AutoCAD
WITH 2D LAB APPLICATIONS

13 RELEASE

FOR DOS® AND WINDOWS™

S. R. Kyles

Addison-Wesley Publishers Limited

Don Mills, Ontario • Reading, Massachusetts
Menlo Park, California • New York • Wokingham, England
Amsterdam • Bonn • Sydney • Singapore • Tokyo • Madrid
San Juan • Paris • Seoul • Milan • Mexico City • Tapei

PUBLISHER: Ron Doleman
MANAGING EDITOR: Linda Scott
EDITORS: Gord Brown, Rodney Rawlings, Michael Cuddy
DESIGNER: Anthony Leung
PAGE LAYOUT: Lexigraf Microsystems
PRODUCTION COORDINATOR: Melanie van Rensburg
MANUFACTURING COORDINATOR: Sharon Latta Paterson
COVER DESIGN: Anthony Leung
PRINTING AND BINDING: Webcom

The author and publisher have taken care in the preparation of this book, but make no expressed or implied warranty of any kind and assume no responsibility for errors or omissions. No liability is assumed for accidental or consequential damage in connection with or arising out of the use of the information contained herein.

Canadian Cataloguing in Publication Data

```
Kyles, S. R. (Shannon R.)
AutoCAD with 2D Lab Applications
2nd ed.
First ed. published under title: AutoCAD with Lab
Applications.
Includes index.
ISBN 0-201-89125-5

1. AutoCAD (Computer file). 2. Engineering design -
Computer programs.
I. Title. II. Title: AutoCAD with lab applications.

T385.K95 1995     620'.0042'02855369     C95-933086-0
```

Copyright © 1996 Addison-Wesley Publishers Limited

All rights reserved. No part of this publication may be reproduced, or stored in a database or retrieval system, distributed, or transmitted in any form or by any means, electronic, mechanical, photocopying, recording or otherwise, without the prior written permission of the publisher.

ISBN 0-201-89125-5

A B C D E -WC- 00 99 98 97 96

TRADEMARKS
AutoCAD, AutoLISP, AutoShade, and AutoSURF are registered trademarks of Autodesk, Inc. dBase is a registered trademark of Ashton-Tate. IBM, IBM/PC/XT/AT, and IBM PS/2 are registered trademarks of International Business Machines. Lotus, Lotus 1-2-3 and 1-2-3 are registered trademarks of Lotus Development Corporation. DOS, MS-DOS, OS/2, and Windows are registered trademarks of the Microsoft Corporation. Ventura Publisher is the registered trademark of Xerox Corporation. WordPerfect is the registered trademark of WordPerfect Corporation.

Preface

AutoCAD is still the world's leading design software, and the software most often chosen in colleges and universities. Architects, engineers and technologists need to be able to use CAD (computer-aided design) as effectively as they use pencils and pens. This book is intended to allow the student to produce both drawings and designs in a 2D (two-dimensional) format using AutoCAD software.

AutoCAD with 2D Lab Applications is designed to introduce technology and engineering students to AutoCAD software. The book can be used as a classroom text, or by an individual in a self-teaching environment.

The main text of each chapter in *AutoCAD with 2D Lab Applications* deals with a new aspect of the software, providing illustrations, examples, and information on how to access the commands in both DOS and Windows. Many students find that software platforms differ among various labs within a college or university. Some also have access to software at home that may be on a different platform. *AutoCAD with 2D Lab Applications* will help students working in either a DOS or Windows environment transparently.

Prelabs Each chapter has a Prelab designed to let the student practice what has just been explained. The Prelabs are detailed examples that explain the functions step-by-step. Students should be able to complete the Prelabs without assistance. In a situation in which there is no lecture time, the student may be requested to read the chapter and complete the Prelab *before* the lab so that any difficult questions can be addressed once the fundamentals of each chapter have been grasped. Where there is ample time for labs and lectures, the Prelabs can be used to get a better understanding of the subject and/or additional help.

Exercises Each chapter closes with several exercises. The Practice Exercise has been developed for those students who have limited time—one hour or less—or merely need a warm-up.

Following the Practice Exercise are exercises that relate to four major disciplines: architectural, civil, electrical and mechanical engineering. Each has been tested in a learning environment to provide a challenging but not impossible exercise for a two-hour lab. In many cases some flexibility is built in; the student can produce all or part of the final example. All commands needed to complete the chapter exercises will have been introduced by the end of the chapter.

The Challenger exercise at the end of each chapter is provided for those students who are eager to become masters of AutoCAD and have the time and equipment to practice; or for those students who have prior knowledge of the software but need to attend classes for necessary credits. It is designed to prevent boredom for those who need a challenge.

Not all of the commands necessary to complete the exercises have been included in the Challenger examples. Students attempting the Challenger will either know what the commands are, or will have a strong sense of what they are and where to find them.

The book is organized into two sections. The first twelve chapters are at the introductory level. At the end of Chapter 12 students should be able to produce a 2D drawing of any given shape. They should also be able to produce a plot of the drawing at the completion of Chapter 12. The next seven chapters are provided for further study in AutoCAD as an intermediate stage. These chapters will help students understand the software better, enabling them to use some of the nongraphic data.

For those who have completed the first twelve chapters, the next seven chapters, plus six or seven of the Challenger exercises, could be considered an intermediate course.

AutoCAD 3D

Another book, *AutoCAD with 2D and 3D Lab Applications*, has been developed to introduce the student to AutoCAD's 3D capabilities. It contains the first nineteen chapters of *AutoCAD with 2D Lab Applications* plus another fifteen on 3D applications, engineering analysis, and development of solid models.

Other Releases

Both these books cover AutoCAD Release 13. Most of the information is also relevant to the earlier Releases, 11 and 12. If a student has Release 11 or 12 at home, most of the commands have remained the same, but the menus are often different. Drawings completed on earlier versions of AutoCAD can be brought up on Release 13. Drawings can be saved to Release 12 by using the command, SAVEASR12, but cannot be taken to earlier releases.

Instructor's Resource Package

For instructors adopting *AutoCAD with 2D Lab Applications*, the publisher will provide, free of charge, the *Instructor's Resource Package*. This contains the *Instructor's Manual* and disks containing AutoCAD files for the discipline-specific exercises at the end of each unit. The disks aid the instructor in marking the student's work on the exercises. The *Instructor's Resource Package* can be requested from your local Addison-Wesley sales representative upon adoption of this book.

Acknowledgments

The exercises in this book went through many years of student testing at Mohawk College. Since the first book (January 1993), there have been many positive suggestions and many bits of constructive criticism. I would like to thank all my students for working with me on the development of new projects, and for proofing tutorials and exercises. I would like to thank Dave Goede of Goede and Associates, for his continuing support and technical genius; Peter Mann of Mohawk College for his encouragement and generous offers of help with architectural examples; Peter Rudyk of Dofasco for his expertise in providing technical support in the mechanical area; Lloyd Mutch of Mohawk College for his help with electrical examples; and Graham Roebuck, Susan Hardy, Jane Zatylny, and Cathy Thompson for technical support.

Illustrations of projects showing industry standards have been generously extended by Sylvia Smith of UMA Engineering Limited in Edmonton, Alberta; Jim Ellis at Micro-Rel in Arizona; Dave Umbach at Tarsons, Benckerhoff, Gore, and Storrie Incorporated; and Andy Slupecki of Dundas, Ontario.

Roger Winn of University College of Cape Breton provided an extensive and in-depth proofing of the final draft, and is also supplying the *Instructor's Manual*. His comments have been of immeasurable value.

I would also like to thank the following people who supplied very useful and constructive comments on the material: Barbara Bang, of North Dakota State College of Science; Jag Mohan of Sheridan College; Curtis Rhodes of the University of Southern California; Tom Hyde of Kirkwood College, USA; Roger Winn of University College of Cape Breton; John Bonacci of the University of Toronto; Dennis Short of Purdue University; and Don Bird of Ricks College.

In addition, I would like to thank Rodney Rawlings, the coordinating editor, Linda Scott, the managing editor, and Ron Doleman, the College Publisher, for their help in producing this book.

It is the responsibility of the author and publisher to correct any problems that users might find within the book. We encourage you to send all of your concerns to the publisher at the following address:

Ron Doleman, Publisher
Addison-Wesley Publishers Limited
26 Prince Andrew Place
Don Mills, Ontario
M3C 2T8

S. R. Kyles
September 1995

Table of Contents

Introduction .... x-xxiv

Chapter 1 Introductory Geometry and Setting Up
Setting UCS, LIMITS, SNAP, and GRID 1
Entry of Points Using Coordinates and Digitizing 6
Coordinate Entry Using Absolute, Relative, and Polar Values 6
Coordinate Entry Using Digitizing 9
Geometry Commands 10
View Commands 13
The REGEN and REDRAW Commands 16
The UNITS Command 17
Prelab 21
Exercises 26
Challenger 34

Chapter 2 Help Files, CHAMFER, OSNAP, ERASE, TRIM, and BREAK
Help Files 37
SNAP and GRID 39
Point Entry and OSNAP 41
The ERASE Command 48
The CHAMFER Command 49
Prelab 50
Exercises 55
Challenger 60

Chapter 3 Entity Commands with Width
The TRACE Command 61
The PLINE Command 62
The PEDIT Command 65
The POLYGON Command 67
The DONUT Command 69
Filling Irregular Shapes 69
The TEXT Command 69
The MLINE Command 71
Prelab 72
Exercises 77
Challenger 82

Chapter 4 Object Selection and Editing
Selecting Objects Within the EDIT Command 83
Editing Commands 86
Editing with Grips 94
Setting LINETYPEs 97
Changing LTSCALE 99
Prelab A 100
Prelab B 102
Exercises 105
Challenger 110

Chapter 5 STRETCH, TRIM, EXTEND, OFFSET, and ARRAY
Removing and Adding Objects 113
Editing Commands 115
Prelab 129
Exercises 133
Challenger 141

Chapter 6 Entity Properties: Layers, Colors, and Linetypes

- About LAYERs 143
- Creating a New Layer 144
- Changing LTSCALE 148
- CHPROP and CHANGE with Layers 149
- Changing the State of the Layers 150
- The LAYER Command 151
- Prelab 153
- Exercises 158
- Challenger 164

Chapter 7 Dimensioning

- About Dimensioning 165
- Entering Dimensions 167
- Dimension Styles 171
- The Geometry Dialog Box 172
- The Format Dialog Box 173
- Annotation 175
- Using Dimension Style Families 177
- Editing Dimensions 180
- When All Else Fails 180
- Prelab 181
- Exercises 188
- Challenger 194

Chapter 8 Text

- Linear Text 195
- Paragraph Text 199
- Text Styles and Fonts 201
- Editing TEXT and DTEXT 204
- Making Isometric Lettering 206
- Using LEADER to Create Notations 207
- Prelab 209
- Exercises 215
- Challenger 220

Chapter 9 HATCH and SKETCH

- The HATCH Command 221
- The BHATCH Command 222
- Editing Hatches 228
- The SKETCH Command 229
- Point Filters 230
- *X, Y* Filters—An Exercise 232
- Prelab A 234
- Prelab B 236
- Exercises 240
- Challenger 247

Chapter 10 Blocks and Wblocks

- Introduction 251
- The BLOCK Command 252
- The INSERT Command 253
- The WBLOCK Command 255
- The MINSERT Command 257
- Editing Blocks 258
- Compiling Drawings with BLOCK 260
- Blocks, Wblocks, Color, and Layers 260
- Prelab 264
- Exercises 269
- Challenger 274

Chapter 11 Setting Up Drawings and PSPACE

- Set Up and Scale for Simple 2D Drawings 275
- Using Blocks to Compile Drawings 277
- Using Paper space to Compile Drawings 280
- Paper space and Tilemode 281
- Scaling Views Within a Drawing 284
- The VPLAYER Command 284
- Dimensioning in Paper space 286
- The MVSETUP Command 286
- Prelab A 288
- Prelab B 297
- Exercises 303
- Challenger 308

Chapter 12 2D Review and Final Drawings

- Review 309
- Problems 310
- Quiz 312
- Final Drawings 327

Chapter 13 POINTS, DIVIDE, MEASURE, INQUIRY, and System Variables

- Point Display or PDMODE Options 333
- Using Divide and Measure 334
- The SPLINE Command 337
- Inquiry Commands 338
- Creating Multiline Styles 344
- Editing Multilines 346
- Prelab 348
- Exercises 353
- Challenger 357

Chapter 14 Creating Attributes

- Introduction 359
- Attributes for Title Blocks and Notations 359
- Defining the Attributes 360
- Editing Attribute Definitions 361
- Displaying Attributes 364
- Creating Attributes for Data Extraction 364
- Prelab A 366
- Prelab B 369
- Exercises 374
- Challenger 378

Chapter 15 Editing and Extracting Attributes

- Editing Attributes Attached to Blocks 379
- The ATTEDIT Command 379
- Data Extraction 383
- Prelab 387
- Exercises 391
- Challenger 395

Chapter 16 Isometric and Orthographic Drawings

- Isometric Views 397
- Prelab A 400
- Creating Orthographic Views 404
- Prelab B 407
- Exercises 410
- Challenger 414

Chapter 17 File Formats and Management

What Are Slides? 415
The MSLIDE Command 416
The VSLIDE Command 417
Script Files 420
Exporting AutoCAD Files 422
Importing Files to AutoCAD 423
Prelab A 424
Managing Larger Files 426
Groups 427
Prelab B 429
Exercises 432
Challenger 436

Chapter 18 Advanced Blocking and Xrefs

Advanced Blocking 437
Xrefs or External Reference Files 439
The XBIND Command 445
Prelab 446
Exercises 450
Challenger 457

Chapter 19 Final Tests and Projects

Section 1: On-Screen Problems 459
Section 2: Review 464
Final Drawings 474
Challenger 480

Appendix A   Glossary of Terms 481
Appendix B   File Extensions 489
Appendix C   Abbreviations and Aliases 491
Appendix D   Plotting and Printing 492

Index 499
Windows Icons 505
DOS Pull-down Menus 509

Introduction

AutoCAD is a very popular, flexible software system that allows the user to create both 2-dimensional and 3-dimensional models and drawings.

For those who are familiar with computers, learning AutoCAD will be easy, simply because they are aware of the typical response structure and the format of their system. This introduction is provided for those approaching AutoCAD and computers for the first time. It will familiarize them with computers, and how computers store information. It will also cover some of the technical terms referred to in this book.

If you are computer-literate, glance through the next few pages to see if there is anything in AutoCAD that looks different from the programs with which you are familiar. If you are not used to working with computers, read the next few pages carefully before starting on Chapter 1.

Using This Book

System Prompts and User Responses

In this book, the system command information will be shown in this style:

```
Command:
From point:
To point:
```

The user responses, (what you should type in), will be shown in bold:

```
Command: LINE
From point: 0,0
To point: 5,3
```

The Enter or Return Key

At the end of each command or entry on the command line, use the Enter key (symbolized visually by ⏎) to signal the end of:

- A command entry:

  Command:**LINE**↵

- A coordinate entry:

  From point:**2,4**↵

- A value:

  New fillet radius.0000:**3**↵

- Or text:

  Text:**All Holes 2.00R Unless Noted**↵

To make the text look neater, the ↵ will not be shown at the end of every entry; it will only be used when the user should press ↵ rather than entering any other response.

Floppy Disks

In order to access AutoCAD in a learning environment, you may need a floppy disk in the A: drive. And in order to store your files, you will definitely need floppy disks—usually two—one for the file itself and one for a backup file.

The two standard sizes for floppy disks are 3 1/2 inches and 5 1/4 inches. Both are certainly acceptable but the smaller floppies can store more information and seem to be more durable.

To see if your floppy disk is formatted, at the prompt enter:

C:**DIR A:**

If your readout says that there are no files found:

No Files Found

Danger
Formatting a disk will erase all data on the disk. Make sure that the disk contains no necessary information before formatting.

then the floppy disk is formatted and ready to go. If your readout says that the disk cannot be read, try formatting it.

Format your floppy in one of the following ways:

Danger
If your floppy has already been formatted and has information on it, do not reformat, unless you want to erase all the information on the disk.

Command
FORMAT A:	On a 1.2 meg drive, this will format a 1.2 meg disk.
FORMAT A:/4	On a 1.2 meg drive, this will format a 360 K disk.
FORMAT A:	On a 1.2 meg drive, this will format a 740 K disk.

If you are using a B: drive, the commands may be the same; just substitute B: for A:.

Starting AutoCAD in DOS

Once the system has been turned on, you will find yourself at a prompt of the C:, D:, G:, H: or some other drive. To enter the AutoCAD Drawing Editor, type the *sign-on code*, **ACAD** or **ACAD13**:

G:**ACAD**

ACAD is usually your sign-on code. If there are two or more versions of AutoCAD running on the network, check the notation on the main menu for specific package information; ACAD13 may mean AutoCAD Release 13; ACAD12 may mean AutoCAD Release 12; ACAD13M may mean AutoCAD Release 13 with a mouse.

Note your sign-on code here:_____

The DOS Drawing Editor

Your sign-on code will get you directly into the Drawing Editor.

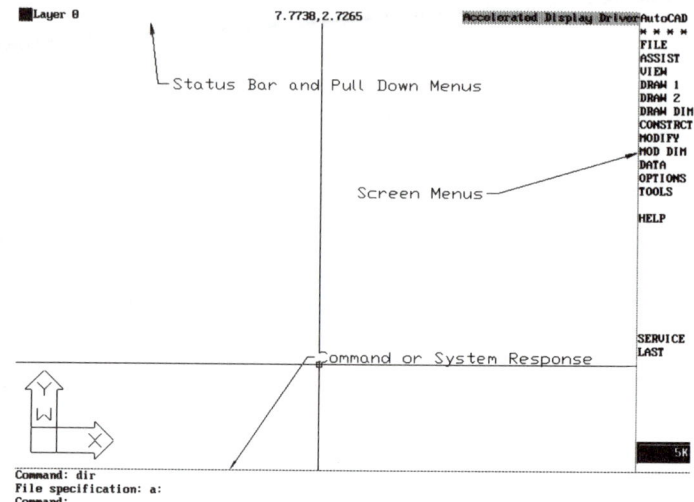

The Drawing Editor divides the screen into four sections.

- **System response area** Three text lines at the bottom of the screen. This will indicate the status of your commands and the system. Watch this area carefully to avoid frustration; it will tell you what command you are in. "Command:" is the system prompt. It will tell you that a new command can be entered. If you do not have the "Command:" prompt, it means you are still using another command. Use Ctrl-C to get back to the "Command:" prompt.

- **Graphics display area** This is the working space where the file or model will be displayed.

- **Status bar** This is located at the top of the screen. It displays the position of the cursor, the first eight characters of the current Snap, and the settings of modes or options. When the cursor is placed on the status bar, the status bar changes to a menu bar containing the pull-down menus.

- **Screen menu** This, located on the right of your screen, will let you access the on-screen menus.

Starting AutoCAD in Windows

Start AutoCAD by clicking on the application icon.

Double-click the application icon to get the program started.

You will automatically be placed in the Drawing Editor.

The Windows Drawing Editor

The initial Windows screen contains the menu bar, the status bar, the drawing window or graphics area, and several toolbars. Toolbars contain icons that represent commands.

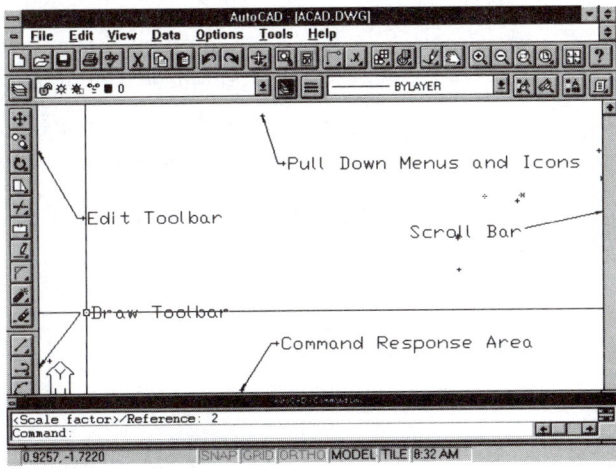

The menu bar contains the pull-down menus. The status bar displays the cursor coordinates and the status modes such as Grid and Snap. Mode names are always visible in the status bar as selectable buttons. Double-click the buttons to toggle the modes. The command line in Windows is "floating," that is, it may be dragged to any location on the screen.

Keyboard and Mouse Functions

Pointing Devices for Both Windows and DOS

There are many different kinds of pointing devices or "mice" on the market. Some have two or three buttons, others have as many as twenty. If you are purchasing a mouse, choose one that has a minimum of three buttons.

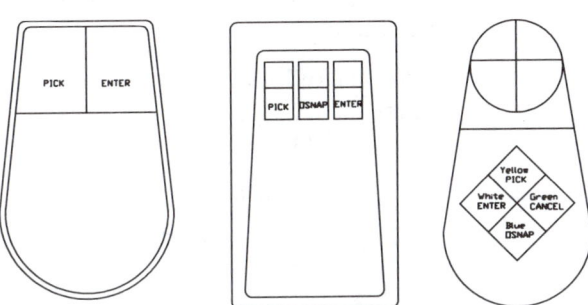

The Pick Button

On all mice there will be a point or command indicator or **pick button**: on a three-button mouse, it is usually on the left side of the device.

The pick button is used to indicate the command you want to access either from the on-screen menu or from the digitizer tablet. It is also used to indicate point positions.

The Enter Button

Another button on the mouse will have the function of the ⏎ key (Enter or Return) on the keyboard. This signals the end of a command.

INTRODUCTION xiii

Other Buttons

Other buttons can mean **Esc** or **Ctrl-C**, **OSNAP**, or other commands.

The mouse (or "puck") is intended to work in conjunction with the keyboard for the entry of commands.

Before entering commands, be sure you have a "Command:" prompt. *To return to the command prompt use Ctrl-C.*

Function Buttons for Both Windows and DOS

To become familiar with the function keys, move the mouse around the screen noting the movement of the crosshairs. Find the F buttons or function keys on the top or side of your keyboard.

DOS		*Windows*
Now press **F6**.	You will notice that the numbers on the status line are moving as you move the mouse. These are the *X* and *Y* values.	
Now press **F9**.	You will notice that the numbers are snapping to even values. In DOS, the word Snap is in the status line; in Windows, the SNAP will be highlighted.	SNAP
Now press **F7**.	You will see a grid on the screen.	GRID

F8 is used with linear commands to create lines at right angles. Once you are in the LINE command, try using **F8** or Ortho within the command to get lines vertical and horizontal.

Type in **LINE** at the command prompt. With the pick button, pick a spot on the screen. Then use ORTHO.

```
Command: LINE
From point: (pick the screen)
```

Now press **F8**. Your lines will only be created on a vertical or horizontal.　　　ORTHO

Press the ⏎ key to exit from the line command.

```
To point: ⏎
```

Many commands invoke a system response that shows an alphanumeric screen with related data. To exit from the alphanumeric screen onto the graphics screen, you can use **F1** in DOS or pick the Command control box at the top of the alphanumeric overlay.

Type in the word **HELP** at the command prompt:

```
Command: HELP
```

To exit from Help, pick OK or the command control box:

Many systems provide *toggle switches* by means of function keys. These allow the user to turn a specific function on and off simply by pressing the key. In Windows these functions can be accessed through icons as well. The function switches are:

Flip Screen	COORDS	GRID	ORTHO	SNAP
F1	F6	F7	F8	F9

When entering commands, you can use the function keys at any time either before or within a command.

Entering Commands and Coordinates in DOS

Once you have entered the Drawing Editor, you can enter information through either the keyboard, the mouse or a digitizer.

You can enter a command either by typing it in at the command prompt or by using the pointing device to pick up commands from either:

- The on-screen menu, found on the right side of the screen
- The pull-down menus, found under the status line at the top of the screen, or
- The Tablet menu on the digitizing board (where available)

Once you have entered the command, you can enter the coordinates or position of objects either by entering it through the keyboard or through the pick button on the mouse.

Entering Commands and Coordinates in Windows

As in the DOS format, you can enter information either through the keyboard or through your mouse or pointing device. With Windows, however, there are also toolbars and icons that help to access the information. You can enter a command by typing it in at the command prompt or you can use the pointing device to pick up commands from either:

- The pull-down menus in the menu bar
- The icons from the toolbars
- The Tablet menu on the digitizing board (where available), or
- The screen menu where loaded

Windows Toolbars

Toolbars contain tools that represent commands. When you move the pointing device over a tool, Tooltips below the cursor display the name of the tool.

Flyouts

Tools that have a small black triangle have *flyouts* similar to the pull-down menus that contain subcommands like the LINE command shown here.

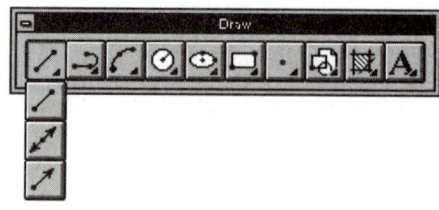

To access the *flyout*, hold the pick button down with the cursor over the tool icon until the flyout appears. Click a tool on the toolbar or flyout to start the command, then select the options from a dialog box or follow the prompts on the command line.

Placing Toolbars

The Standard toolbar is visible by default. It carries frequently used tools such as Zoom, Redraw, and Undo. You can display multiple toolbars on-screen at once, change their contents, resize them, and dock or float them. A **docked toolbar** attaches to any edge of the graphics window. A *floating toolbar* can lie anywhere on the application screen, and it can be resized and does not overlap with the drawing window.

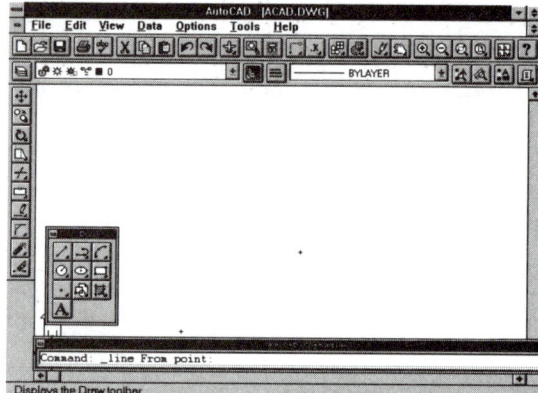

Showing a Toolbar

From the tool windows flyout on the standard toolbar, choose the icon for the tool you want to show.

The command line equivalent is **TOOLBAR**.

Alternatively, from the Tools menu, choose Toolbars; then choose the name of the toolbar you want to show.

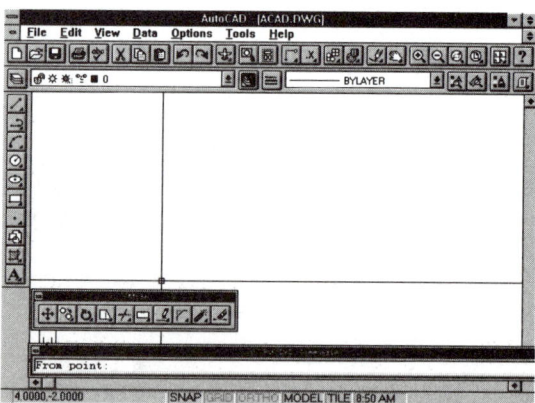

To Dock a Toolbar

1. Position the cursor on the caption, and press the pick button on the pointing device.

2. Drag the toolbar to a dock location at the top, bottom, or either side of the drawing window.

3. When the outline of the toolbar appears in the docking area, release the pick button.

To place the toolbar in a docking region without docking it, hold down the Ctrl key as you drag.

Releasing a Toolbar

To get the toolbar off the screen, choose the control menu box on the top left corner.

The Windows Command Window

Like the toolbars, the Windows command line can be moved and docked as well. By default the command window is a floating window with a caption and frame.

You can move the floating command window anywhere on the screen and change its width and height with the pointing device. This window is also dockable within the AutoCAD application window. Dock the command window by dragging it until it is over the top or bottom dock regions of the application window. Undock the window by selecting any part of its border and dragging it away from the dock region until it has a thick outline. Drop it to make it a floating window.

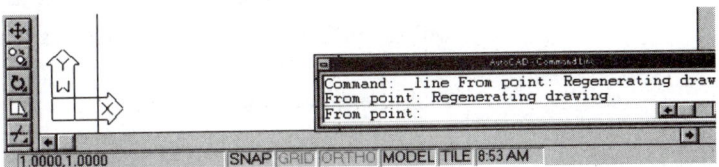

You can resize the commmand window vertically and horizontally, both with the pointing device and with the splitter bar located at the top edge of the window when docked on the bottom and on the bottom edge when docked at the top. Resizing and docking the command window can help you to create more space for your drawings on-screen.

Scroll Bars in Windows and DOS

In both Windows and DOS there are **scroll bars** that advance the file you are viewing. Each scroll bar has arrows that indicate a move up or down. To access an area not displayed, either pick on the up or down arrow until the information is displayed or pick the box within the scroll bar and move it quickly up and down the screen.

Scroll bars can be either vertical or horizontal. In Windows, the scroll bars on the top and bottom move the file across the screen in the same way that a PAN does.

Menus

In Windows, AutoCAD divides the command set by default between menus and toolbars. You can work exclusively with menus by loading a menu file that contains the full command set.

In DOS, the commands are all available either in the screen menu or in the pull-down menus.

Screen Menus in DOS

The screen menu is a set of submenus to the right of the drawing window. It contains the full AutoCAD command set.

To load the screen menu in Windows:

1. At the command prompt, enter **MENU**.

Command: **MENU**

Notes

If you are taking drawings from a Windows environment to a DOS environment, often the menu is loaded with the drawing. Thus, when loading a Windows drawing in DOS, you may need to load the menu file as well. Type **MENU** at the command prompt.

2. In the Select Menu File dialog box, enter the menu file name or select a name from the list. Often the menus are under the SUPPORT subdirectory.
3. Choose OK to load the menu.

Menus are often in the SUPPORT subdirectory of the ACAD software.

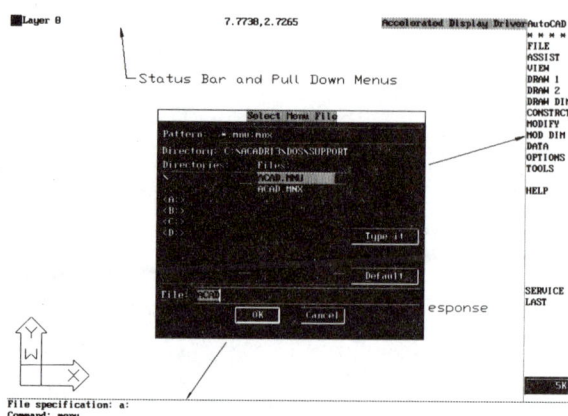

Pull-down Menus for Both DOS and Windows

Pull-down menus are used to access many of the same commands that are found on the screen menu. Place your cursor on the status bar in DOS or on the menu bar in Windows. Pick the menu that you want, then pick the command or submenu that you want.

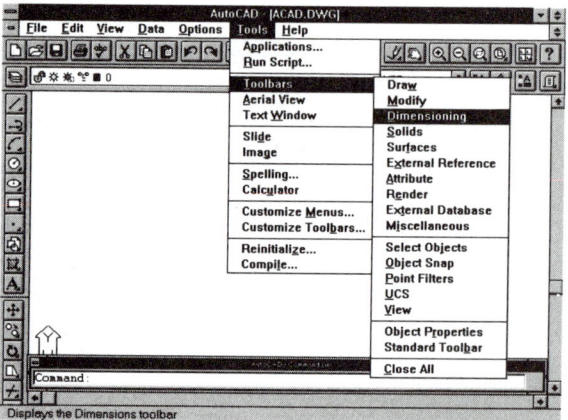

Reading the Pull-down Menus

Under each separate menu you can access commands, subcommands and dialog boxes. For example, in the View menu, if you choose Redraw All, you are simply invoking that command. The command REDRAW will take place in all viewports. Often with the commands from the pull-down menus a repeat is added, so you may need to use **Ctrl-C** or **ESC** to exit the command if it repeats.

Also under the View menu are the options with triangles following them. The triangle indicates that there is another "flyout" menu (as with Zoom in the illustration at right).

When you have invoked the ZOOM command, you will reach a secondary or submenu that will offer you several choices within that command. You are prompted to decide the amount of Zoom that you require: In, Out, All, Window, etc.

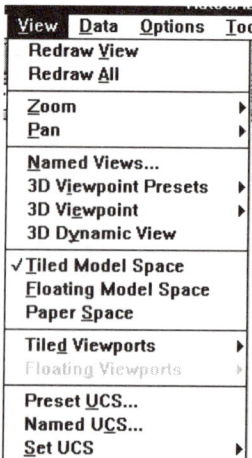

The third type of pull-down menu choice has three dots following the entry, meaning a dialog box is invoked either immediately or after an object is selected. In View, the dialog box menu choices include Named Views, which brings up the Named Views dialog box where you can modify the amount of named views on your file.

Switching from DOS to Windows

Those switching from DOS to Windows or Windows to DOS will find that many automatic functions are different from one platform to the other. These functions are outlined below.

Function	DOS	Windows
Cancel	**CTRL-C**	**ESC**
Directory of files	Command:**Dir A:**	Command:**FILES**
Text editor	Command:**EDIT**	Notepad

Opening or Accessing Drawings

Once you have accessed the Drawing Editor, you can start drawing and later save your work under a specified name in a specified directory.

If you have a drawing started in AutoCAD Release 13 or some earlier version, you can use OPEN to work on it. Only one drawing can be active at any given time. If you are already in a drawing, AutoCAD will prompt you to save it before opening the next drawing.

Opening Existing Drawings

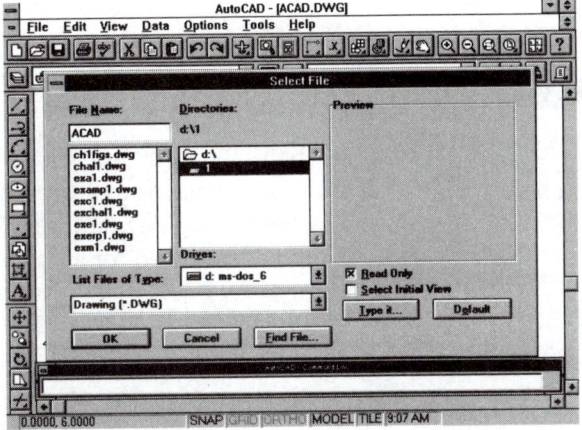

WINDOWS From the Standard Toolbar or the File menu, choose Open.

DOS From the File menu, choose Open.

The command line equivalent is **OPEN**.

In Windows under File Name, double-click the file name in the list of files. Use the scroll bars to access other files. To access other directories, pick the desired directory under Directories. You can also type in the drawing name by picking the long white box, then typing in the name of the file. If you prefer to type in both the directory and the name choose the Type it...option.

In DOS, the Open Drawing dialog box appears. The functions are the same: choose the directory, then the file name, or use Type it ... to simply type in the desired file name.

Starting a New File

If you would like to start a new file, access the same File menu and choose New. You will be prompted to save the current file before the new file is opened.

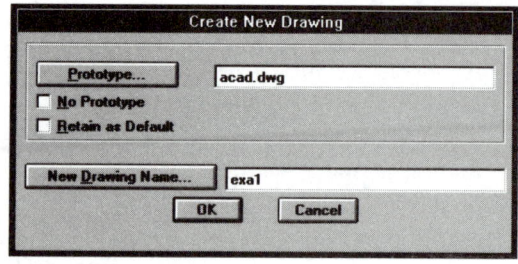

Before the new file is created, you can choose a default drawing file environment and/or enter the name of the file that you wish to create.

The default file environment can be either the ACAD.DWG standard file or a prototype file that contains all the settings for a specific application.

Recovering Files

If you have a problem with retrieving a file in the Open command, you may need to Recover the file. Usually these problems are caused by either bad diskettes or removing the floppy disk from the drive before AutoCAD has completely exited from the file. If you need to Restore a file, simply choose Recover from the main File menu and enter the name of the drawing file.

You do *not* need to Recover the file every time you load it.

Saving Files

Notes

To save a file so it can be read in Release 12, type in SAVEASR12, and enter the file name. You may lose some hatch or text objects.

Computers have a tendency to lose information at the worst possible times. It is suggested that when you are using AutoCAD you save your files at least every hour.

The first time you save a drawing, you will be prompted for the name of the file before it saves. If you have already entered the name of the current file under the New option in file, then AutoCAD simply saves the file under the given name and directory and you will not be prompted for a name.

To save a named file, use SAVE.

1. Start the QSAVE (Quick Save) command using one of the following:

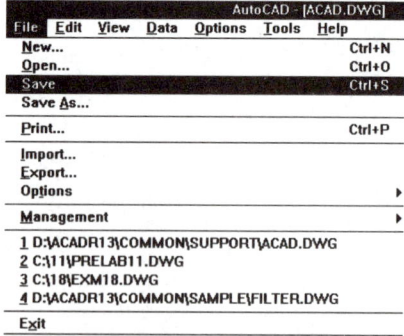

WINDOWS From the Standard Toolbar or the File menu, choose Save.

DOS From the File menu, choose Save.

The command line equivalent is **SAVE**.

2. In the Save Drawing As dialog box, enter the new drawing name. Then choose OK.

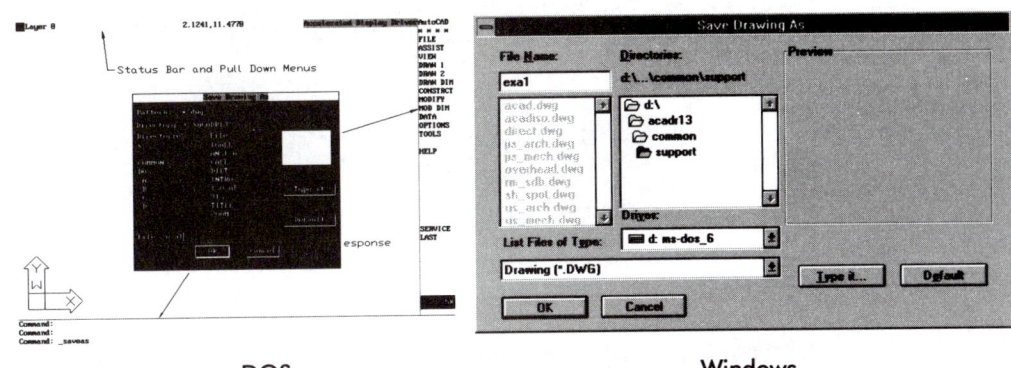

DOS   Windows

Choose Save every subsequent time you would like to save the drawing, and the drawing will automatically be saved under this specified file name.

If you specified a directory and file when you signed on, use SAVE to save the file under this name.

To save the file under a new name or on a different directory, choose Save As from the File pull-down menu.

To change the directory, double-click on the directory listing that you want. The line reading Directory must reflect the directory chosen.

To enter the name of a file, pick the File area and type it in. To Exit the file and Save it at the same time, type in End at the command line.

Changing the Drawing Name or Directory

If you want to change the drawing name or directory, use Save As. If you have been addressing C: while creating your drawing, you can save the file onto a floppy disk before exiting the file by using Save As; then pick A: or B: for the directory or drive.

Browsing Through Files

The Browse/Search dialog box displays small images of drawings in the directory you specify. Use the options on the dialog box to sort images by file type and change the size of the images. Click an image to select it.

To access the dialog box, pick the Find File button from the Select File dialog box. Use Open to find the Select File dialog box.

File names should be less than eight characters long and without dots (.), backslashes (\), or spaces.

While you are creating an AutoCAD file, there is a temporary file being created of your changes. If you are addressing A: or B:, you cannot remove your diskette from the drive without saving the file back to C:.

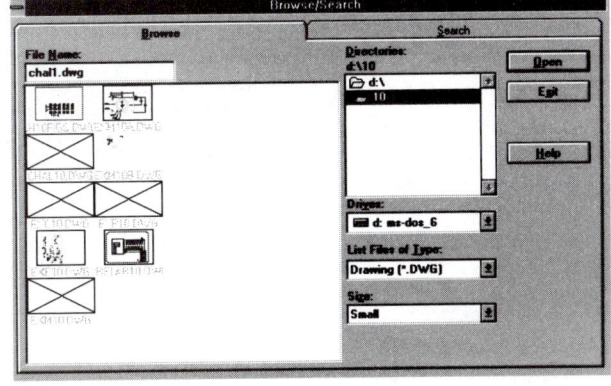

Exiting AutoCAD

Once you have saved the file, you can exit AutoCAD.

> **WINDOWS** Pick the Control menu box, then pick **Close Alt-F** or simply type in **C**.
>
> **DOS** From the File menu, pick Exit AutoCAD.

The command line equivalent is **QUIT**.

Do not remove your floppy disk from the drive before you have completely exited from AutoCAD.

Entering Commands

When you enter commands at the command prompt, AutoCAD displays a set of options or a dialog box. Many commands have a series of options, for example the FILLET command:

```
Command:FILLET
Polyline/Radius/Trim/<Select first object>:
```

The default is to choose the first object that you would like to have filleted. If you would like to choose one of the options, simply type in the first letter of the option at the command line and the option will be invoked. For example, if you would like to change the Radius of the fillet, simply type in **R** and you will be prompted for the size of the radius.

```
Polyline/Radius/Trim/<Select first object>:R
Radius <1.00>:6
```

Then enter the size of radius that you require.

Values Versus Coordinates

In many commands you will be prompted for a coordinate or position on the screen. This is an X, Y (and, in 3D, Z) coordinate.

```
Command:LINE
From point:2,3
To point:4,3
```

If you are asked to specify a radius or diameter, or a distance such as the length of a line or the width of a linetype, this is a value that is expressed with only one numeral.

```
Radius<1.000>:3
Offset distance <0.5000>:12
```

Object Selection Prior to Command Entry

Once you are in the Graphics Editor, commands are entered either through the menus or icons with the pick button or by the keyboard. If you should pick the graphics screen without first invoking a command, you will create an object selection box that can be used for editing commands. (See Chapters 4 and 5.)

Window

If you pick anywhere on the screen and then pick to the right, either above or below the first pick, you will be creating a window. Every object that is fully within that window will be selected. (See Chapter 4.)

Crossing

If you pick anywhere on the screen and then for the second time to the left of the first point, you will be creating a crossing window. This selection window selects all objects fully or partially within its borders. (See Chapter 4.)

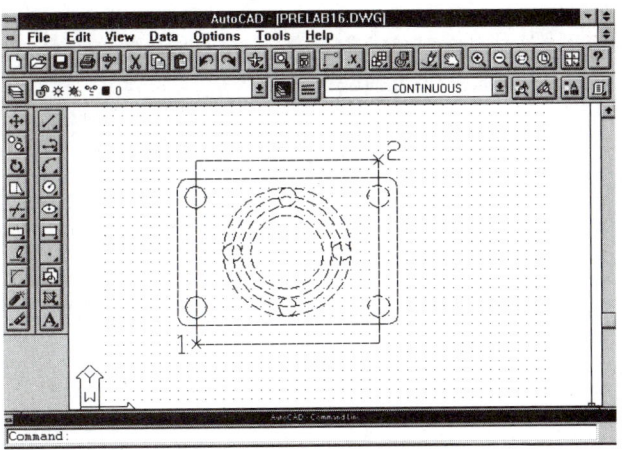

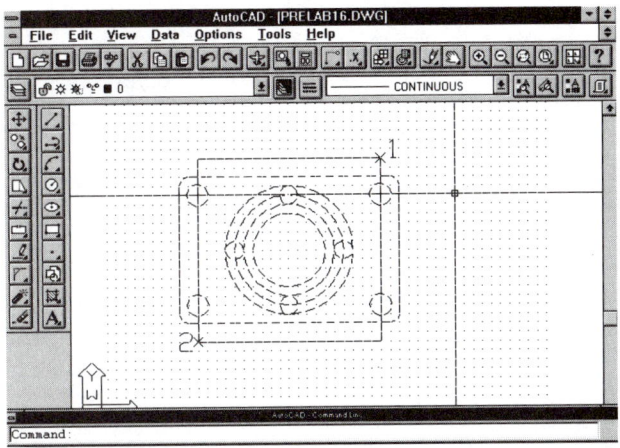

Objects

If you select an object on screen without invoking a command first, small squares called **grips** will appear on the object or objects you select. These are for editing the object. (See Chapters 4 and 5.)

If you have selected objects either through a window, through a crossing window, or as an individual object selection, and you do not want to edit the objects at the time, simply continue to enter the next command and the selection will be ignored.

Help Files

When first entering AutoCAD, many people have a tendency to press the space bar before starting work on the file. This will invoke the Help files. If you are not familiar with computers, this can be rather disconcerting.

To exit the Help files and continue with the drawing:

> **WINDOWS** Press the Command Control box on the top left of the help file menu. Then pick Close **Alt-F** or simply type in **C**.
> You can also simply press **ESC** on the keyboard.
>
> **DOS** Press OK at the bottom of the menu.

Accessing Commands

Typing in the Commands

You can type in commands instead of picking them from menus, but they must be spelled correctly! Any commands on-screen that are followed by a colon can be typed in.

You cannot access the menus by typing in the menu titles (note that menu selections are not followed by colons). In DOS, if you type in EDIT by mistake, it will convert your screen to alphanumeric format because you have entered Edit, the DOS editor. Press the command control box or File Exit to return to the graphics screen.

Speeding Up Your Entries

The space bar on the keyboard and the enter button on the pointing device, when used with a "Command:" prompt, will bring back the last command entered. Effective use of this technique can greatly enhance your speed when entering LINEs etc.

Command Aliases

Many commands can be accessed by simply typing the first letter of the command.

L = LINE	E = ERASE
A = ARC	Z = ZOOM
C = CIRCLE	P = PAN

Exiting

SAVE	QUIT or EXIT
Makes a copy	Exits without saving

When you use AutoCAD, the system is generating a temporary file on the drive that you are accessing. *Do not remove your floppy disk until you have saved and exited the Drawing Editor.*

When working in an environment where the computer is accessible to others, there is always the possibility that someone, for whatever reason, will QUIT or EXIT your file if you leave it unattended. It is also possible that a severe storm or power surge could cut off the electricity temporarily, thereby erasing your file. So it is good practice to perform a SAVE every half-hour or hour, to reduce the amount of work you may accidentally lose.

The only way to improve your AutoCAD skills is by practicing. There is a wide variety of exercises in this book. The more you do, the better you will get.

Introductory Geometry and Setting Up

Upon completion of this chapter, you should be able to:

1. Set the UCS origin
2. Change the screen LIMITS, SNAP, and GRID
3. Use coordinate entry methods
4. Create simple geometry using LINE, ARC, CIRCLE, and FILLET
5. Use the Display commands ZOOM and PAN
6. Set up the UNITS

This book describes both WINDOWS and DOS environments. If you are not familiar with the platform — WINDOWS or DOS — you are using, read the preface for your platform before continuing with the specific AutoCAD commands in this chapter.

The text is designed for AutoCAD Release 13; however, much of the information is relevant to Release 12 and most of the commands are the same as in the earlier releases if typed in.

Setting UCS, LIMITS, SNAP, and GRID

Once you have entered the Drawing Editor, AutoCAD establishes a default working environment. This includes LIMITS set at 12 × 9 inches, GRID set at one dot per unit, and SNAP set at one-unit increments. This drawing environment is stored in a drawing called ACAD.DWG which is loaded as a default prototype for new files.

You will probably want to change these for every design you do. To make a drawing on paper, the paper size can be set when you are ready to plot.

AutoCAD uses Cartesian coordinates for point entry. The points are set around a determined origin at $X0$, $Y0$, $Z0$. All points to the right of the 0,0 are positive X; all points to the left are negative X. All points above 0,0 are positive Y; all points below are negative Y.

Moving the cursor around the screen with the Coordinates ON (**F6** in DOS, **COORDS** in Windows), you will notice that the 0,0 position is at the bottom left corner of your screen.

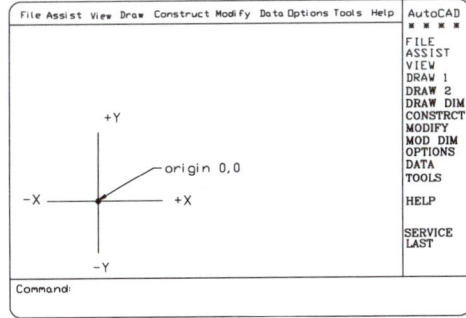

Figure 1-1

Choosing the Origin with UCS

The origin should be the most easily accessible point on the design. If a large percentage of the dimensions on a model stem from one point, this should be made the origin. The reason for this is that the coordinate readout on the top of the screen is there to help you find your position. The placement of the origin is important to establish a base for your readouts.

It is advisable to have the drawing or part start at 0,0 so that the dimensions of the drawing match the coordinate readout on the screen. To move 0,0 from the bottom of the screen use the UCS command with the Origin option as follows:

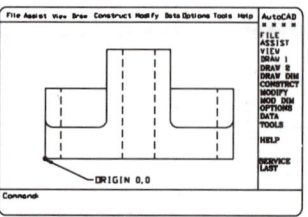 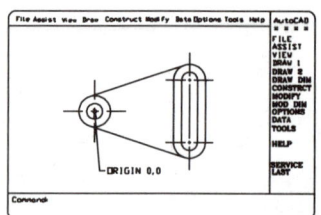 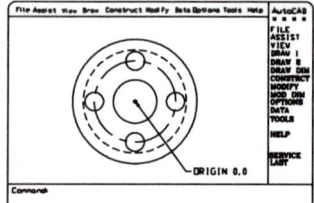

Figure 1-2

UCS stands for **user coordinate system**. This system allows placement of the origin anywhere on the screen for easy identification of 0,0. To place the origin at the positions shown in the above illustration, use UCS Origin and pick the position on the screen as shown. Because many options on the pull-down menus are not relevant at this stage, simply type in UCS at the command prompt.

```
Command:UCS
Origin/ZAxis/3point/Entity/View/X/Y/Z/Prev/Restore/Save/Del/?/
    <World>:O
Origin point<0,0,0>: (pick a point on the screen)
```

For a rectangular part, a point further up on the bottom left is a good idea. For a symmetrical part, the center of the screen is a good idea.

If the model or drawing is larger than 12 inches, you must set the screen to the size needed.

All drawings and models in AutoCAD should be created at a full size scale of 1:1. Plotting to a scale factor other than 1:1 is done when you want to scale a drawing to fit a piece of paper.

Danger

If you press the space bar without entering a command, the system will offer you the help files. If you don't want the help files, use the cancel button or **Ctrl-C** to cancel, then press **F1** to flip the screen back to the graphics mode.

The LIMITS Command

The LIMITS command is used to set up the screen to accept the lengths of lines and arcs you need. LIMITS sets the size of your screen and the area covered by the screen grid.

Here are some examples of LIMITS which you may set:

● = the position of the origin

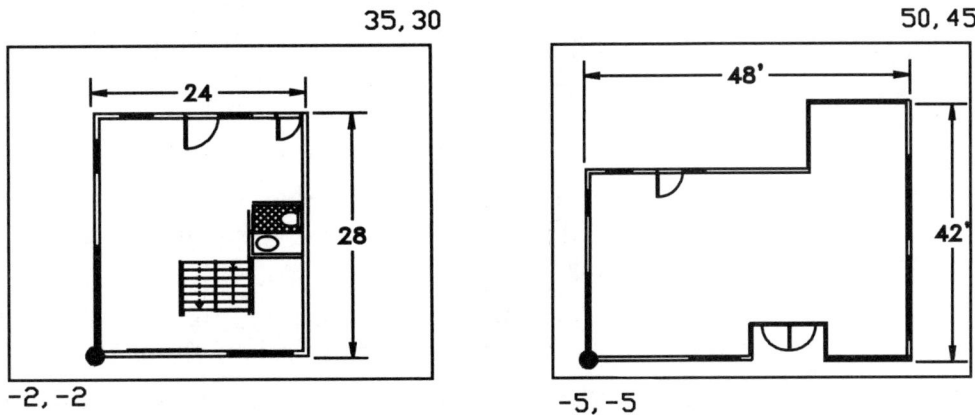

In architectural applications, dimensions are either interior or exterior. The origin is often at the bottom left.

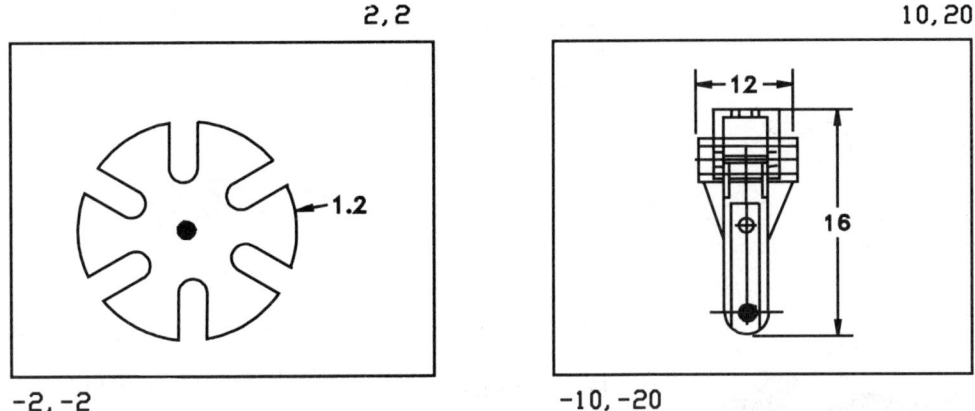

In mechanical applications, the parts are often symmetrical so the origin is found on the part for easy measuring.

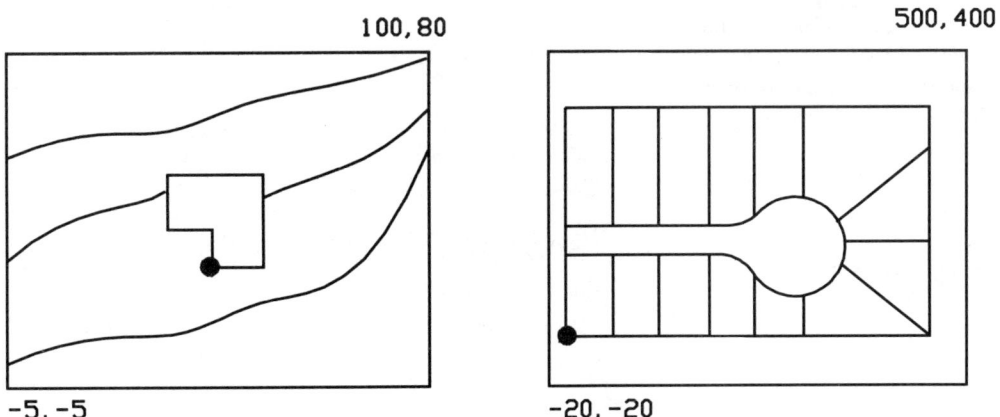

In civil applications, large sizes are often needed. The origin or 0,0 is usually on the bottom left.

Figure 1-3

LIMITS sets a flexible general size for your drawing.

To find LIMITS:

> **Windows** From the Data pull-down menu, choose Drawing Limits.
>
> **DOS** From the Data screen menu, choose Drawing Limits.

The command line equivalent is **LIMITS**.

Setting LIMITS does not limit your model; it merely lets you determine how big the finished product might be. Should the design change, you can change the LIMITS at any time.

If you set square limits, e.g. 10 × 10, you will use only part of the screen, because the screen's X value is larger than the Y.

If you choose the ON option you cannot work outside the area defined by the LIMITS command. You can choose OFF to have the grid extend to the edges of your screen.

Once the LIMITS are set, you may want to set up a GRID and SNAP.

ZOOM All allows you to view the size you have chosen. To make sure the limits and zoom are correct, press **F6** in DOS or **COORDS** in Windows and move the cursor around. The coordinates should change.

SNAP will set an increment that the cursor will move by.

GRID will set a visual aid to help you place objects. The GRID is often set to twice the SNAP value. The grid will extend over the area given by the LIMITS command. If the grid setting is too small for the screen limits, an error message will be issued: "Grid too dense to be displayed." Change the GRID spacing to a larger value.

> **Notes**
> If your GRID is not covering your screen, simply invoke the LIMITS command and pick the bottom left and top right of your screen.

To find GRID and SNAP

> **Windows** From the Tools pull-down menu, choose Drawing Aids.
>
> **DOS** From the Assist screen menu, choose Snap or Grid.

The command line equivalent is **SNAP** or **GRID**.

Setting a Drawing Boundary

Setting LIMITS sets up an invisible boundary that helps fit the drawing to paper at the scale chosen.

Example A1: A house that is 40′ × 36′.

```
Command:LIMITS
ON/OFF/Lower left corner<0'-0",0'-0">:-5',-5'
Upper right corner<12.0000,9.0000>:45',40'
Command:ZOOM
All/Center/Dynamic/Extents/Left/Previous/Vmax/Window/<Scale(X/
   XP)>:ALL
Command:SNAP
ON/OFF/Aspect/Rotate/Style/<1.0000>:1
Command:GRID
ON/OFF/Snap/Aspect/<0>:5
```

These settings will provide plenty of viewing space on either side of the drawing and allow you to enter the units by decimal point. The smallest integer picked will be one unit. This can be changed at any time.

Example M1: A template that is 220 mm × 160 mm.

```
Command:LIMITS
ON/OFF/Lower left corner <0.0000,0.0000>:-5,-40
Upper right corner<12.0000,9.0000>:240,180
Command:ZOOM
All/Center/Dynamic/Extents/Left/Previous/Vmax/Window/<Scale
   (X/XP)>:ALL
Command:SNAP
ON/OFF/Aspect/Rotate/Style/<1.0000>:5
Command:GRID
ON/OFF/Snap/Aspect/<0>:10
```

These settings will provide plenty of viewing space on either side of the model and allow a minimum five-unit integer entry.

Example C1: An intersection that is 15 m × 12 m.

```
Command:LIMITS
ON/OFF/Lower left corner <0.0000,0.0000>:-5,-5
Upper right corner<12.0000,9.0000>:20,15
Command:ZOOM
All/Center/Dynamic/Extents/Left/Previous/Vmax/Window/<Scale
   (X/XP)>:ALL
Command:SNAP
ON/OFF/Aspect/Rotate/Style/<1.0000>:1
Command:GRID
ON/OFF/Snap/Aspect/<0>:2
```

These settings provide plenty of viewing space and allows a minimum .5 unit integer entry.

Entry of Points Using Coordinates and Digitizing

All parts of geometry are entered by means of points. **Lines** have two points each. **Circles** have a center point and a point determining the radius. **Arcs** have a center point, a radius point, a start point, and an end point.

There are three ways of entering points:

- Entering them by coordinates: absolute values, relative values, or polar values
- Picking them on the screen, with or without SNAP, also called *digitizing*
- Entering them relative to existing geometry

In this chapter we will look only at the first two methods of point entry. The LINE command will be used to illustrate coordinate entries.

The LINE Command

Find LINE as follows:

> **WINDOWS** From the Draw toolbar, choose the Line button:
>
> **DOS** From the Draw pull-down menu, choose LINE.

The command line equivalent is **LINE** or the *command alias* **L**.

To create a LINE, you will need to know where it starts, and where it ends. Pick two or more points on the screen or enter the coordinates. Terminate the command by pressing ⏎ (the Return or Enter key).

Coordinate Entry Using Absolute, Relative, and Polar Values

The coordinates of an item, the *X*, *Y*, and *Z* values, can be entered either relative to the origin — the absolute value of the line — or relative to the last point entered — the incremental value.

Absolute Value Entries

In this method, the origin of the model or drawing does not change: the objects are placed relative to the origin. To enter the absolute value of an item, type in the *X* value, then the *Y* value, separated by a comma. Press ⏎ to signal the end of the coordinate entry.

```
Command:LINE
From point:0,0
To point:4,0 (from the absolute position of 0,0 to the
   absolute position of 4,0)
```

Relative Value Entries

To enter an incremental or relative value, type the **@** symbol (**Shift-2**) before the number. @ means "from the last point."

```
Command:LINE
From point:2,3
To point:@4,0 (from the absolute position of 2,3 to a position
   4 units in positive X from this point)
```

Try this example:

Absolute	*Relative*
Command:**LINE**	Command:**LINE**
From point:**0,0**	From point:**5,5**
To point:**4,0**	To point:**@4,0**
To point:**4,4**	To point:**@0,4**
To point:**0,4**	To point:**@-4,0**
To point:**0,0**	To point:**@0,-4**
To point:⏎	To point:⏎

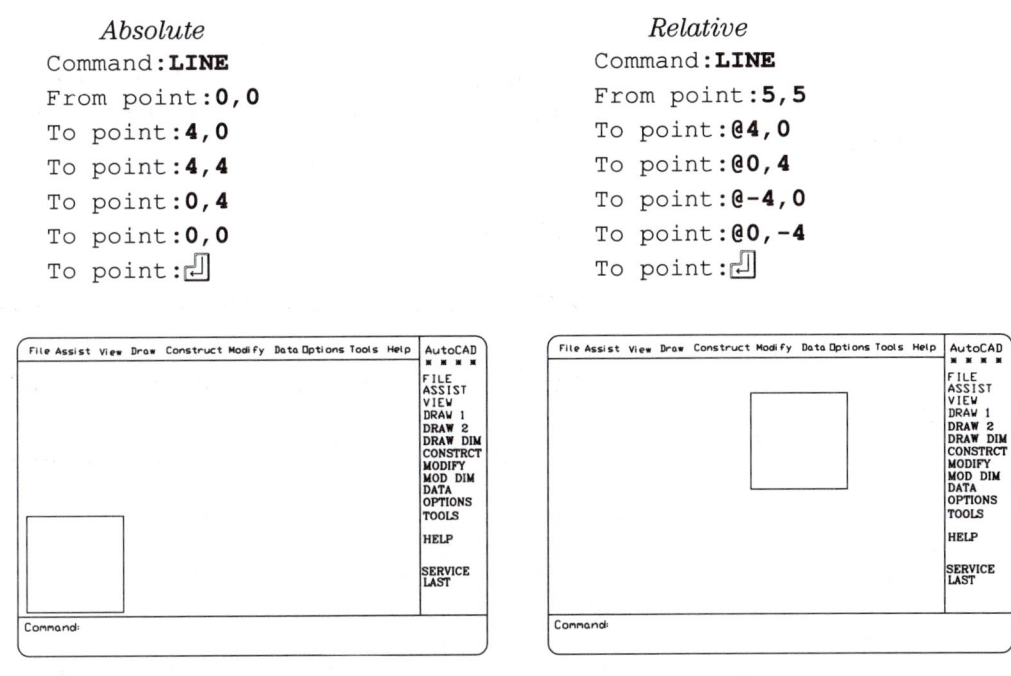

Absolute   Relative

Figure 1-4

The first example will give you a four-unit square starting at 0,0. The second will give you a four-unit square starting at 5,5.

To draw a line from point 5,6 to point 8.37,6 use either of the following.

Absolute	*Relative*
Command:**LINE**	Command:**LINE**
From point:**5,6**	From point:**5,6**
To point:**8.37,6**	To point:**@3.37,0**

In choosing between the absolute and the incremental method, the deciding factor is what you know. If you know that the final point is going to be 8.37,6, use the absolute value. If you know that the line is going to be 3.37 units in positive *X* from the last point, then enter the incremental coordinates.

Try these examples:

<pre>
        Absolute   Relative
Command:LINE   Command:LINE
From point:0,0   From point:7,5
To point:2,0   To point:@2,0
To point:2,3   To point:@0,3
To point:4,3   To point:@2,0
To point:4,2   To point:@0,-1
To point:6,2   To point:@2,0
To point:6,4   To point:@0,2
To point:0,4   To point:@-6,0
To point:0,0   To point:@0,-4
To point:⏎   To point:⏎
</pre>

The objects should look like this, one starting at 0,0 and the other starting at 7,5:

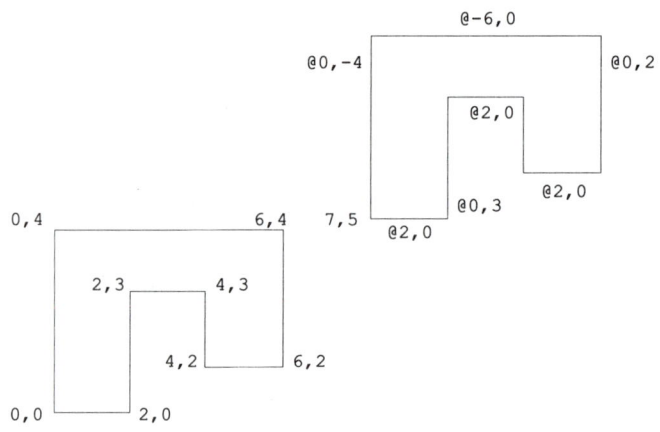

Figure 1-5

Polar Value Entries

Polar coordinates allow you to enter an item, relative to the last item, at a specified length and angle.

<pre>
Command:LINE
From point:3,4
To point:@4<45>
</pre>

Figure 1-6

Where: **@** = relative to the last point
4 = the length of the line
< = angle
45 = the angle that the line will be drawn at; all angles are calculated counterclockwise

Try this example:

```
Command: LINE
From point: 6,0
To point: @2<0>
To point: @3<90>
To point: @2<0>
To point: @1<270>
To point: @2<0>
To point: @2<90>
To point: @6<150>
To point: @1<210>
To point: C (for close)
```

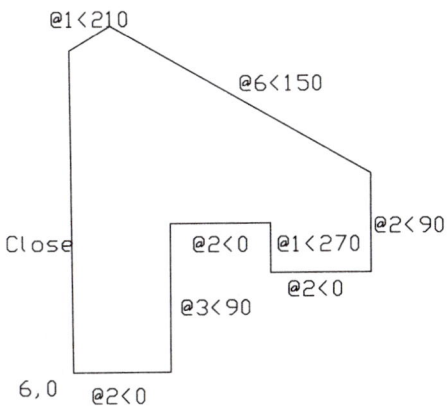

Figure 1-7

As noted above, angles are calculated counterclockwise from the furthest point in positive *X*.

Coordinate Entry Using Digitizing

The "pick" button on the mouse will enter a point every time you press it while in a geometry command. You can make your digitizing or picking of points much easier and much more accurate by using the SNAP function.

With SNAP you can draw lines, arcs or circles at preset integers. If you set the snap to .25, all entries will be to the nearest .25 interval as follows:

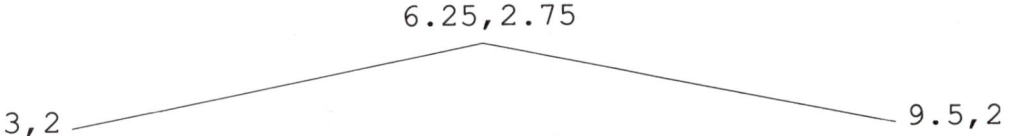

Figure 1-8

Try repeating the previous exercises using the mouse and setting the SNAP value to 1. If the coordinate readout does not move, press **F6** or the COORDS button.

If you set SNAP to 1, all the points you digitize or pick from the screen will be accurate to one-unit integers. You cannot be accurate without using SNAP.

Do not make the mistake of thinking that the grid will help to place items on the screen. The grid is for visual reference only.

ORTHO Mode

With the ORTHO command or option (F8 in DOS, the ORTHO button in Windows), lines can only be drawn vertically or horizontally. Assume SNAP is set at .25 and enter the following:

Figure 1-9

You will notice that the cursor only goes vertically and horizontally. By pressing **F8**, you will be able to draw diagonal lines again.

The GRID (**F7**) will give you a visual display of distance.

You can use any of the above methods in combination at any time.

The LINE Command

The LINE command is as simple as the above examples indicate. Indicate, with either a pick on the screen or a coordinate position, where the point should be. Any combination of points is accepted. Use ⏎ at the end of any coordinate entries.

> **Windows** From the Draw toolbar, pick Line.
>
> **DOS** From the Draw pull-down menu, choose Line.

The command line equivalent is **LINE** or **L**.

```
Command: LINE
From point: (pick a point)
To point: @3<250
To point: (pick another point)
To point: ⏎
```

When drawing lines, you are creating objects that are described by two points: a beginning and an end. Any number of points can be entered in the LINE command with each point joined to the last by a separate line. If you have entered five or six points in a single command, any of the lines can be erased.

LINE Options

C will close the string of lines with a line from the last point to the first point.

U will undo the last entered point.

.X, .Y, .Z are dealt with in Chapter 9.

Pressing ⏎ at the "From point:" prompt will attach the new line to the last point given.

Entering **U** at the command prompt will undo the last command.

The CIRCLE Command

Drawing a CIRCLE you are also describing an object that has two points; a center and a radius. The ARC has four points: a center, a radius, a start, and an end. The CIRCLE command will prompt you for the information needed to complete the circle. The command is:

> **Windows** From the Draw toolbar, choose the Circle flyout.
>
> **DOS** From the Draw menu, choose Circle.

The command line equivalent is **CIRCLE** or **C**.

```
Command:CIRCLE
3P/2P/TTR/<Center point>: (pick a point)
Diameter/<Radius>: DRAG (pick another point or type in a
   radius value)
```

Where:
3P	=	a circle fit through three points
2P	=	a circle fit through two points
TTR	=	a circle that is tangent on its diameter to two selected objects indicated with a specified radius
<Center point>:	=	the default circle which is described by a center point and a radius, in that order
Diameter/<Radius>:	=	the default circle specified by a radius; or type **D** to specify a diameter
DRAG	=	the drag mode; the circle will expand and contract following the movement of the cursor

Options appear when you type in **C** or **CIRCLE**. When picking CIRCLE from either the screen menu or the pull-down menu, you will be prompted for one of the options listed above. All of the options for creating circles are shown on page 24.

Notes: If you hold down the pick button on CIRCLE on the Draw toolbar, options will be offered.

The ARC Command

Arcs are also created by using options to control how the ARC is entered. The default is to define the first, then the second, then the third or final point of an arc. The command is:

> **Windows** From the Draw toolbar, choose the Arc flyout.
>
> **DOS** From the Draw 1 menu, choose Arc.

The command line equivalent is **ARC** or **A**.

```
Command:ARC
Center/<Start point>: (pick a point)
Center/End/<Second point>: (pick another point)
End point: (pick another point)
```

Introductory Geometry and Setting Up

Where: **Center** = an option to pick a center point
       \<Start point>: = the default first point to create an arc stretched through three points
       End = an option to pick the end point
       \<Second point>: = the default for a three-point arc
       End point = the end of a three-point arc

> **Notes**
> The key to using ARC is to determine what is not known and choose the options that work.

There are many variations on the ARC command, illustrated on page 25.

Should you want to enter the options at the command line, simply type in the option that you want as follows:

```
Command:ARC
Center/<Start point>: (pick 1)
Center/End/<Second point>:E
End point: (pick 2)
Radius/<Diameter>:R
Radius: (pick 3)
```

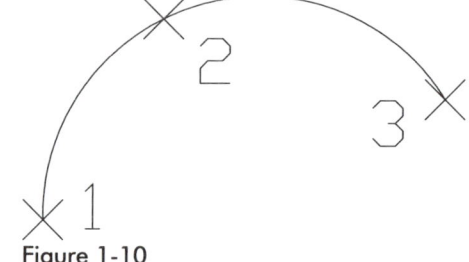
Figure 1-10

The FILLET Command

The FILLET command provides an easy way to place an arc between two existing objects, usually lines. FILLET can also be used with Radius 0 to clean up corners and connect lines to an apex.

> **Windows** From the Modify toolbar, choose the Fillet flyout.
>
> **DOS** From the Construct menu, choose FILLET.

The command line equivalent is **FILLET**.

```
Command:FILLET
(TRIM mode) Current fillet
   radius = <0.0000>
Polyline/Radius/Trim/<Select
   first object>: (pick 1)
Select second object: (pick 2)
```

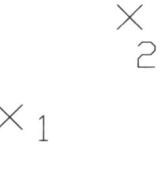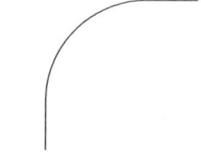

Figure 1-11

To change the radius choose **R**.

> **Notes**
> To restart the previous command at the command prompt, press ↵.

```
Command:FILLET
TRIM mode) Current fillet radius = <0.0000>
Polyline/Radius/Trim/<Select first object>:R
Radius:12
```

View Commands

The View menu and toolbar offer commands which will change the display of the model or drawing relative to the screen. Commands from the View menu will not change the coordinates or position of the model or the database. They only change the way you look at it.

The following commands appear in the View menu:

ZOOM	= magnifies a section of the screen
PAN	= moves the model across the screen without changing the magnification factor (zoom)
REDRAW	= updates the view and clears off blips (pick marks — see later) and erase marks
REGEN	= recomputes the file

The ZOOM Command

The following is the list of options for the ZOOM command. ZOOM is accessed by typing it on the command line, using the side menu, or using the pull-down menus or toolbars.

Note that the choices for the ZOOM command on the pull-down menu are limited.

```
Command: ZOOM
All/Center/Dynamic/Extents/Left/Previous/Vmax/Window/<Scale
   (X/XP)>
```

Where:		
	A	= All. Expands or shrinks the model or drawing to fit onto the screen relative to your limits
	C	= Center. Centers the model on the screen; you must enter a magnification factor or "height"
	D	= Dynamic. Creates a dynamic display of the item for zooming
	E	= Extents. Expands or shrinks the model or drawing to fit all of the objects on screen
	P	= Previous. Returns you to the Previous zoom factor
	V	= View maximum. Zooms the object as far out as possible without causing a REGEN
	W	= Window. Describes by two diagonal points a rectangle around the area you want to view
	Scale X	= specifies a percentage of the existing size
	Scale XP	= specifies a size relative to paper space

Scale works like this: .8x will display an image at 80% of its current size; .5x will display an image at half the current size; and 2x will display an image twice the size of the current size.

Zooming In and Out

Zooming in doubles the size of the image, zooming out reduces the image by half.

Zoom Limits shows the screen limits.

Introductory Geometry and Setting Up

Windows From the standard toolbar, choose In or Out.

DOS From the View menu, choose Zoom, then choose In or Out.

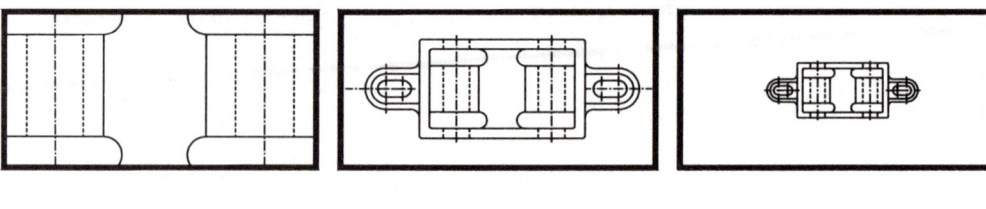

Zoomed in   Original   Zoomed out

Figure 1-12

ZOOM Window

To zoom into an area by specifying its boundaries, use Window.

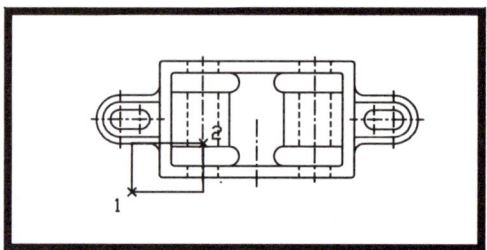 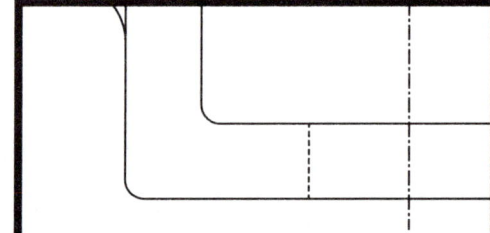

Figure 1-13

```
Command:ZOOM
All/Center/Dynamic/Extents/Left/Previous/Vmax/Window/
  <Scale (X/XP)>:W
First corner: (pick 1)
Other corner: (pick 2)
```

ZOOM All

To display the entire drawing, use ZOOM All.

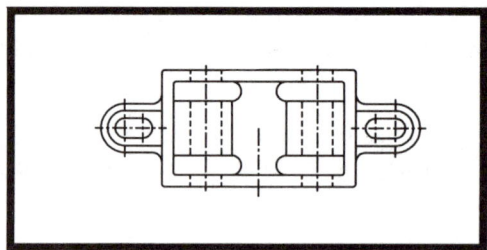 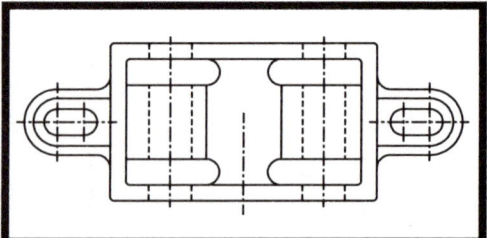

Figure 1-14

```
Command:ZOOM All
All/Center/Dynamic/Extents/Left/Previous/Vmax/Window/
  <Scale(X/XP)>:A
```

Zoom Relative to the Current Size

To scale a view relative to the current size, use the x. To double the size of the current view use 2x; to reduce the view to 75% use .75x.

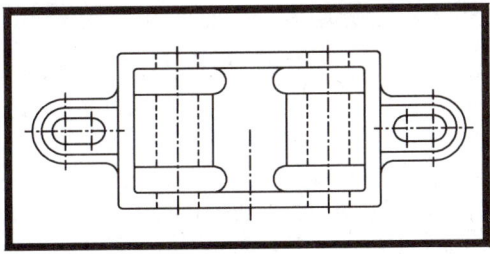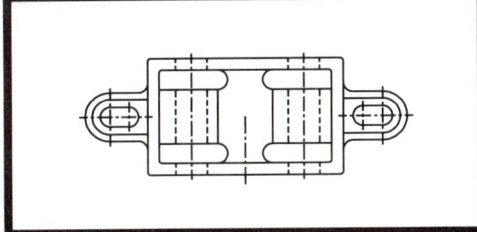

Figure 1-15

```
Command:ZOOM
All/Center/Dynamic/Extents/Left/Previous/Vmax/Window
   <Scale(X/XP)>:.75X (this is the same as Zoom Out)
```

Zoom Relative to the Drawing Limits

To scale the view to twice the drawing limits use 2. To display the image at half the full size use .5.

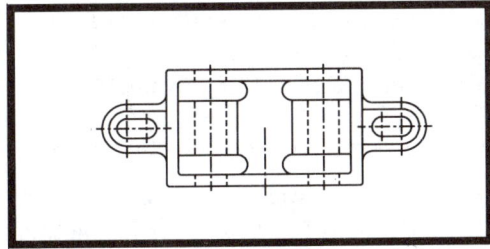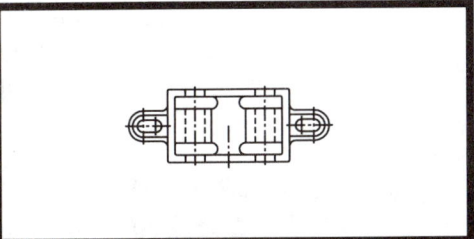

Figure 1-16

```
Command:ZOOM
All/Center/Dynamic/Extents/Left/Previous/Vmax/Window
   <Scale(X/XP)>:.5
```

The PAN Command

To move the view across the screen without changing the display size, use PAN.

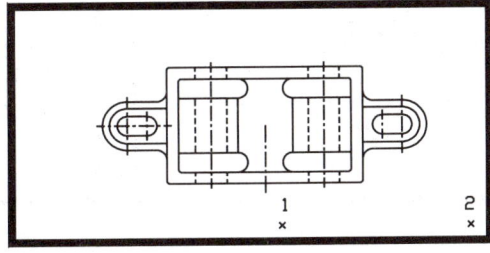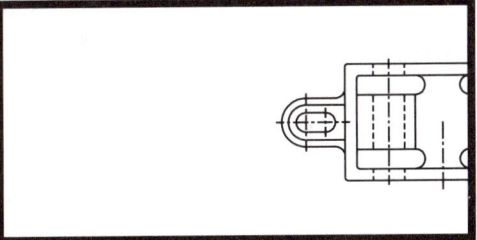

Figure 1-17

```
Command:PAN
Displacement: (pick 1)
Second point: (pick 2)
```

Windows Pick the icon or the slide bars on the side and bottom of the screen.

DOS From the View pull-down menu, pick PAN.

The command line equivalent is **PAN** or **P**.

PAN and ZOOM

The PAN command moves the model across the screen, while ZOOM magnifies the model within the screen. The database, i.e. the 0,0 point and associated coordinate points, remains the same.

If you want to see an area not currently in view but with the same magnification factor, use PAN to translate the data across the screen without changing the magnification. Often this does not require a screen regeneration, which can be quite time-consuming depending on the amount of information in the database. Past a certain point in either PAN or ZOOM, a screen regeneration may be required.

Transparent View Commands

ZOOM and PAN can be entered at the command line, or they can be accessed through the pull-down menus. If accessing the commands from the pull-down menus, ZOOM and PAN can be used within a command sequence. If you are in the middle of a line sequence and need to access a certain area for accuracy, choose ZOOM Window from the pull-down menu without exiting from LINE.

The REGEN and REDRAW Commands

While your data is always available, to save space in RAM, it is not always completely generated. Thus, the use of ZOOM or PAN sometimes results in a screen regeneration.

The REDRAW command cleans the screen and redraws any information that may be absent due to previous erasing or changes in the position of objects. It refreshes the screen from the available generated data.

While you are creating a model, the screen will fill up with a series of small items called **blips**, which look like tiny crosses. These indicate where you have digitized or picked a spot on the screen.

To clean the screen, you can use the REDRAW command by typing in **R** or using REDRAW on the View menu. If your grid is properly set, pressing **F7** twice (GRID on, GRID off) provides another method.

In the illustration at the right, the blips have been removed:

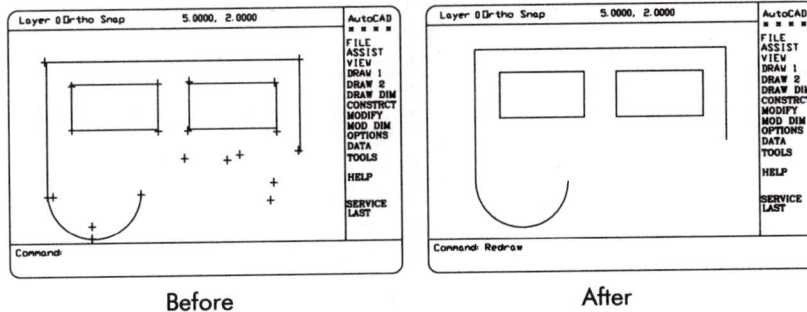

Before   After

Figure 1-18

The REGEN command actually regenerates all the data, and takes much longer. REGEN is used to update arc and circle displays to make the objects look more rounded. If you pick up an arc or circle in a ZOOM Window command, it is not rounded but octagonal, or at least "squared" rather than rounded at the corners. To have a better image of the circle, try REGEN. This will update the display of an arc or circle relative to the current magnification factor and display a superior image.

The UNITS Command

In AutoCAD it is suggested that you draw everything at full scale or 1:1 scale, and plot the drawing at the required scale factor later. The type of units chosen determines how AutoCAD interprets coordinate and angle entries.

AutoCAD offers various types of units of measure for entering into drawings. Before setting up the parameters of the drawing, first set up the units so that the readout displays the required units. Decimal mode may be used for metric units and for inch or foot units. When plotting to paper, if you want the number to represent an inch on the final plot, choose inches in the PLOT command. If you want the decimal to represent millimeters, choose mm in the PLOT command.

The scientific, decimal, and fractional modes simply equate one drawing unit to one displayed unit. The engineering and architectural modes assume that one drawing unit equals one inch. The decimal mode makes no assumptions.

Use these settings in UNITS to set up your drawing and they will be reflected properly in the final plot.

When you install AutoCAD Release 13 you are given the option of using Inches or Metric units as default units.

The UNITS command can be accessed either through the command line or through the DDUNITS dialog box.

> **Windows** From the Data menu, choose UNITS.
>
> **DOS** From the Data menu, choose UNITS.

The command line equivalent is **UNITS** or **DDUNITS**.

The dialog box offers the same parameters as the screen command, but in a slightly different format.

Introductory Geometry and Setting Up

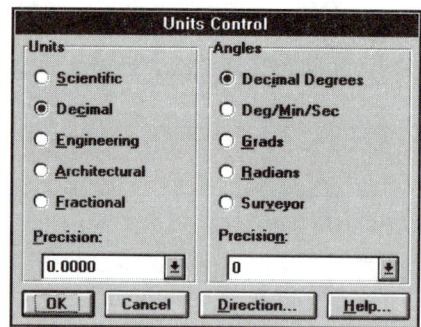
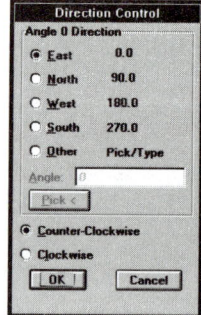

Figure 1-19

From the dialog box, choose the units readout that you require by picking the appropriate button. Choose the precision by picking the 0.0000 box; then fill in the desired accuracy. For surveying, the direction is important, so a direction dialog box is provided. Choose the appropriate readout for your discipline. Once completed, pick OK.

For the same information at the command line, use the following:

```
Command: UNITS
Report formats:
    1. Scientific      1.55E+01
    2. Decimal         15.50
    3. Engineering     1-3.58"
    4. Architectural   1'-3 1/2"
    5. Fractional      15' 1/2"
```

Decimal (Option 2)

This option changes the amount of decimals shown on the status line and in your dimensions. Once the type of unit you want to use is selected, you are then prompted for the precision.

```
Number of digits to the right of the decimal point (0 to 8):
```
2 = 1.00

6 = 1.000000

Architectural (Option 4)

When Architectural (i.e. imperial) units are chosen, the following prompt will appear:

```
Denominator of smallest fraction to display (1, 2, 4, 8, 16,
   32, or 64):
```
1 = 1″

2 = 1/2″

16 = 1/16″

When using architectural measurements, each numerical entry is accepted as an inch. The ′ and ″ symbols must be used to differentiate between feet (′) and inches (″). Following is an example of an architectural unit input using the LINE command.

```
Command: LINE
From point: 2,3 (X 2 inches, Y 3 inches)
To point: @4',0 (4 feet in X and none in Y)
To point: @0,2'3 (none in X and 2 feet 3 inches in Y)
To point: @3'4-1/2",0 (3 feet 4 1/2 inches in X and none in Y)
```

All angles will be measured in decimal degrees in a counterclockwise direction. If you want to change this option, use the following:

```
Systems of angle measure:
  1. Decimal degrees        45.0000
  2. Degrees/minutes/seconds 45deg0'0"
  3. Grads                  50.0000g
  4. Radians                0.7854r
  5. Surveyor's units       N45d0'0"E
```

> **Notes**
> The hyphen (-) in the architectural point entry is not the same as in the architectural readout on the status line.

Usually the default is decimal degrees. The next few options are mainly for surveyors; keep accepting the defaults until the command prompt is reached, then press **F1** in DOS or **F2** in Windows to toggle back to the graphics screen.

Do not worry about the paper size until you are ready to plot.

Scientific (Option 1)

In scientific notation, the E is the exponent factor of 10 and is shown with a plus sign. The quantity of 4.1325E+6 is read as 4132500. This notation is used when numbers are very large, such as in astronomy or physics. Its use reduces the need for a large number of trailing zeros.

Engineering (Option 3)

Often in civil engineering there is a need to express feet and inches without fractions. With this option, you can do exactly that.

$$3'\ 4\ 3/4'' = 3'\ 4.75''$$

In fact, with engineering units, the inch symbol is not required. Therefore, you are able to place objects in inches with good precision.

```
Command: LINE
From point: 4'3.5,0 (4 feet 3 1/2 inches in X and 0 in Y)
To point: @4.4,0 (4 4/10 inches in X and none in Y)
```

Introductory Geometry and Setting Up

Fractional (Option 5)

With fractional units the information is entered as a fraction with a slash (/). For mixed numbers a hyphen (-) must be added with the slash (/).

```
Command:LINE
From point:1-1/2,2-3/4
To point:@3/4,0
```

Choose the required precision from the next prompt.

Surveying

The surveyor's "compass rose" is much the same as a ship's compass — divided into four parts with the top being north, the left being west, etc. Angles are expressed in 90 degree quadrants.

The quadrant between north and east, for example, starts at 0 degrees due east and progresses 90 degrees to due north. To express 25 decimal degrees using AutoCAD's default origin for angles, enter **N25d0'0"E**. You may omit null minutes and/or seconds and enter **N25dE**.

When entering this measurement, do not use spaces.

```
Command:L
From point:0,0
To point:@38<S44d14'9"W
```

Measuring Angles

AutoCAD's default setting for angles is zero degrees at due east. You may change this zero-degree reference point to due north, due west, or due south. These are the only four positions offered by the UNITS command. To orient the zero reference at an angle other than those specified, you can change the user coordinate system (UCS) as explained later.

CLOCKWISE OR COUNTERCLOCKWISE These two angle rotations can be applied to all angles. The rotation always starts at the zero reference point.

UNTRANSLATED ANGLES If the UNITS command is set to a nondecimal angular mode (e.g. radians), an angle can be preceded by a **<** to enter a measurement counterclockwise from three o'clock.

If an angle measurement direction or origin has been changed, enter <before an angle measurement to have the angle measured counter clockwise from three o'clock.

USING UNITS EFFECTIVELY Remember that the UNITS command does not operate like the SNAP command. The status line will indicate the position relative to the units that were specified, but points will not be entered exactly at the position specified without setting the SNAP.

The units can be changed when beginning a model or drawing, or when entering all or part of a model. For example, in a floor layout at 1:1 decimal units, you can switch to architectural units and then scale the entire model by 12. See Chapter 11 for plotting scales.

Prelab 1 Using LINE, ARC, CIRCLE and FILLET

Step 1 Start a new file and set the LIMITS to 100,60, GRID to 10, SNAP to 5. Use the screen menu to do this. Then pick ZOOM All from the View pull-down menu or the button.

At the command line, type **LIMITS**, then **SNAP**, then **GRID**.

```
Command:LIMITS
ON/OFF/Lower left corner<0.0000,
   0.0000>:⏎
Upper right corner<12.0000,9.0000>
   :100,60
Command:SNAP
ON/OFF/Aspect/Rotate/Style/
   <1.0000>:5
Command:GRID
ON/OFF/Snap/Aspect/<1.0000>:10
Command: (from the View menu, pick ZOOM All)
```

Step 2 Start by drawing circles. With the first one, use the screen menu.

With the second, use the command alias. With the third, use the pull-down menu.

> **Windows** From the Draw toolbar, choose the Circle flyout.
>
> **DOS** From the Draw 1 menu, choose Circle. (If the Draw 1 menu is not displayed, pick AutoCAD at the top right of your screen.)

The command line equivalent is **CIRCLE** or **C**.

```
Command:CIRCLE
3P/2P/TTR/<Center point>:25,30
Diameter/<Radius>:10
Command:C
3P/2P/TTR/<Center point>:70,40
Diameter/<Radius>:(pick a point 5
   units straight over)
Command:⏎ (retrieves the previous
   command)
3P/2P/TTR/<Center point>:70,20
Diameter/<5.0000>:⏎
```

Introductory Geometry and Setting Up

Step 3 Now use ARC with the option St,E,Rad to create an arc. Use the screen menu. Make sure that SNAP is on. The word SNAP should be on the status line. If not, press **F9** or the Snap button.

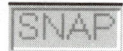

> **Windows** From the Draw toolbar, choose the Arc flyout.
>
>
>
> **DOS** From the Draw 1 menu, choose Arc.

The command line equivalent is **ARC**.

```
Command:ARC St,E,Rad
Center/<Start point>: (pick 1)
Center/End/<Second point>:
End point: (pick 2)
Radius/<Diameter>:
Radius: (pick 3)
```

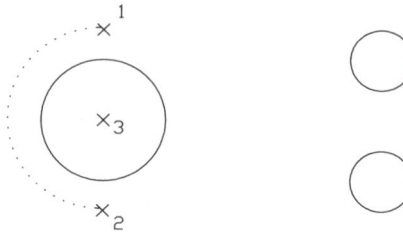

Step 4 With the SNAP still on, add the other two arcs. Type in **A** at the command prompt to access the command.

```
Command:A
Center/<Start point>: (pick 4)
Center/End/<Second point>:E
End point: (pick 5)
Radius/<Diameter>:R
Radius: (pick 6)
Command:⏎
Center/<Start point>: (pick 7)
Center/End/<Second point>:E
End point: (pick 8)
Radius/<Diameter>:R
Radius: (pick 9)
```

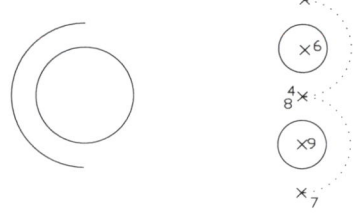

Step 5 Now draw in the lines from the ends of the arcs.

```
Command:LINE
From point: (pick 10)
To point: (pick 11)
To point:⏎
Command:L
From point: (pick 12)
To point: (pick 13)
To point:⏎
```

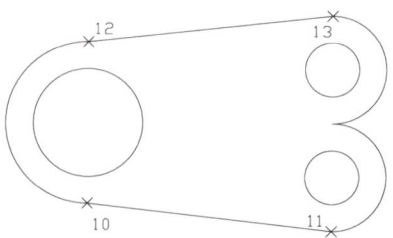

Step 6   Use the FILLET command to complete the drawing.

> **Windows** From the Modify toolbar, choose the Feature flyout, then Fillet.
>
>
>
> **DOS** From the Construct menu, choose Fillet.

```
Command: FILLET
FILLET polyline/Radius/<Select first
    object>:R
Current radius <0.0000>:15
Command:FILLET
FILLET polyline/Radius/<Select first
    object>: (pick 14)
Select second object: (pick 15)
```

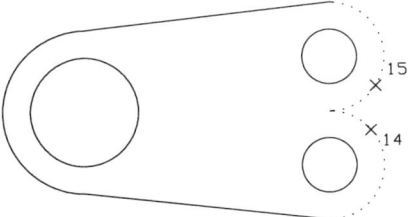

Your drawing should look like this.

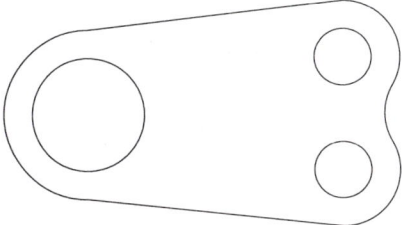

To save the file, access the File pull-down menu. Choose SAVE. See Introduction, page xx for details.

Introductory Geometry and Setting Up   **23**

Circle Draw Menu

Center Radius
If you know the center and the radius

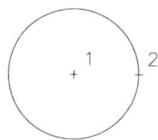

Command:CIRCLE
3P/2P/TTR/<Center point>: **(pick 1)**
<Radius>/Diameter: **(pick 2)**

Center Diameter
If you know the center and the diameter

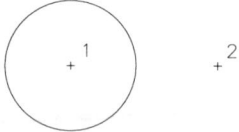

Command:**CIRCLE**
3P/2P/TTR/<Center point>: **(pick 1)**
Radius/<Diameter>: **(pick 2)**

2-Point

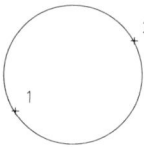

Command:**CIRCLE**
3P/2P/TTR/<Center point>:**2P**
First point on diameter: **(pick 1)**
Second point on diameter: **(pick 2)**

3-Point

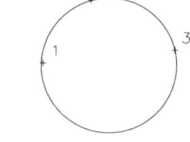

Command:**CIRCLE**
3P/2P/TTR/<Center point>:**3P**
First point: **(pick 1)**
Second point: **(pick 2)**
Third point: **(pick 3)**

TTR

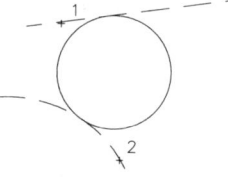

Command:**CIRCLE**
3P/2P/TTR/<Center point>:**TTR**
Enter Tangent spec: **(pick 1)**
Enter second Tangent spec: **(pick 2)**
Radius:**2**

Fillet

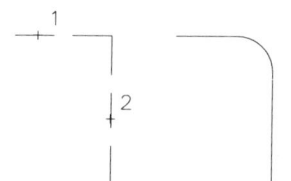

Command:**FILLET**
FILLET Polyline/Radius/<Select first object>:**R**
Current radius<0.0000>:**15**
Command:**FILLET**
FILLET Polyline/Radius/<Select first object>: **(pick 1)**
Select second object: **(pick 2)**

TTT

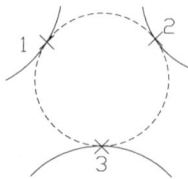

Command:**CIRCLE**
3P/2P/TTR/<Center point>:**3p**
First point:_tan to **(pick 1)**
Second point:_tan to **(pick 2)**
Third point:_tan to **(pick 3)**

Arc Draw Menu

3-point

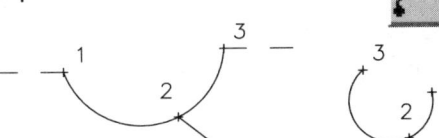

```
Command:ARC
Center/<start point>: (pick 1)
Second point: (pick 2)
Endpoint: (pick 3)
```

St,C,End   Start, Center, End

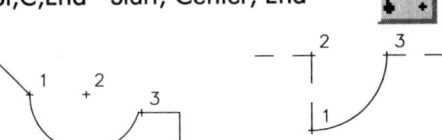

```
Center/<start point>: (pick 1)
Center: (pick 2)
Endpoint: (pick 3)
```

St,C,ANG   Start, Center, Angle

```
Command:ARC
Center/<start point>: (pick 1)
Center: (pick 2)
Angle:90
```

St,C,Len   Start, Center, Length

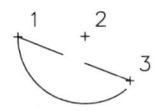
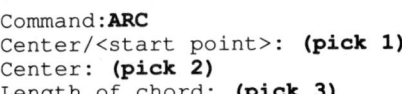

```
Command:ARC
Center/<start point>: (pick 1)
Center: (pick 2)
Length of chord: (pick 3)
```

St,E,Ang   Start, End, Angle

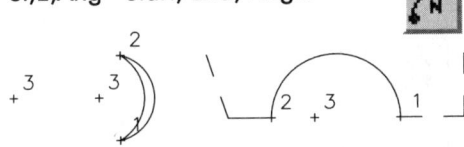

```
Command:ARC
Center/<start point>: (pick 1)
Endpoint: (pick 2)
Angle:90
```

St,E,Rad   Start, End, Radius

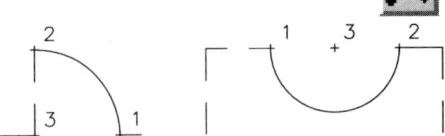

```
Command:ARC
Center/<start point>: (pick 1)
Endpoint: (pick 2)
Radius:.5
```

St,E,Dir   Start, End, Direction

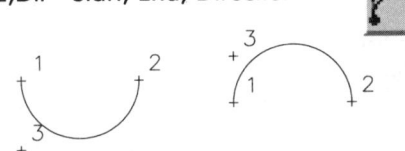

```
Command:ARC
Center/<start point>: (pick 1)
Endpoint: (pick 2)
Direction:9
```

Ce,S,End   Center, Start, End

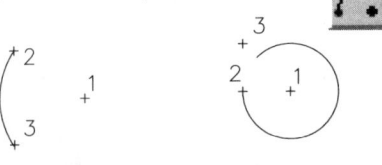

```
Command:ARC
Center: (pick 1)
Start point: (pick 2)
Endpoint: (pick 3)
```

Ce,S,Ang   Center, Start, Angle

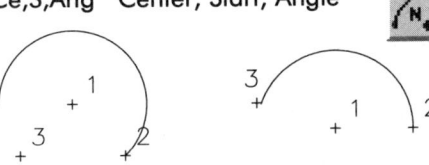

```
Command:ARC
Center: (pick 1)
Start point: (pick 2)
Angle:9
```

Ce,S,Len   Center, Start, Length

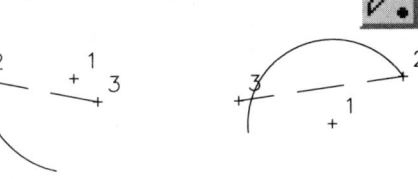

```
Command:ARC
Center: (pick 1)
Start point: (pick 2)
Length of chord: (pick 3)
```

Introductory Geometry and Setting Up

Practice Exercise 1

For these exercises you will not need to change your limits, but you may need to change your SNAP and GRID. File each as a different drawing.

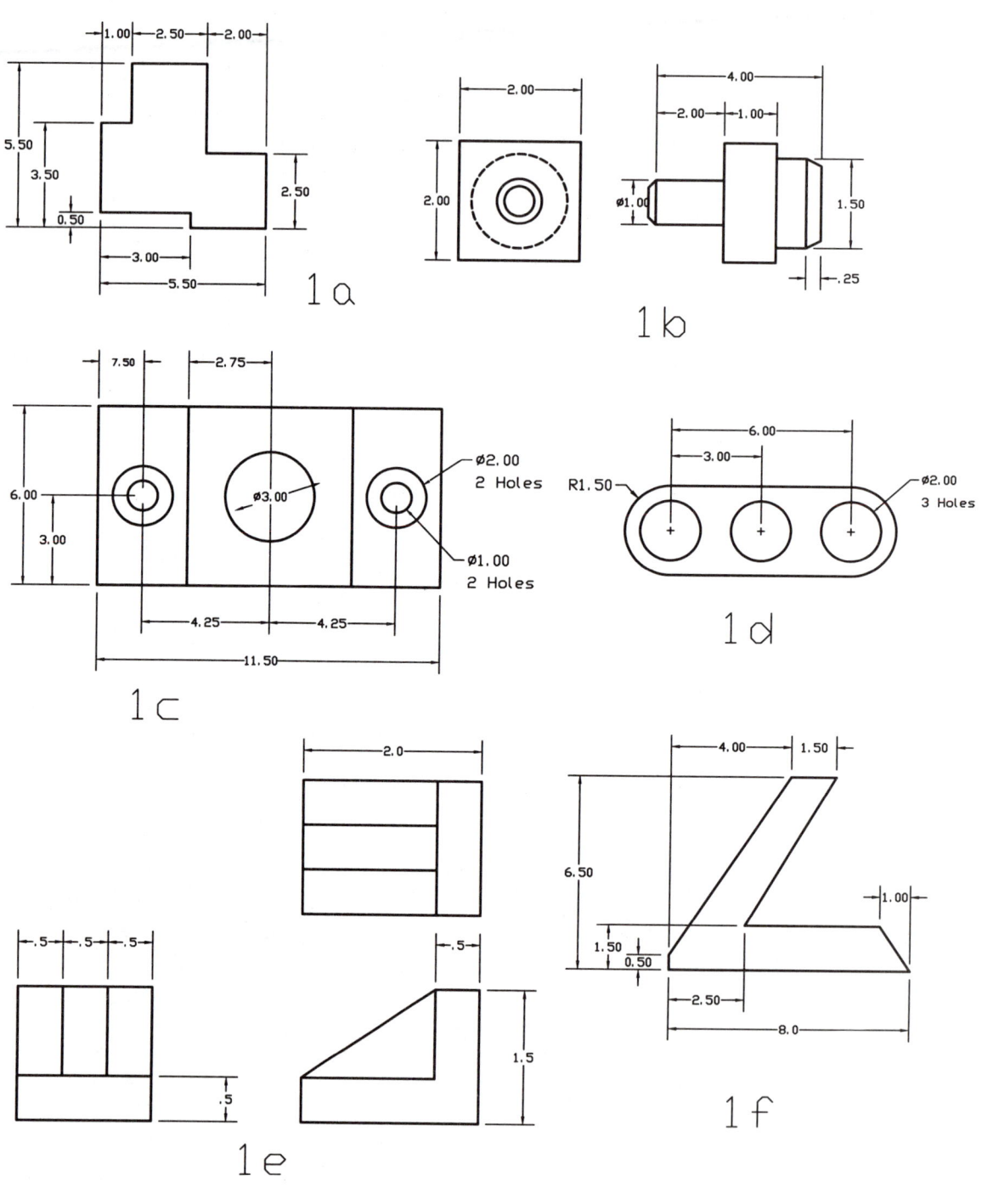

Exercise A1

Set up your screen using UNITS, LIMITS, GRID, and SNAP, then use LINE and ARC to add in the floor plan outline as shown. If you get finished early, the remainder of the interior walls are on Exercise A5.

```
Command:UNITS, 4, 1
  (press ↵ accepting the default on the other settings until
    you return to the command prompt)
```

Use **F1** to flip your screen back to the graphics screen.

```
Command:LIMITS -3',-3'    45',38'
Command:ZOOM A
Command:SNAP 1
Command:GRID 6
```

Use **F6** to get your coordinates to move.

When you are finished, SAVE the file as SECONDFL.

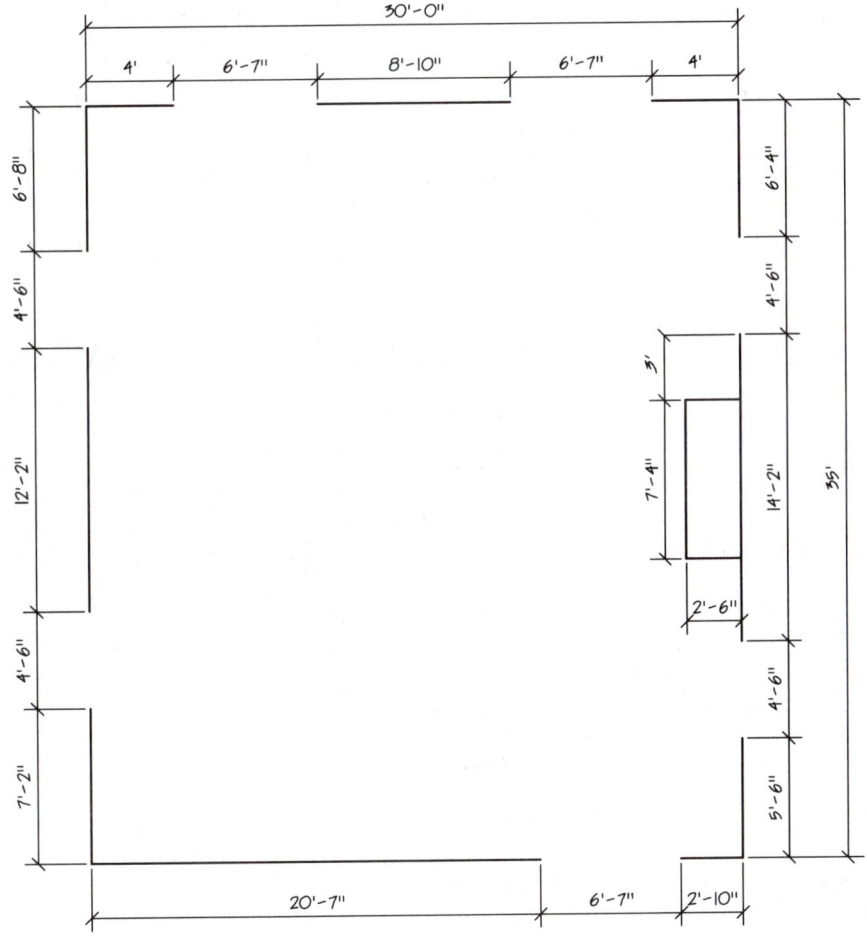

Introductory Geometry and Setting Up   **27**

Exercise C1

Settings Menu

Command:**LIMITS**
Bottom left corner<0.0000,0.0000>: **-5,-5**
Upper right corner<12.0000,9.0000>:**35,25**

Display Menu

Command:**ZOOM**
All/Center/Dynamic/Extents/Left/Previous/Vmax/Window:**ALL**

Settings Menu

Command:**SNAP**
On/Off/Value/Aspect/Rotate/Style/<1.0000>:**1**
Command:**GRID**
On/Off/Value<0.0000>:**5**

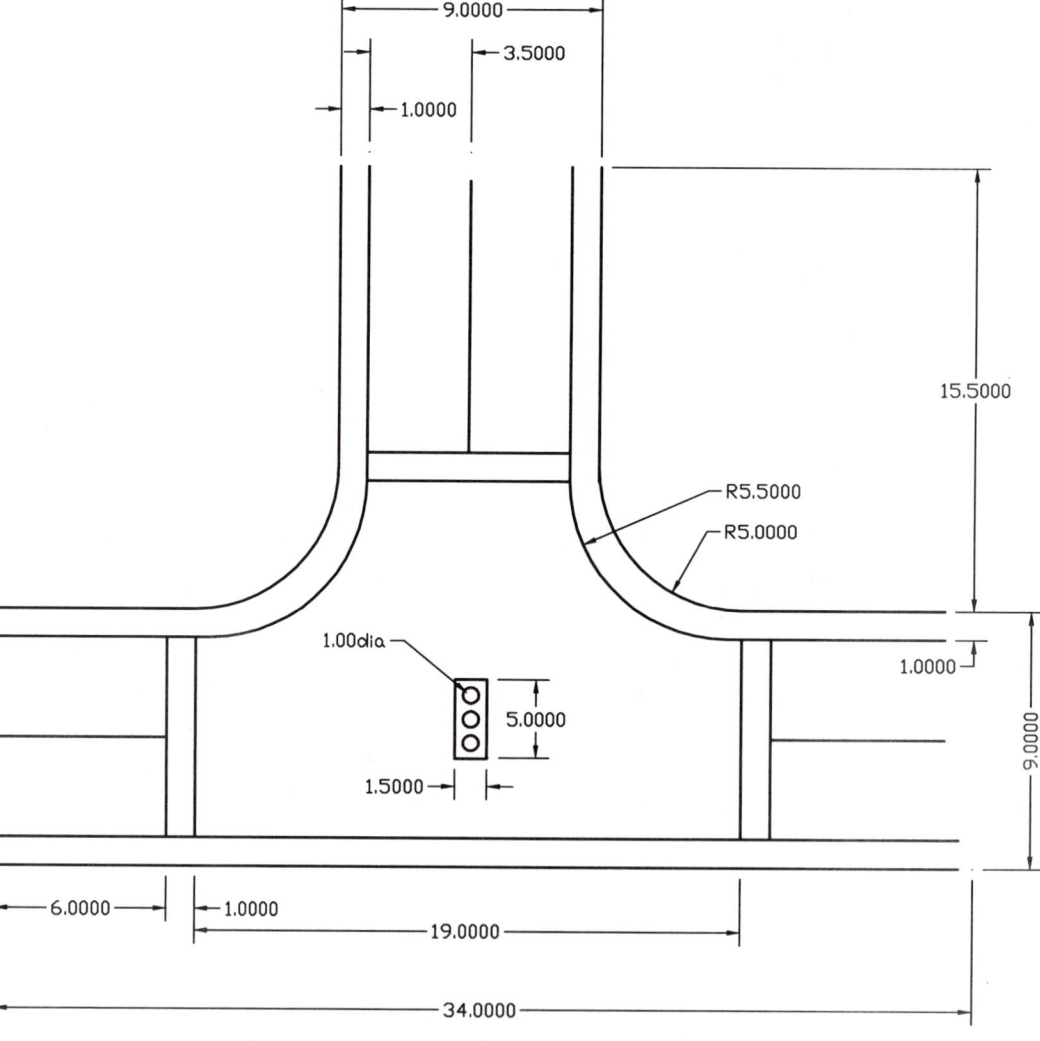

Do the geometry commands only. Do not put in the dimensions.

Hints for C1

Use the following commands to complete the exercise.

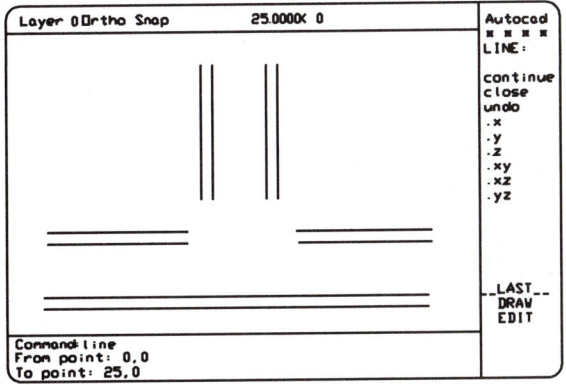

Lines
With your snap on draw the lines

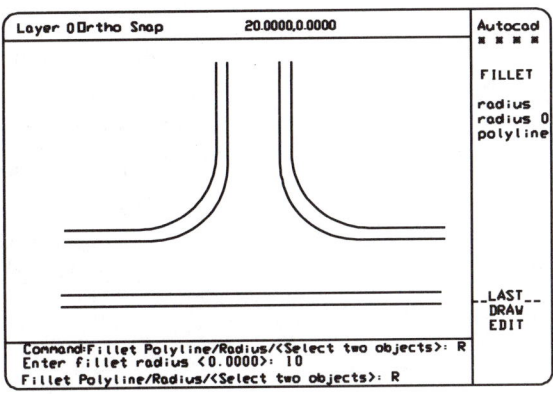

Fillets
Fillets are under the EDIT menu
Change the Radius

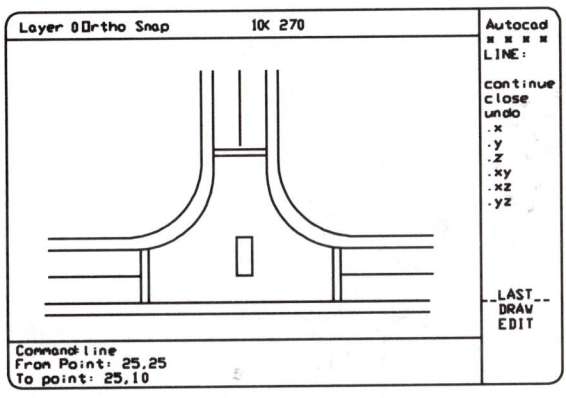

Lines
Change your snap for more accuracy

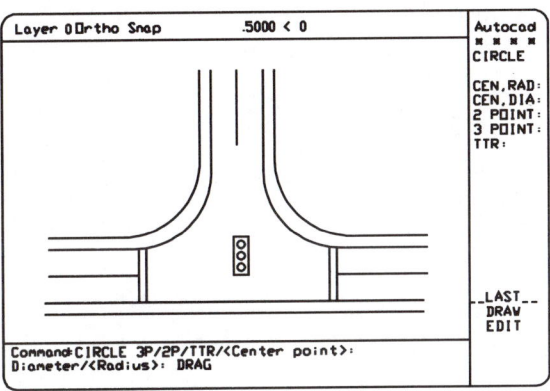

Circles
Use CENtre RADius for circles
Use drag to change the radius

Make sure you have a "Command:" prompt before starting a new command. If you do not, use **Ctrl-C** or the cancel button.

ARC, LINE, and CIRCLE are under the Draw menu. FILLET is under the Edit menu.

When you are finished, type **END**. This will save it on your floppy disk.

Introductory Geometry and Setting Up   29

Exercise E1

Settings Menu

```
Command:LIMITS
Bottom left corner<0.0000,0.0000>:-1,-1
Upper right corner<12.0000,9.0000>: ⏎
```

Display Menu

```
Command:ZOOM
All/Center/Dynamic/Extents/Left/Previous/Vmax/Window:ALL
```

Settings Menu

```
Command:SNAP
On/Off/Value/Aspect/Rotate/Style/<1.0000>:.5
Command:GRID
On/Off/Value<0.0000>:1
```

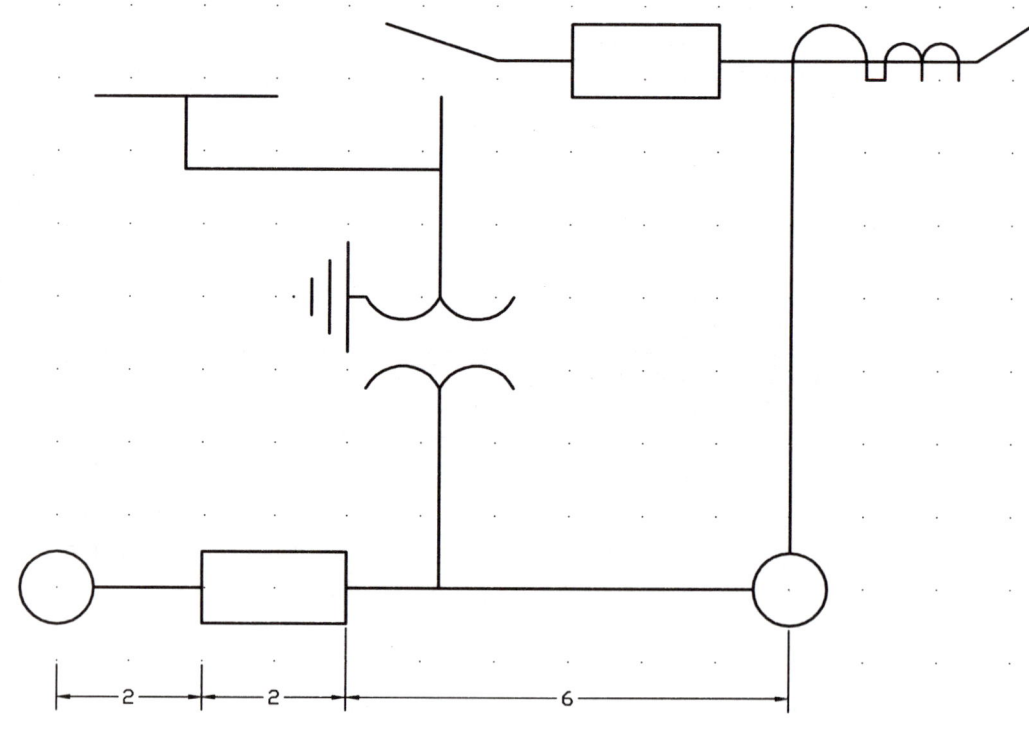

Do the geometry commands only. Do not put in the dimensions.

Hints for E1

Use the following commands to complete the exercise.

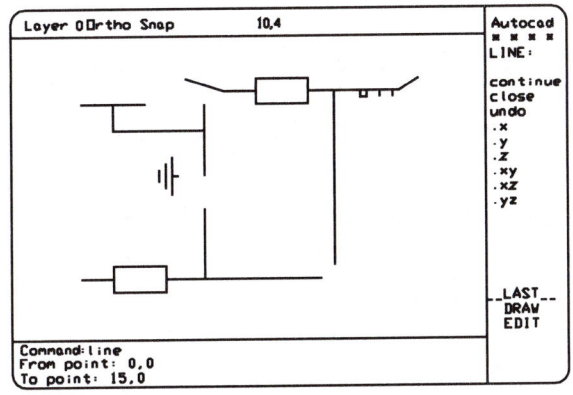

Lines

With your snap on draw the lines
Use only LINE, not X, Y etc.

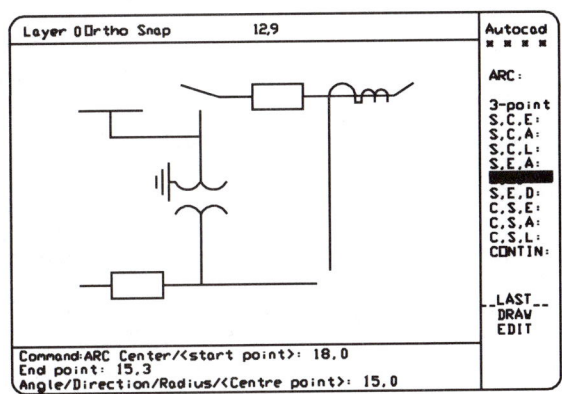

Arcs

Use Start End Radius for arcs
Remember, only counter clockwise

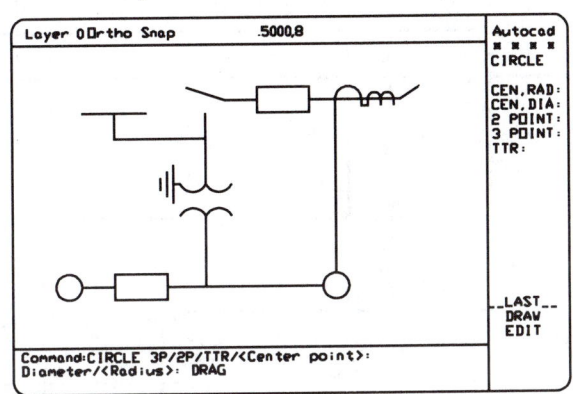

Circles

Use CENtre RADius for circles
Drag the radius over, use snap

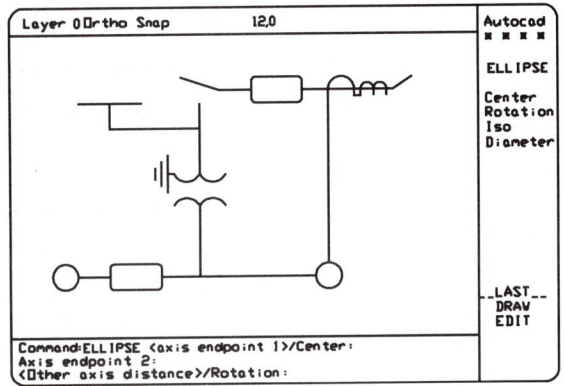

Fillets

Fillets are under the EDIT menu
Change the Radius

Make sure you have a "Command:" prompt before starting a new command. If you do not, use **Ctrl-C** or the cancel button.

ARC, LINE, and CIRCLE are under the Draw menu. FILLET is under the Edit menu.

When you are finished, type **END**. This will save it on your floppy disk.

Introductory Geometry and Setting Up

Exercise M1

Settings Menu

```
Command:LIMITS
Bottom left corner<0.0000,0.0000>:-10,-40
Upper right corner<12.0000,9.0000>:220,110
```

Display Menu

```
Command:ZOOM
All/Center/Dynamic/Extents/Left/Previous/Vmax/Window:ALL
```

Settings Menu

```
Command:SNAP
On/Off/Value/Aspect/Rotate/Style/<1.0000>:5
Command:GRID
On/Off/Value<0.0000>:10
```

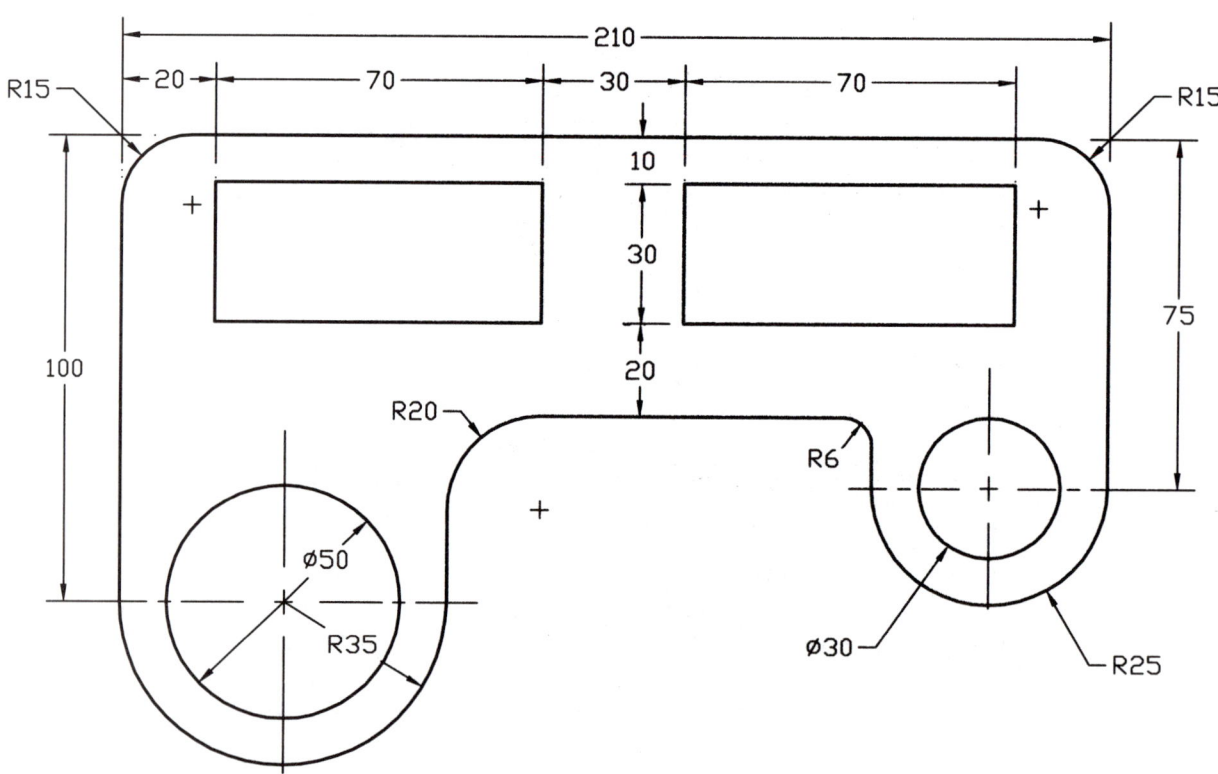

Do the geometry only. No dimensions are needed yet.

Hints for M1

Use the following commands to complete the exercise.

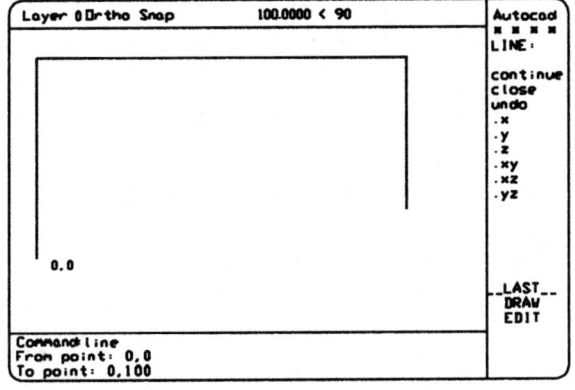

Lines
With your snap on draw the lines
Watch the top line for position

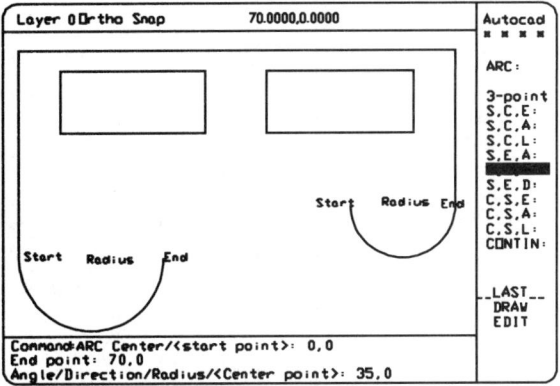

Arcs
Use Start End Radius for arcs
Remember to go counter clockwise

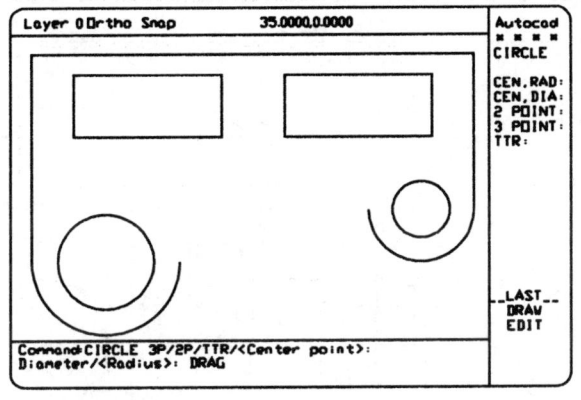

Circles
Use CENtre RADius for circles
DRAG means drag out the radius

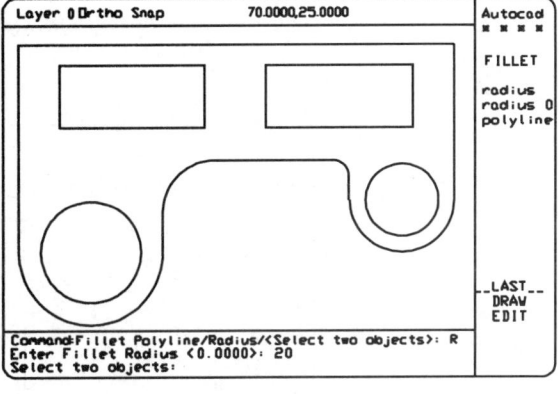

Fillets
Fillets are under the EDIT menu
Change the Radius

Make sure you have a "Command:" prompt before starting a new command. If you do not, use **Ctrl-C** or the cancel button.

ARC, LINE, and CIRCLE are under the Draw menu. FILLET is under the Edit menu.

When you are finished type in **END**. This will save the drawing on your floppy disk.

Challenger 1A

If you have taken AutoCAD before, try these.

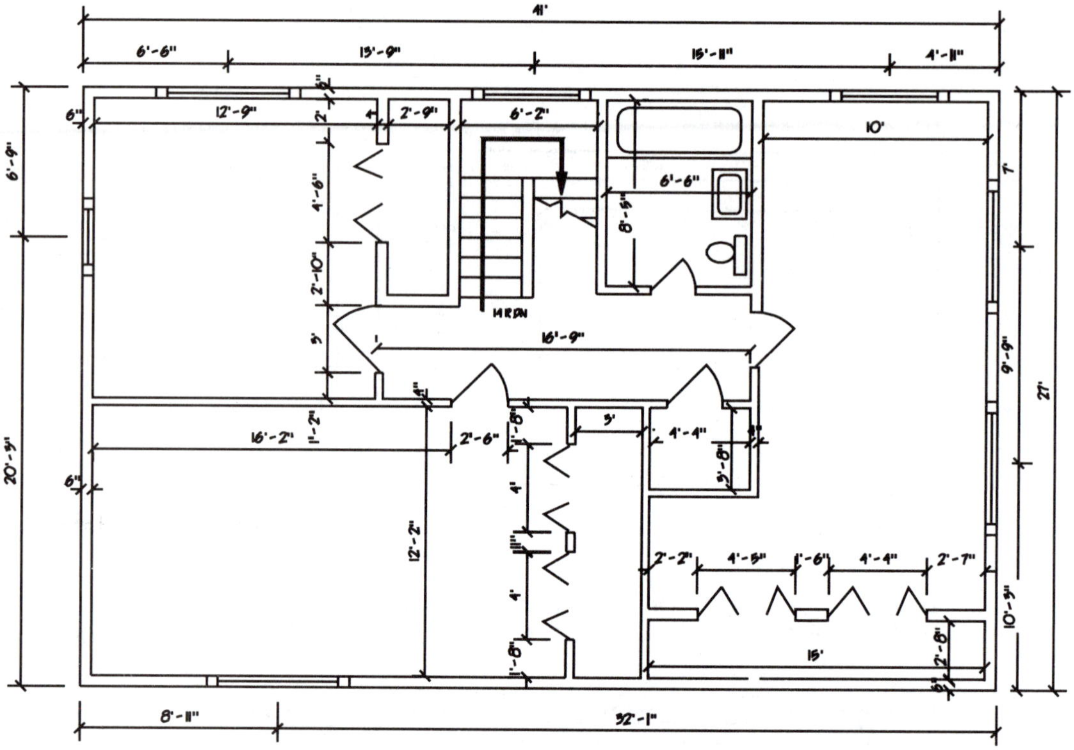

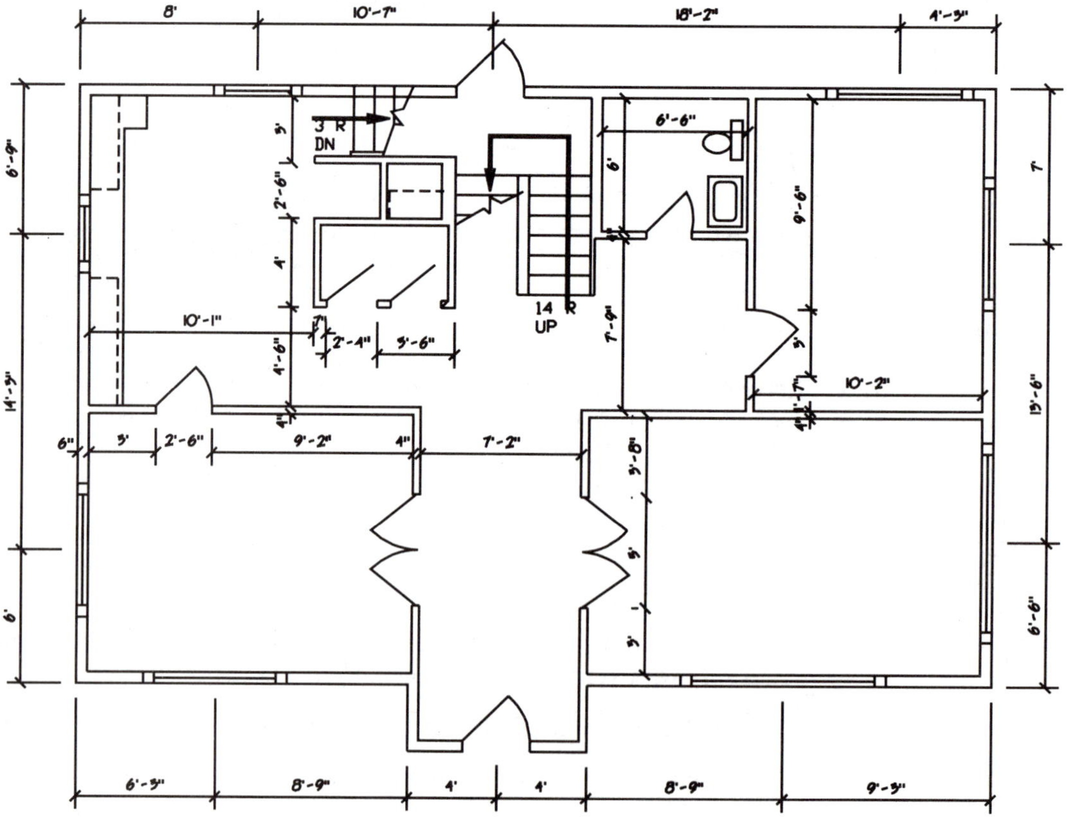

Challenger 1B

These exercises will help you with your skills.

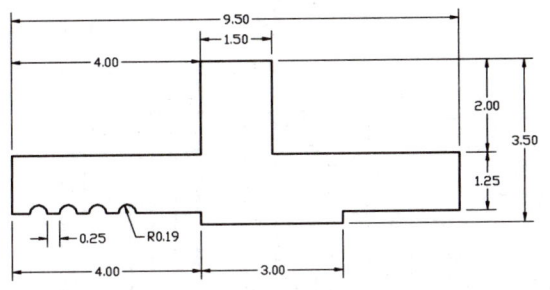

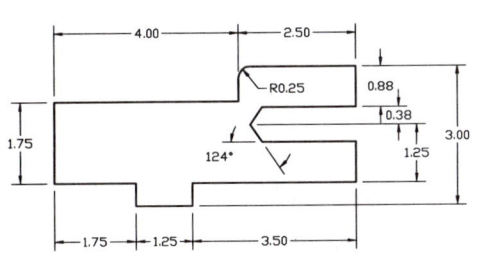

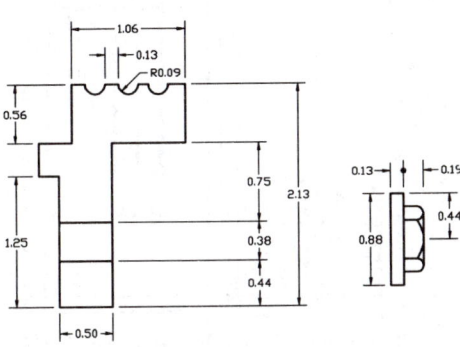

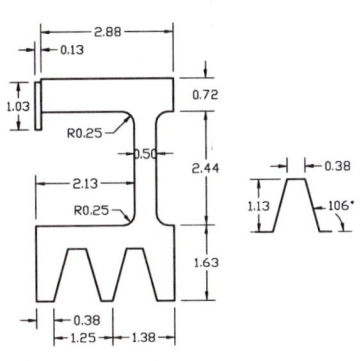

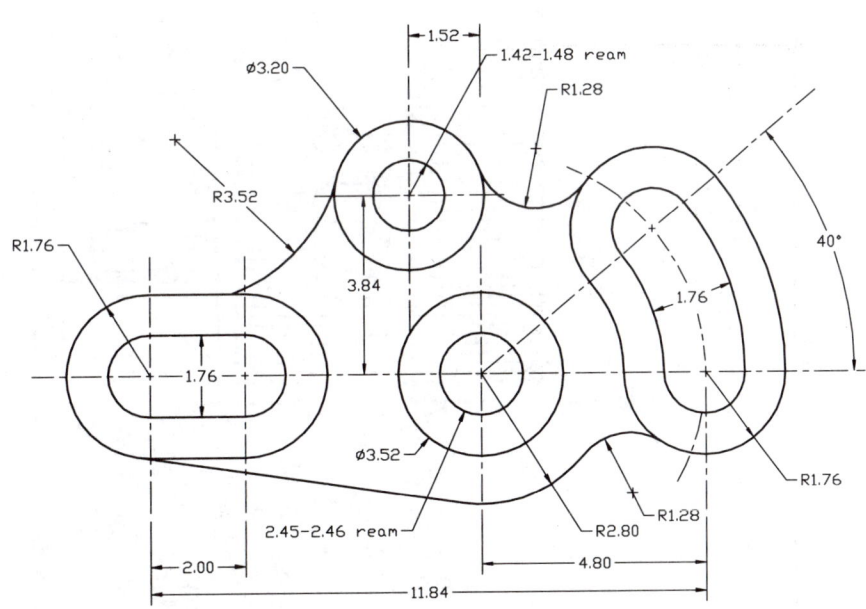

Introductory Geometry and Setting Up

Challenger 1C

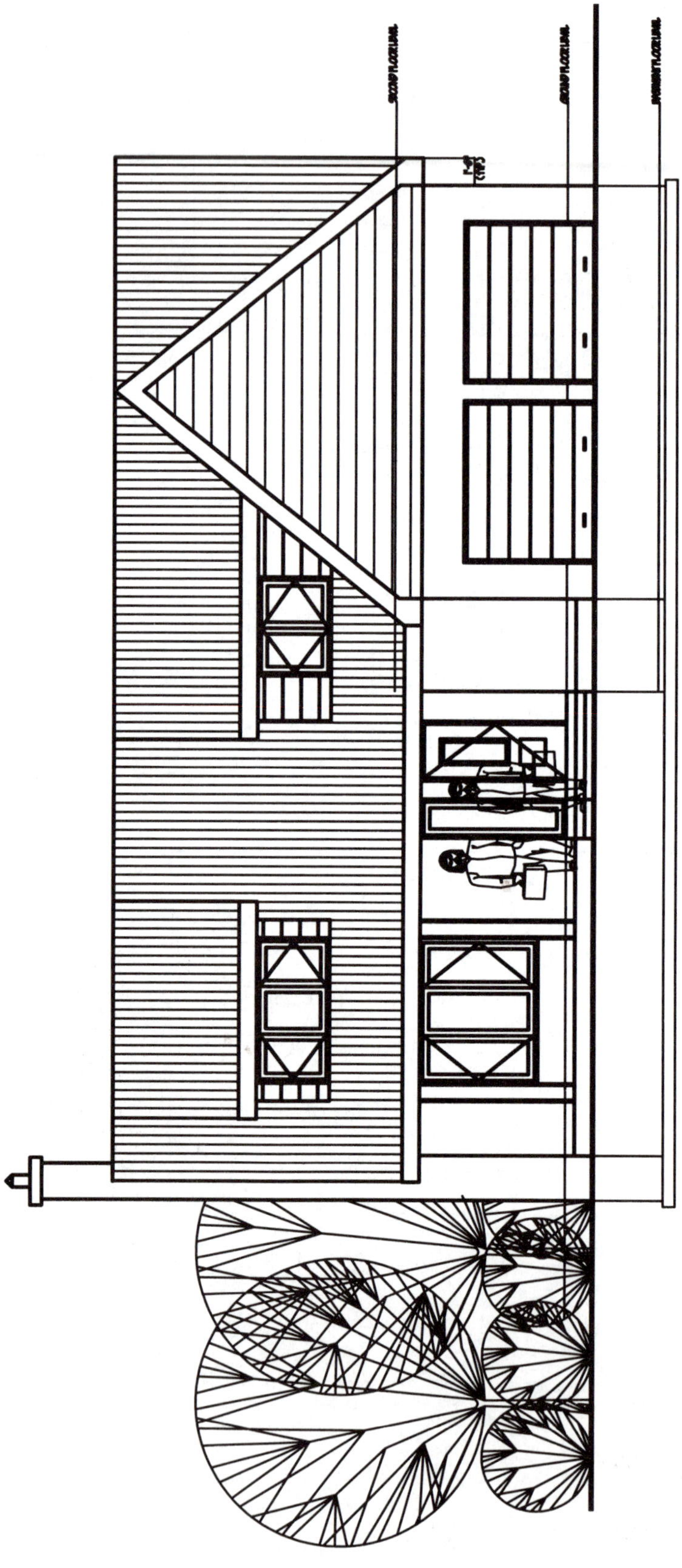

36 CHAPTER ONE

Help Files, CHAMFER, OSNAP, ERASE, TRIM, and BREAK

Upon completion of this chapter, you should be able to:

1. Retrieve on-line documentation or Help files
2. Use the commands GRID and SNAP effectively
3. Use OSNAP both within a command and as a setting
4. Use TRIM and BREAK to erase portions of objects
5. Use ERASE to remove whole objects
6. Use the CHAMFER command

Help Files

Once you have learned how to sign on to the system and have located all the menus, you can learn the system from on-screen documentation from this point on.

The Help files have two main functions.

Function #1

The first function serves as an index of commands. When looking for a command that will change the magnification of the data on the screen, you may be able to spot it by using the following.

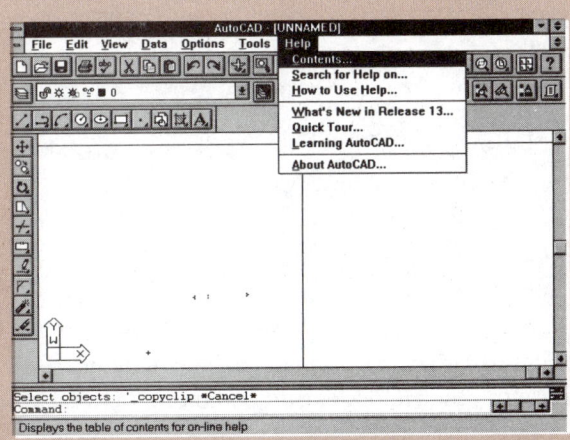
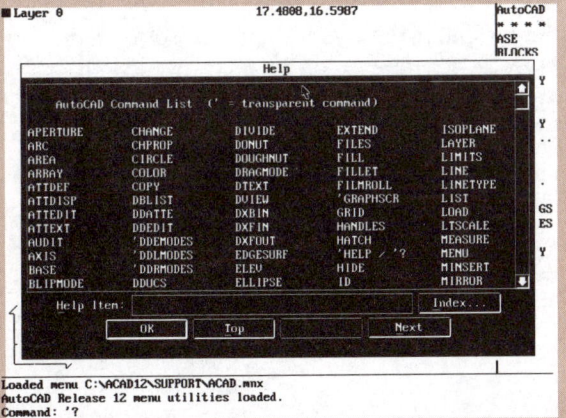

Windows Choose Help from the standard toolbar, then Contents or type in **?** as follows:

DOS From the Help pull-down menu, choose Help.

```
Command:?
Command name or return for list:⏎
```

To exit from the Help files:

Windows Pick the graphics screen or **F2**.

DOS Pick OK or **F1**.

This will give you a listing of the various commands which are available on the system. Use the scroll bar to look at the commands (the scroll bar is described in the Introduction, page xvii). By reading the various command names, you may be able to identify a word that is similar in meaning to the one you are after.

If you have completed what you are expected to complete for the day, use Help to become familiar with more of the commands.

If the command is shown with an apostrophe before it, as in 'ZOOM, it may be used within a current command. For example, if you are in the LINE command and would like to ZOOM Window on an area, choose ZOOM from the pull-down menu, perform the operation, and continue with the LINE command. Commands marked in this way are called *transparent commands*.

' = transparent command

Function #2

The second function of the Help files permits you to quickly retrieve information about a specific command. AutoCAD responds to any command by a series of prompts. The Help files will explain those prompts.

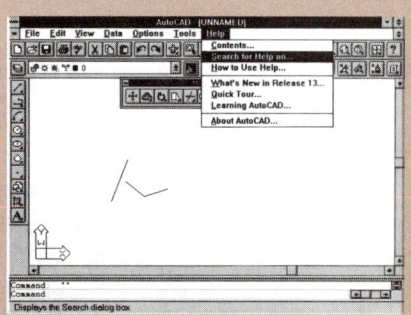 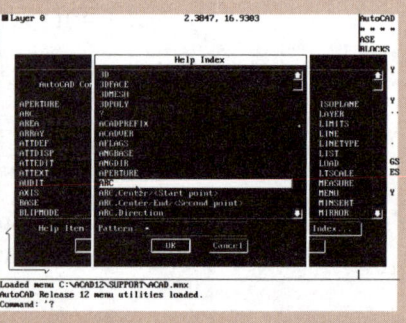

Windows Choose Help, Search for help on ... For example, to find out about the ARC command, find the ARC submenu on the slide bar. Pick ARC, then Go To, then pick Show Topics.

To exit from the Help files:

Windows Pick the graphics screen or **F2**.

DOS From the Help pull-down menu, choose Search for help on ...

DOS Pick OK or **F1**.

On both DOS and Windows versions of AutoCAD, the Help files will explain the syntax or command line sequence and the prompts and options of each command. With a little practice, you will be able to understand the Help files and use them for commands that you have not already used.

Angle Brackets

Within the command strings are the options associated with the commands. You will notice that one of the options is contained in angle brackets. In any prompt the information offered in angle brackets, < >, is the default. This is what the command will do if you do not specify something else.

In the case of the ARC command, it will accept a Start point unless you tell it that you want something else.

```
Command: ARC
Center/<Start point>:
```

Exiting from the Help Menu

In the first two or three weeks of classes, students often press the space bar before they enter any command. If you do this immediately after you have signed on, the system will offer you the Help files. If this should happen, press OK to exit from Help.

Also, when trying to reach the AutoCAD prompt on the top of the side menu, the student may pick Help by mistake because the commands are very close together.

Help files are also useful once you have learned AutoCAD and you switch to a system using a different release of the program.

SNAP and GRID

In AutoCAD, always draw the objects at a scale of 1:1. The LIMITS command is one way to set up the screen.

LIMITS defines the general size that you want the model to be.
ZOOM All allows you to view the size chosen. To make sure it has changed, press **F6** and move the cursor around.
SNAP will set an increment that the cursor will move to.
GRID will set a visual aid to help place objects.

If you prefer, you can turn LIMITS off and work with no set space. SNAP can be used without setting LIMITS.

The SNAP command sets a spacing for point entries and restricts cursor movements to the preset integer. The GRID command places dots on the screen, providing a visual framework.

The SNAP (**F9**) and GRID (**F7**) functions act as toggles, but can be modified through the command system. Type in your choice of GRID or SNAP, then type the desired spacing length. The Aspect Ratio (X vs. Y) will be equal ($X = Y$) unless you change it.

The SNAP Command

SNAP allows you to indicate points or positions on the screen at preset regular integers. It also allows a rotated or isometric drawing to be entered. In the following example of a simple lug section, we will set a snap as a guide to draw in both the horizontal and the angled section of the part.

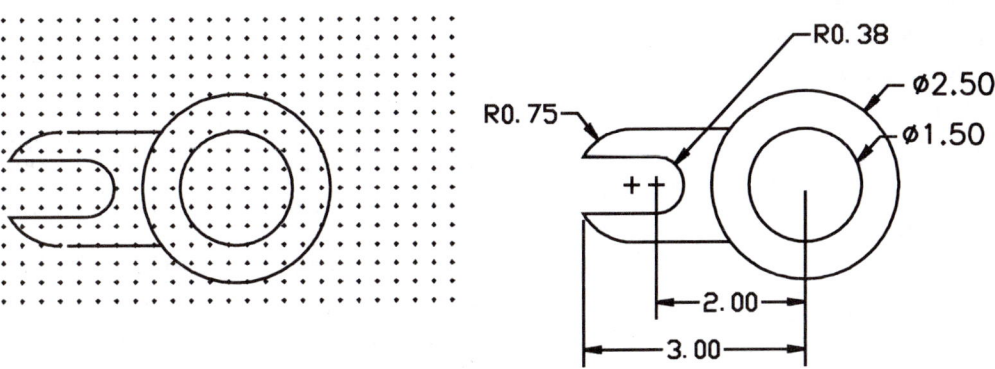

Figure 2-1

First set SNAP and GRID to .25 units. Then draw in Figure 2-1.

Next rotate SNAP by 45 degrees and set the base point in the center of the large circle. The GRID will follow the SNAP angle. The crosshairs will remain perpendicular, but are seen at an angle.

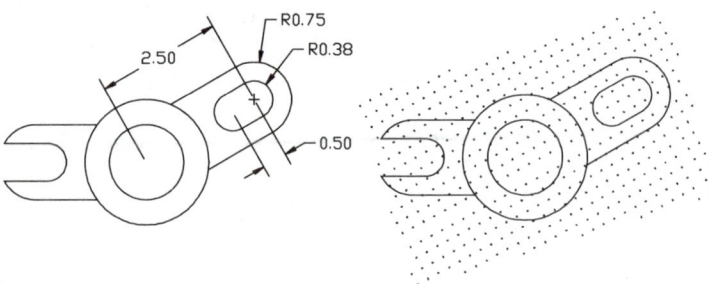

Figure 2-2

```
Command:SNAP
Snap spacing or ON/OFF/Aspect/Rotate/Style/<0.000>:R
Base point<0'-0.00",0'-0.00">:CEN of (pick the circle)
Rotation angle<0.00>:45
```

Where: **On** = SNAP activated or turned on
      Off = SNAP turned off
      Aspect = the option of changing the X value with relation to the Y value
      Rotate = the option of rotating SNAP for drawing on a given angle
      Style = the option of changing the style from Standard to Isometric, thus allowing isometric drawing
      Value = the spacing of SNAP entered by number

The GRID Command

The GRID command sets a visual aid for drawing. The grid size will follow the snap size unless changed using GRID. To change the relative X, Y value, use Aspect.

```
Command:GRID
Grid spacing ON/OFF/Snap/Aspect<0.000>:A
Value of X for square grid:1
Aspect of Y for rectangular grid:2
```

Notes
GRID and SNAP can be changed at any time during the creation of a drawing.

Where: **On** = GRID has been activated or turned on
      Off = GRID has been turned off
      Value = the spacing
      Aspect = the option of changing the relative X and Y values

Using the Drawing Aids Dialog Box

The dialog box can be used to rotate SNAP and GRID as well. The box is transparent, and therefore can be used to change the SNAP in the middle of another command. To change the snap angle and base point in the dialog box:

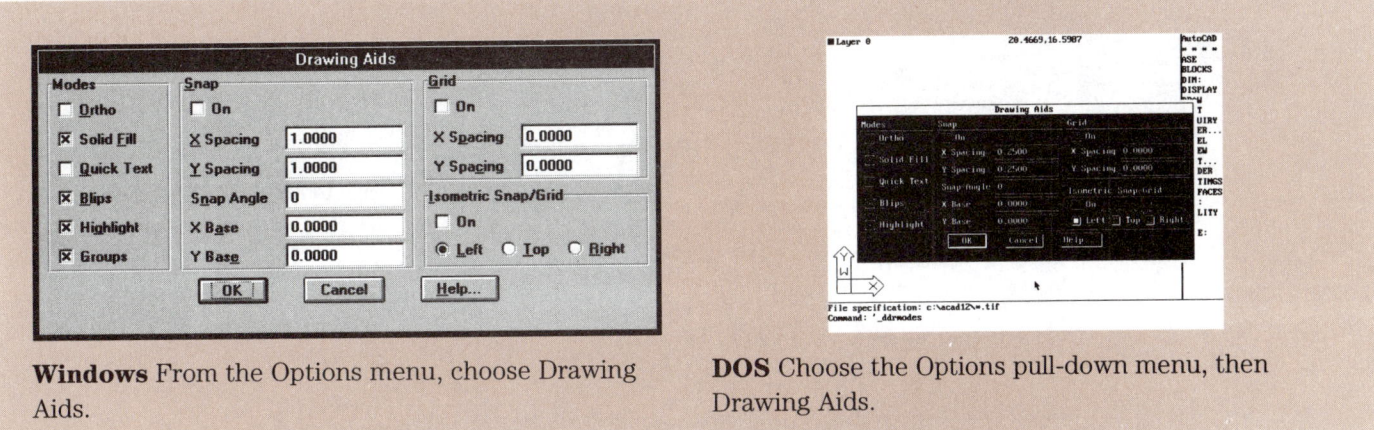

Windows From the Options menu, choose Drawing Aids.

DOS Choose the Options pull-down menu, then Drawing Aids.

The command line equivalent is **DDRMODES**.

1. In the Snap Angle box, enter 45, for 45 degrees.
2. Change the *X* and *Y* coordinate values in the X Base and Y Base boxes to change the base point for the rotation.
3. Choose OK to make the changes happen.

To return to standard snap and grid values, choose 0 degrees and 0,0 for the angle and base points. You can pick the Snap, Grid, and Ortho modes on or off by picking the box On. Blips can also be switched off here.

Point Entry and OSNAP

Point Entry

In most CAD systems, there are three basic ways to enter points:

PICKING If you pick or digitize a point on the screen, the object will be placed at exactly the point indicated. The SNAP command can be used to access a point at a preset integer to increase your accuracy. The GRID and ORTHO commands can also help the entry process.

COORDINATE ENTRY You can enter absolute, relative, or polar coordinates in any order.

ENTITY SELECTION, OBJECT SNAP, OR OSNAP This allows you to use existing objects on screen to create your file. Accessing points on existing objects is called using OSNAPs.

OSNAP: Object SNAP

Object Snaps allow you to specify precise points on objects in order to create or edit objects. Osnaps are used with any command that requires a "point."

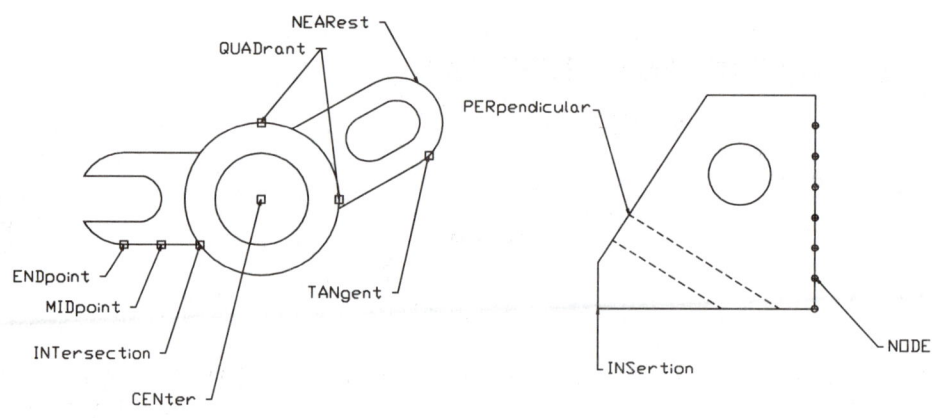

Figure 2-3

AutoCAD has 13 Object Snap modes that allow you to specify precise points on objects such as circles or lines. Following is a short description of the available object snaps. The capitalized letters are those needed when typing in the option.

APPINT	(apparent intersection) snaps to a real or imaginary intersection of two objects.
CENter	snaps to the center of an arc or circle.
ENDpoint	snaps to the closest end of any object.
FROM	establishes a temporary reference point from the parameters of an existing objects.
INSertion	snaps to the insertion point of a block.
INTersection	snaps to the intersection of two items.
MIDpoint	snaps to the midpoint of a selected item.
NEARest	snaps to a point on an object nearest to the digitized point.
NODE	snaps to a point created by POINT, DIVIDE, or MEASURE.
NONE	turns the object snap mode off.
PERpendicular	snaps to a 90 degree angle to an existing line.
QUADrant	snaps to the 0, 90, 180, or 270 degree point on an arc, circle, or ellipse.
Quick	snaps to the first snap point found; it must be used in conjunction with other OSNAPs.
TANgent	snaps to the tangent of an arc or circle.

Accessing OSNAPs

OSNAPs can be accessed in one of 4 ways:

1. There may be a designated button on your mouse.
2. In DOS, choose the "****" under the word AutoCAD on the top right of your screen.
3. Type in the first few letters of the OSNAP mode.
4. Choose Object Snaps from the Assist menu.
5. Choose DDOSNAP under the Options menu.

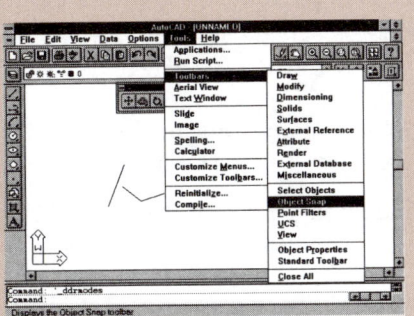

Windows From the Tools menu, choose Toolbars, then Object Snap.

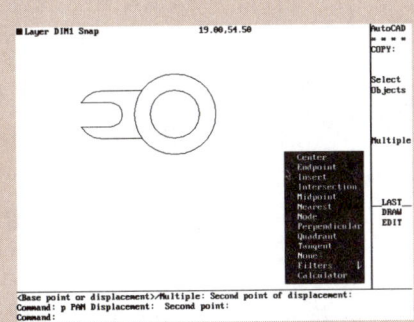

DOS Choose Assist, then Object Snap, "****," or Shift plus enter on the mouse.

Using OSNAPs

This lifting socket will illustrate how to use Object Snaps within the command line. First, two circles are drawn and then lines are added TANgent to the circles and from the QUADrant of the circles.

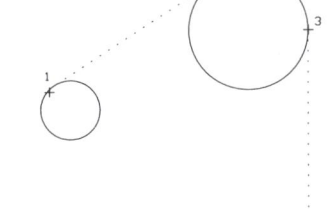

```
Command:(pick AutoCAD, Draw)CIRCLE CEN RAD or
3P/2P/TTR/<Center point>:0,0
Diameter/<Radius>:1.5
Command:C
3P/2P/TTR/<Center point>:9,4
Diameter/<Radius>:3
Command:PAN (pan the objects onto the
    screen)
```

Figure 2-4

TANgent creates a tangent to the identified object from the last object. This creates an object tangent to a circle or arc.

```
Command:LINE
From point:TANgent to (pick 1)
To point:TANgent to (pick 2)
```

AutoCAD will calculate the tangent for you; all you do is indicate which side of the object the tangent should be calculated on.

Figure 2-5

QUADrant takes the top, the bottom, or the far right or left of a specified circle or arc.

```
Command:LINE
From point:QUADrant of (pick 3)
To point:@0,-10
Command:ZOOM .5x
```

CENter takes the center of an arc or circle. Only items with a defined radius can have a center point.

ENDpoint takes the closest end point. All lines are made up of two end points. Circles have one end point, and arcs have two.

When finding the CENter of the arcs and circles, pick the object itself, rather than the spot where the center might be; the system will calculate that for you.

```
Command:CIRCLE
3P/2P/TTR/<Center point>:CENter of (pick 4)
Diameter/<Radius>:.55
Command:CIRCLE
3P/2P/TTR/<Center point>:CENter of (pick 5)
Diameter/<Radius>:1.5

Command:ARC
Center/<Start point>:-1.5,-5
Center/End/<Second point>:E
End point:ENDpoint of (pick 6)
Angle/Direction/Radius/<Center point>:R
Radius:15
```

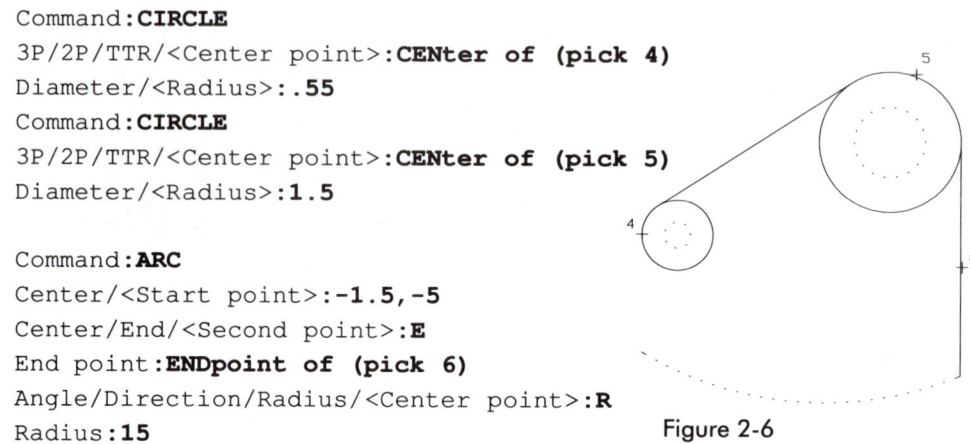

Figure 2-6

PERpendicular forms a normal to an identified object.

```
Command:LINE
From point:ENDpoint of (pick 7)
To point:QUADrant of (pick 8)
To point:↵
Command:LINE
From point:QUADrant of (pick 9)
To point:@0,-2.5
To point:PERpendicular to (pick 10)
To point:↵
```

Figure 2-7

Erase the short line between points 9 and 11. Add an arc going from point 11 tangent to the lower quadrant of the circle.

```
Command:ARC
Center/<Start point>:ENDpoint of (pick 11)
Center/End/<Second point>:E
End point:TANgent to (pick 12)
Angle/Direction/Radius/<Center point>:R
Radius:10
```

Then use TRIM to trim off the unnecessary pieces.

The TRIM Command

TRIM is used to remove portions of existing objects between specified cutting edges. Notice that TRIM goes to the closest intersection. TRIM will be discussed at length in Chapter 5.

```
Command:TRIM
Select cutting edges (Projmode = UCS,
   Edgemode = No extend):
Select objects: (pick 13)
Select objects: (pick 14)
Select objects: (pick 15)
Select objects: (pick 16)
Select objects:⏎ (no more cutting edges
   are needed)
<Select object to trim>Project/Edge/Undo:
   (pick the dotted lines)
```

Figure 2-8

With the exception of APPint, QUICK, NONE, and FROM, the OSNAP modes not illustrated above are fit into the command strings exactly as those shown — before the object selection digitize.

APParent Intersection

APPint (apparent intersection) snaps to a real or visual 3D intersection formed by objects you select or by an extension of those objects. (A *visual intersection* occurs when objects appear to meet in the current view but do not actually intersect in 3D space.)

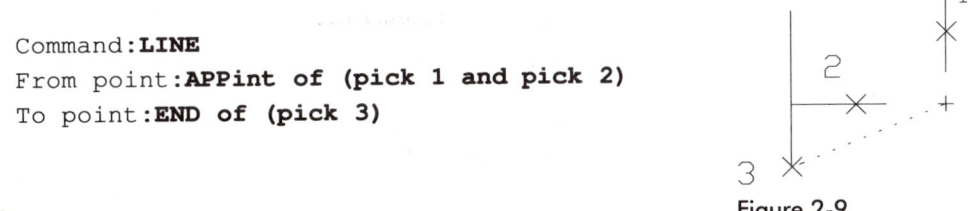

```
Command:LINE
From point:APPint of (pick 1 and pick 2)
To point:END of (pick 3)
```

Figure 2-9

From

From establishes a temporary reference point as a basis for specifying subsequent points. It is usually used in combination with other object snaps and relative coordinates.

To establish a point that is 2 units in the X direction and 3 units in the Y direction from the middle of an existing line, use the following.

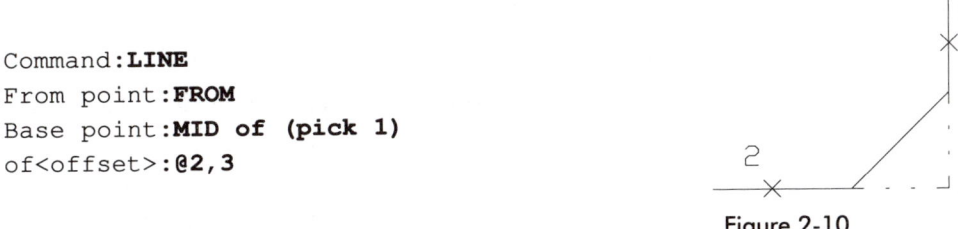

```
Command:LINE
From point:FROM
Base point:MID of (pick 1)
of<offset>:@2,3
```

Figure 2-10

None

None is used to turn off the current running object snap. The aperture will disappear from the cursor and no object snap will take place.

Quick

Quick, when used with other object snaps, snaps to the first eligible point it finds. When Quick is off, AutoCAD snaps to the point nearest to the center of the crosshairs. In complex drawings, the Quick object snap shortens time otherwise required to analyze which point is closest.

OSNAPs and Layers

You cannot snap to objects on layers that have been turned off, or on the blank portions of dashed lines, but you can snap to objects in blocks, on locked layers, solids, polyline segments, and floating viewport boundaries.

Using OSNAPs with the BREAK Command

Like TRIM, BREAK is used to remove portions of existing objects.

Once BREAK is invoked, you are prompted to choose the object to BREAK. You can break the object from the point chosen on the object and a second point, taking the selection digitize as a first point. Or you can choose the object and then specify **F** to choose a different break point.

OSNAPs are used to define points, often intersections, between any two objects: arcs, lines, circles, etc. The objects must be identified in the area of the intersection, preferably where the objects actually intersect.

The BREAK Command

BREAK is found under the Modify menu in both Windows and DOS.

```
Command:BREAK
Select object: (pick 1)
Enter second point (or F for first
  point):F
Enter first point:CENter of (pick 2)
Enter second point:INTersection of (pick 3)
```

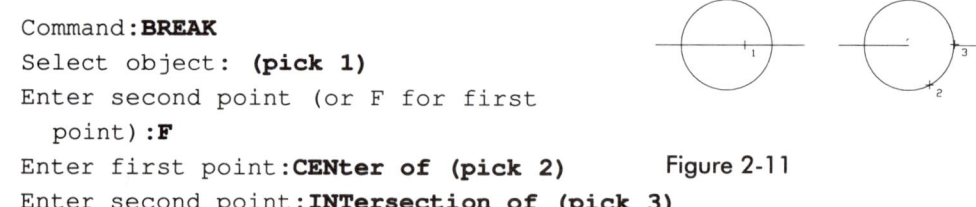

Figure 2-11

The first pick indicates the object itself and should be placed where the system cannot mistake it for any other object.

The second pick indicates the first point at which the item will be broken.

When breaking circles, break points are measured counterclockwise around the pick point.

Running Object SNAP Mode

If you use the OSNAP command or the DDOSNAP Running Object Snap dialog box, AutoCAD understands that you want every following object selection to be that particular portion of the object selection.

```
Command: OSNAP
Object snap modes: ENDpoint
```

This tells AutoCAD that you want only to pick end points.

Once you have indicated an Object Snap mode, your cursor will have a "target" box or aperture in the middle of the crosshairs.

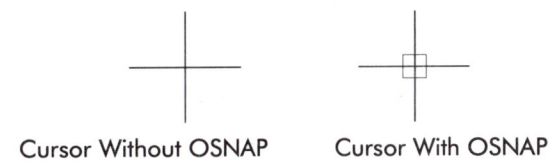

Cursor Without OSNAP   Cursor With OSNAP

Figure 2-12

> **Danger**
> If you do not turn your Object Snap off, it will be set for every following command. If you have a box on your crosshairs, and you don't know why, turn OSNAP off by using OSNAP OFF or OSNAP NONE.

To set Running Object Snap modes, open the Running Object Snap dialog box by one of the following methods:

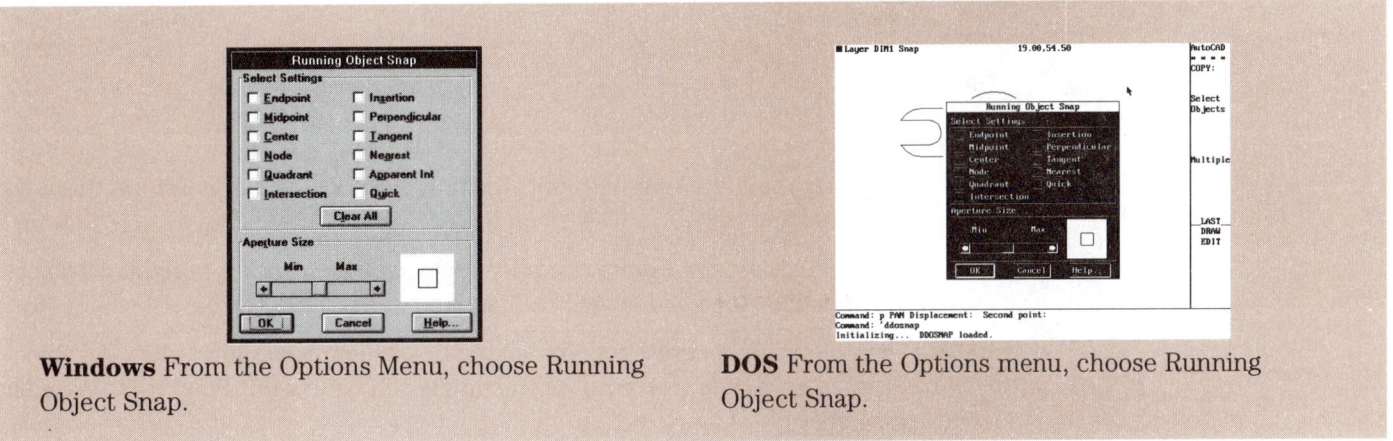

Windows From the Options Menu, choose Running Object Snap.

DOS From the Options menu, choose Running Object Snap.

The command line equivalent is **OSNAP**.

Aperture Size

To change the size of the aperture, use the following:

```
Command: APERTURE
Object snap target height (1-50 pixels)<20>: 6
```

Object Selection Cycling

The selection pick box can touch more than one object at a time. AutoCAD normally selects the most recently acquired object. A new feature in Release 13 is *object selection cycling*, which allows you to cycle through the possible selections and choose any of the possible objects.

At any "Select objects:" prompt, enter a **Ctrl-N** before picking, then repeatedly pick at a location where multiple objects lie within the selection aperture.

Grips

If you pick an object on the screen before picking a command, there are places on the object that are identified for object snap purposes. These are called **grips** and are discussed in Chapter 4.

The ERASE Command

The ERASE command will take objects out of the file. There are several options to help you choose items. You can pick the objects individually, or pick either a window or a *crossing* to indicate your selection set.

    Command:**ERASE**
    Select objects: **(use either Window, Crossing, or individual
        items)**

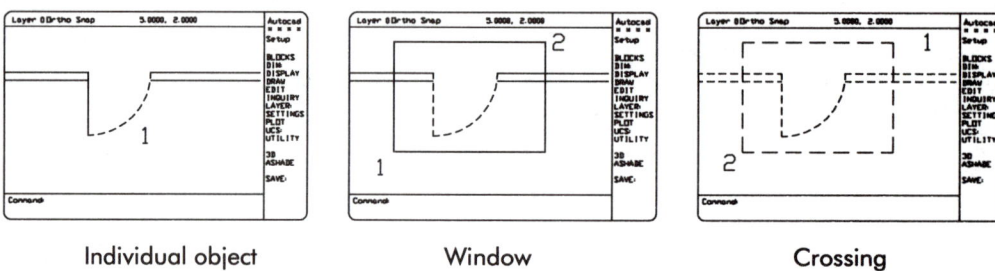

Individual object   Window   Crossing

Figure 2-13

In the first illustration, the user has indicated only the door swing; therefore, this is the only object to be erased.

In the second illustration, a window is indicated by picking from left to right. All objects *totally contained* in the window will be erased.

In the third illustration, a crossing is indicated by picking from right to left. All objects *either entirely or partly crossing* the rectangle will be erased. Press ⏎ to have the objects disappear.

Other options are discussed in Chapter 4.

Undoing an ERASE

If you have just erased an object or group of objects and then realize that you should not have, you can use the OOPS command to bring back the erased items.

    Command:**ERASE**
    Select objects: **(pick 1, pick 2)**
    Select objects:⏎
    Command:**OOPS (all erased objects will reappear)**

The CHAMFER Command

The CHAMFER command is very similar to the FILLET command. Both commands modify the connection between two objects. With FILLET you can create a curved fillet ending relative to a defined radius. With CHAMFER you can create a chamfered edge relative to a defined distance.

To find CHAMFER:

> **Windows** From the Modify toolbar, choose the Feature flyout, then Chamfer.
>
> **DOS** From the Construct menu, choose Chamfer.

The command line equivalent is **CHAMFER**.

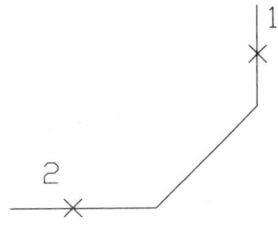

Figure 2-14

```
Command: CHAMFER
(Trim mode) Current chamfer Dist1 = 1.00, Dist2 = 1.00
Polyline/Distance/Angle/Trim/Method/<Select first line>: D
Enter first chamfer distance<1.00>: 2
Enter second chamfer distance<2.00>: ⏎
Command: ⏎ (restarts command)
(Trim mode) Current chamfer Dist1 = 1.00, Dist2 = 1.00
Polyline/Distance/Angle/Trim/Method/<Select first line>: (pick 1)
Select second line: (pick 2)
```

Where: **Polyline** = can create chamfered edges on all sides of a closed polyline
      Distance = allows you to set a distance
      Angle = allows you to set different angles for the two sides
      Trim = allows you to have the objects either trimmed when chamfered or not trimmed when chamfered
      Method = allows you to choose either two distances or an angle and a distance

Prelab 2 Using OSNAP and TRIM

Step 1 Start a new file and set the screen size to 35,25 with either LIMITS or Zoom scale. No GRID or SNAP are needed for this example.

```
Command: (pick Data)LIMITS
Lower left corner<0.0000,0.0000>:-5,-10
Upper right corner<12.0000,9.0000>:30,20
Command:  (pick View, Zoom All)
```

Step 2 Start by drawing circles. With the first one, use the screen menu. With the second, use the command alias. With the third, use the pull-down menu.

```
Command: (in DOS pick AutoCAD, Draw1)CIRCLE CEN RAD
```

Or in Windows:

```
3P/2P/TTR/<Center point>:0,0
Diameter/<Radius>:1
Command:C
3P/2P/TTR/<Center point>:13,2
Diameter/<Radius>:1
Command:↵
3P/2P/TTR/<Center point>:13,-2
Diameter/<Radius>:1
Command:Z
All/Center/Dynamic/Extents/Left/Previous/Vmax/Window/X/XP:
   (pick 1)
Other corner: (pick 2)
```

Step 3 Now use LINE command with the object snaps to access the appropriate points on the identified objects. Access the OSNAPs by the mouse buttons, from the "****" menu, from the Assist menu, or by typing them in. Use the screen menu to first find the LINE command.

```
Command: (pick AutoCAD, Draw1)LINE
From point:TANgent to (pick 3)
To point:TANgent to (pick 4)
To point:↵
Command:L
From point:QUADrant of (pick 5)
To point:PERpendicular to (pick 6)
To point:↵
Command: (space bar)
From point:QUADrant of (pick 7)
To point:PERpendcular to (pick 8)
To point:↵
```

Step 4   Use TRIM to remove the portions of the existing objects not wanted on the final drawings. TRIM can be found under the Modify menu.

```
Command:TRIM
Select cutting edges:
Select objects: (pick 9)
Select objects: (pick 10)
Select objects:↵
Select object to trim: (pick 11)
Select object to trim:↵
```

Step 5   Use the PAN command to translate the view to the left, and add some more circles.

```
Command:PAN
Displacement: (pick 12)
Second point: (pick 13)

Command: (pick Draw)CIRCLE CEN RAD
3P/2P/TTR/<Center point>:26,-5
Diameter/<Radius>:1
Command:C
3P/2P/TTR/<Center point>:26,5
Diameter/<Radius>:1
```

Use PAN to get the objects fully on screen as they are in this example.

Step 6   Now using the two new circles, create lines between the circles. First use the OSNAP command under the Assist menu to set the OSNAP to TANgent. Notice that the aperture is already on the crosshairs.

```
Command:OSNAP Osnap mode:TANgent
Command: (pick AutoCAD, Draw)LINE
From point: (pick 14)
To point: (pick 15)
To point:↵
Command:L
From point: (pick 16)
To point: (pick 17)
To point:↵
Command: (space bar)
From point: (pick 18)
To point: (pick 19)
To point:↵
```

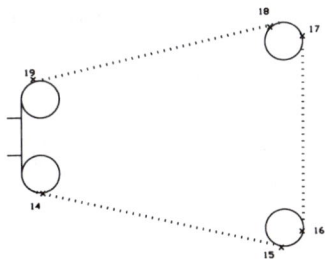

Set your OSNAP to the OFF or NONE option.

```
Command:OSNAP Osnap mode:NONE
```

Help Files, CHAMFER, OSNAP, ERASE, TRIM, and BREAK   **51**

Step 7   Use the TRIM command to remove portions of the circles.

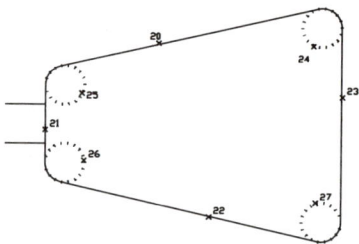

```
Command:TRIM
Select cutting edges:
Select objects: (pick 20, 21, 22, 23)
Select objects:↵
Select object to trim: (pick 24, 25, 26, 27)
Select object to trim:↵
```

Step 8   Use the Zoom command to reduce the drawing to 50% of the current size and create lines from the midpoints of the identified lines at the angles shown.

```
Command:Z
All/Center/Dynamic/Extents/Left/Previous/Vmax/Window/<Scale
   (X/XP)>:.5X
Command:L
From point:MID of (pick 28)
To point:@10<120
To point:↵
Command:(space bar)
From point:MID of (pick 29)
To point:@10<240
To point:↵
```

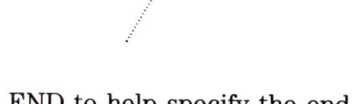

Step 9   Create circles on the ends of the identified lines using END to help specify the end point.

```
Command:(pick Draw)CIRCLE 2Point
First point on diameter:END of (pick 30)
Second point on diameter:@-2,0
Command:C
3P/2P/TTR/<Center point>:2P
First point on diameter:END of (pick 31)
Second point on diameter:@-2,0
```

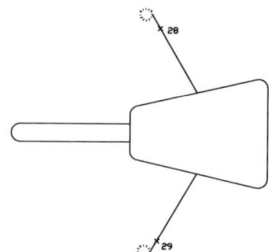

Step 10 This next command is extremely useful and difficult to do by hand. You will be making two lines tangent from the two circles at a specified angle and length.

```
Command:L
From point:TAN of (pick 32)
To point:@10<300
To point:↵
Command:(space bar)
From point:TAN of (pick 33)
To point:@10<60
To point:↵
```

The lines are parallel to the first lines drawn.

Step 11 Object snaps are also sometimes part of the command as in CIRCLE TTR. Use this option to place the two circles as shown.

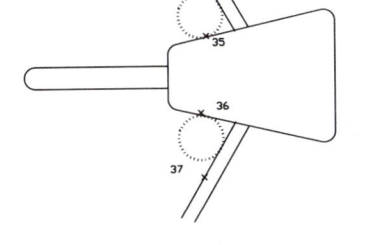

```
Command:(pick Draw)CIRCLE TTR
First tangent: (pick 34)
Second tangent: (pick 35)
Radius:2
Command:C
3P/2P/TTR/<Center point>:TTR
First tangent: (pick 36)
Second tangent: (pick 37)
Radius:2
```

Step 12 Finally we will use the TRIM command to trim off all of the unwanted geometry. Many cutting edges can be used at once. TRIM using the cutting edges shown in Figure (a) to remove the portions of objects in Figure (b).

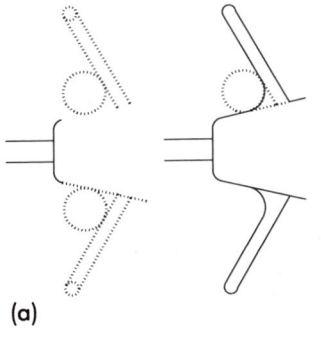

```
Command:TRIM
Select cutting edges:
Select objects:(pick objects in
   (a) with crossing)
Select objects:↵
Select object to trim:(pick
   objects in (b))
Select object to trim:↵
```

(a)

When completed, your object should look like this. It is a plan view of a crowd control unit.

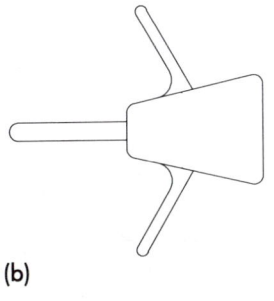

(b)

Help Files, CHAMFER, OSNAP, ERASE, TRIM, and BREAK   53

Command and Function Summary

BREAK is used to erase a portion of an existing object.

CHAMFER is used to trim two intersecting lines at a specified distance from their intersection and connect the intersecting lines by a new line segment.

ERASE erases selected objects.

HELP files are used to access information about a selected command.

Object SNAPs are used to access particular points on existing objects during other commands.

PAN is used to shift the view of your drawing without changing the magnification.

REDRAW refreshes the screen display and removes blips.

REGEN regenerates the entire file.

TRIM trims an object or series of objects according to an identified cutting line.

ZOOM is used to change the magnification of your views.

Practice Exercise 2

These exercises will help you with object snaps and TRIM.

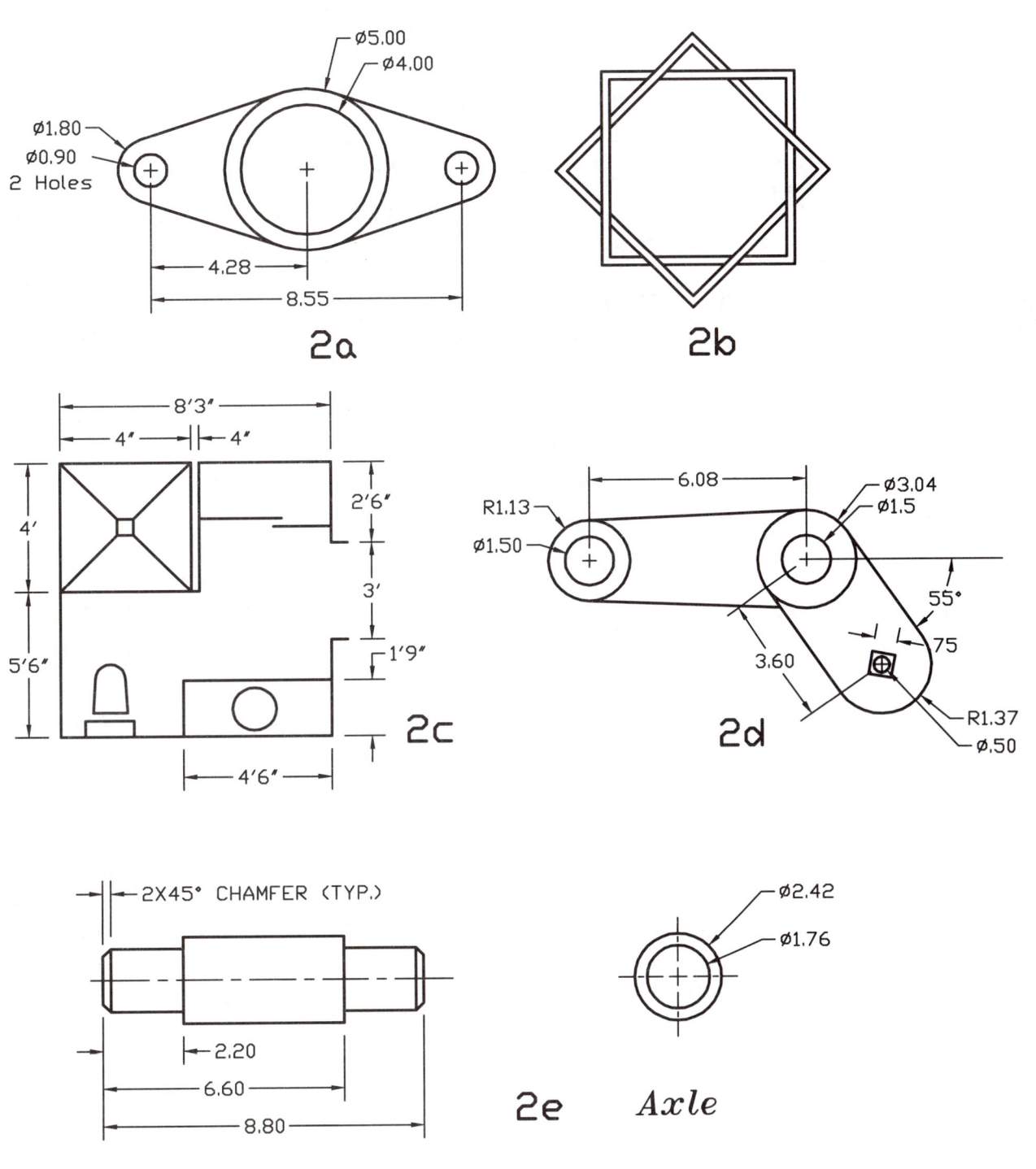

Exercise A2

You will need PAN and ZOOM to place these interior walls.

```
Command: UNITS, 4, 1 (press ⏎, accepting the default on the
    other settings until you return to the command prompt)
Command: LIMITS -3',-3' 45',38'
Command: SNAP 1
Command: GRID 6
```

Use **F6** to get your coordinates to move.
When you are finished, SAVE the file as FIRSTFL.

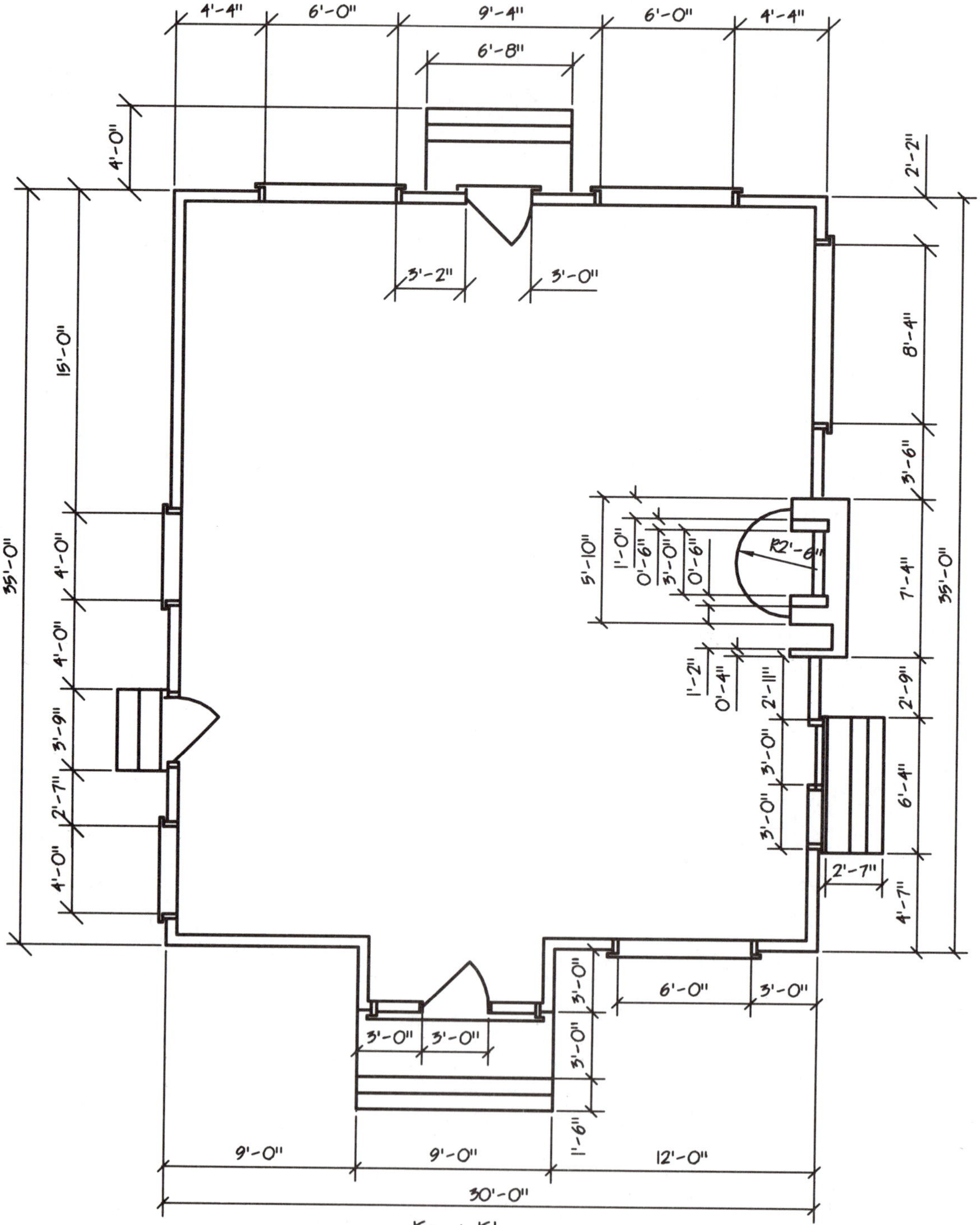

First Floor

Exercise C2

Set up your file as follows:

    LIMITS -5,-5
               60,40
    SNAP 1
    GRID 5
    ZOOM All

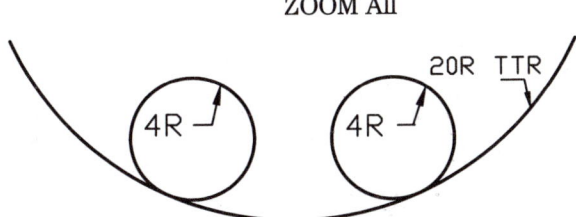

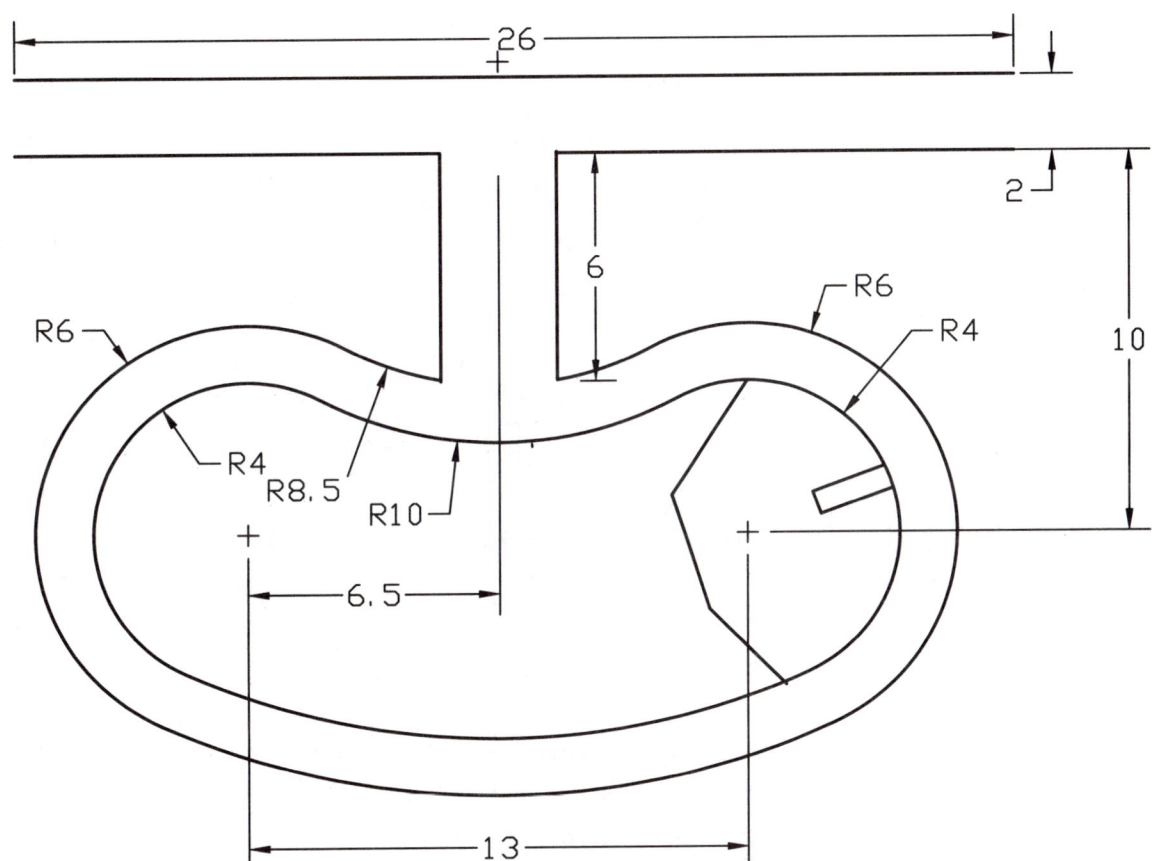

Use OSNAP to enter objects correctly in this swimming pool design.

Place two circles at 4R, then two circles at 6R. You will need CIRCLE TTR (tangent tangent radius) to connect them.

Use NEAR to place the diving board.

Exercise E2

Set up a file as follows:

```
LIMITS -5,-5
       45,30
SNAP 1
GRID 5
ZOOM All
```

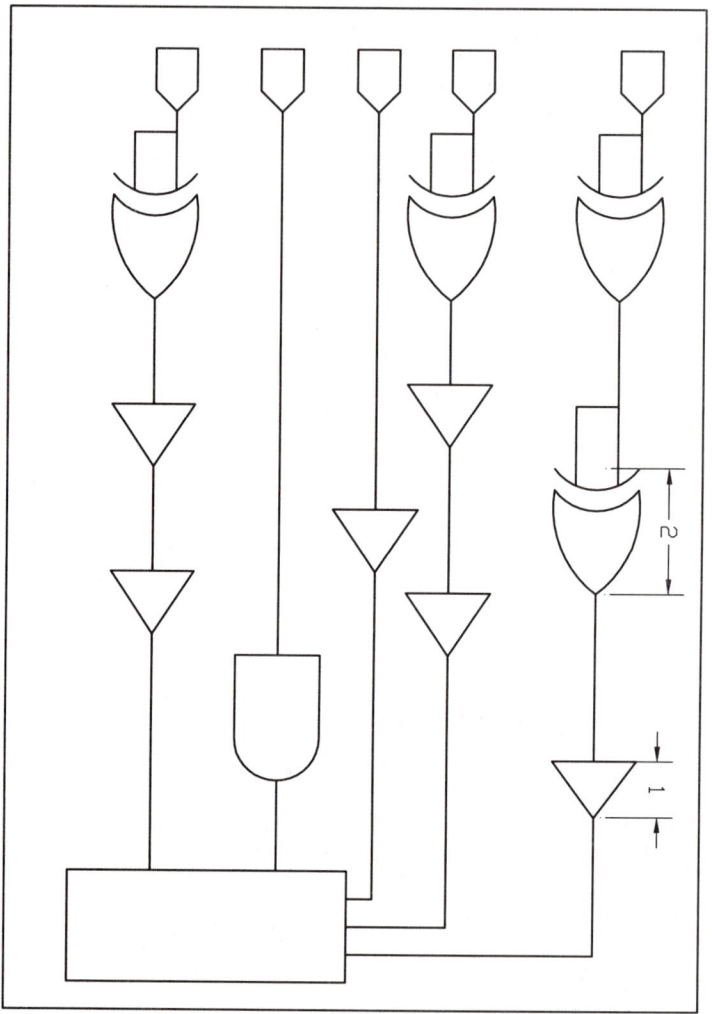

Use LINE for the outline of the schematic. Then change the SNAP to .5 and enter the lines for the top line. Use the space bar to speed up entry. Put the arcs in, overlapping the lines, and use BREAK to create a good corner.

Use ZOOM to get a good look at your work. PAN the drawing over to complete an adjacent section.

Practice on Exercise M2 if you finish early.

Exercise M2

Set up a separate file for each of the following. No changes in limits, snap, or grid are required for the first one; set the limits to top right 150,100 for the second. Use ZOOM Window, ZOOM All, and PAN to access the geometry. Place the objects using the OSNAPs. Use TRIM to remove unwanted lines.

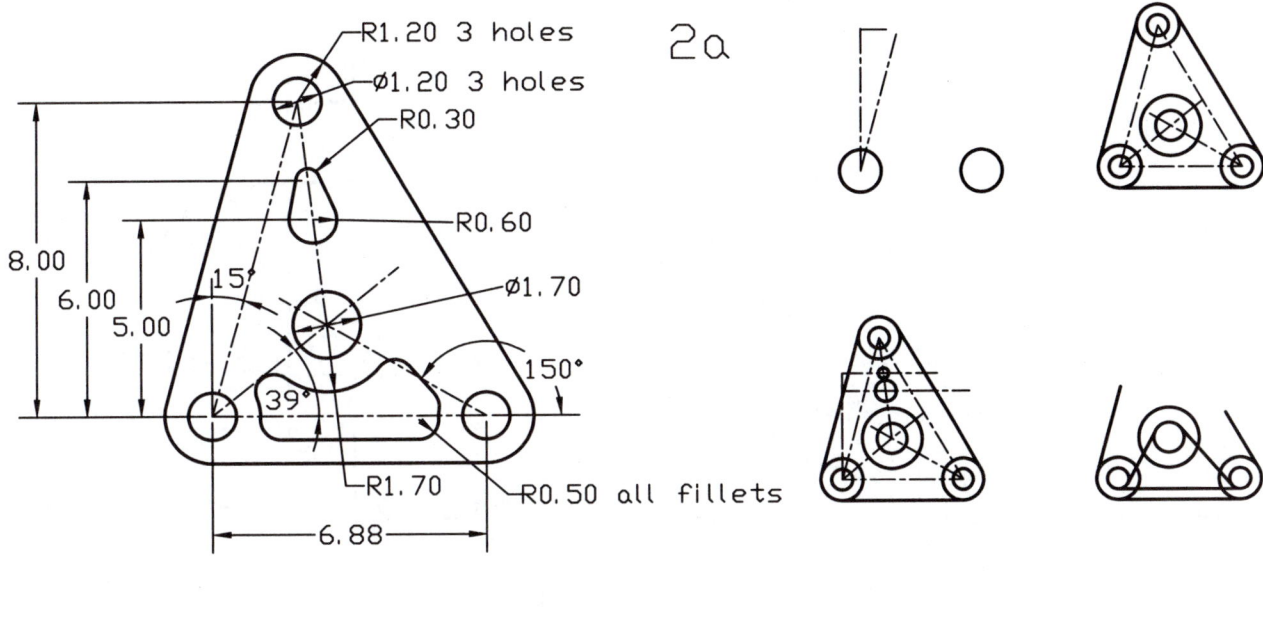

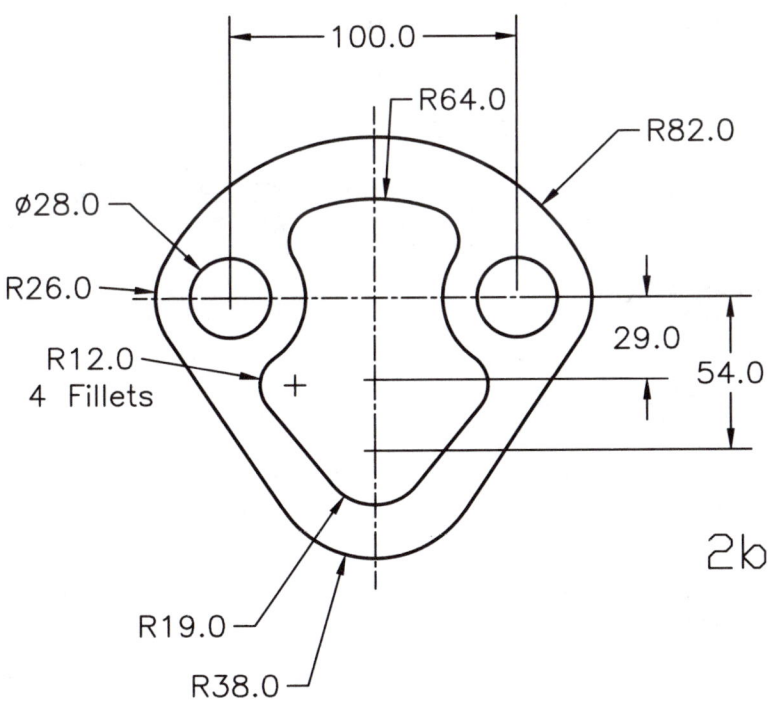

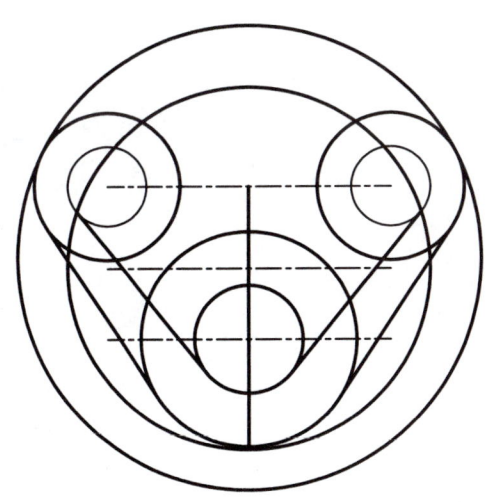

Challenger 2

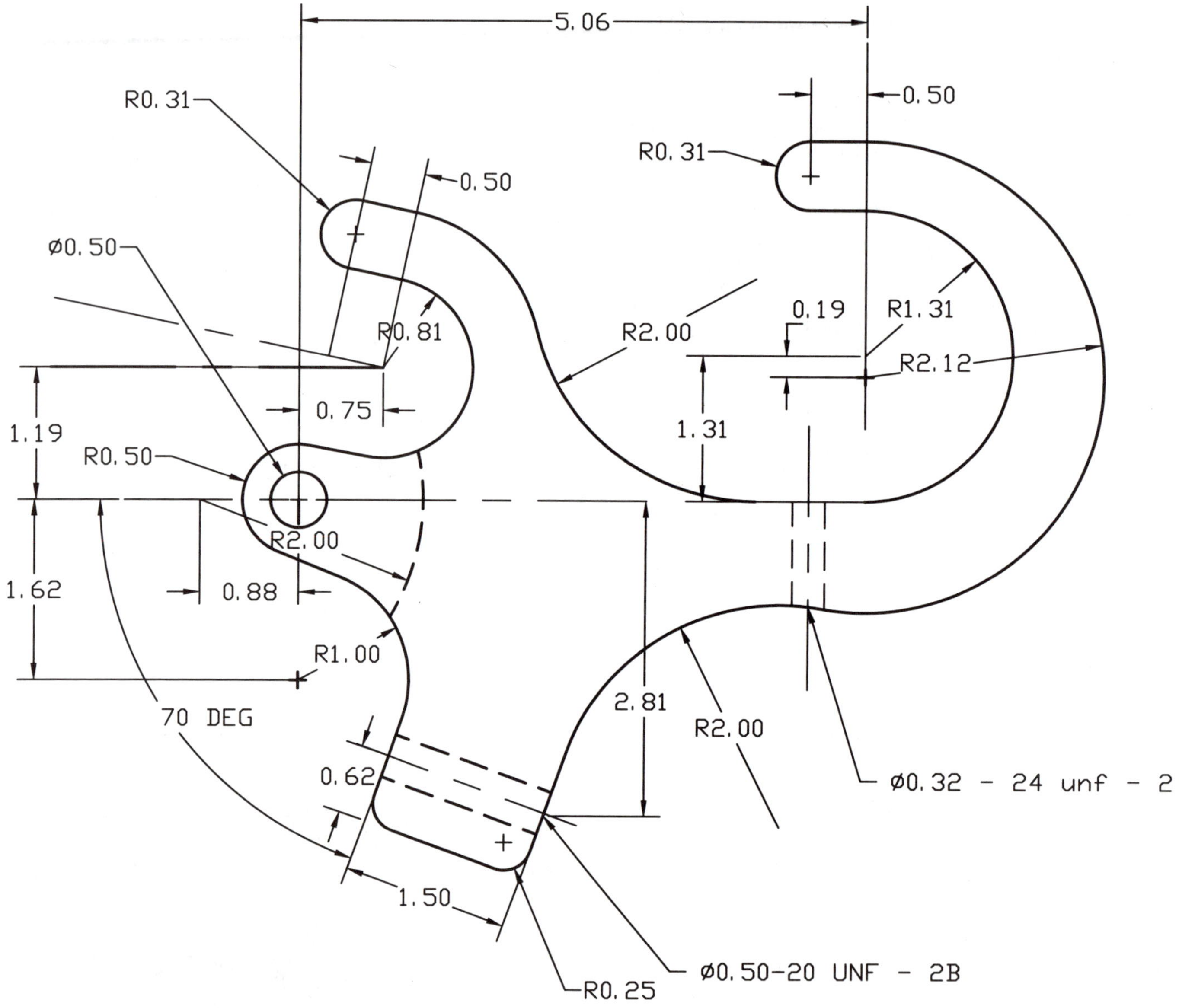

This is an excellent OSNAP exercise.

Use LINE, ARC, CIRCLE, FILLET, OFFSET (Chapter 4), and BREAK to create this model. Your origin should be the center of the small circle.

3 Entity Commands with Width

Upon completion of this chapter, you should be able to:

1. Create a TRACE
2. Create a polyline PLINE with acceptable corners and widths
3. Edit a pline using PEDIT to change the width and curve factors
4. Create a POLYGON
5. Create a SOLID
6. Create a DONUT with a specific width
7. Enter simple TEXT
8. Add Multilines

The TRACE Command

The TRACE command makes a thickened line with bevelled or mitred corners. Traces are drawn in the same way as lines, from point to point; the difference is that a width must be specified.

> **WINDOWS** From the Miscellaneous menu, pick Trace.
>
> **DOS** From the DRAW 2 screen menu, pick Trace.

The command line equivalent is **TRACE**.

```
Command:TRACE
Trace width<0.0500>:.1
From point:5,.5
To point:11,.5
```

The line segment created will show up one pick behind because AutoCAD is calculating the mitre to the next edge. If you want to keep this mitre, draw in an extra segment to have the mitre calculated, then erase it as in the following examples.

```
Command:TRACE
Trace width<0.1000>:.2
From point: (pick 1)
To point: (pick 2)
To point: (pick 3)
To point: (pick 4)
To point: (pick 5)
To point: (pick 6)
To point:↵
```

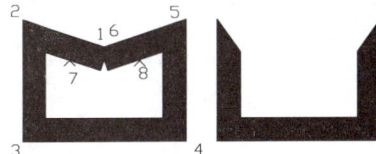

Figure 3-1

Each segment of the trace is a separate entity once it is placed, and thus the segments adjacent to the mitred edges can be erased to provide a final mitred edge.

```
Command:ERASE
Select objects: (pick 7, 8)
Select objects:⏎
```

To create a closed square with TRACE, start halfway down the side of the square and continue around so that all sides have square corners.

The disadvantage of TRACE is that it cannot be edited. To create thick lines that can be edited, use PLINE.

The PLINE Command

A *pline* is a single-drawing entity that includes line and curve sections that may vary in thickness, and may be edited using PEDIT for the Spline and Fit curve options to create contour lines or airfoils. The individual segments are connected at vertices; the direction, tangency, and line width are stored at each vertex.

The PLINE command can create rectangles as single entities as well as curved segments of varying thickness.

> **Windows** From the DRAW toolbar, pick Polyline.
>
> **DOS** From the Draw 2 menu, pick Polyline.

The command line equivalent is **PLINE**.

```
Command:PLINE
From point: (pick a point)
The current line-width is 0.0000
Arc/Close/Halfwidth/Length/Undo/Width/<Endpoint of line>:
```

Where: **Arc** = a change from line entry to arc entry
        Close = a closed pline, in which the first point will be joined to the last entered point in the pline to make a closed object; more than two points are needed to have a closed pline
        Halfwidth = a specified halfwidth on either side of the pline vector
        Length = the length of the pline
        Undo = an undo of the last point entered
        Width = a specified width of the line or arc segments on either side of the pline vector

The first PLINE command prompt asks for a point at which the pline will start.

```
Command:PLINE
From point: (pick a point)
```

Figure 3-2

You *must* enter the first point, after which you can choose one of the options (Width, Arc, etc.).

The next command default is to enter the second point, "<Endpoint of line>." If you pick a second point, this assumes a straight segment or a line. If you continue picking points, the object created will look like a series of lines, but it will be a single object that can be edited using PEDIT or other editing commands.

To change the width of a pline, enter the first point and then specify width by entering **W** or picking Width from the side menu. The system will prompt you for both the beginning and the end width. If these are the same, press ⏎ to accept the default.

```
Command:PLINE
From point:0,0
The current pline is 0.0000 units
Arc/Close/Halfwidth/Length/Undo/Width/<Endpoint of line>:W
Start width<0.0000>:.25
End width<0.2500>:⏎
Arc/Close/Halfwidth/Length/Undo/Width/<Endpoint of line>:11,0
```

You can continue drawing with this line at the current thickness or change it at any time.

```
Arc/Close/Halfwidth/Length/Undo/Width
   /<Endpoint of line>:11,8.5
Arc/Close/Halfwidth/Length/Undo/Width
   /<Endpoint of line>:0,8.5
Arc/Close/Halfwidth/Length/Undo/Width
   /<Endpoint of line>:C
```

Figure 3-3

To achieve a perfect corner on a box or rectangle, use the Close option. This will attach the first point to the last entered point and create a clean, bevelled corner.

```
Command:PLINE
From point:2,2
The current pline is 4.00 units
Arc/Close/Halfwidth/Length/Undo/Width/<Endpoint of line>:W
Start width<4.00>:.25
End width<0.25>:⏎ (don't pick)
```

Do not pick! If you pick, you will create an object similar to that of Figure 3-4. There, the second pick was assumed to be the width of the end point of your pline, measured from the first entered point.

Danger

You must enter both the beginning and the end width of the pline, or the system will create a triangle. When prompted for the second point, if you pick a point in space, the PLINE command assumes that the distance between the first and second points entered is the ending width.

Figure 3-4

Entity Commands with Width

PLINE with Varying Width

You can create objects such as arrows using one PLINE with a series of different segments with varying widths. You can change the width at every vector.

```
Command:PLINE
From point: (pick 1)
The current pline is 4.00 units
Arc/Close/Halfwidth/Length/Undo/Width/<Endpoint of line>:W
Start width<4.00>:0
End width<4.00>:.35
Arc/Close/Halfwidth/Length/Undo/Width/<Endpoint of line>:
   (pick 2)
Arc/Close/Halfwidth/Length/Undo/Width/<Endpoint of line>:W
Start width<0.35>:.10
End width<0.10>:.10
Arc/Close/Halfwidth/Length/Undo/Width/<Endpoint of line>:
   (pick 3)
```

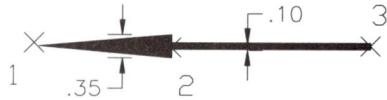

Figure 3-5

Using PLINE Effectively

There are a few tricks with PLINE that will make the entered plines perfect. First, let us look at creating good corners.

Remember that the points entered for a pline are considered to be vertices. This means the pline is to be calculated as a series of points entered with a specific sequence in mind.

While creating a PLINE, you are creating only one object that is fit through a series of points at a defined width. As the pline changes direction, the end of each pline is calculated relative to the points or vertices used to create it. If a thickened pline is created with just two points, its ends are calculated perpendicular to the pline vector. If three points are used to create a pline, the end of the pline that attaches to a segment going in another direction is calculated to create a sharp corner. As shown in Figure 3-6, the ends of the plines are perpendicular to the direction of the pline itself. When entering plines, do not pick a point twice in the same spot; this will create a corner that has two perpendicular ends as shown in Figure 3-7. To get the corners to close properly, only one pick per corner is needed.

Figure 3-6

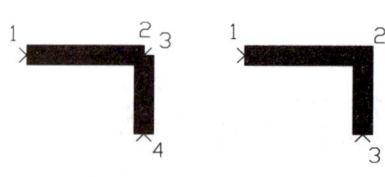

Figure 3-7

The Fill Option

If a drawing is becoming too dark, or is taking a long time to regenerate after ZOOM commands, you may want to use the FILL command. This allows the lines to show the edges of your pline, but the pline will not be filled in.

Figure 3-8 demonstrates how the Fill option can change the appearance of a pline. On the left, the Fill is off; on the right, the Fill is on.

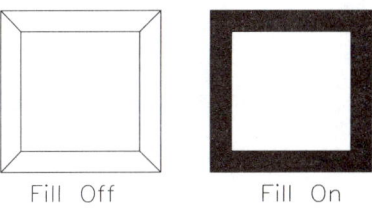

Figure 3-8

You need to use the REGEN command to see the effect of turning Fill on and off. Turn fill off for quicker draft plots.

Fillet and Chamfer with Polylines

Both the FILLET and the CHAMFER command have polyline options that apply to all the vertices or corners.

Polyarcs

The PLINE command can be used to create arcs within polyline segments or on their own. The Arc options are:

Notes

If your Polyarc is not working, create an arc using the ARC command and then make it a polyarc with PEDIT.

```
Command:PLINE
From point: (pick a point)
The current pline width is 0.1000
Arc/Close/Halfwidth/Length/Undo/Width/
   <Endpoint of line>:A
Angle/Center/Close/Direction/Halfwidth/
   Line/Radius/Second point/Undo/Width/<Endpoint of Arc>:
```

Figure 3-9

The command default is to create a two-point arc. Once the first arc segment is in, the system assumes that you want to continue with a series of arcs until the **L** option is entered, which will return you to a line segment.

The PEDIT Command

One of the great advantages of PLINE is that, once the pline is entered, it can be modified using PEDIT (Polyline Edit).

In the introductory stages, this command sequence is used most often to change the width of a pline. In Figure 3-10, the border is edited from .25 units to .10 units.

The PEDIT command changes the width of all the segments of the identified pline.

> **Windows** From the Edit toolbar, choose Polyline Edit.
>
> **DOS** From the Modify menu, choose Edit Polyline.

The command line equivalent is **PEDIT**.

```
Command:PEDIT
Select polyline: (pick the polyline)
Close/Join/Width/Edit vertex/Fit
   curve/Spline curve/Decurve/Undo/
   eXit<X>:W
New width for all segments:.10
Close/Join/Width/Edit vertex/Fit
   curve/Spline curve/Decurve/Undo/
   eXit<X>:↵
```

Figure 3-10

More segments can be added to the pline by using the Join option. This will add lines or arcs to a pline which can then be edited for width. Below, the arc and the two lines identified by picks 2, 3, and 4 are added to the pline selected with the first pick before the command option Join. Only segments with a common end point can be joined.

```
Command:PEDIT
Select polyline: (pick 1)
Close/Join/Width/Edit vertex/Fit curve/
   Ltype gen/Spline curve/Decurve/Undo/
   eXit<X>:J
Select object: (pick 2)
Select object: (pick 3)
Select object: (pick 4)
Select object:↵
Close/Join/Width/Edit vertex/Fit curve/Ltype gen/Spline
   curve/Decurve/Undo/eXit<X>:↵
```

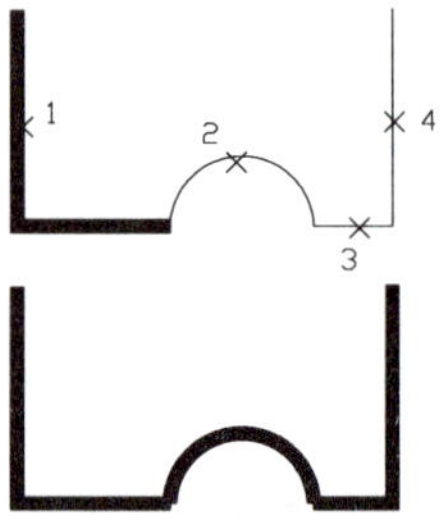

Figure 3-11

You can change polylines back to regular lines by using the command EXPLODE. This will also remove the width given.

If you apply PEDIT to an object that is not a polyline, you will be given the option of turning it into one. Then use Join.

PEDIT with Spline and Fit Curve

To create contour lines and other items used in surveying among other fields, etc., a pline can be modified to become a spline or fit a curve through a series of points. Splines and Spline editing will be dealt with in more detail in Chapter 13. For purposes of this chapter, use PEDIT to fit a curve through a series of vertices.

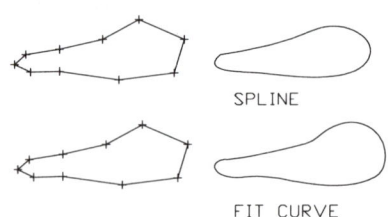

The Spline curve option of PEDIT will edit the pline according to the series of points used to create it. Make sure you use the Close option when entering the pline to get a continuous spline.

Figure 3-12

The Fit curve option of PEDIT will create continuous curves through a series of points.

Figure 3-12 shows two "airfoil" shapes. The plines are closed and were drawn without a width.

In the upper example, the Spline option was used to create a smooth curve using the points as a guide.

In the lower example, the Fit curve option was used, and the resulting curve is much more choppy.

The Spline option, on the other hand, uses the points as a guide but has a degree of continuity through each of the adjacent points.

If you need to be quite specific about spline generation, you can change the curve of the spline with the SPLINETYPE system variable.

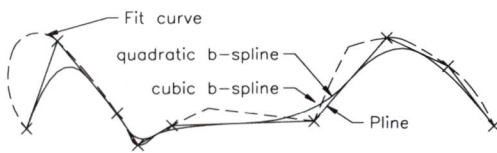

Figure 3-13

```
Command:SPLINETYPE
New value for the SPLINETYPE<6>:5
```

The default is a cubic B-spline <6>. A quadratic B-spline can be achieved by changing the variable to 5.

The POLYGON Command

The POLYGON command is like a PLINE in that you are creating an object that has many vertices.

> **Windows** From the Draw toolbar, choose the Polygon flyout, then Polygon.
>
> **DOS** From the Draw 2 menu, choose Polygon.

The command line equivalent is **POLYGON**.

```
Command:POLYGON
Number of sides<4>:5
Edge/<Center of polygon>: (pick a
   center point)
Inscribed in circle/Circumscribed
   in circle (I/C)<I>:I
Radius of circle:1.5
```

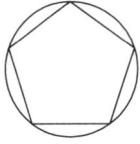 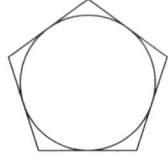

Inscribed   Circumscribed

Figure 3-14

Hexagons

In mechanical drawing, hexagons (six-sided regular polygons) are quite common.

If you need to draw a hexagon that is measured by *the distance across the flats* choose the option Circumscribed in circle. The diameter across the circle will equal the distance across the flats.

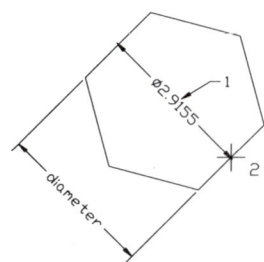

Figure 3-15

```
Command:POLYGON
Number of sides<4>:6
Edge/<Center of polygon>: (pick center)
Inscribed in circle/Circumscribed in circle (I/C)<I>:C
Radius of circle: (pick 2)
```

You can also use the polygon command to draw regular polygons by specifying *the length of an edge*.

```
Command:POLYGON
Number of sides<4>:6
Edge/<Center of polygon>:E
First endpoint of edge: (pick 1)
Second endpoint of edge:@4<0
```

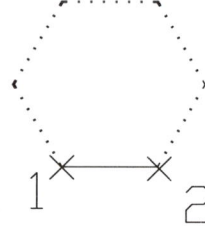

Figure 3-16

The SOLID Command

The SOLID command creates a polygon filled with the currently selected color. The Fill On option will fill in the solid with each screen regeneration, while the Fill Off option will display only an outline for quicker redraws.

The command line equivalent is **SOLID**.

> **Notes**
> Do not confuse SOLID with the SOLIDS option given under the Draw menus.

Windows From the Draw toolbar, choose SOLID.

DOS From the Draw 1 screen menu, choose SOLID.

The command line equivalent is **SOLID**.

```
Command:SOLID
First point: (pick 1)
Second point: (pick 2)
Third point: (pick 3)
Fourth point: (pick 4)
Third point:
```

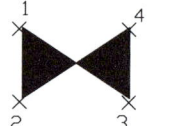 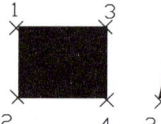 

Figure 3-17

If you continue to digitize after the fourth entry, the system will continue to add to the original entity until you terminate with ⏎. To create a triangle, use ⏎ after the third point prompt.

The order in which points are entered is very important. Figure 3-18 provides some examples.

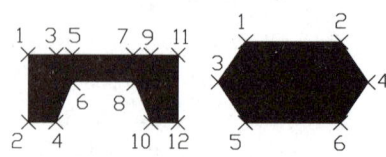

Figure 3-18

The DONUT Command

The DONUT command is used to create a thick or solid circle. The inside diameter is used to determine the hole of the doughnut. Use an inside diameter of zero to create a solid circle; use a larger diameter to create a ring. Once DONUT is active, a doughnut will be drawn every time you digitize until you press ↵.

The DONUT command can be accessed through either of the Draw menus or by typing it in.

```
Command:DONUT
Inside diameter<.5>:0
Outside diameter<1.0>:2
Center of doughnut: (pick 1)
Center of doughnut: (pick 2)
Center of doughnut:↵
```

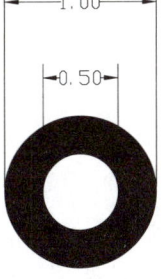

Figure 3-19

Filling Irregular Shapes

The HATCH command can be useful when you are trying to fill an irregular shape. The problem with HATCH, however, is that every given space is filled with a series of vectors. This can take a lot of space. Try creating an irregular shape with PLINE and changing the width with PEDIT to fill in the given space.

The TEXT Command

Chapter 8 deals with many aspects of text, including text style, editing, fonts, and a wide variety of text alignments. Here we introduce simple, one-line text entry. If you prefer to use DTEXT, the command is the same except that DTEXT will allow you to see the text as it is written on the screen.

Both the commands TEXT and DTEXT will place strings of characters on your drawing. DTEXT will allow you to see the text being entered onto the screen. When entering text, AutoCAD will prompt you to choose a height for each character, a rotation angle for your string, and a point at which to place the text string on the model or drawing.

To access TEXT:

> **Windows** From the Draw toolbar, pick either Text or DTEXT. (Do *not* pick MTEXT.)
>
> **DOS** From the Draw menu, pick Text, then Single-Line Text.

The command line equivalent is **TEXT**.

```
Command:TEXT
Justify/Style/<Start point>:J
Align/Center/Fit/Middle/Right/TL/TC/TR/ML/MC/MR/BL/BC/BR:
```

Where: **Justify** = placement of the text
Style = switches the style of the letters; the styles must be loaded in AUTOCAD to be accessible
Align = an alignment by the end points of the baseline; the aspect ratio (*X* versus *Y*) will correspond to the preset distance
Center = the center point of the baseline; this option will center the text on the point indicated
Fit = an adjustment of width only of the characters that are to be fit or "stretched" between the indicated points
Middle = a placement of the text around the point; i.e. the top and bottom of the text are centered as well as the sides
Right = an alignment with the right side of the text

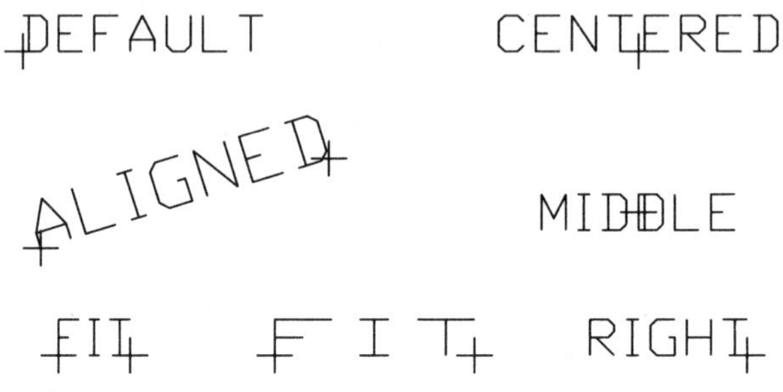

Figure 3-20

The initials TL, TC, etc. stand for other alignments. See Chapter 8.

The examples in Figure 3-20 demonstrate the standard justifications.

The default is left justification at the baseline of the text string.

Once you have chosen a point at which to place your text, the command will prompt you for the height of the letters, the rotation angle, and the text or string of characters itself.

```
Command:TEXT
Justify/Style/<Start point>: (pick 1)
Height <.2000>:.15
Rotation angle <0>:↵ (to accept the default)
Text:Scale:
Command:↵
Justify/Style/<Start point>: (pick 2)
Height<.2000>:.25
Rotation angle<0>:↵
Text:1/4"=1'0"
```

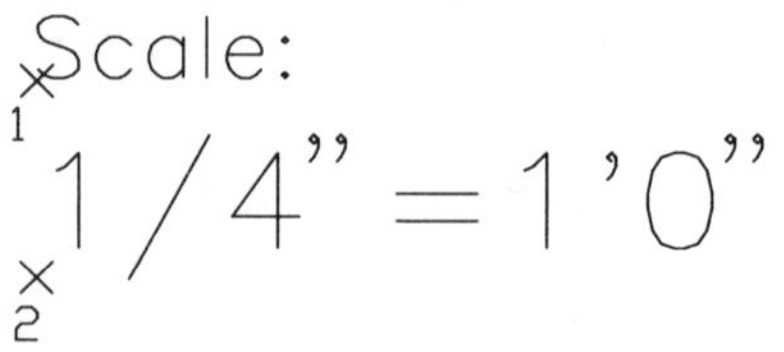

Figure 3-21

Multilines

The MLINE or multiline command is used to create multiple, parallel lines.

> **Windows** From the Draw toolbar, choose the Polyline flyout, then
>
> **DOS** From the Draw menu, choose Multiline.

The command line equivalent is **MLINE**.

MLINE is particularly useful for drawing walls and other architectural features. From a specified point, AutoCAD draws a multiline segment using the current multiline style, and continues to prompt for other points. Like LINE, using Undo undoes the last vertex point on the multiline. If you create a multiline with two or more segments, the Close option will be included in the command string.

```
Command:MLINE
Justification  = Top, Scale = 1.00,
   Style = Standard
Justification/Scale/STyle/
   <From point>: (pick 1)
<To point>: (pick 2)
Undo/<To point>: (pick 3)
Close/Undo/<To point>: (pick 3)
Close/Undo/<To point>: (pick 4)
Close/Undo/<To point>:C
```

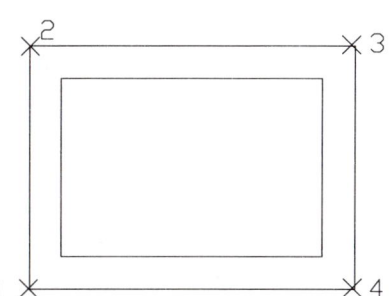

Justification

This option determines how the multiline is drawn between the points you specify.

```
Top/Zero/Bottom<Top>:
```

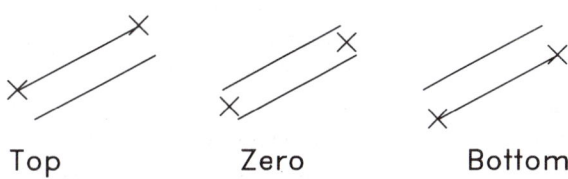

Top   Zero   Bottom

Scale

This option determines the distance between the two lines of the multiline.

The scale is based on the width established in the multiline style.

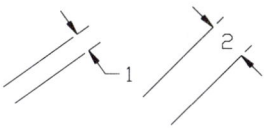

Style

Specifies a style. For more information see Chapter 13.

Prelab 3 Using PLINE and SOLID

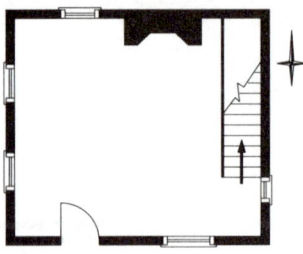

By using PLINE and SOLID effectively, we can create a presentation view of a small cabin with a staircase and a fireplace.

Step 1 First change the units to Architectural. At the command prompt type **UNITS**.

```
Command:UNITS
Report formats:4
Denominator of smallest fraction to display:4
```

Keep pressing ⏎ until you have returned to the command prompt.

Step 2 Make your LIMITS -2′, -2′ by 22′, 18′. Put the SNAP at 12″ or 1′ to enter the outline of the cabin as shown.

```
Command: (pick Last from screen menu, then LIMITS)
Lower left corner<0′0",0′0">:-2′,-2′
Upper right corner<1′0",0′9">:22′,18′
Command:(pick Next from screen menu, then SNAP)
```

Step 3 Add the PLINEs in, as shown on the illustration.

Things really get complicated when you start to dimension this.

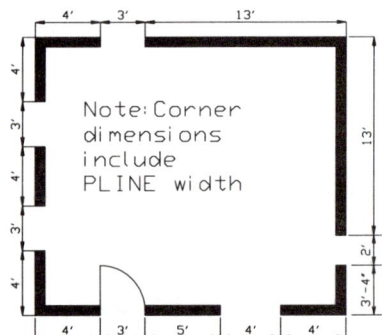

> **Notes**
> Since PLINEs are created along a center line with width added to *both* sides, be sure to deduct half the wall thickness to obtain the correct wall length.

Windows From the Draw menu on the standard toolbar, choose Polyline.

DOS From the Draw menu, choose Polyline.

```
Command:PLINE
From point:4′,0
The current pline is 0.00 units
Arc/Close/Halfwidth/Length/Undo/Width/<Endpoint of line>:W
Start width<0.00>:8
End width<8.00>:⏎
Arc/Close/Halfwidth/Length/Undo/Width/<Endpoint of line>:0,0
Arc/Close/Halfwidth/Length/Undo/Width/<Endpoint of line>:0,4′
Arc/Close/Halfwidth/Length/Undo/Width/<Endpoint of line>:⏎
```

Step 4   Now use Zoom Window to get a closer look at the northeast corner of the building. Add the lines as shown, with PLINE width at 2″.

> **Windows** From the standard toolbar, pick Zoom Window.
>
> **DOS** From the View menu, choose Zoom Window.

Once sufficiently zoomed, enter the staircase with PLINE.

PLINEs are entered from their MIDDLE point, so calculations will need to be adjusted.

```
Command:PLINE
From point:16'8",5'
The current pline is 8.00 units
Arc/Close/Halfwidth/Length/Undo/Width/
   <Endpoint of line>:W
Start width<8.00>:2
End width<2.00>:↵
Arc/Close/Halfwidth/Length/Undo/Width/
   <Endpoint of line>:16'8", 17'6"
```

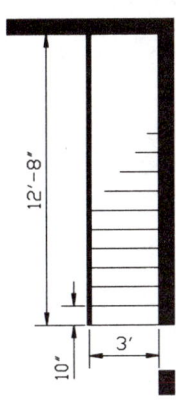

Step 5   Now use PLINE to place an arrow pointing up the stairs.

```
Command:PLINE
From point:18'2",4'6"
The current pline is 2.00 units
Arc/Close/Halfwidth/Length/Undo/Width/
   <Endpoint of line>:18'2",7'
Arc/Close/Halfwidth/Length/Undo/Width/
   <Endpoint of line>:W
Start width<4.00>:7
End width<7.00>:0
Arc/Close/Halfwidth/Length/Undo/Width/
   <Endpoint of line>:18'2",8'
Arc/Close/Halfwidth/Length/Undo/Width/
   <Endpoint of line>:↵
```

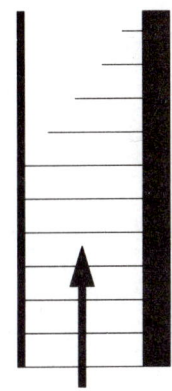

Step 6   Now use PLINE to create a break line on the staircase.

```
Command:PLINE
From point: (pick 1)
The current pline is 0.00 units
Arc/Close/Halfwidth/Length/Undo/Width/
   <Endpoint of line>: (pick 2, 3, 4,
 5, 6, 7 in sequence)
```

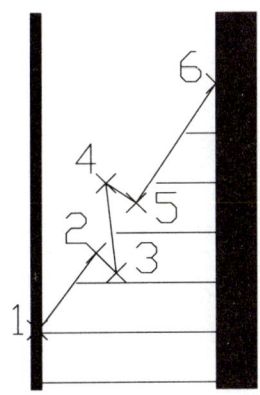

Step 7 Now use PAN to move the drawing to the right, at the same scale factor, and then use SOLID to create a fireplace.

> **Windows** From the standard toolbar, choose PAN. Then from the Draw toolbar, choose Solid.
>
> **DOS** From the View menu, choose Pan. Then from the Draw menu, choose Solid.

```
Command: SOLID
First point: (pick 1)
Second point: (pick 2)
Third point: (pick 3)
Fourth point: (pick 4)
Third point: (pick 5)
Fourth point: (pick 6)
Third point: (pick 7)
Fourth point: (pick 8)
Third point: (pick 9)
Fourth point: (pick 10)
Third point: (pick 11)
Fourth point: (pick 12)
Third point: ⏎
```

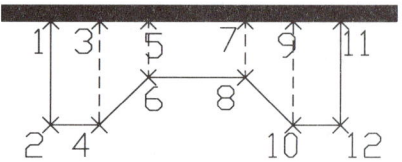

Step 8 Use PAN again to move the drawing over to the left so that you can create a "north arrow" outside of the building. Draw in a vertical line at 3 feet, and a horizontal line at 1.5 feet across the lower section as shown.

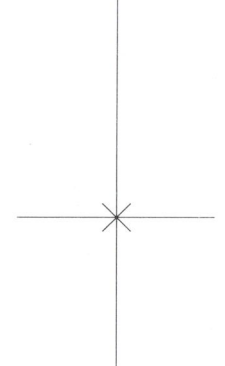

Step 9 Then, with your SNAP still set to 2, draw in two diagonal lines through the intersection. Now use SOLID to fill in the sides of the north arrow.

```
Command: SOLID
First point: (pick 1)
Second point: (pick 2)
Third point: (pick 3)
Fourth point: ⏎
Third point: ⏎
```

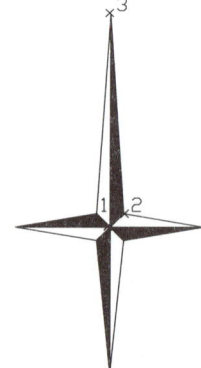

Step 10   Finally, add lines for windows, and your simple floor plan is complete.

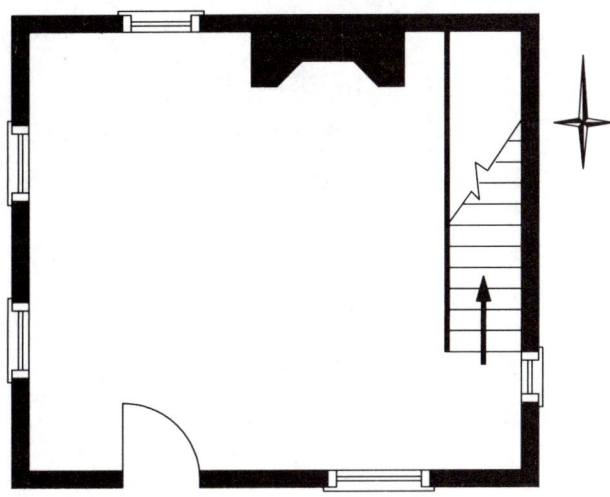

Step 11   Now PAN your view over and, using the same overall dimensions, add a second floor plan with MLINE. The MLINE will act like a polyline; the lines are one unit. Change the Scale of the exterior walls to 8″ and the interior walls to 4″.

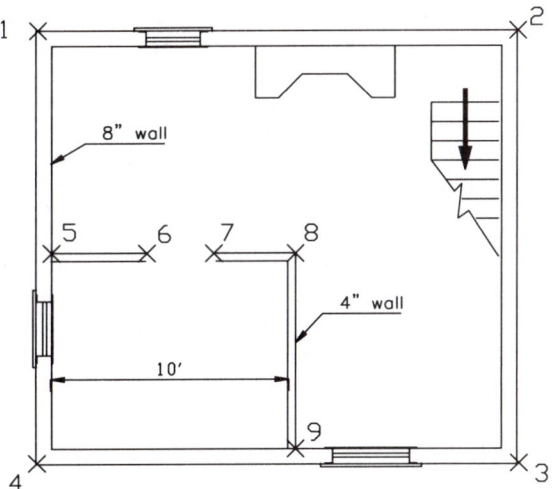

Entity Commands with Width   **75**

Command and Function Summary

DONUT creates filled circles or rings.

PEDIT is used to edit the PLINEs.

PLINE creates lines with multiple segments in varying widths.

POLYGON is used to create polygons with equal sides, having anywhere from 3 to 1024 sides.

SOLID creates filled polygons.

TEXT is used to place text or lettering on drawings.

TRACE is used to make thick lines with mitred corners.

Practice Exercise 3

Use PLINE, SOLID, and DONUT to create this map. Then use DONUT and BREAK with the OSNAP QUADrant to create the elevation marker.

POLYGON can be used to create the bolt.

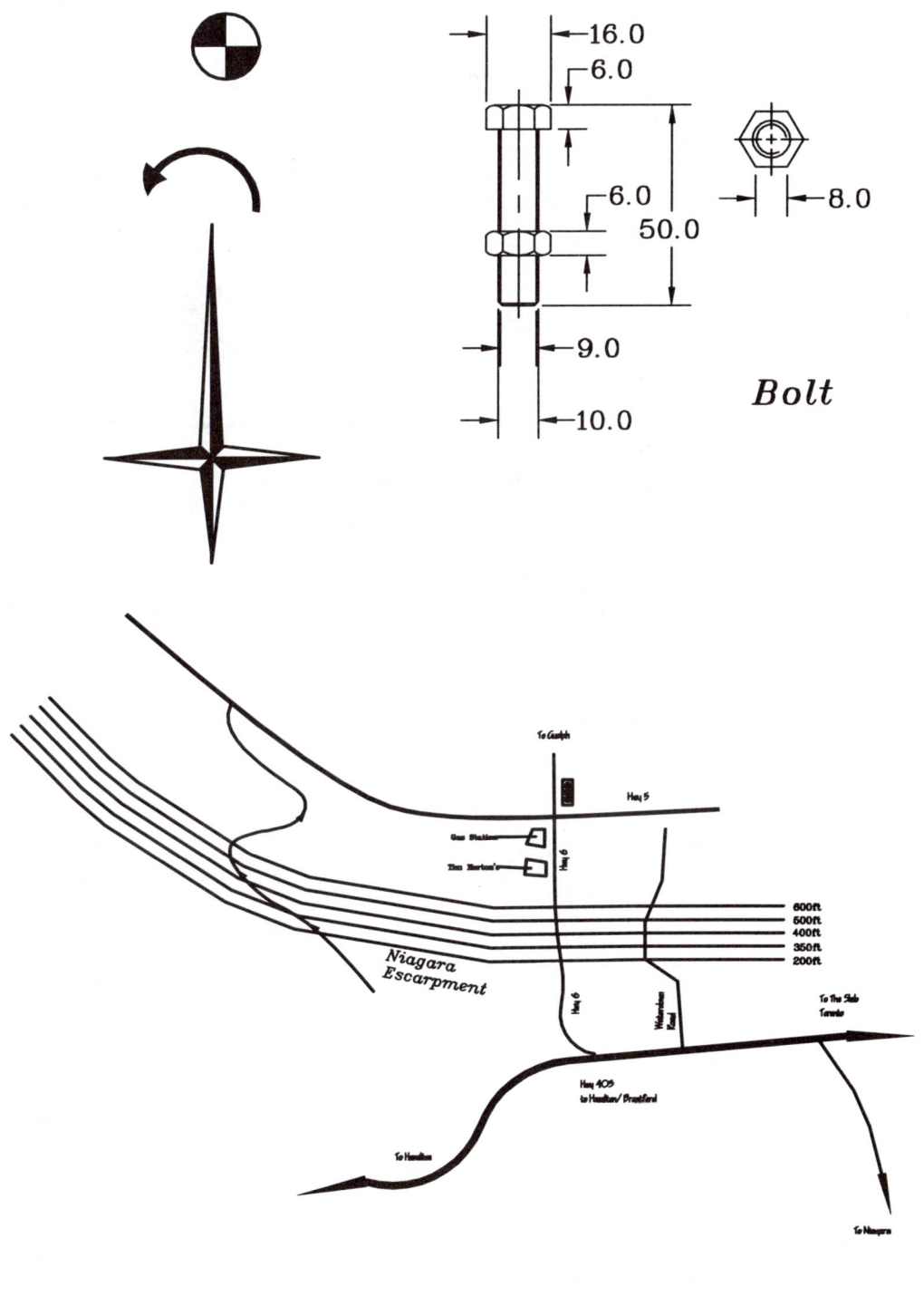

Exercise A3

Using the dimensions from the floor plan in Chapter 2, use PLINE to create this presentation floor plan. Make the outside walls 8″ thick and the inside walls 5″ thick. Add a fireplace using SOLID, and columns to support the terrace roof with DONUT. Save the file as PRESENTFP. Open a new file to create the title block below. Save it under the name of ATITLE.

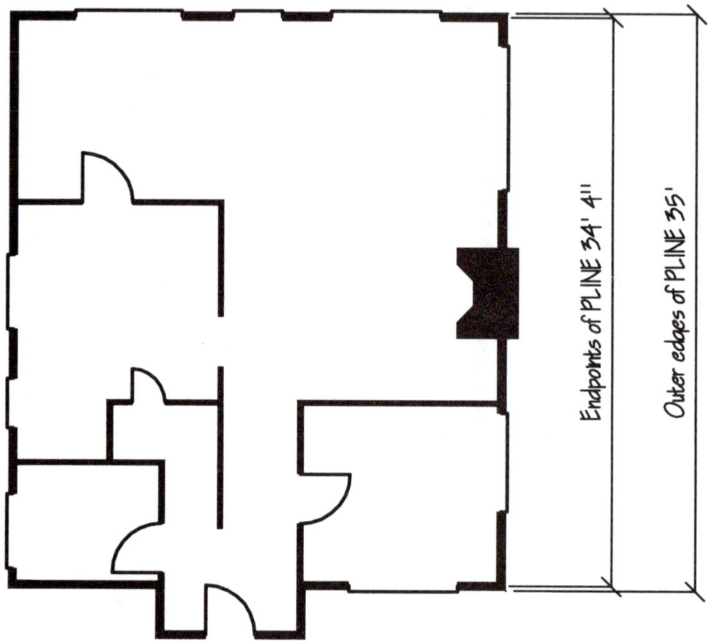

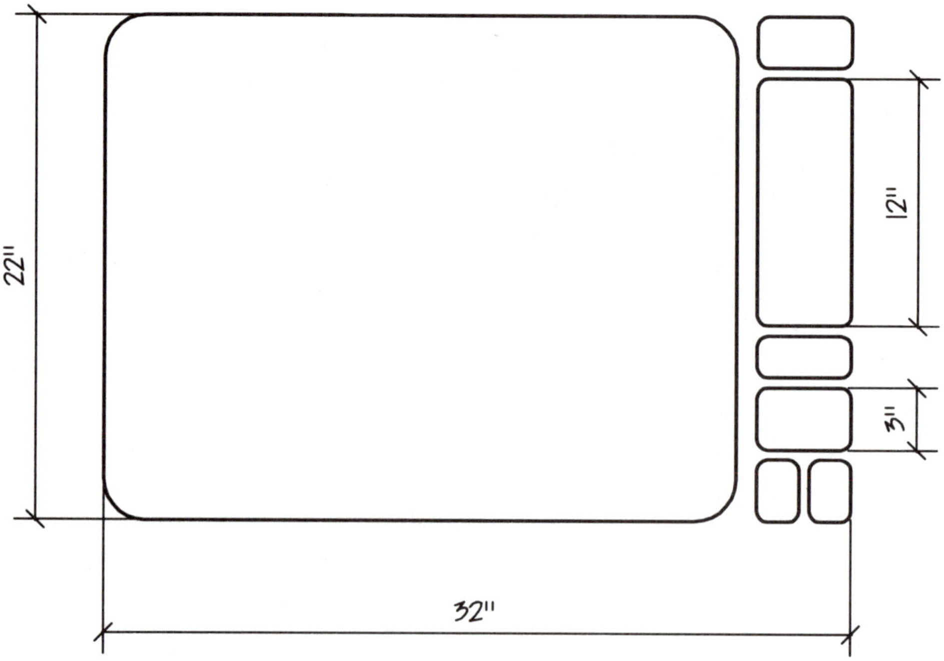

Exercise C3

Use surveyor's units to create the lot plan shown below. Save this as file LOT. Then create the titleblock and call it CTITLE. Finally make a file called NORTH for the north arrow. For help on accessing surveyor's units, see Chapter 1.

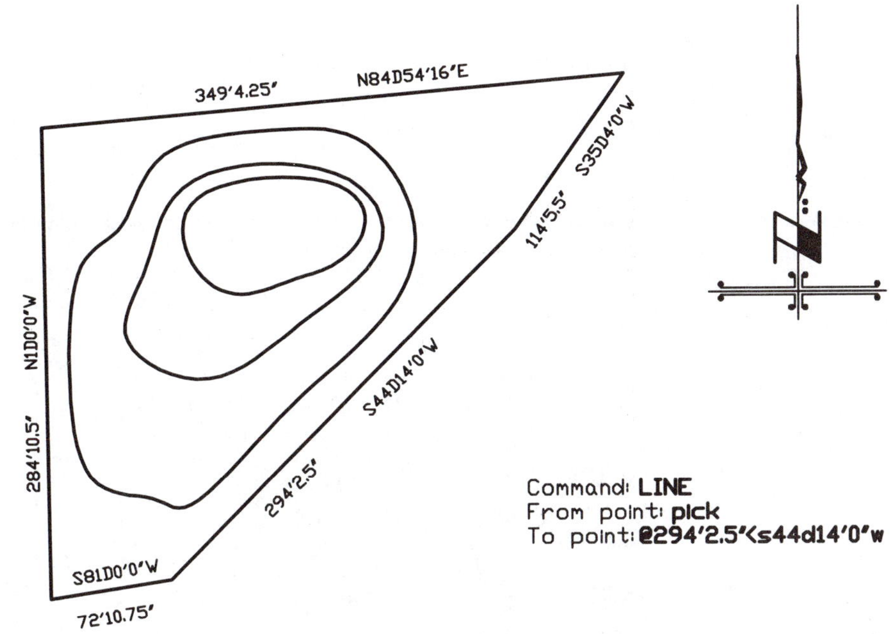

```
Command: LINE
From point: pick
To point: @294'2.5"<s44d14'0"w
```

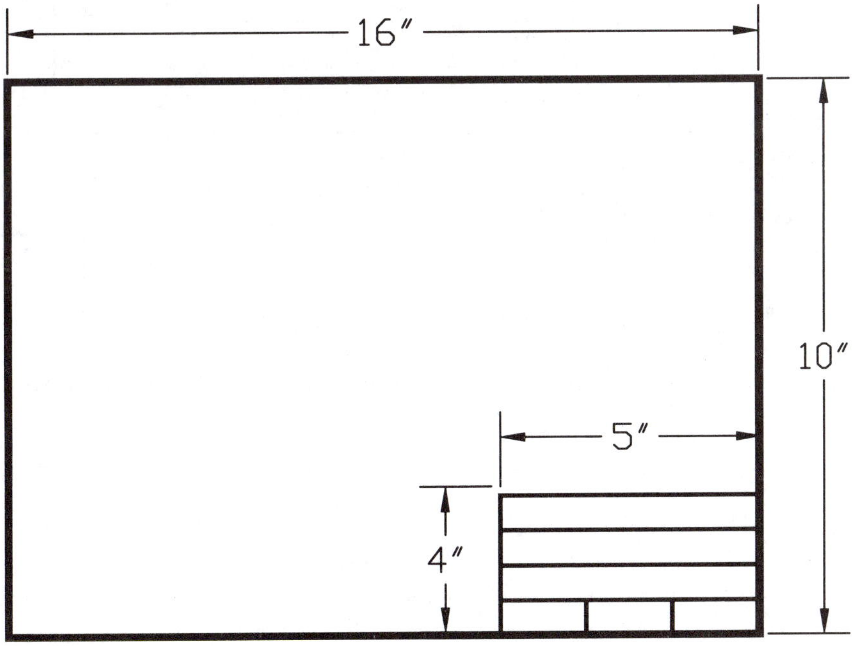

Exercise E3

Create the schematic below and save it as E3. Then use PLINE to draw the title block and save it as ETITLE.

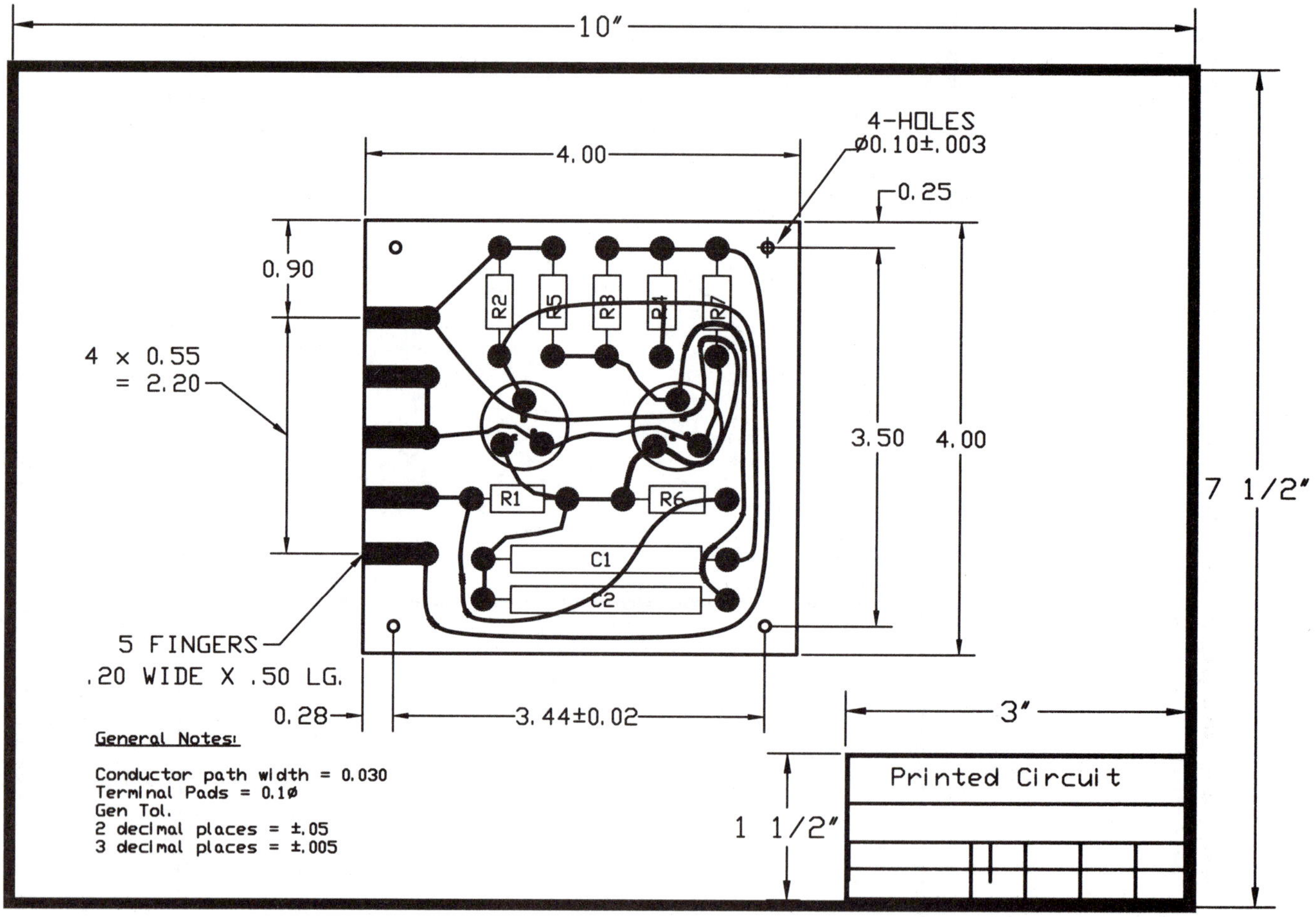

Exercise M3

Draw the lifting screw assembly shown below using PLINE and SOLID. Save the file as M3. Then open a new file and draw the title block and border shown below. Save this as MTITLE.

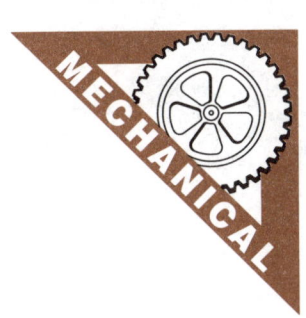

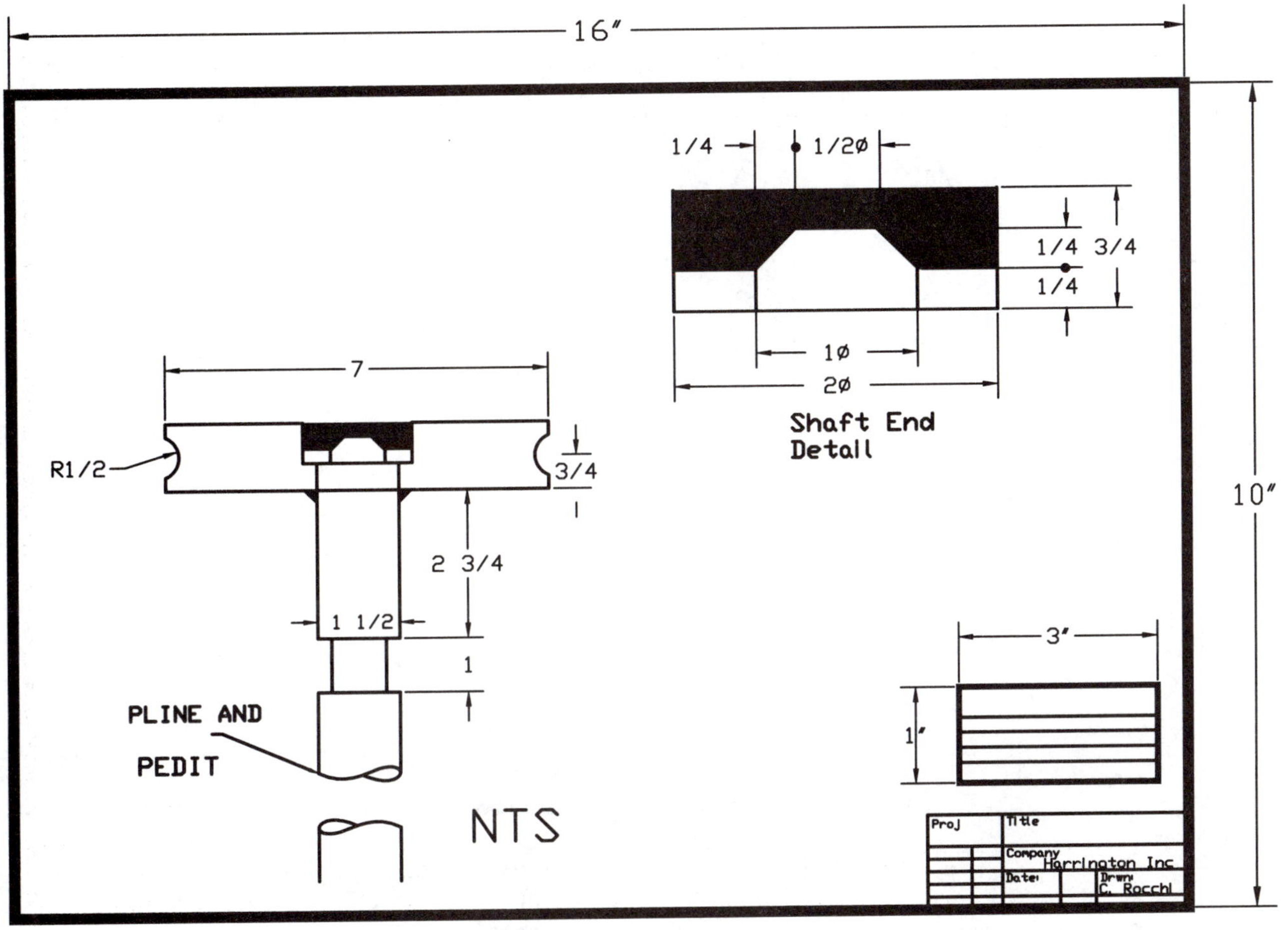

Entity Commands with Width

Challenger 3

Draw the ball bearing shown below using PLINE, DONUT, and SOLID. Save the file as CH3.

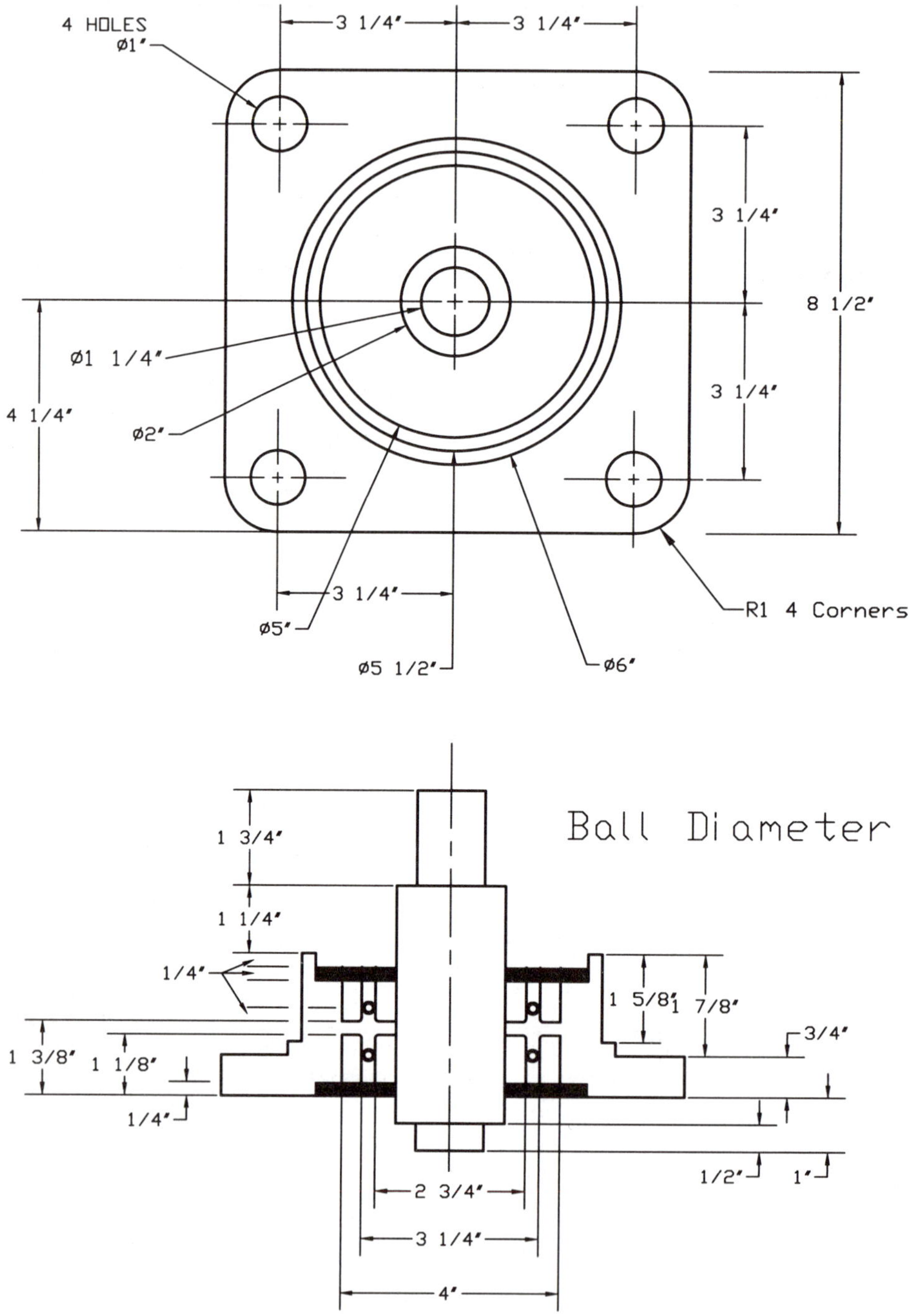

Object Selection and Editing

Upon completion of this chapter, you should be able to:

1. Use the various options for selecting objects
2. Edit objects with the MOVE command
3. Edit objects with the COPY command
4. Edit objects with the MIRROR command
5. ROTATE objects
6. SCALE objects
7. Use grips to edit objects
8. Change the LINETYPE

Once objects are drawn on the screen, editing commands (Modify and Construct) are used to cut down on drawing time. When editing, OSNAP is particularly important to get the exact position of objects when they are moved, copied, or rotated.

Selecting Objects Within the Edit Command

In virtually every editing command, you will be prompted to "select objects." This makes sense, because the editing command will change the position or the parameters of objects, and thus the system needs to know which objects you want to change. As you select items, they become highlighted (dotted lines).

Objects can be selected in a variety of ways. If you are selecting the object after you have invoked the command you can select the objects by:

- Digitizing the desired item with the cursor "pick box"
- Indicating a group of items with the Crossing option
- Indicating a group of items with the Window option
- Indicating a series of items with Fence
- Indicating a group of items with CPolygon
- Indicating a group of items with WPolygon
- Indicating the last entered item using the Last option
- Indicating the previous selection set with Previous
- Indicating all objects by typing in **ALL**

Once you have identified your selection set or chosen the objects that are to be edited, AutoCAD will keep prompting you to select objects to be added or removed from the selection set. To continue with the Edit command after selecting objects, press ⏎ to signal the end of object selection.

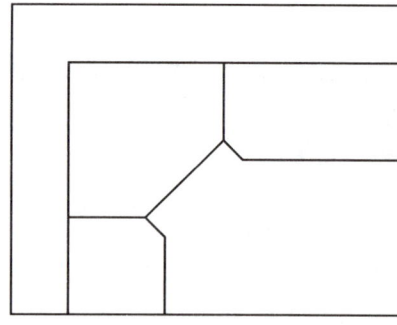
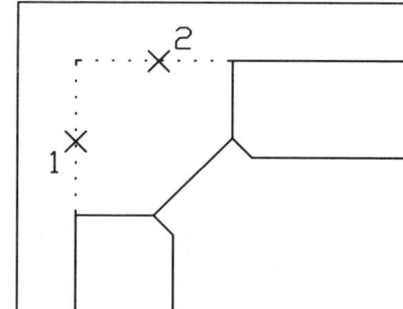

Figure 4-1

The object selection default is to select every item one by one.

```
Select objects: (pick 1, 2)
```

Only the two objects selected will be affected by the editing command.

Selection Windows

You can also select the objects by drawing a rectangular window area in response to the "Select objects:" prompts.

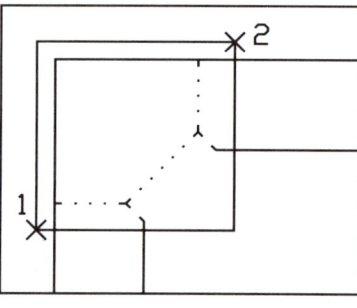
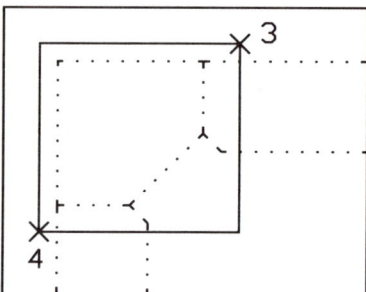

Figure 4-2

```
Select objects: (pick 1)
Other corner: (pick 2)
```

```
Select objects: (pick 3)
Other corner: (pick 4)
```

In Figure 4-2 the first selection is a Window; the second is a Crossing box. Dragging the window from left to right (Window selection box) identifies only the objects completely enclosed in the rectangle. Dragging the window from right to left (Crossing selection box) identifies objects within the window as well as any objects that cross into the selection area.

Sometimes while trying to indicate a Window or Crossing, your selection picks up an object rather than starting the window or crossing. To specify that you want a Window you can also specify Window from the menu or by typing in **W**. You can specify Crossing from the menu or by typing in **C**. In both cases you will be prompted for the other corner once your selection pick doesn't select an object.

Selection Polygons

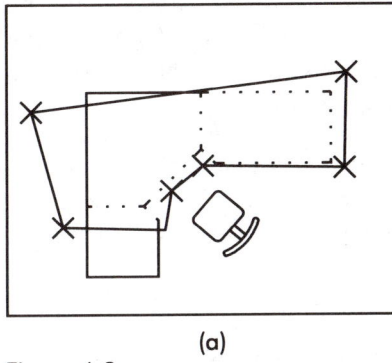 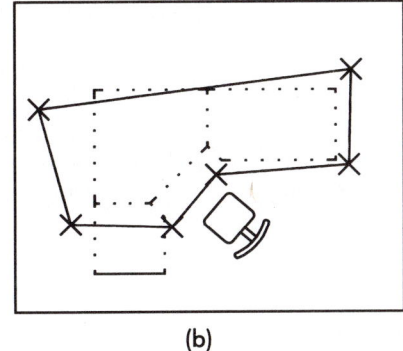

(a)   (b)

Figure 4-3

To select objects in an irregularly shaped area, use Window Polygon or Crossing Polygon. With the WPolygon option (Figure 4-3(a)) the objects contained within the polygon are chosen. With the CPolygon option (Figure 4-3(b)), all objects touching the three- or four-sided polygon will be picked up. Up to 16 points can be used on the polygon.

```
Select objects:WP (Window Polygon) (pick points as in
   Figure 4-3(a))
Select objects:CP (Crossing Polygon) (pick points as in
   Figure 4-3(b))
```

Selection Lines or Fence

With the Fence option, the objects touching the fence line will be picked up. Fence is useful for nonadjacent objects.

```
Select objects:F (Fence) (pick points)
```

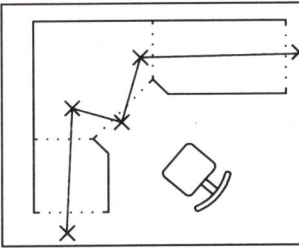 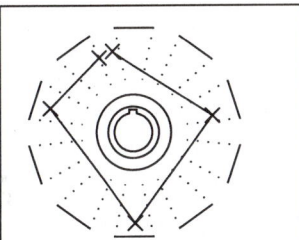

Figure 4-4

Last and Previous

To identify the same selection set just used, type in **P** for Previous. To identify for editing the last object entered, type in **L** for Last.

```
Select objects:L (Last)
Select objects:P (Previous)
```

Object Selection Cycling

In cases where a selection pick box touches more than one object, AutoCAD normally selects the most recent object. To avoid picking many times in an area with many overlapping objects, enter **Ctrl-N** before picking, then repeatedly pick at the location where multiple objects lie.

You can use any of the above methods for object selection in any of the editing commands. The following examples show how to choose first the command, and then the selection set. Choosing the selection set first is covered on page 94.

The COPY Command

The COPY command takes an item or group of items and places a copy at another location or at multiple locations.

> **Windows** From the Modify toolbar, choose Copy.
>
> **DOS** From the Construct menu, choose Copy.

The command line equivalent is **COPY**.

```
Command:COPY
Select objects: (pick 1, 2)
Select objects:↵
<Base point> or Multiple:END of (pick 3)
Displacement:END of (pick 4)
```

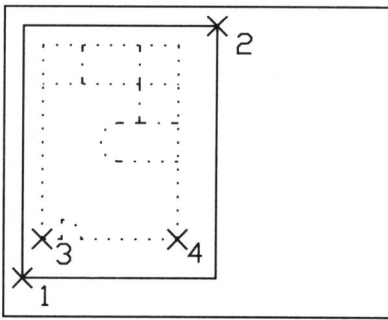 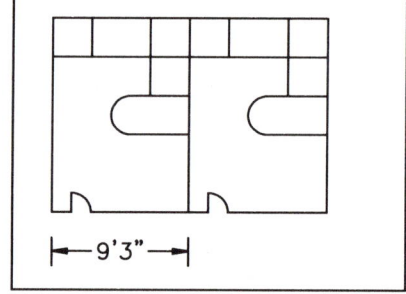

Figure 4-5

In the example of Figure 4-5, an incremental value could be used provided you know the exact location of the displacement in relation to the base point. For example:

```
Displacement:@9'3",0
```

COPY Multiple

The COPY command can be used to place multiple copies of objects at random spacing. Once the objects have been selected, the command prompts for either the base point or the Multiple option.

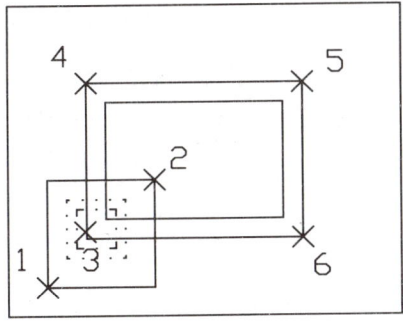
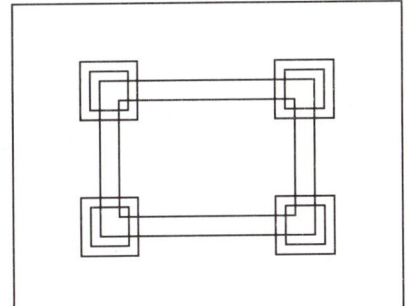

Figure 4-6

```
Command:COPY
Select objects: (pick 1)
Other corner: (pick 2)
Select objects:↵
<Base point> or Multiple:M
<Base point>:END (pick 3)
Displacement:END (pick 4)
```

All displacements are relative to the first base point, so choose that point carefully. In Figure 4-6 the end point of the bottom left corner is chosen as the base point. This point is referenced each time the object is to be placed. SNAP can also be used to place them accurately.

The MOVE Command

The MOVE command moves an object or series of objects from one point to another, relative to a defined point on the object or a base point.

> **Windows** From the Edit toolbar, choose Move.
>
> **DOS** From the Modify menu, choose Move.

The command line equivalent is **MOVE** or **M**.

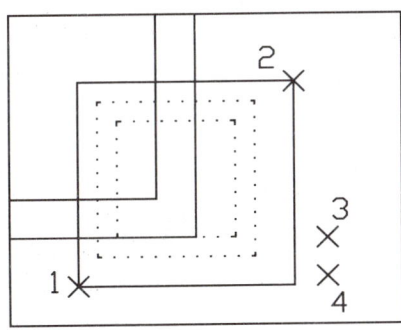
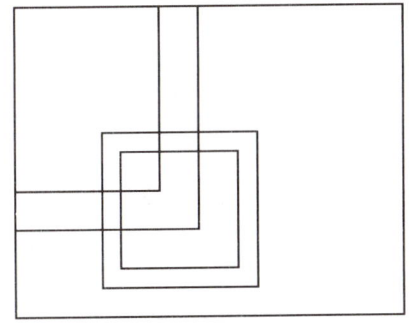

Figure 4-7

```
Command:MOVE
Select objects: (pick 1)
Other corner: (pick 2)
Select objects:↵
Base point: (pick 3) (from where)
Displacement: (pick 4) (to where)
```

Object Selection and Editing

In Figure 4-7, pick 3 is where the objects are coming from, and pick 4 is where they are going to.

MOVE Using Incremental Values

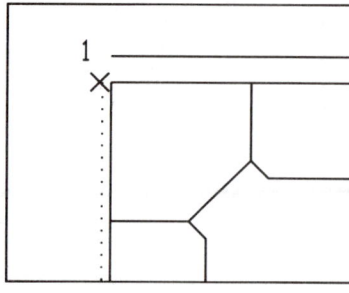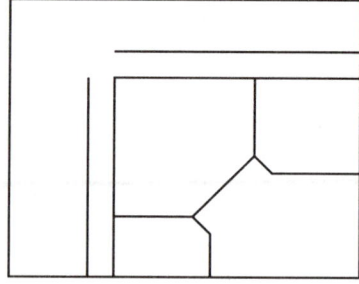

Figure 4-8

In Figure 4-8, the base point could have been picked up anywhere on the screen. The second point, @-2,0, describes a point two units in negative *X* from the current position, wherever that may be. The point given is a relative or incremental point.

```
Command:MOVE
Select objects: (pick 1)
Base point:0,0
Displacement:@-2,0
```

Moving Objects to 0,0

To move a selected group of objects from their current position to 0,0, first identify the selection set. Identify which point will be finally 0,0. Use 0,0 as the displacement. Since the exact end point of that object must be at 0,0, use END (end point). The displacement is given as an absolute value.

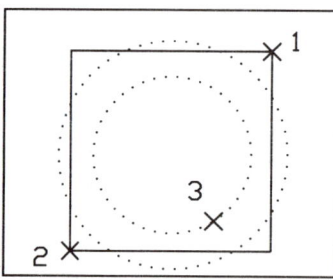

Figure 4-9

```
Command:MOVE
Select objects: (pick 1)
Other corner: (pick 2)
Select objects:↵
Base point:CEN of (pick 3)
Displacement:0,0
```

You may wonder: What is the difference between the PAN command and the MOVE command?

PAN moves the screen viewing area. It is a display command. After the PAN command is finished, the parameters or the coordinates of each object will be the same: even though they appear to have moved, they have only moved relative to the screen. MOVE, on the other hand, actually moves the items from one point to another so that the coordinates of the objects change.

The MIRROR Command

The MIRROR command creates a mirror image of an item or group of items through a specified mirroring plane selected by a real or imaginary line.

After you have executed this command, you are prompted to either delete the old objects or keep them. Thus, to "flip" a series of objects, delete the old objects and you will only retain the mirror image.

Windows From the Edit toolbar, choose Mirror.

DOS From the Construct menu, choose Mirror.

The command line equivalent is **MIRROR**.

In Figure 4-10, the object is mirrored through a *mirroring-plane object* or *mirroring plane* — picks 3 and 4 — placed at the halfway point of the final object. The mirroring plane was deleted after the operation.

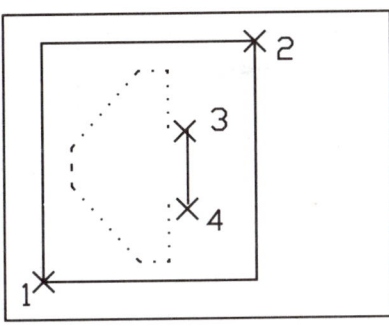
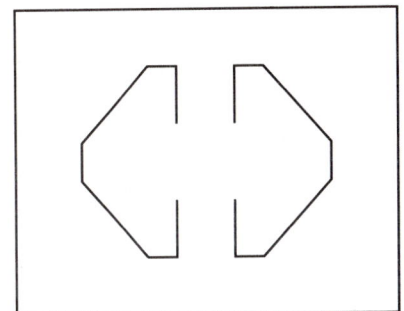

Figure 4-10

```
Command:MIRROR
Select objects: (pick 1)
Other corner: (pick 2)
Select objects:↵
First point of mirroring plane:END of (pick 3)
Second point of mirroring plane:END of (pick 4)
Delete old objects?<N>:↵
```

If existing objects can be used to describe the mirroring plane, use them.

In Figure 4-11, the mirroring plane is calculated from the center of the circle, in the center at the bottom. With ORTHO on, the mirroring plane would be easier to identify.

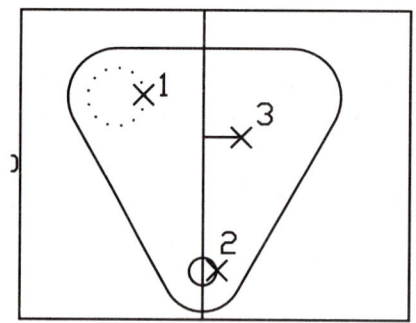 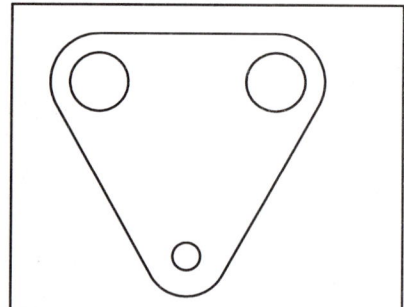

Figure 4-11

```
Command:MIRROR
Select objects: (pick 1)
Select objects:↵
First point of mirroring plane:CEN of (pick 2)
Second point of mirroring plane: (pick 3) (use F8 Ortho)
Delete old objects?<N>:↵
```

Mirroring Using Absolute Values

If no object is accessible as a mirroring-plane base point, one can calculate the distance using the coordinates, and pick them on the screen. The mirroring plane can be indicated by picking points or by entering values, as demonstrated in Figure 4-12.

When MIRROR prompts for a point, any kind of point may be entered.

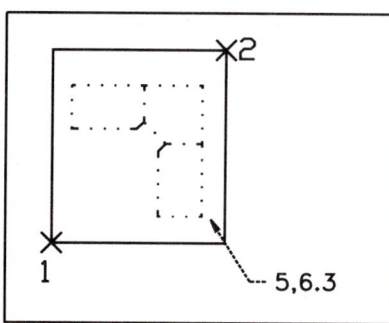 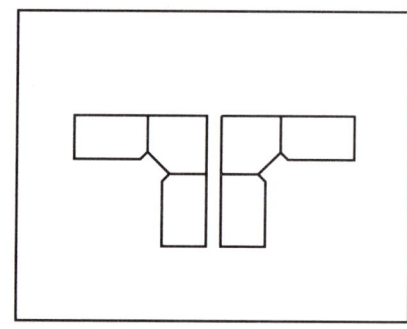

Figure 4-12

```
Command:MIRROR
Select objects: (pick 1, 2)
Select objects:↵
First point of mirroring plane:5,6.3
Second point of mirroring plane:@0,3
Delete old objects?<N>:↵
```

In Figure 4-12, the point 5,6.3 was chosen because it is 1.5 units in the X direction past the desk, allowing for a 3-unit space between the desks.

Mirrored Text

The MIRRTEXT variable allows you to set text while either creating mirror-image (backwards) text, or simply copying it.

```
Command:MIRRTEXT
0 Retains text direction
1 Mirrors the text
```

The ROTATE Command

The ROTATE command rotates an object or series of objects around a specified base point. Using the command sequence of Figure 4-13, the objects are rotated 45 degrees around a point in the middle of the object itself. To view the rotation of the object, move the cursor in a circle around the base point.

> **Windows** From the Edit toolbar, choose Rotate.
>
> **DOS** From the Modify menu, choose Rotate.

The command line equivalent is **ROTATE**.

```
Command:ROTATE
Select objects: (pick 1)
Other corner: (pick 2)
Select objects:
Base point: (pick 3) (use SNAP for accuracy)
<Rotation angle>/Reference:45
```

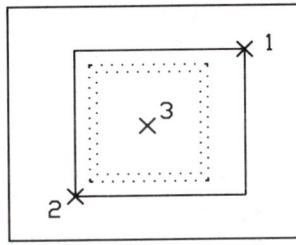

Figure 4-13

In Figure 4-14, the objects were first copied from the position on the front wall, then rotated relative to the base point at the ENDpoint of the window so the selection fits perfectly into the space provided.

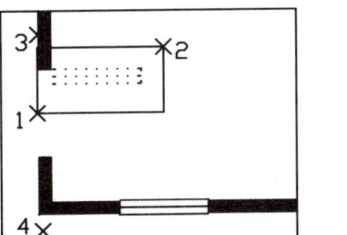

Figure 4-14

Object Selection and Editing 91

```
Command:ROTATE
Select objects: (pick 1)
Other corner: (pick 2)
Select objects:↵
Base point:END of (pick 3)
<Rotation angle>/Reference: (pick 4) (use ORTHO F8 for
   accuracy)
```

ROTATE and COPY to Create a New Object

In Figure 4-15, the objects are copied in the same place, then rotated around a point identified as the middle of the object.

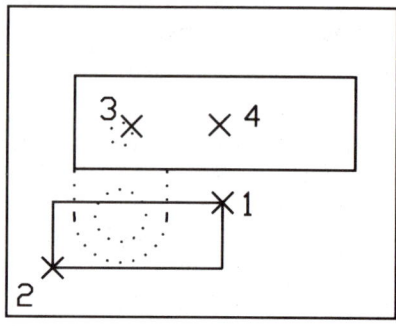
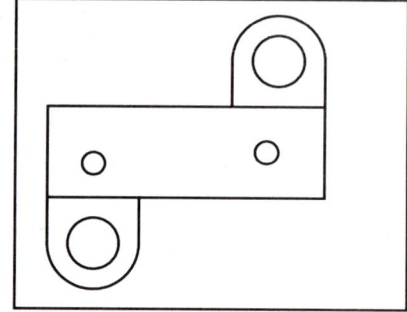

Figure 4-15

```
Command:COPY
Select objects: (pick 1)
Other corner: (pick 2)
Select objects: (pick 3)
Select objects:↵
<Base point>/Multiple:0,0
Second point:0,0
```

This makes a duplicate copy of the arc, two circles, and two lines at the same spot.

```
Command:ROTATE
Select objects:P (this will take the previous selection set)
Select objects:↵
Base point: (pick 4)
<Rotation angle>/Reference:180
```

Use Previous to identify the first selection set *but not the copy*.

Rotating Using a Reference

Use the reference option to rotate something when you are not sure what the angle is, but you know the desired final angle.

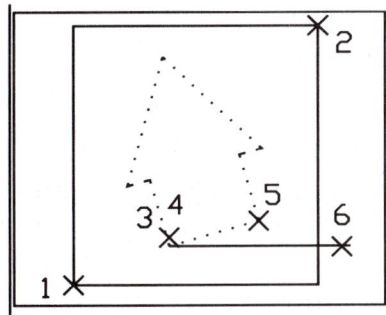 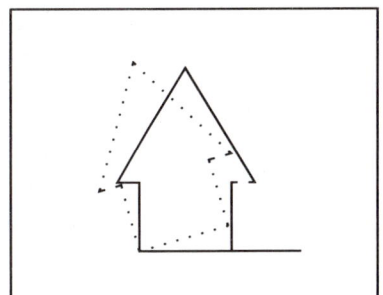

Figure 4-16

```
Command:ROTATE
Select objects: (pick 1)
Other corner: (pick 2)
Select objects:↵
Base point:(END of pick 3)
<Rotation angle>/Reference:R
Reference angle<0>:(END of pick 4)
Second point:(END of pick 5)
New angle:(END of pick 6)
```

See Prelab 4B for ROTATE/COPY using grips.

The SCALE Command

The SCALE command is very similar to those described above. You select objects, pick a base point, then enter a scale factor. Not surprisingly, a scale factor more than 1:1 will make the image bigger. A scale factor less than 1:1 will make it smaller.

> **Windows** From the Modify toolbar, choose the Resize flyout, then:
>
> **DOS** From the Modify menu, choose Scale.

The command line equivalent is **SCALE**.

```
Command:SCALE
Select objects: (use an object selection method)
Base point: (pick a point, usually on the object)
<Scale factor>/Reference:3
```

Problems with Editing

How do you easily remove objects that have been moved, copied, or mirrored to the wrong place?

The Undo command negates the previous command. Almost any type of command can be undone. Typing **U** ↵ will undo any number of commands back to the beginning of the current editing session. All the commands are stored in the order in which they were performed.

```
Command:U ↵ Group
Command:U ↵ Group
```

Repeat until you have removed all the objects that are incorrect.

UNDO is similar to U, but UNDO offers more power and is more dangerous. UNDO will prompt for the number of commands to be undone and undo them in a single operation. For example, to undo the last five entries, type UNDO 5.

To override an UNDO, use REDO. This command must directly follow the UNDO command.

```
Command:U Group
Command:REDO (this will restore the information just undone)
```

What happens when you move or copy a group of objects and they disappear?

Use **U** ↵ to undo the command, or **ZOOM A** to view the spot to which it was moved or copied.

During the COPY or MOVE command, you may have pressed ↵ when the system was prompting for the base point. Remember to read the prompts, then give the needed information.

It is a good idea to save the model or drawing with SAVE before doing extensive editing. If you undo more than expected, you can quit and reload the original file.

Getting Good Results the First Time

Always use SNAP, OSNAP, or actual coordinates to pick base points. If a reference object is needed to complete an editing command (for example, a line in the MIRROR command), insert it, reference it with OSNAP, then erase it.

Editing with Grips

The commands demonstrated above were chosen before the objects were selected. With the aid of grips, you can use the SELECT command to select objects, then use the editing commands or the grip modes to edit them.

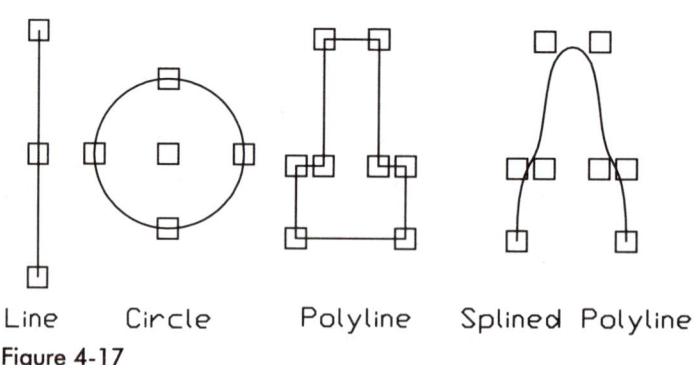

Line   Circle   Polyline   Splined Polyline

Figure 4-17

Figure 4-17 shows where the grips would be placed on a variety of objects. Grips allow you to combine objects and command selection and thereby edit more quickly.

Grip Modes

Object and command selection can be combined by using the grip modes — Stretch, Move, Rotate, Scale, and Mirror. Once you have identified the selection set, you identify a base point or **hot grip**.

When the hot grip is chosen, it turns a solid color. The grip modes will then be loaded and you can then edit using one of the five modes.

Clear grips from the selection set by using **ESC** twice in Windows or **Ctrl-C** twice in DOS, or by selecting a new command.

Use the space bar or Enter key to toggle through the modes, or type in the first two letters of the mode. For example, to reach Scale mode from another mode, keep pressing ⏎ until Scale appears, or enter **SC**.

If the grip mode does not work on your system, type **GRIPS** and then enter **1**.

Grips Example

First draw in a series of lines as shown.

At the command prompt pick up the two objects on the right. The grips will appear in blue on the set of objects.

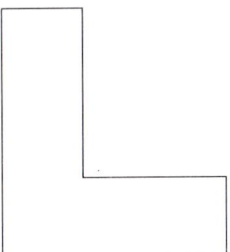

Figure 4-18

```
Command: (pick 1, 2)
```

The cursor snaps to any grip over which it is moved. Select the top right grip to act as the base point for the edit.

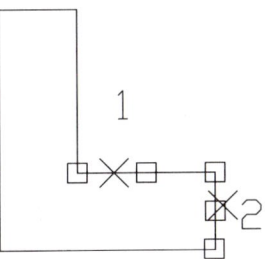

Figure 4-19

```
Command: (pick 3) (pick the right
   corner of the object)
```

The base grip will be identified with a solid color, probably red.

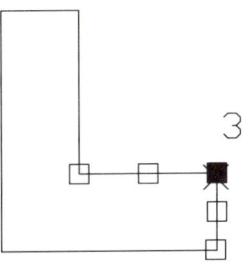

Figure 4-20

Now that you have the base point highlighted, the grip modes appear and you can choose one. First use Stretch. Pick the point for the displacement of the specified objects.

```
Command:**STRETCH**
<Stretch to
   point>/Base/Copy/Undo/eXit:
   (pick 4)
```

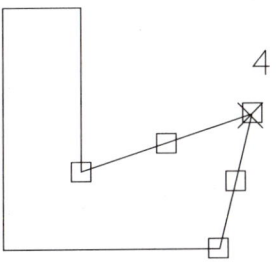

Figure 4-21

Object Selection and Editing **95**

Now pick up the upper left objects to be moved plus the base point.

```
Command: (pick 5, 6) (the objects
  to be moved)
Command: (pick 7) (the base point
  or hot point)
```

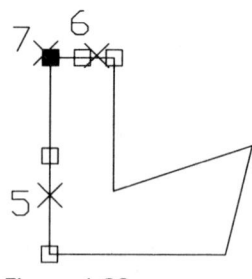

Figure 4-22

Once you have chosen the base point, the grip mode is invoked, starting with Stretch. Use ⏎ to advance to the next mode, which is Move. Then pick the displacement point for the objects to move to.

```
Command:**STRETCH**⏎
Command:**MOVE**
<Move to point>/Base point/Copy/
  Undo/eXit: (pick 8)
```

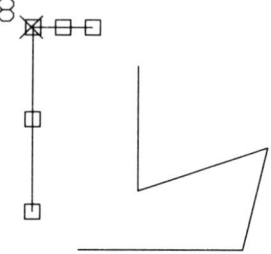

Figure 4-23

Again use **Ctrl-C** or **ESC** to remove the grips from the objects.

Now we will rotate one object.

Pick the lowest horizontal line once to pick the object and then at the corner to identify the base point.

```
Command: (pick 9) (the object to
  be rotated)
Command: (pick 10) (the base
  point)
```

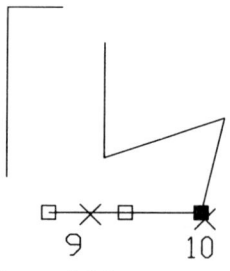

Figure 4-24

You will be in the grip mode. Use ⏎ to toggle to rotate.

```
Command:**STRETCH**⏎
Command:**MOVE**⏎
Command:**ROTATE**
<Rotation angle>/Base point/Copy/
  Undo/eXit:-10
```

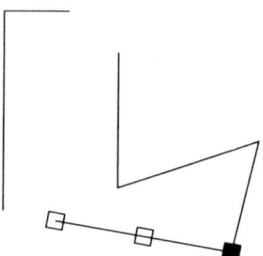

Figure 4-25

For Scale, use the same sequence. Pick the objects, pick the base point, then enter the scale factor. You can type in **SC** to get Scale rather than toggling through.

```
Command: (pick 11, 12, 13)
Command:**STRETCH**SC
Command:**SCALE**
<Scale factor>/Base point/Copy/
   Undo/eXit:.5
```

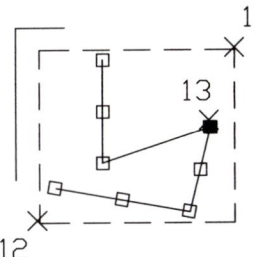

Figure 4-26

Finally, the mirror mode allows you to mirror the objects, but doesn't allow you to keep the original selection set. Only the mirrored image is left.

```
Command: (pick 14, 15)
Command: (pick 16)
Command:**STRETCH**MI
Command:**MIRROR**
<Scale factor>/Base point/Copy/
   Undo/Exit: (pick 17)
```

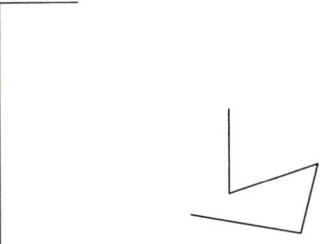

Figure 4-27

Editing with grips should make your editing much faster.

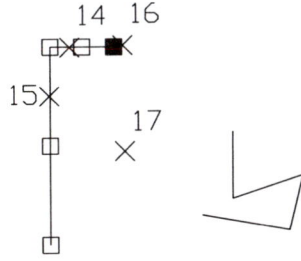

Figure 4-28

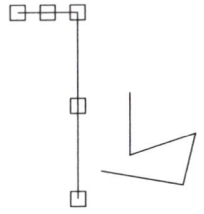

Figure 4-29

Setting LINETYPEs

AutoCAD offers the standard different types of lines — center, hidden, and dashed — in addition to many variations to make drawings more legible and more attractive. The *linetypes* are not loaded with the default AutoCAD drawing, so for you to use them they must be loaded into the file.

> **Windows** Under the Standard toolbar, choose Linetype.
>
> **DOS** Under the Data menu, choose Linetype ... for the DDLtype dialog box.

The command line equivalent is **LINETYPE**.

Object Selection and Editing

To get a listing of the available linetypes from the command line, use the following:

Command:**LINETYPE**
?/Create/Load/Set:**?**

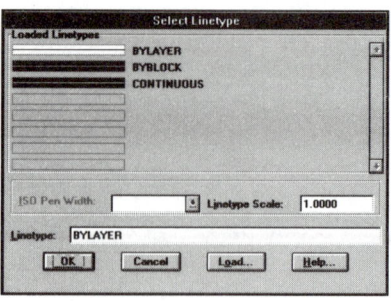

The dialog box is the easiest way to access the linetypes.

From the linetypes listed, pick the one you require. If the linetype you require is not listed, choose Load.

Choose the linetypes that you would like to load, then pick OK.

You will then return to the main menu. If you want to use the new linetype to draw with, use Set. This will set the current linetype for subsequently drawn objects.

For each linetype there are three listings. For example:

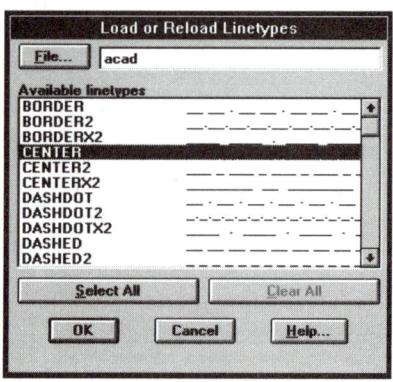

- Hidden = 3 dashes per inch
- Hidden2 = 6 dashes per inch
- HiddenX2 = 1 1/2 dashes per inch

You may use these to modify the appearance of your lines.

The command line LINETYPES works as follows:

Command:**LINETYPE**
?/Create/Load/Set:

Where: **?** = a listing of possible linetypes
Set = a setting of that linetype; only that type of line can be used
Load = a load of that linetype onto the system; once it is loaded, you can use it whenever you want or attach it to a layer
Create = the ability to create new linetype patterns

If you want to load a linetype, pick **L** for Load.

Command:**LINETYPE**
?/Create/Load/Set:**L**
File to search<ACAD>:⏎ **(accepts the default file listing)**
Linetype name:**HIDDEN**
Select Linetype File dialogue box:⏎
Linetype HIDDEN loaded
?/Create/Load/Set:⏎

Notes

If a dialog box appears after the "File to Search" prompt, press ⏎ to accept the file.

Be careful to type the linetype exactly as it is spelled in the listing, or it will not load. (The most common stumbling-block in Canada is the spelling of "centre." Be sure to use the American spelling, *center*.)

If you would like to load all of the available linetypes use the DOS wildcard option or *****.

```
Command:LINETYPE
?/Create/Load/Set:L
File to search<ACAD>:↵    (accepts the default file listing)
Linetype name:*
```

All LINETYPEs will be loaded. Once the linetype is loaded, you can invoke it with the Set option.

```
Command:LINETYPE
?/Create/Load/Set:S
File to search<ACAD>:↵
Linetype name:HIDDEN
```

Danger
Be sure to set your linetype back to BYLAYER when you are finished using Hidden.

All objects that you draw will now be in linetype Hidden. If you cannot see them, change the LTSCALE.

Prototype Drawings

To avoid loading linetypes for every drawing, open ACAD.DWG, the prototype for new drawings. Load all the linetypes. Save the changes, then exit. Linetypes require little memory, so it is easier if they are always loaded.

Changing LTSCALE

Depending on the size of your object, linetypes that are not continuous may show up as such. This is because the scale of the drawing is too large or too small.

All scales are set to be viewed on a 12' 9" screen. This means that a hidden line, for example, will show up with three long dashes for every actual inch of line. If your screen is showing a line which is 200 units in length, this means there are 600 actual segments to that line. Obviously, you can't see 600 segments of a line on a screen that is 14 inches wide, so the line will appear to be continuous.

To see it properly you need to change the scale of the screen display. Pick LTSCALE under the Settings menu on the right side of your screen.

```
Command: LTSCALE
New scale factor:13
```

To determine the LTSCALE of the screen, start with the scale of the drawing. If a floor plan is to be drawn at 1/4"=1'0", then the scale is 1/48: try 48 for the LTSCALE. If you have not determined the scale of a drawing yet and simply want a screen display, find the furthest point in *X* and divide by 12. If you have set up the screen but have not as yet entered any points or geometry, take the furthest limit in positive *X* and divide by 12. If the limits are 200,160, then the LTSCALE will be 200/12, or approximately 15. If your limits are 3, then your LTSCALE will be 3/12, or .03.

Prelab 4A Using MOVE, COPY, and MIRROR

First we'll make a roller arm. Open a new file, keep the settings the same, and draw in two circles — one with a radius of .5, the other with a radius of .75 as shown. Use SNAP to access the same center.

Step 1

```
Command:C
3P/2P/TTR/<Center>: (pick 1)
Diameter/<Radius>:.5
Command:C
3P/2P/TTR/<Center>: (pick 1)
Diameter/<Radius>:.75
```

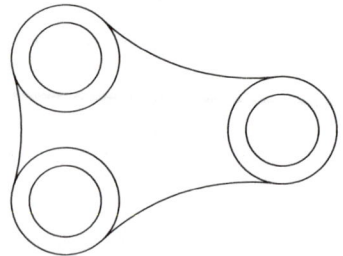

Step 2

Now MOVE these from wherever they are on the screen to 0,0.

```
Command:MOVE
Select objects: (pick 1, 2)
Base point:CEN (pick 3)
Displacement:0,0
```

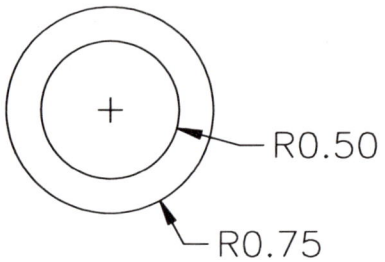

Step 3

Now PAN the objects across the screen so you can see them, and use COPY to place another set of circles.

```
Command:PAN
Displacement: (pick 1)
Second point: (pick 2)

Command:COPY
Select objects: (pick 1, 2)
<Base point or displacement>/
   Multiple:CEN of (pick 3)
Displacement:@3,1
```

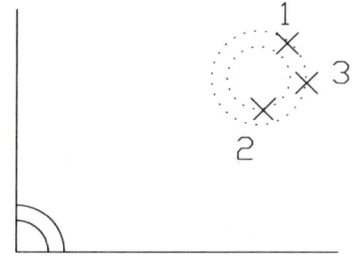
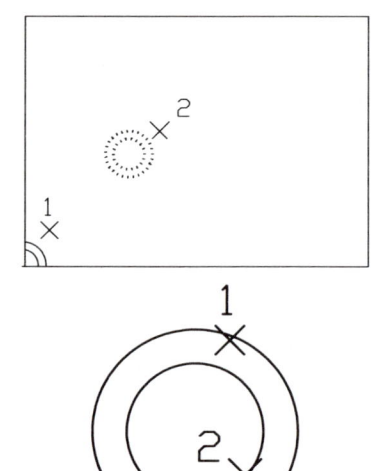

Step 4   Now use MIRROR to get the third arc in.

```
Command:MIRROR
Select objects: (pick 1)
Other corner:(pick 2)↵
First point of mirror line:CEN of
   (pick 3)
Second point: (pick 4)
Delete old objects?<N>:↵
```

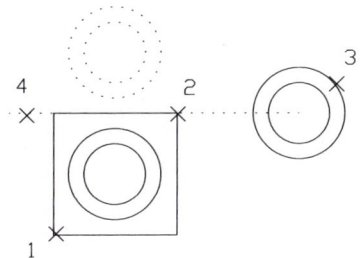

Set ORTHO on if it makes it any easier.

Step 5   Use FILLET to create the arcs between the parts.

```
Command:FILLET
Polyline/Radius<Select two objects>:R
Enter fillet radius:3
Command:FILLET
Polyline/Radius<Select two objects>:
   (pick 1 to 6)
```

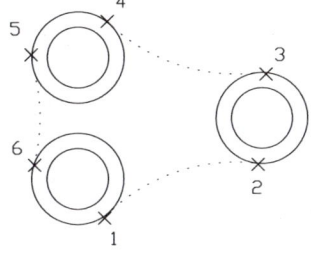

Prelab 4B Using ROTATE, COPY and MIRROR

In this example we will make a bay window using ROTATE, COPY, and MIRROR.

Step 1 First use LINE to draw in the window itself.

```
Command:LINE
From point:0,0
To point:8,0
To point: (continue drawing the window)
Command:ZOOM
All/Center/Dynamic/Extents/Left/Previous/
   Vmax/Window/<Scale X/XP>:A
Command:ZOOM
All/Center/Dynamic/Extents/Left/Previous/
   Vmax/Window/<Scale X/XP>:.8X
```

Step 2 Use PLINE to create the wall section.

```
Command:PLINE
From point: (pick 1)
The current line-width is 0.0000
Arc/Close/Halfwidth/Length/Undo/Width/
   <Endpoint of line>:W
Start width<0.1000>:8
End width<8.000>:↵
Arc/Close/Halfwidth/Length/Undo/Width/
   <Endpoint of line>:@0,-12
Arc/Close/Halfwidth/Length/Undo/Width/<Endpoint of
   line>:@12<315
Arc/Close/Halfwidth/Length/Undo/Width/<Endpoint of line>:↵
```

Step 3 Use COPY to get a copy of the window to the end of the PLINE.

```
Command:COPY
Select objects:W (pick 1, 2)
<Base point or displacement>/Multiple:
   MID of (pick 3)
Displacement:END of (pick 4)
```

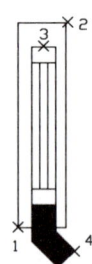

It is important to use the OSNAPs MID and END to get the correct point of reference on the window.

Step 4 Now use ROTATE to rotate the window at 45 degrees.

```
Command:ROTATE
Select objects:W (pick 1, 2)
Base point:MID of (pick 3)
<Rotation
   angle>/Reference:45
```

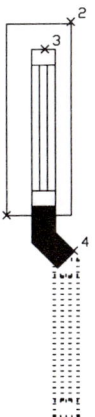

Step 5 Finally use MIRROR to create the other half of the image. Again, make sure that your OSNAPs are used.

```
Command:MIRROR
Select objects:W (pick 1, 2)
First point of mirroring plane:MID of
   (pick 3)
Second point of mirroring plane: (pick 4)
```

Put ORTHO on if it makes it easier.

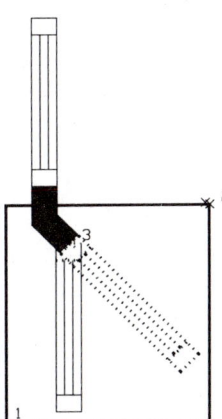

Your window should look like this:

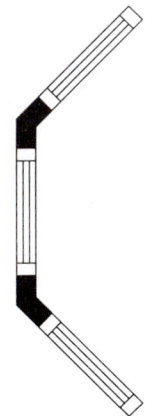

Object Selection and Editing

Command and Function Summary

COPY is used to make one copy or a series of copies of selected objects at random spacing.

Crossing is an object selection method which picks up all objects that cross over into the selected rectangle.

Grips speed up the editing process, allowing you to use the pointing device to combine object and command selection.

Fence is an object selection method that identifies any object that is touched by the Fence line.

LINETYPE is used to create, load, or set to a variety of different linetypes.

LTSCALE is used to change the scale of the linetypes relative to the drawing.

MIRROR is used to make a mirror image of selected objects through a specified mirroring line.

MOVE is used to move selected objects from one position to another along an identified vector.

Crossing Polygon is an object selection method that identifies any object contained within the defined polygon.

Window Polygon is an object selection method that identifies any object that crosses into the defined polygon.

ROTATE is used to rotate objects around an identified base point.

UNDO or **U** is used to undo the commands in reverse order of entry.

Window is an object selection method that identifies any object completely contained within the defined window.

Practice Exercise 4

Use COPY, ROTATE, MIRROR, and MOVE to complete these drawings. No dimensions are needed.

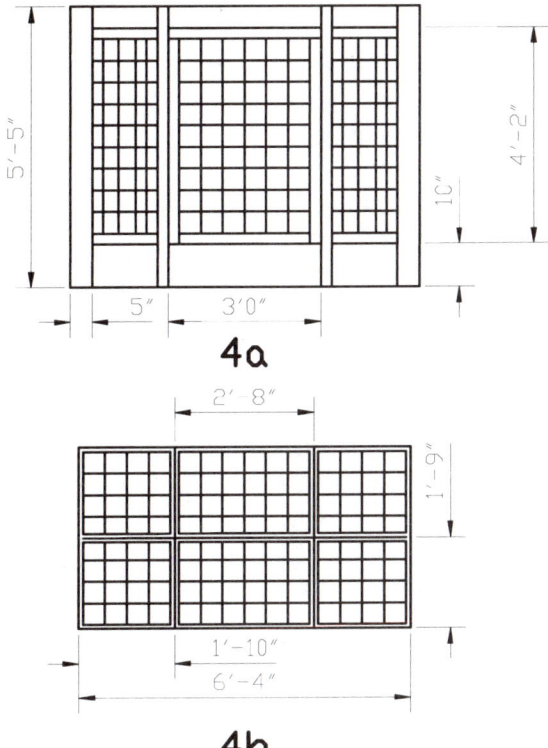

4a

4b

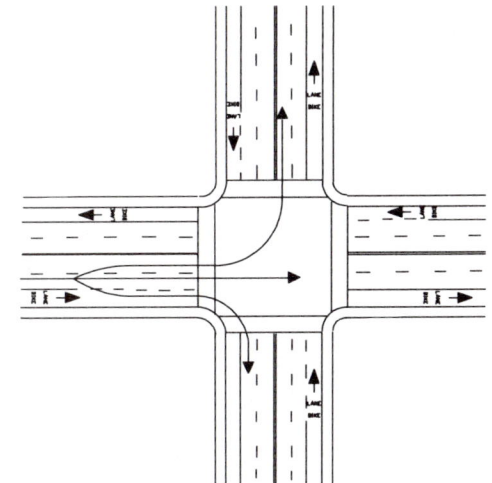

4d

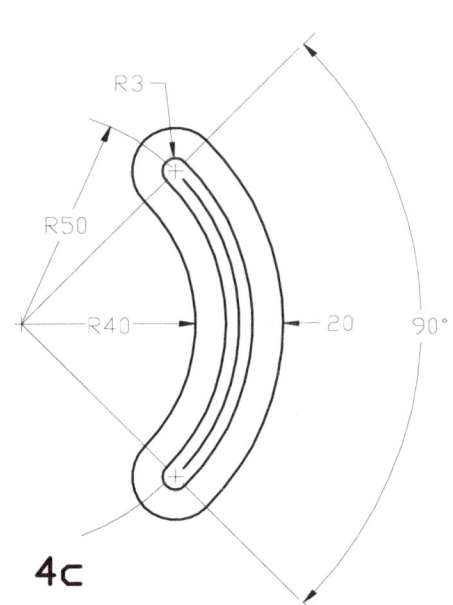

4c

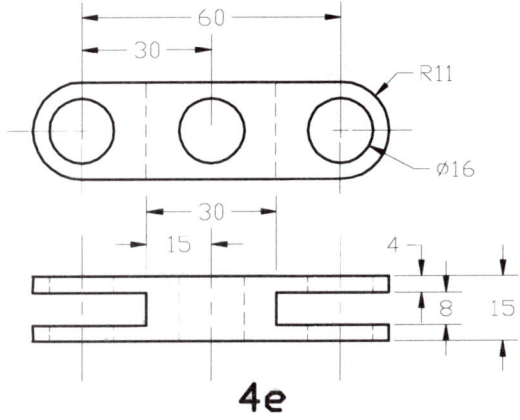

4e

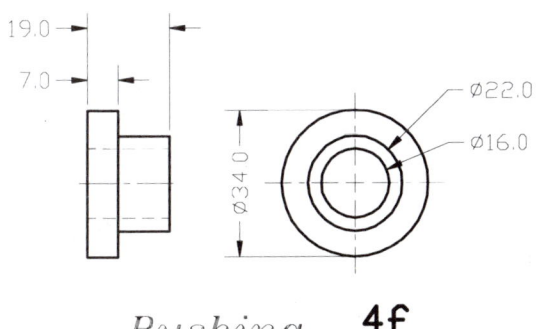

Bushing 4f

Exercise A4

Open FIRSTFL, the file you created earlier. Use LINE to draw in one window. Then use COPY, ROTATE, and MIRROR to place the windows within the walls. SNAP can be useful for drawing the windows. Be sure to use the OSNAPs when placing them.

Leave a 2″ gap between the exterior wall and the ledge of the window. This will be filled in with a veneer hatch in Chapter 8.

When the first floor is completed, start the second.

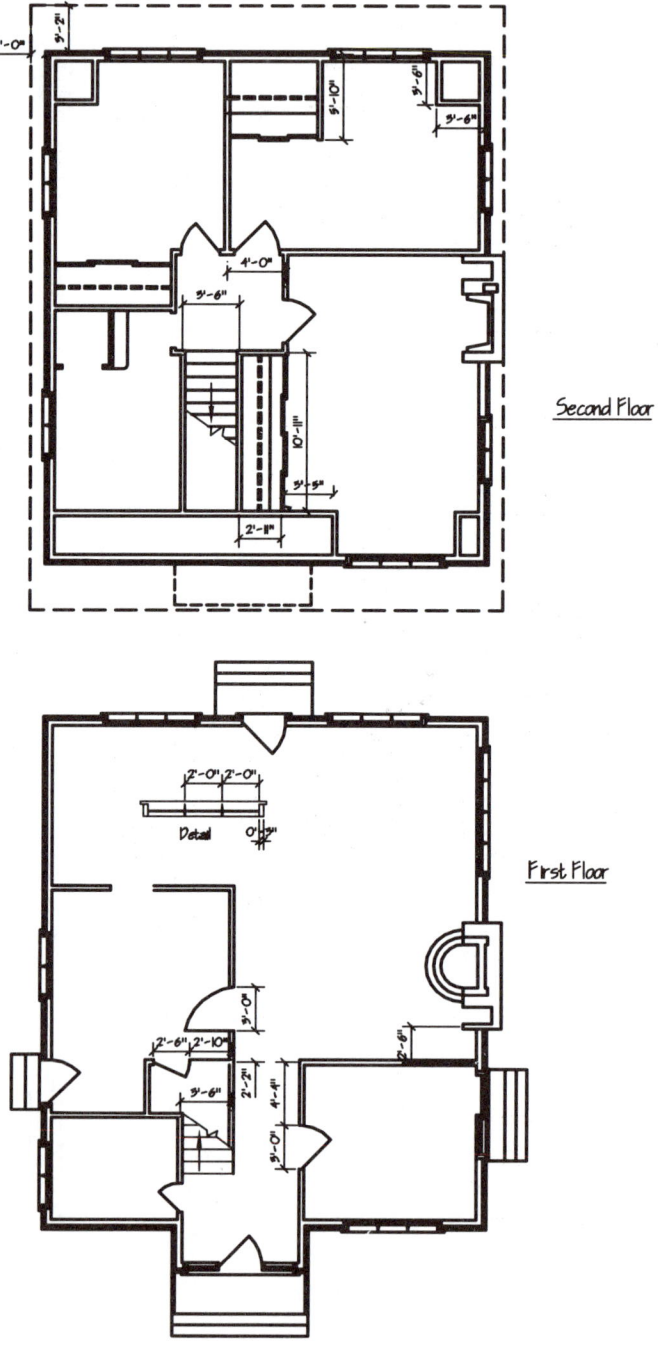

Exercise C4

For this exercise, use MIRROR and COPY for the road, COPY and ROTATE for the trees, and MIRROR for the fire hydrant. The scale in this case is metric.

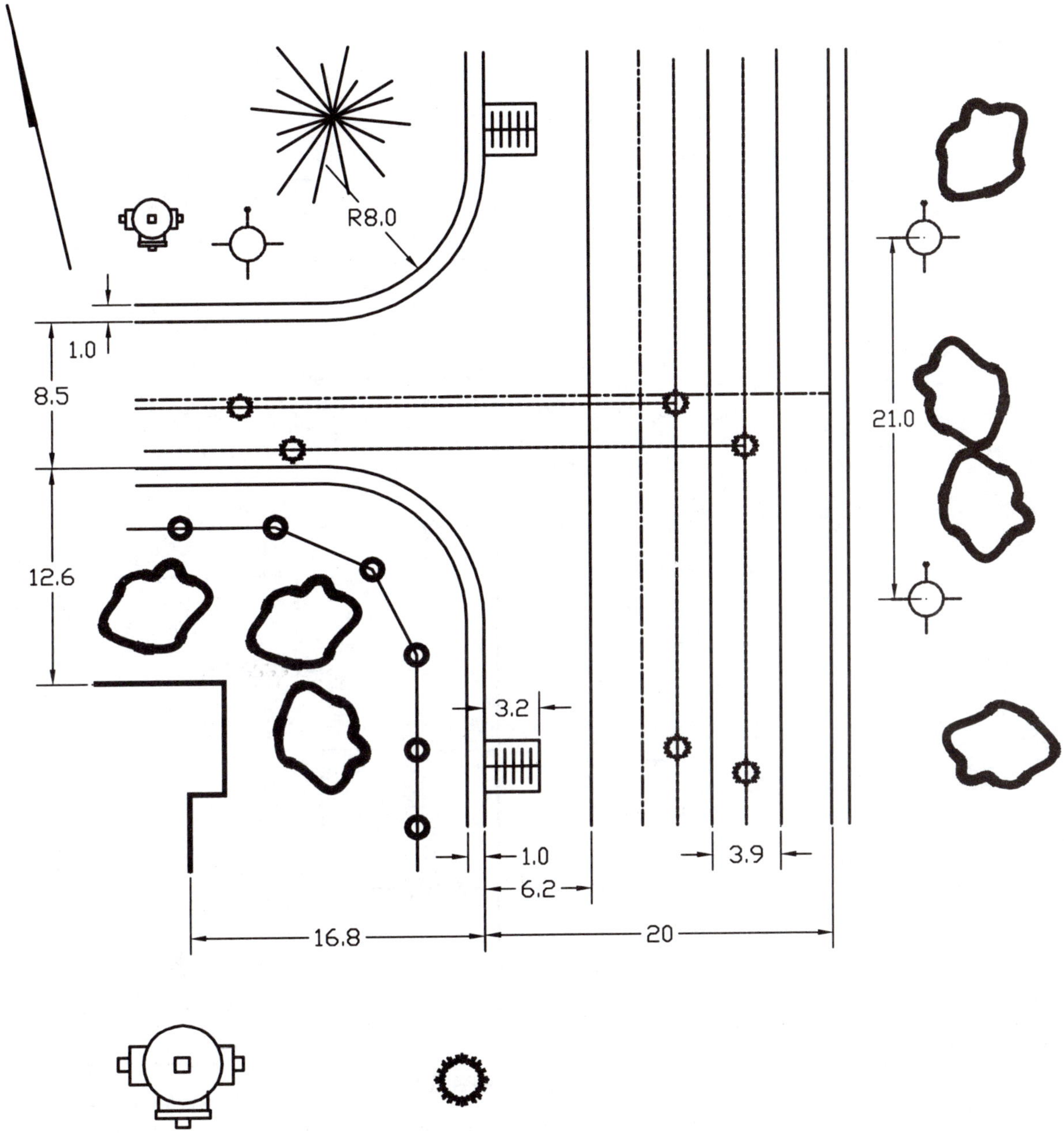

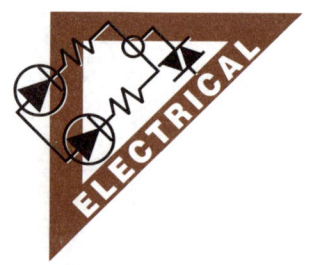

Exercise E4

Use COPY, COPY ROTATE with Grips, MIRROR, and MOVE to complete this schematic. Use OSNAP to access the end points of objects and the midpoints of objects. Create the symbols first, then put them in.

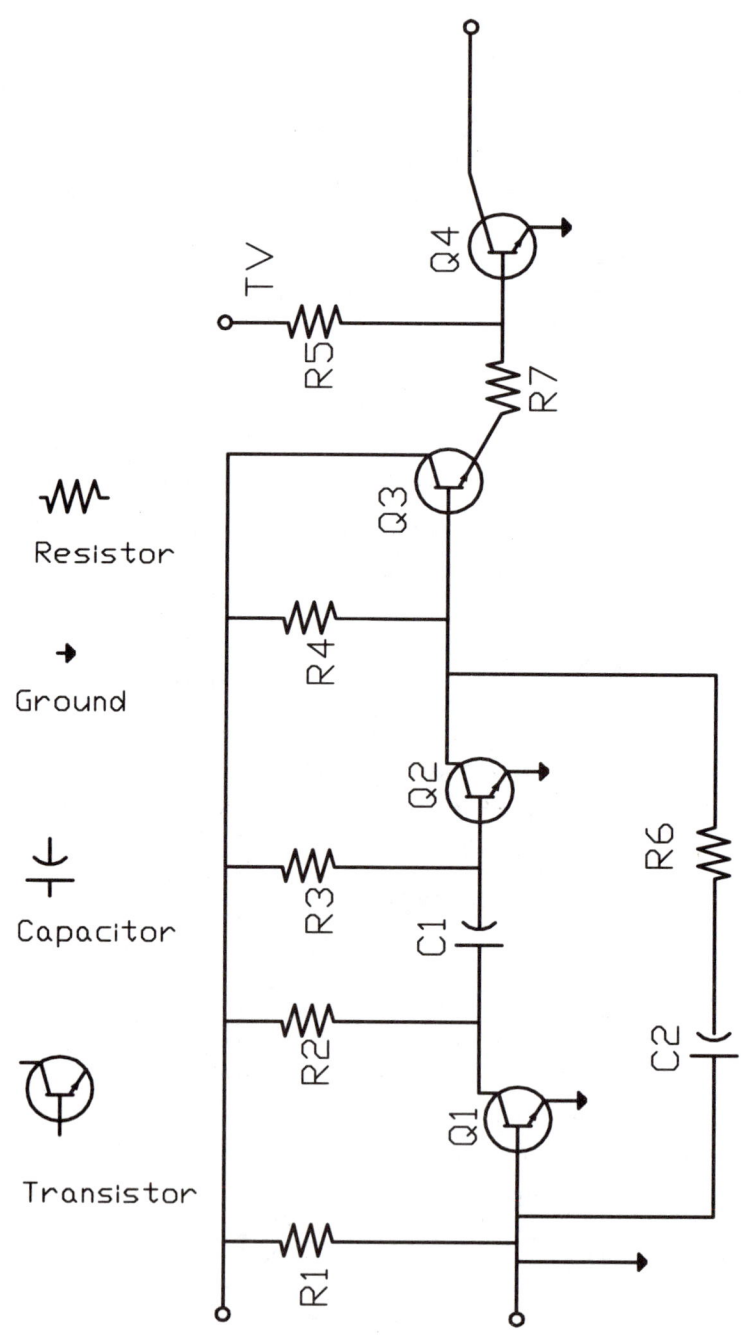

Exercise M4

Use COPY, MIRROR, MOVE, and TRIM to complete these drawings.

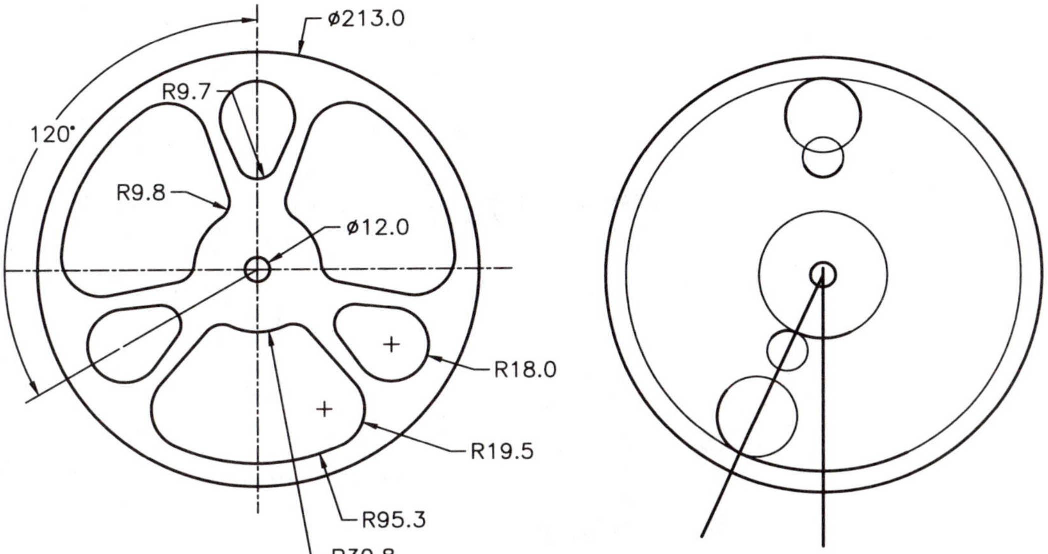

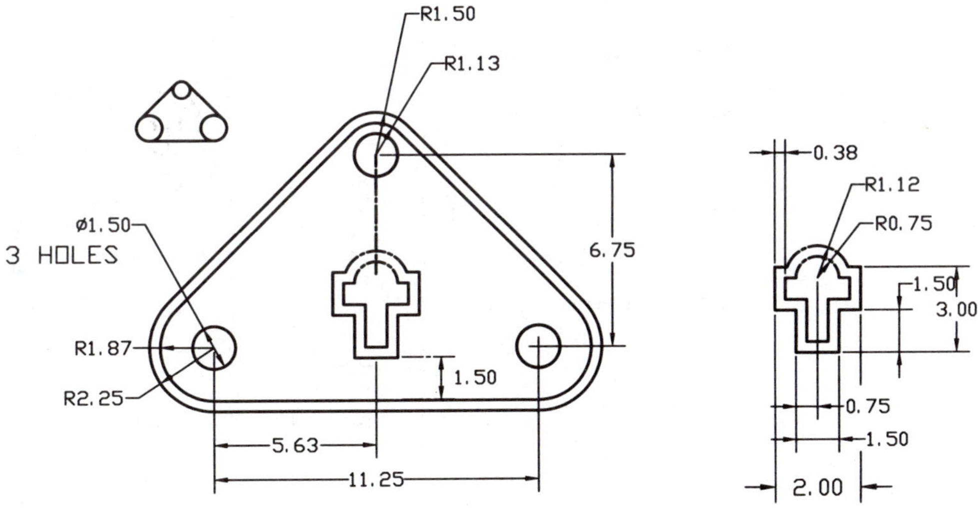

Challenger 4A

View from the side to gain maximum use of the screen.

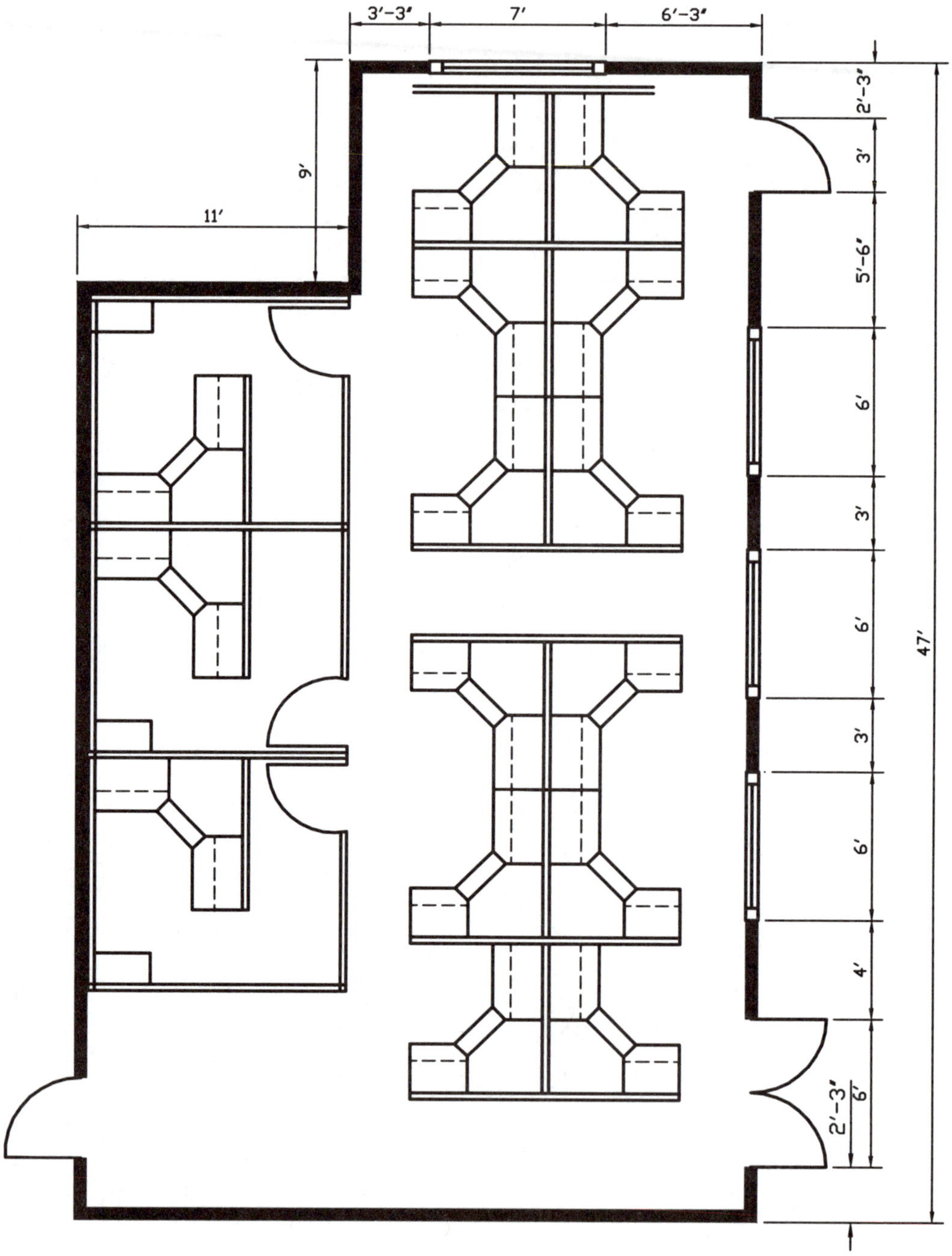

Hints on Challenger 4A

In this example, change to architectural units with a Snap value of 3. Remember, any number entered without a foot (') or inch ('') symbol will be accepted as inches. It should be easy to draw with SNAP and GRID on.

The drawing is of modular furniture. Create the offices by using lines. The panels separating the offices are 3 inches thick. Be sure to leave enough space for them.

Use ROTATE, COPY, and MOVE to create the offices exactly as you see them.

PLINE can be used to make the exterior walls easier to see.

Do not dimension.

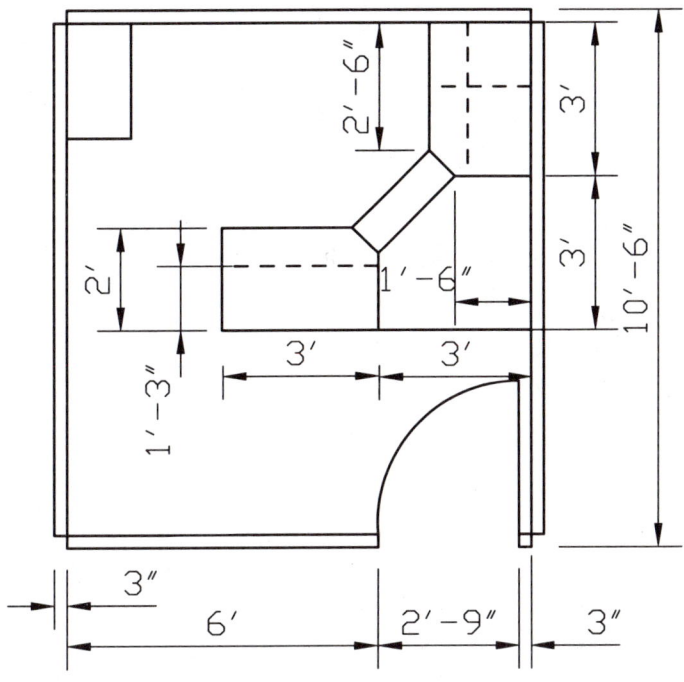

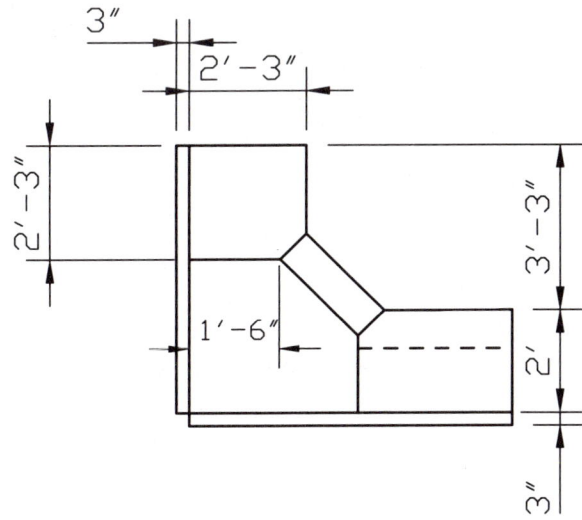

When finished with this drawing, save it so you can edit it later with STRETCH, as well as dimension it as shown in Chapter 7.

Challenger 4B

These commercial windows should give you some experience with editing commands.

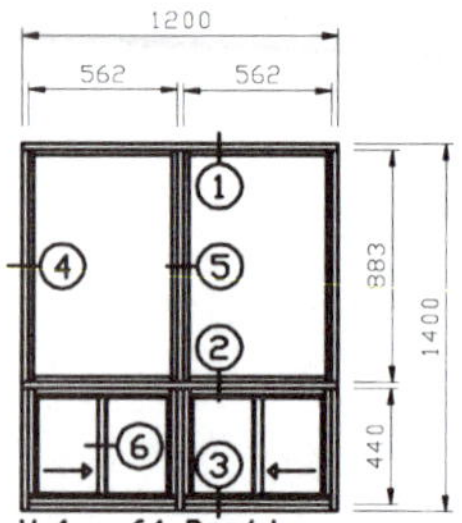

W 1   64 Req'd.
4200 Series

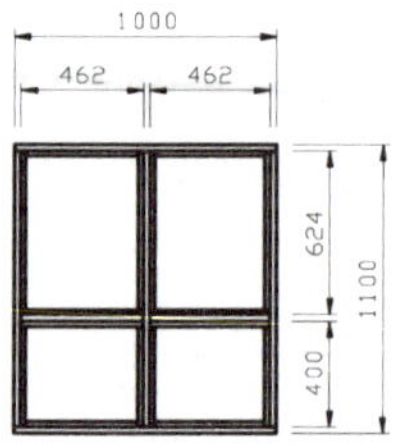

W 3   6 Req'd.
4200 Series

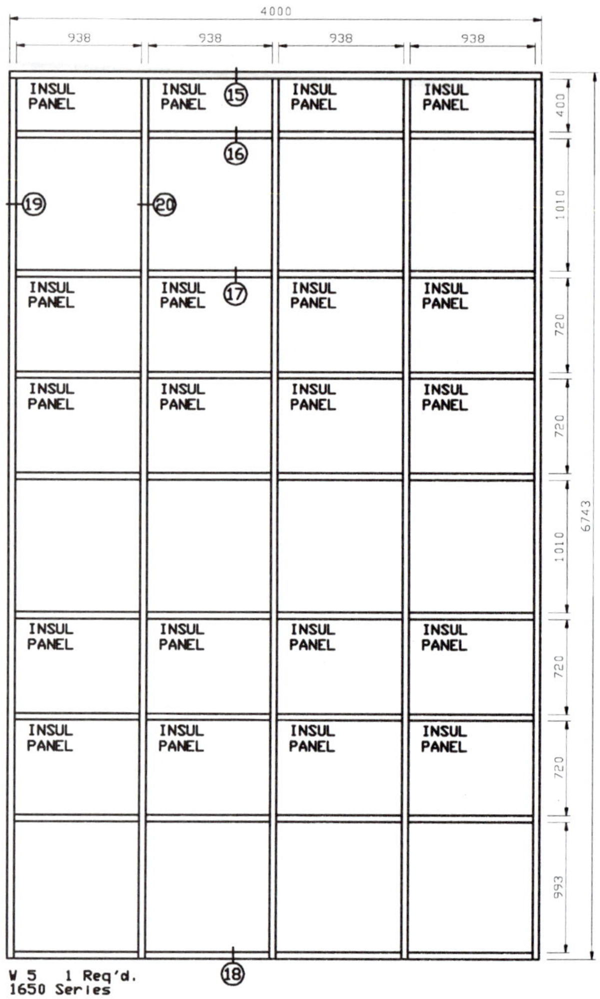

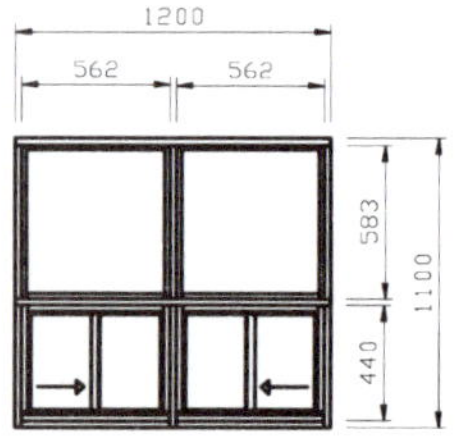

W 4   8 Req'd.
4200 Series

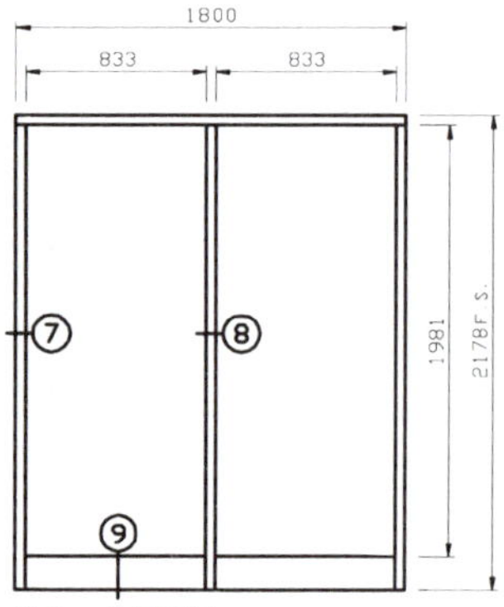

W 6   1 Req'd.
4500 Series

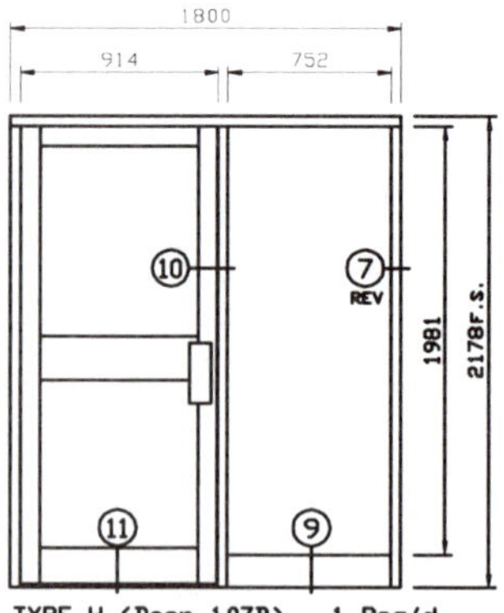

W 5   1 Req'd.
1650 Series

TYPE H (Door 107B)   1 Req'd.
4500 Series
Hardware By Others, As Per List.
EXTERIOR

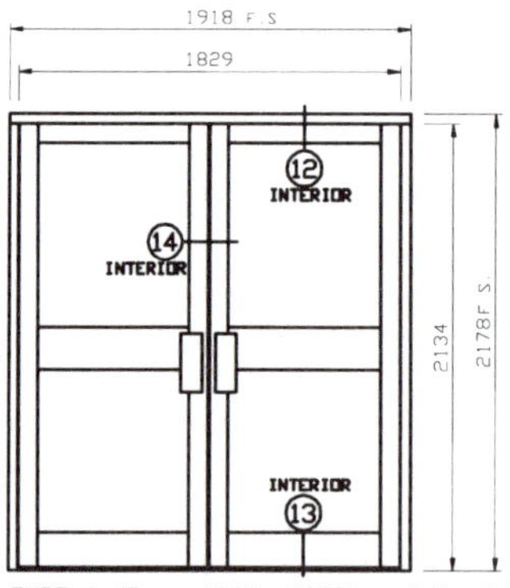

TYPE A (Door 100A, 100B)   2 Req'd.
4500 Series
Hardware By Others, As Per List.
1 EXTERIOR, 1 INTERIOR

112   CHAPTER FOUR

STRETCH, TRIM, EXTEND, OFFSET, and ARRAY

Upon completion of this chapter, you should be able to:

1. Use the Remove and Add Selection Set options
2. Edit objects using STRETCH, TRIM, EXTEND, OFFSET, and ARRAY
3. Use the ALIGN command

Removing and Adding Objects

Chapter 4 looked at identifying objects for a selection set using Window and Crossing. Now we will discuss removing objects from the selection set and adding information that has been removed.

You can find the selection set options as follows:

> **Windows** From the Tools menu, choose Toolbar, then Select objects.
>
>
>
> **DOS** Once the editing command is chosen, from the Screen menu, choose Select objects.

The command line equivalent is **REMOVE** or **R** and **ADD** or **A**.

Remove Option

The Remove option allows for objects to be removed from the selection set. Use Remove when a group of objects has been identified with Window or Crossing and a few need to be deleted from this selection set. Once this option is chosen, the "Select objects:" prompt changes to a "Remove objects:" prompt. Note that you can use Window and Crossing within the Remove option.

To copy just the lines and solids in this electrical symbol you could pick each object separately (four picks) or identify the whole set with Window, then remove the circle (three picks).

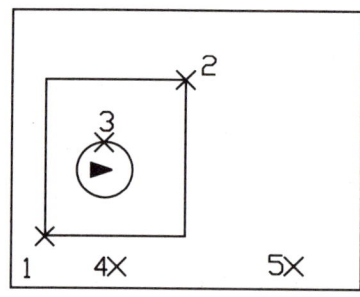
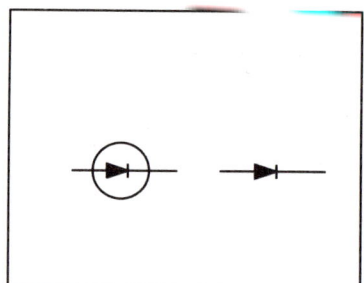

Figure 5-1

```
Command:COPY
Select objects: (pick 1)
Other corner: (pick 2)
Select objects:R
Remove objects: (pick 3)
Remove objects:⏎
<Base point or displacement> or Multiple: (pick 4)
Displacement: (pick 5)
```

Add Option

If objects have been removed and should be added back into the selection set, use Add. Use Add to change the prompt from "Remove Objects:" to "Select objects:", then continue to add objects to the selection set.

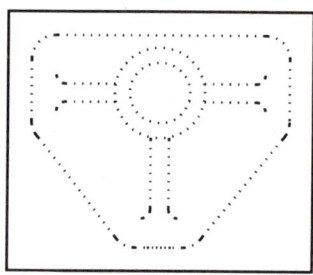
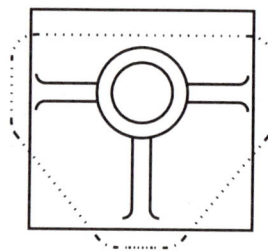
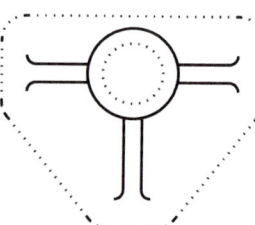

Figure 5-2

```
Command:COPY
Select objects: (pick 1)
Other corner: (pick 2)
Select objects:R
Remove objects: (pick 3)
Other corner: (pick 4)
Remove objects:A
Select objects: (pick 5)
Select objects:⏎
<Base point or displacement> or Multiple: (continue with
    command)
```

You may alternate between Remove and Add until you are satisfied with the objects in the selection set.

Editing Commands

These editing commands are similar to those discussed in Chapter 4, but have a more intricate structure. Keep reading the prompts for maximum efficiency.

Like the commands offered in Chapter 4, these can be found under the pull-down menus Modify and Construct in DOS, under the Edit screen menu in DOS, under the Edit toolbar in Windows, and by typing them in with aliases in both systems.

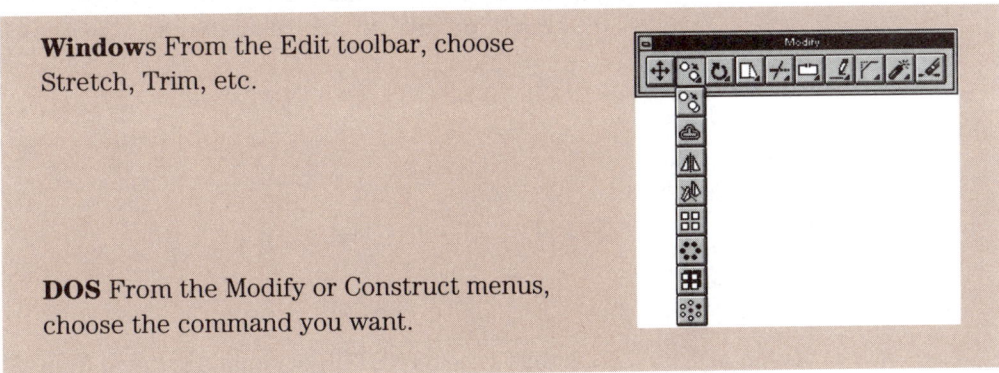

Windows From the Edit toolbar, choose Stretch, Trim, etc.

DOS From the Modify or Construct menus, choose the command you want.

The STRETCH Command

The STRETCH command is used to make lines, plines, and other linear objects either shorter or longer. It repositions the selected side of an existing object or group of objects relative to a new point or position.

First select which *portion* of an object you would like to redefine with a Crossing Window or CPOLY, then place it in the new position. The object will be redrawn relative to the new end point while the unselected point will remain at its original position, acting as an anchor for the selected adjoining lines.

> **Notes**
> You must use a Crossing Window with STRETCH.

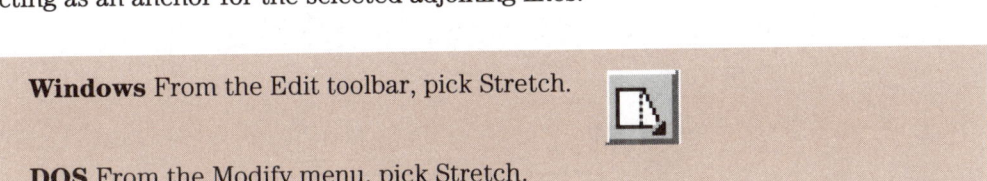

Windows From the Edit toolbar, pick Stretch.

DOS From the Modify menu, pick Stretch.

The command line equivalent is **STRETCH**.

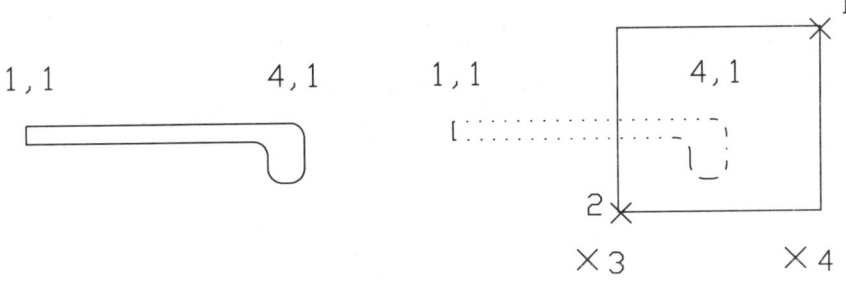

Figure 5-3

```
Command:STRETCH
Select objects to stretch by crossing-window or polygon ...
Select objects: (pick 1)
Other corner: (pick 2) (the right side of the objects was
   picked up)
Select objects:
Base point or displacement: (pick 3)
Second point of displacement: (pick 4)
```

1,1   7,1

Figure 5-4

Note that if the STRETCH command is taken from the side menu, the pull-down menu, or the toolbar, the Crossing option is automatic. All objects in the Crossing are highlighted when they are chosen.

Only the right side of the object is affected by the command above, because only the right-side point was picked up with the Crossing option.

The LINE can also be made shorter or diagonal through changing the position of the second point.

```
Base point:7,1                          Base point:7,1
New point:@-4,0                         New point:@-4,3
```

Figure 5-5

The first point has not changed; the item has been stretched by identifying one point and moving it in relation to the original point.

Stretching Circles

Circles are moved as full circles — stretching a circle does not create an ellipse. One end of an arc can be picked up and stretched into a different radius.

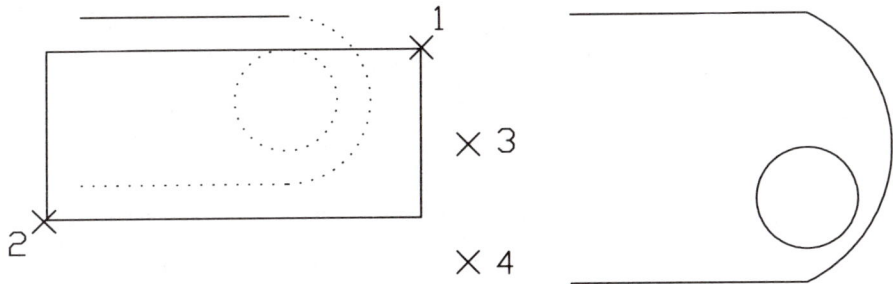

Figure 5-6

```
Command:STRETCH
Select objects to stretch by crossing-window or polygon ...
Select objects: (pick 1)
Other corner: (pick 2)
Select objects:⏎
Base point or displacement: (pick 3)
Second point of displacement: (pick 4)
```

Stretching Using Remove

Objects removed from the selection set will not be stretched.

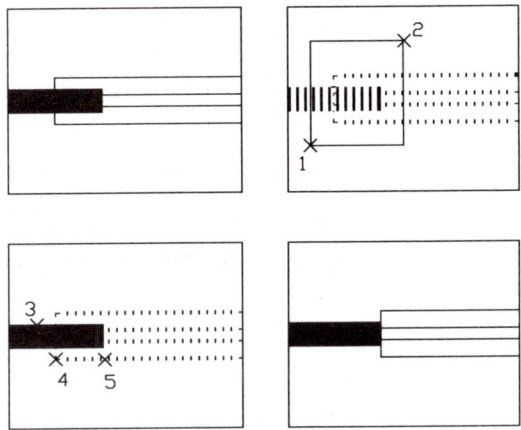

Figure 5-7

```
Command:STRETCH
Select objects to stretch by crossing-window or polygon ...
Select objects: (pick 1)
Other corner: (pick 2)
Select objects:⏎
Select objects:R (Remove)
Remove objects: (pick 3) (indicates that pline is not to be
   stretched)
Remove objects:⏎ (indicates no more items on the list)
Base point or displacement: (pick 4)
Second point of displacement: (pick 5)
```

The objects identified will be "loaded on the cursor" and will move along with the cursor while identifying the new point.

In this example, notice that the door opening retains its width when the walls on either side are selected and stretched.

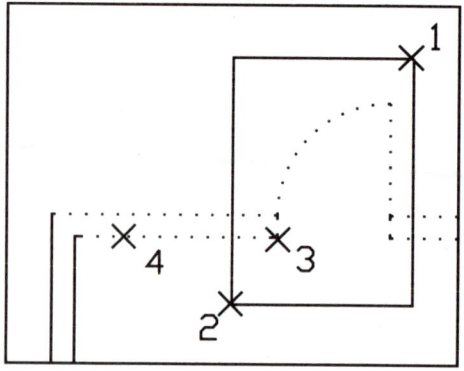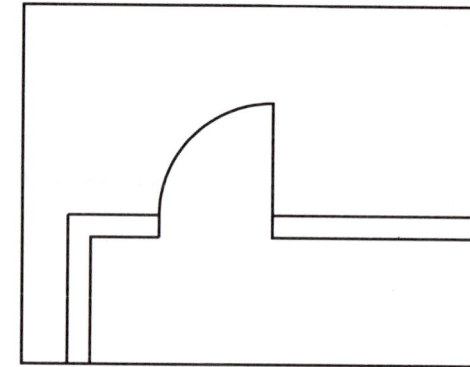

Figure 5-8

Danger

If you pick all the objects with your Crossing selection, you will MOVE the objects instead of STRETCHing them. Choose only the side of the object that you would like to stretch.

```
Command: STRETCH
Select objects to stretch by crossing-window or polygon ...
Select objects: (pick 1)
Other corner: (pick 2)
Select objects:↵
Base point or displacement: (pick 3)
Second point of displacement: (pick 4)
```

Use ORTHO to make sure that the walls remain straight.

It is advisable to use STRETCH or EXTEND when making an object fit into a larger space than was originally intended. It is *not* a good idea to add extra lines.

The TRIM Command

The TRIM command is used to cut off an object or a series of objects at their intersection with a boundary or cutting edge. While you have already used TRIM on individual objects, you can also use it on multiple objects.

The cutting edge or boundary must already exist on the drawing before you can use TRIM.

In this example, the horizontal lines will be trimmed to the diagonal line which has been defined as the cutting edge. Notice that the cutting line is to be chosen first. It becomes highlighted when chosen.

> **Windows** From the Edit toolbar, pick Trim.
>
> **DOS** From the Modify menu, pick Trim.

The command line equivalent is **TRIM**.

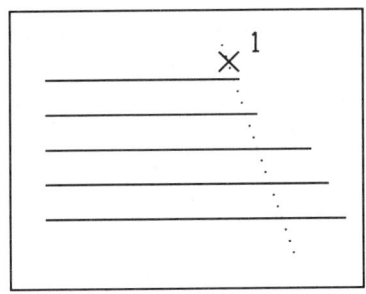
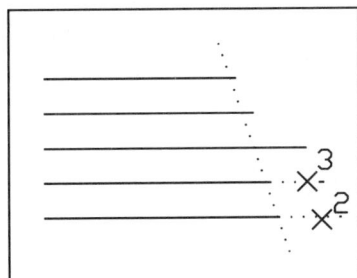

Figure 5-9

```
Command: TRIM
Select cutting edges (Projmode = UCS, Edgemode = No extend):
Select objects: (pick 1)
Select objects: ⏎ (no more cutting edges are needed)
<Select object to trim>Project/Edge/Undo: (pick 2)
<Select object to trim>Project/Edge/Undo: (pick 3)
<Select object to trim>Project/Edge/Undo: (select the
   remaining objects)
```

The first pick(s) indicates the cutting edge. Use ⏎ to continue when the cutting edges are selected. The following picks are those items which will be trimmed to this edge. Any number of objects can be trimmed. Using Fence at the "Select objects to trim:" prompt can greatly speed the process.

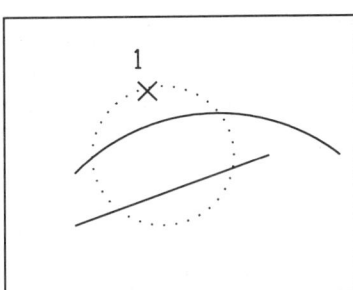
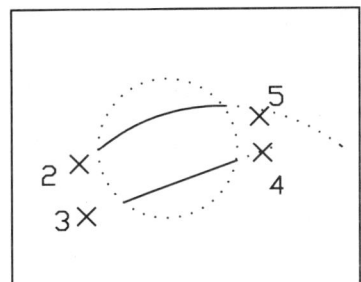

Figure 5-10

The objects are trimmed from the end chosen to the closest cutting edge. Choose the segments to be trimmed away.

```
Command: TRIM
Select cutting edges (Projmode = UCS, Edgemode = No extend):
Select objects: (pick 1)
Select objects: ⏎ (no more cutting edges are needed)
<Select object to trim>Project/Edge/Undo: (pick 2)
<Select object to trim>Project/Edge/Undo: (pick 3 to 5)
```

You can also use Window to select the cutting edges.

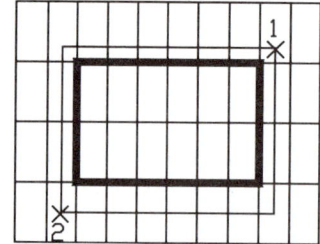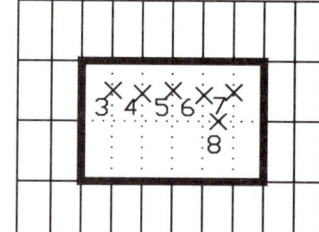

Figure 5-11

```
Command: TRIM
Select cutting edges (Projmode = UCS, Edgemode = No extend):
Select objects: (pick 1)
Select objects:⏎ (no more cutting edges are needed)
<Select object to trim>Project/Edge/Undo: (pick 2 to 8)
```

You can also choose multiple objects as the cutting edges. In the following example, all of the objects are considered cutting edges.

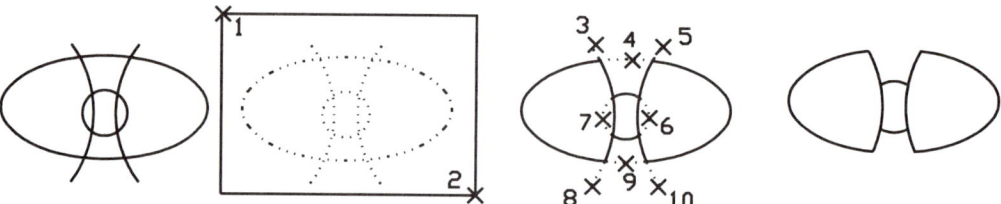

Figure 5-12

```
Command: TRIM
Select cutting edges (Projmode = UCS, Edgemode = No extend):
Select objects: (pick 1)
Other corner: (pick 2)
Select objects:⏎
<Select object to trim>/Project/Edge/Undo: (pick 3 to 10)
<Select object to trim>/Project/Edge/Undo:⏎
```

The EXTEND Command

The EXTEND command also uses a boundary (or boundaries), but uses it as the item to extend objects to.

In this command, the boundary is clearly the diagonal line. Once picked, it will be highlighted.

> **Windows** From the Edit toolbar, pick Extend.
>
> **DOS** From the Modify menu, pick Extend.

The command line equivalent is **EXTEND**.

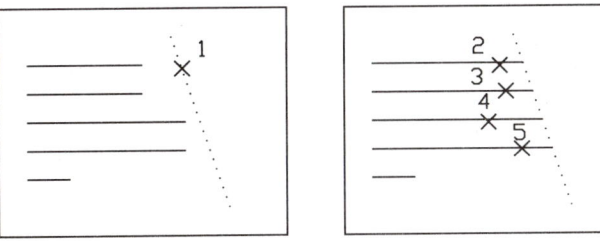

Figure 5-13

```
Command: EXTEND
Select boundary edges (Projmode = UCS, Edgemode = No extend):
Select objects: (pick 1)
Select objects:↵
Select object to extend/Project/Edge/Undo: (pick 2)
Select object to extend/Project/Edge/Undo: (pick 3 to 5)
```

Notes

The Undo option lets you undo the previous pick while still in the command.

In this example, the LINEs are extended to the closest boundary selected. More than one boundary can be chosen for objects to extend to, and the closest end point will always be extended to the boundary chosen. The Fence option is also useful here.

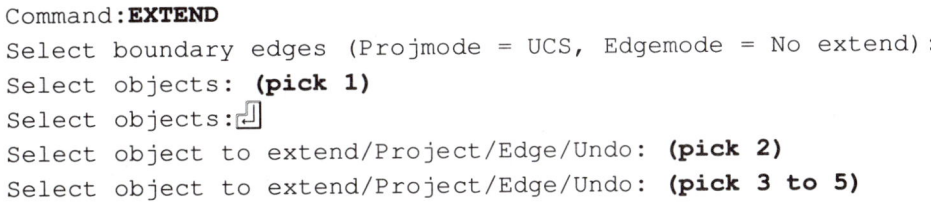

Figure 5-14

```
Command: EXTEND
Select boundary edges (Projmode = UCS, Edgemode = No extend):
Select objects: (pick 1)
Select objects:↵
Select object to extend>Project/Edge/Undo: (pick 2 to 5)
```

TRIM and EXTEND with Implied Intersections

Implied intersections are the points where two objects would intersect if they were either extended or trimmed. You can trim objects using their implied intersections as cutting edges for the trim, and extend objects using their implied intersection as the boundary for the extend.

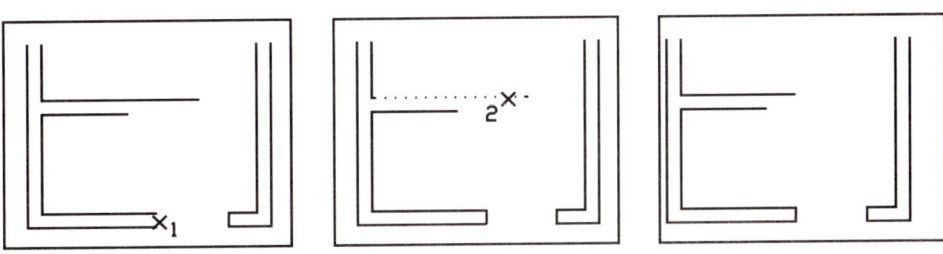

Figure 5-15

STRETCH, TRIM, EXTEND, OFFSET, and ARRAY

```
Command:TRIM
Select cutting edges (Projmode = UCS, Edgemode = No extend):
Select objects: (pick 1)
Select objects:⏎ (no more cutting edges are needed)
<Select object to trim>Project/Edge/Undo:E
Extend/<No extend>:E
<Select object to trim>Project/Edge/Undo: (pick 2)
<Select object to trim>Project/Edge/Undo:⏎
```

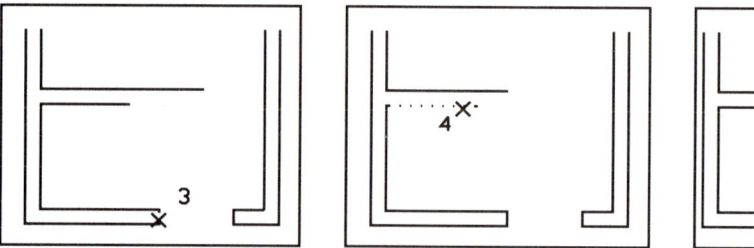

Figure 5-16

```
Command: EXTEND
Select boundary edges (Projmode = UCS, Edgemode = No extend):
Select objects: (pick 3)
Select objects:⏎
Select object to extend>Project/Edge/Undo:E
Extend/<No extend>:E
<Select object to extend>Project/Edge/Undo: (pick 4)
<Select object to extend>Project/Edge/Undo:⏎
```

What Can Go Wrong with TRIM and EXTEND

If AutoCAD responds with:

No edges selected = you have chosen an object that cannot be referenced as a boundary or cutting edge. TRACE, TEXT, and hatching are not valid boundaries and will be rejected.

Cannot extend this entity = the object chosen cannot be extended. For example, if you choose a complete circle, this cannot extended.

Entity does not intersect an edge = the object does not line up with the edge selected. Try using FILLET RAD 0. Arcs continue their radius and may curve away from the boundary.

The OFFSET Command

OFFSET is arguably the most often used editing command on the system. It is so useful that it is the only command in both the Draw and the Edit menu.

OFFSET copies an object parallel to an existing object at a given distance. Any amount of objects can be offset at the specified distance within one command string. Note that the OFFSET command works only in the *X-Y* plane.

Windows From the Edit toolbar, pick Offset.

DOS From the Construct menu, pick Offset.

The command line equivalent is **OFFSET**.

Offset first requires the offset distance — the distance that all of the objects will be offset by. It doesn't matter how far away the pick point is from the offset object; only the side is important.

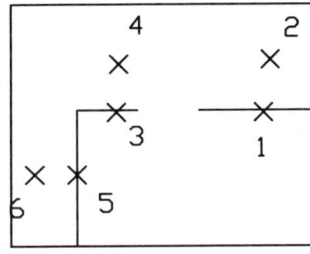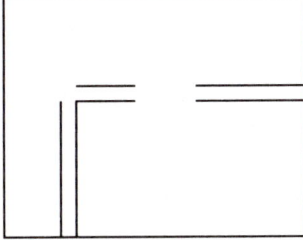

Figure 5-17

```
Command: OFFSET
Offset distance or Through<Through>: 8
Select object to offset: (pick 1)
Side to offset?: (pick 2)
Select object to offset: (pick 3)
Side to offset?: (pick 4)
Select object to offset: (continue to pick points and sides)
```

The prompts for the objects to offset and the side to offset will continue in a paired sequence until you press ⏎. After the walls are created, clean up the corners with FILLET RAD 0.

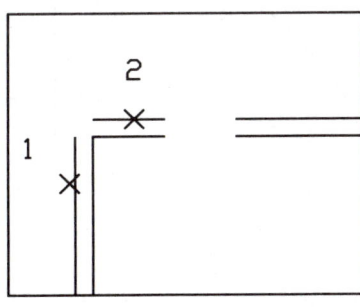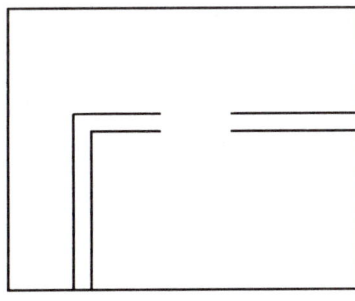

Figure 5-18

```
Command: FILLET
Radius/Diameter<First object>: (pick 1)
Other object: (pick 2)
```

STRETCH, TRIM, EXTEND, OFFSET, and ARRAY

The Through option in OFFSET is used to specify a point through which you would like the offset calculated, as opposed to having a specified distance.

Offsetting a circle or an arc will result in the second object having the same center, but the radius will be larger or smaller by the offset distance. The center point does not change; only the size of the circle or arc changes.

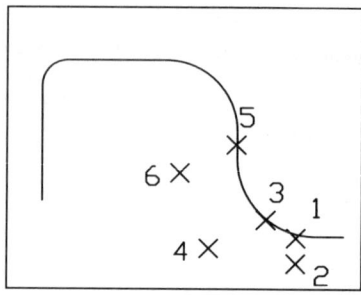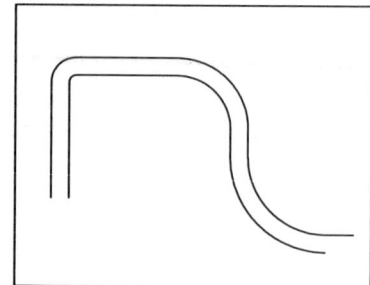

Figure 5-19

```
Command:OFFSET
Offset distance or Through<Through>: (pick 1)
Through point: (pick 2)
Select object to offset: (pick 3)
Side to offset?: (pick 4)
Select object to offset: (pick 5)
```

Similarly, with plines that are fit into curves or splines, the resulting object is fit through the same series of points at an offset distance. The inside offset spline will be defined by the same number of vertices, but the distance between the vertices will be different.

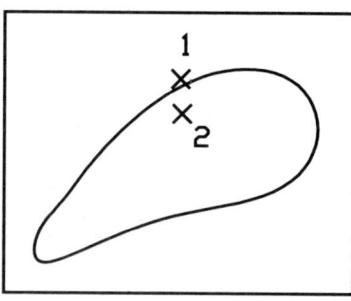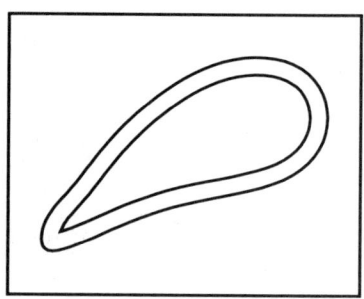

Figure 5-20

```
Command:OFFSET
Offset distance or Through<Through>:1
Select object to offset: (pick 1)
Side to offset: (pick 2)
Select object to offset:⏎
```

The ARRAY command makes circular (polar) or rectangular patterns of selected objects by copying them along or around an identified point. You are prompted first for the objects that are to be arrayed, then for the type of array — polar or rectangular. Finally you are asked to specify how many and how they are to be spaced.

In this polar array example, the object (a chair) is arrayed around the center of the larger circle (a table) at equal distances through 360 degrees, rotating the copies in line with the center point.

Windows From the Edit toolbar, pick Polar Array.

DOS From the Construct menu, pick Array, then Polar.

The command line equivalent is **ARRAY**.

```
Command:ARRAY
Select objects: (pick 1)
Other corner: (pick 2)
Select objects:⏎ (indicates that no more are needed)
Rectangular/Polar array (R/P)<R>:P
Center point of array:CENter of (pick 3)
Number of items:8
Angle to fill (+=CCW,-=CW)<360>:⏎ (accepts default of 360)
Rotate objects as they are copied?<Y>:⏎ (accepts default)
```

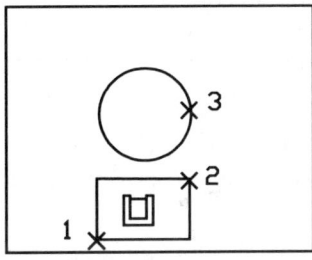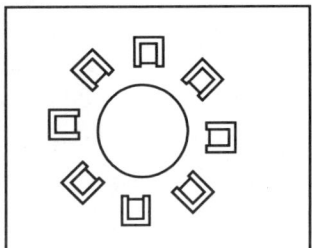

Figure 5-21

With a polar array, you can specify the incremental angle between each object rather than the total distance if you prefer.

You must know both the angle that you need between items and the number of items.

Respond with a **0** when prompted for the angle to fill, and you will be asked for the incremental angle.

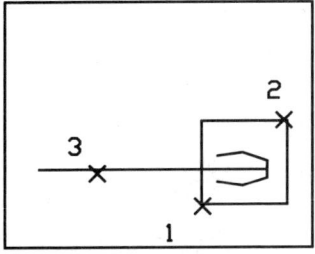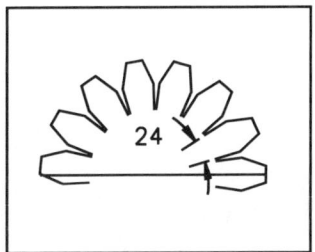

Figure 5-22

```
Command:ARRAY
Select objects: (pick 1)
Other corner: (pick 2)
Select objects:⏎ (indicates no more objects are needed)
Rectangular/Polar array (R/P)<R>:P
Center point of array:MIDpoint of (pick 3)
Number of items:8
Angle to fill (+=CCW,-=CW)<360>:0 (indicates incremental angle)
Angle between items (+=CCW, -=CW):24
Rotate objects as they are copied?<Y>:⏎ (accepts default)
```

If you don't know the number of items but you know what the final angle and the angle between items is, use:

```
Center point of array:MIDpoint of (pick)
Number of items:⏎
Angle to fill (+=CCW,-=CW)<360>:180
Angle between items (+=CCW, -=CW):24
Rotate objects as they are copied?<Y>:⏎
```

The Rectangular option of the ARRAY command prompts for the number of rows (copies in the *Y* direction) and the number of columns (copies in the *X* direction). Then you are prompted for the distance between the copies in rows and columns. This is the center-to-center distance, which is particularly useful for both building trades and mechanical engineering since the measurements are center-to-center of wall studs, holes, slots, etc.

In the following diagram, the objects are ARRAYed along *X* or in columns. Note the distance between the columns includes the halfwidth of the item itself.

Windows From the Edit toolbar, pick Rectangular Array.

DOS From the Construct menu, pick Array, then Rectangular.

The command line equivalent is **ARRAY**.

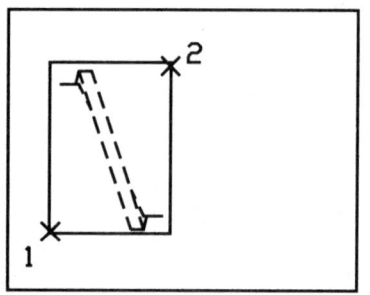

Figure 5-23

```
Command:ARRAY
Select objects: (pick 1)
Other corner: (pick 2)
Select objects:↵
Rectangular/Polar array (R/P)<R>:↵
Number of rows(--)<1>:↵ (accepts the default of 1)
Number of columns(|||)<1>:7
Distance between the columns:.5625
```

When doing layouts that contain many objects in both the *X* and the *Y* direction, specify both rows and columns. In specifying the number of rows and columns, the total number is indicated.

If you have a chair that is 24 inches wide and you want an aisle of 18 inches between it and the next chair, the spacing would be the width of the chair plus the width of the aisle (24 + 18 = 3'6'').

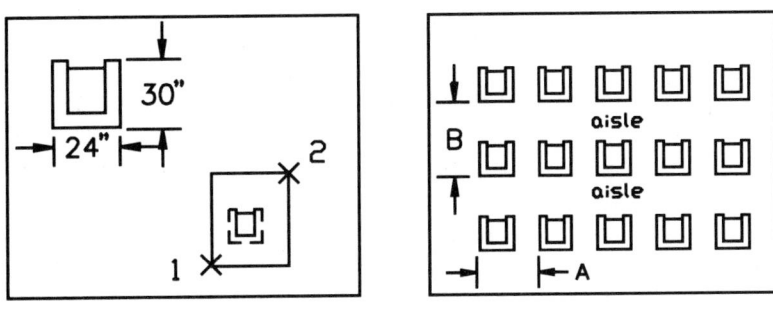

Figure 5-24

```
Command:ARRAY
Select objects: (pick 1)
Other corner: (pick 2)
Select objects:↵
Rectangular/Polar array (R/P)<R>:↵
Number of rows(--)<1>:3
Number of columns(|||)<1>:5
Distance between the rows:60 (distance B)
Distance between the columns:42 (distance A)
```

A negative distance will place the array in a negative direction: to the left if columns, and down from the original if rows.

Unit cell satisfies the distance between rows and columns with a resultant distance of *X* and *Y*. The object arrays the directions the unit cell window has selected.

STRETCH, TRIM, EXTEND, OFFSET, and ARRAY

```
Command:ARRAY
Select objects: (pick 1)
Other corner: (pick 2)
Select objects:↵
Rectangular/Polar array (R/P)<R>:↵
Number of rows(--)<1>:3
Number of columns(|||)<1>:5
Distance between the rows<--->: (pick 3)
Other corner:@60,42
```

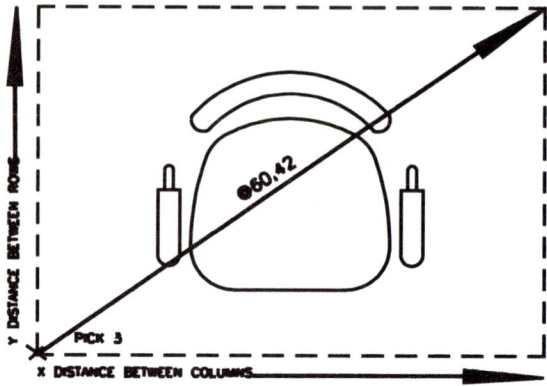

Figure 5-25

Align

Align moves and rotates objects to align with other objects, either in 2D or 3D. Align uses up to three pairs of source and destination points. The first set of points defines a move for the objects. The second set of points can define either a 2D or 3D transformation and rotation of the objects. The third set of points defines an unambiguous 3D transformation of the objects.

Windows From the Modify toolbar, choose the Rotate flyout, then

DOS From the Modify menu, choose Align.

The command line equivalent is **ALIGN**.

```
Command:ALIGN
Select objects: (pick the upper set
  of objects)
Select objects: ↵
1st source point: (pick 1)
1st destination point: (pick 2)
2nd source point: (pick 1)
2nd destination point: (pick 2)
3rd source point: ↵
3rd destination point: ↵
```

Prelab 5   Editing Commands

This example will illustrate how to use ARRAY, TRIM, EXTEND, and OFFSET to create a backyard pool.

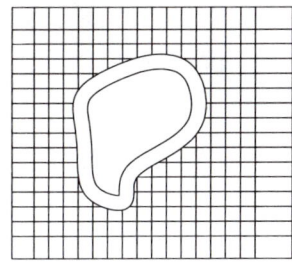

Step 1   Set LIMITS to -1,-1 and 40,30 or place your UCS origin on screen.

Set GRID to 2 and SNAP to 1. ZOOM All.

Step 2   Create a line from 0,0 to 0,28, and ARRAY it to make 16.

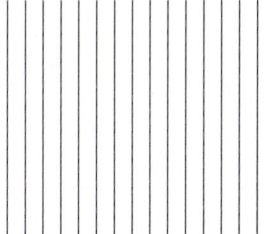

```
Command:L
From point:0,0
To point:0,28
To point:⏎
Command:ARRAY
Select objects:L
Select objects:⏎
Rectangular or Polar (R/P)<R>:R
Number of rows (---)<1>:⏎
Number of columns (|||)<1>:16
Distance between columns (|||):2
```

Step 3   Create a line from 0,0 to 30,0 and ARRAY it by 15 to make a rectangular grid (the patio stones).

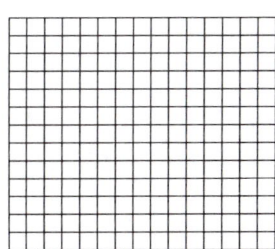

```
Command:L
From point:0,0
To point:30,0
To point:⏎
Command:ARRAY
Select objects:L
Select objects:⏎
Rectangular or Polar (R/P)<R>:⏎
Number of rows (---)<1>:15
Number of columns (|||)<1>:⏎
Distance between columns (|||):2
```

STRETCH, TRIM, EXTEND, OFFSET, and ARRAY   **129**

Make a PLINE and use PEDIT to create a pool shape.

```
Command:PLINE
From point: (pick 1)
Current width is 0.0000
Arc/Close/Halfwidth/Length/Undo/Width/
   <Endpoint of line>: (pick points 1 to
   whatever)
Arc/Close/Halfwidth/Length/Undo/Width/
   <Endpoint of line>:C
Command:PEDIT
Select polyline:L
Close/Join/Width/Edit vertex/Fit curve/Ltype gen/Spline
   curve/Decurve/Undo/eXit/<X>:S
Close/Join/Width/Edit vertex/Fit curve/Ltype gen/Spline
   curve/Decurve/Undo/eXit/<X>:⏎
```

Step 4 Now OFFSET the pool shape to create a pool border.

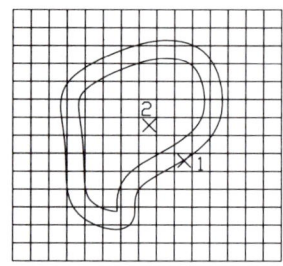

```
Command:OFFSET
Offset distance or through<0.0000>:2
Select object to offset: (pick 1)
Side to offset ?: (pick 2)
Select object to offset:⏎
```

Step 5 Now use trim to remove the lines from the patio stones within the pool.

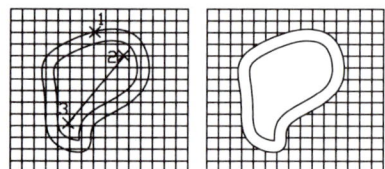

```
Command:TRIM
Select cutting edges: (pick 1)
Select objects:⏎
<Select object to trim> Undo:FENCE
   (pick points 2 to whatever)
```

Step 6 Use OFFSET to create a set of larger patio stones on the outside of the original stones as a border.

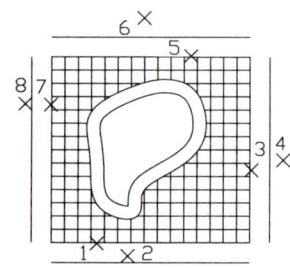

```
Command:OFFSET
Offset distance or through<0.0000>:3
Select object to offset: (pick 1)
Side to offset ?: (pick 2)
Select object to offset: (complete
   picks indicated)
```

Use FILLET to make the corners square.

```
Command:FILLET
Polyline/Radius<Select two objects>: (pick one line)
Second object: (pick another line)
```

Now use EXTEND to extend the lines from the existing stones to the new boundary.

```
Command: EXTEND
Select boundary edges (Projmode = UCS,
    Edgemode = No extend):
Select objects: (pick 1 to 4)
Select objects:⏎
Select object to extend>Project/Edge/
    Undo:(pick 5 and so on)
```

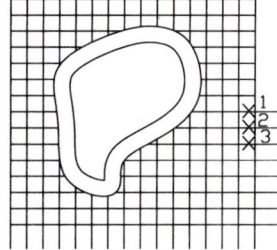

Step 7 Now STRETCH the stones on the right just one unit further.

```
Command:STRETCH
Select objects to stretch by crossing-
    window or polygon ...
Select objects:C
First corner: (pick 1)
Other corner: (pick 2)
Select objects:⏎
Select objects:R (Remove)
Remove objects: (pick 3) (indicates that pline is not to be
    stretched)
Remove objects:⏎ (indicates no more items on the list)
Base point or displacement: (pick 3)
Second point of displacement:@1,0
```

Now you have a completed exterior pool area with 2×2 stones, 2×3 stones and 2×4 stones.

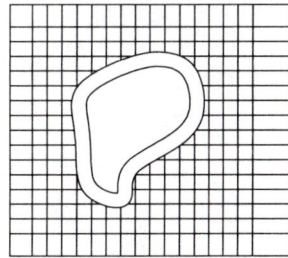

STRETCH, TRIM, EXTEND, OFFSET, and ARRAY **131**

Command and Function Summary

Add is an object selection option that allows you to keep adding to your selection set after the Remove option has been used.

ARRAY makes copies of an entity in either a polar or a rectangular pattern.

EXTEND is used to extend an object or a series of objects to reach a selected boundary.

OFFSET makes a copy of an entity parallel to and at a specified distance from that entity.

Remove is an object selection option that allows you to remove objects from the selection set once they have been identified.

STRETCH distorts an object or a series of objects along a specified vector.

Note: You are now at the point where you can draw the geometry for almost any part. The only way to speed your understanding is through practice, not only in your own discipline but in all.

Do as many of the exercises in this chapter as possible. Remember to save your files!

Practice Exercise 5A

Use the editing commands to complete these mechanical examples.

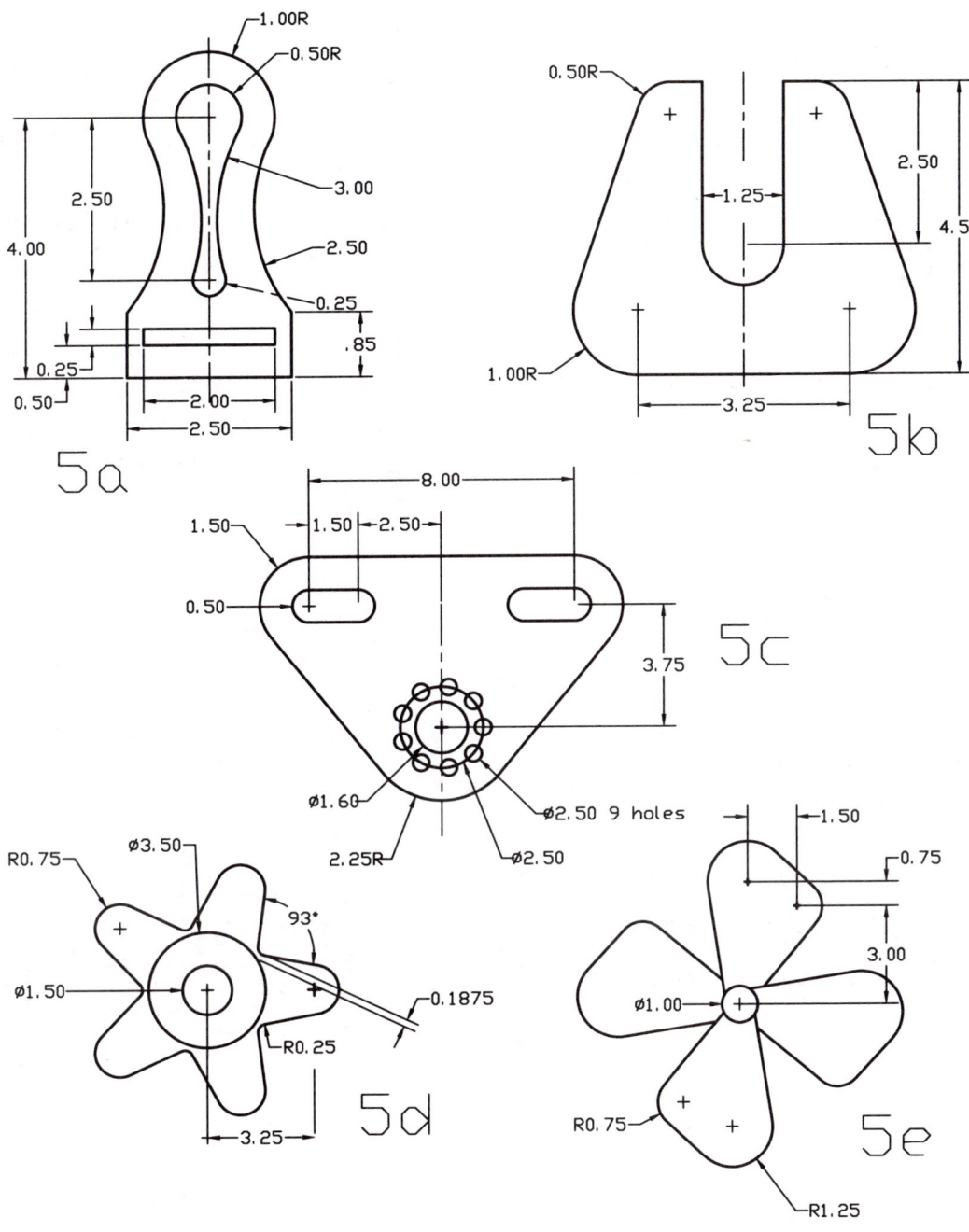

Practice Exercise 5B

In the digital display use SOLID to make the numerals, then DONUT, CIRCLE, LINE, and FILLET to complete the drawing.

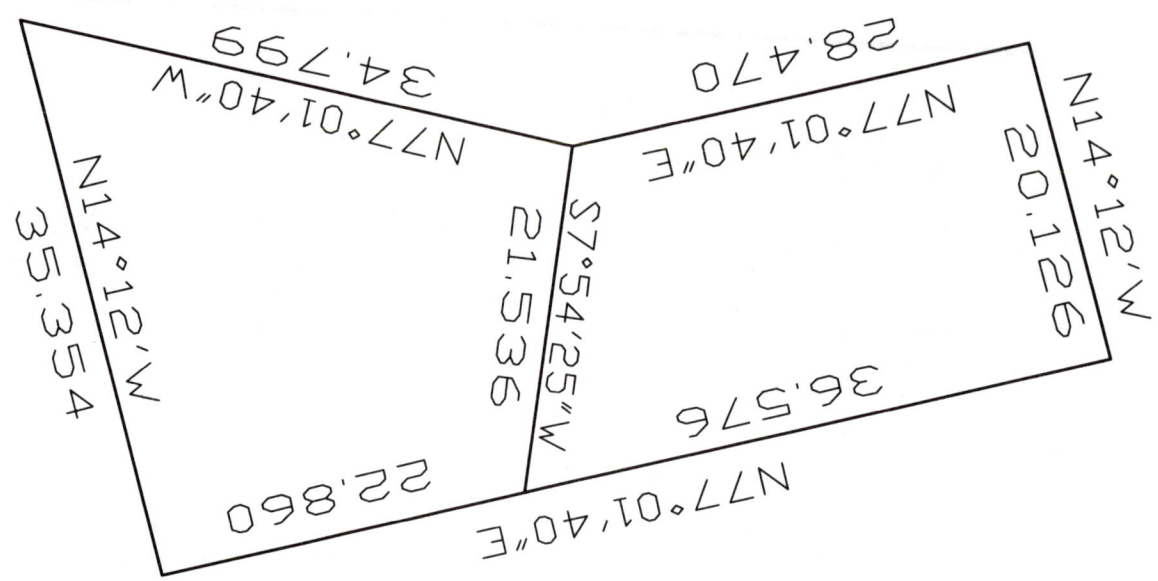

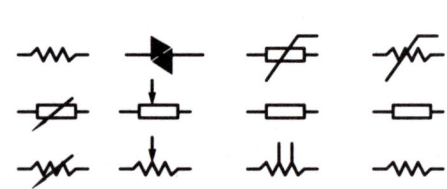

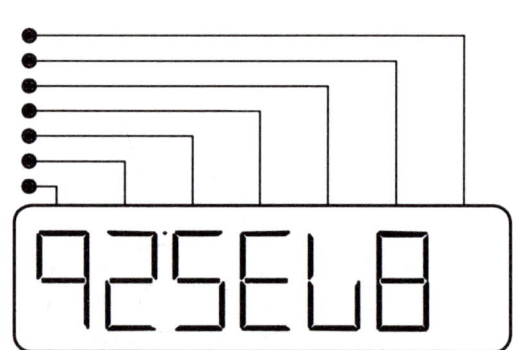

Practice Exercise 5C

OFFSET and ARRAY will be useful for these examples.

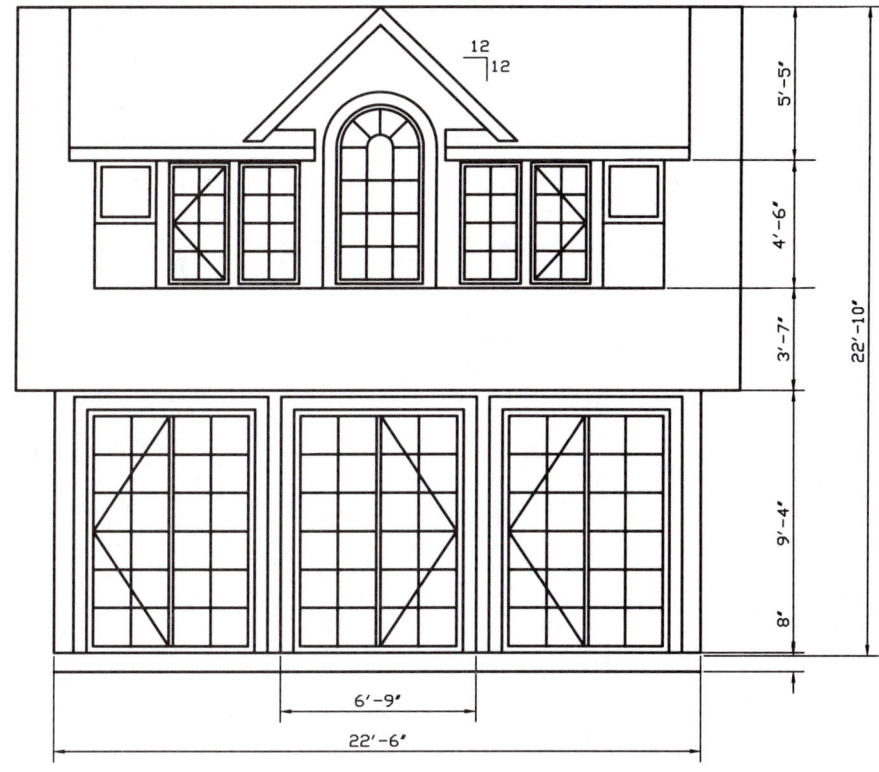

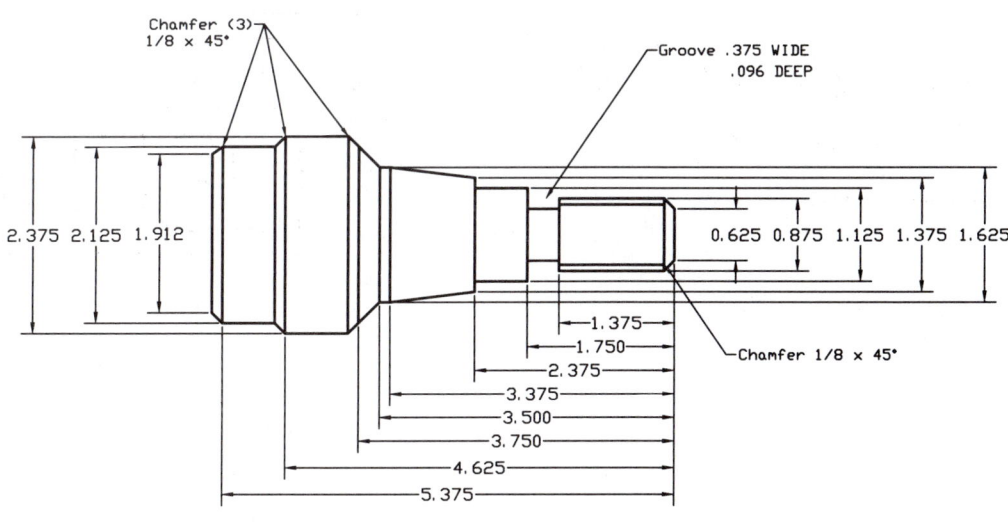

Exercise A5

Using the files from the previous chapters, for the interior walls of the second floor use OFFSET and FILLET. If you are not happy with some of the design features, use STRETCH to move things around. Use ARRAY to place the upper stairs.

If time permits, draw in the bathrooms on this floor and on the first floor.

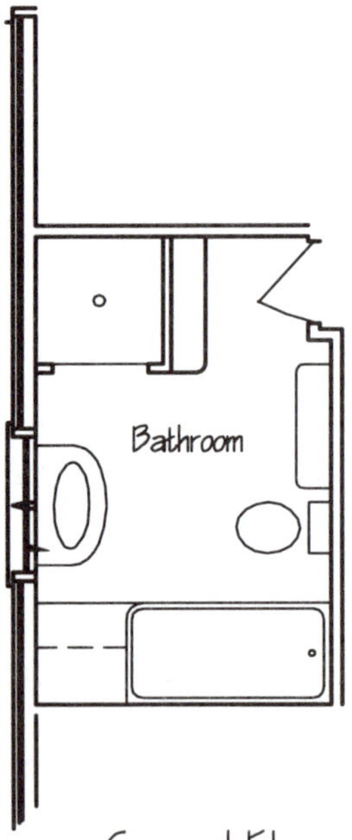

Second Floor

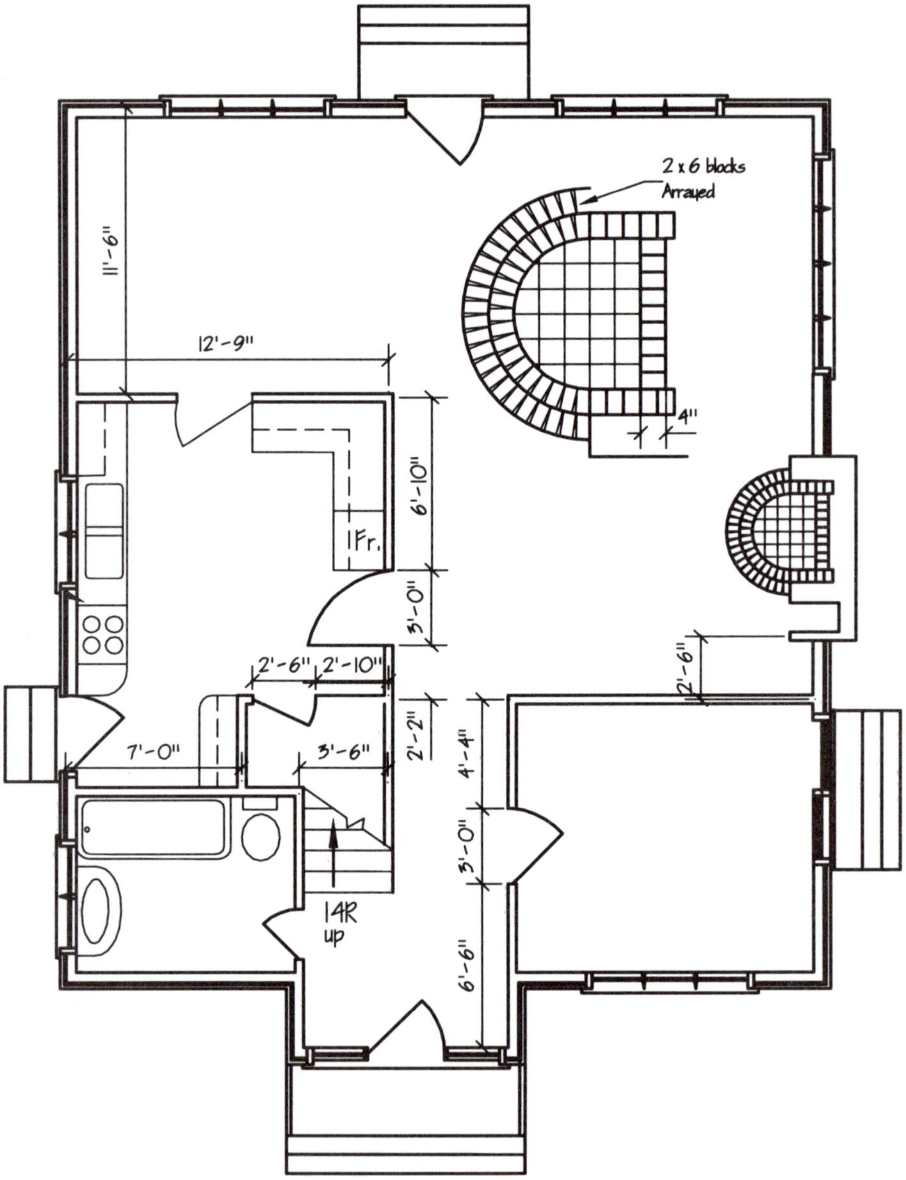

First Floor

Exercise C5

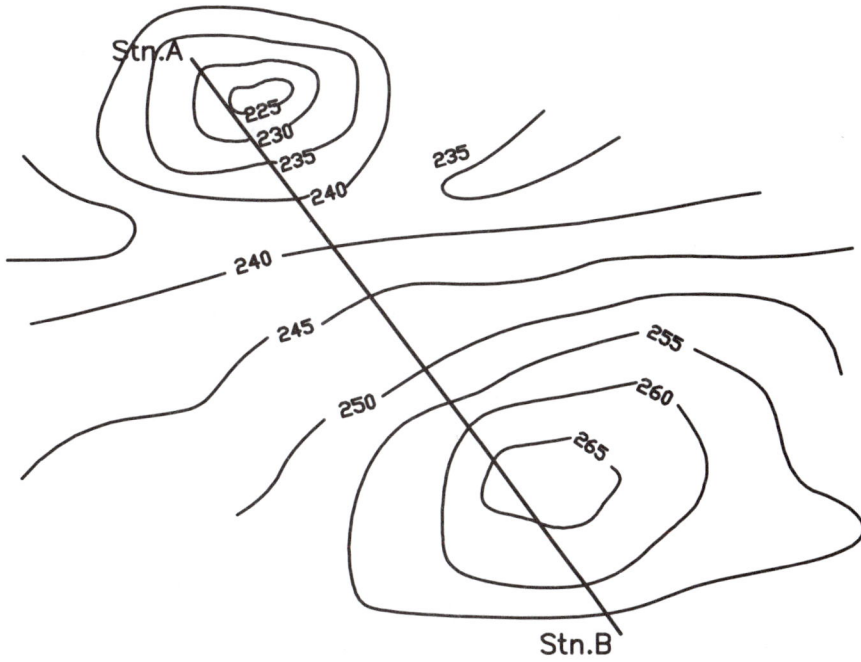

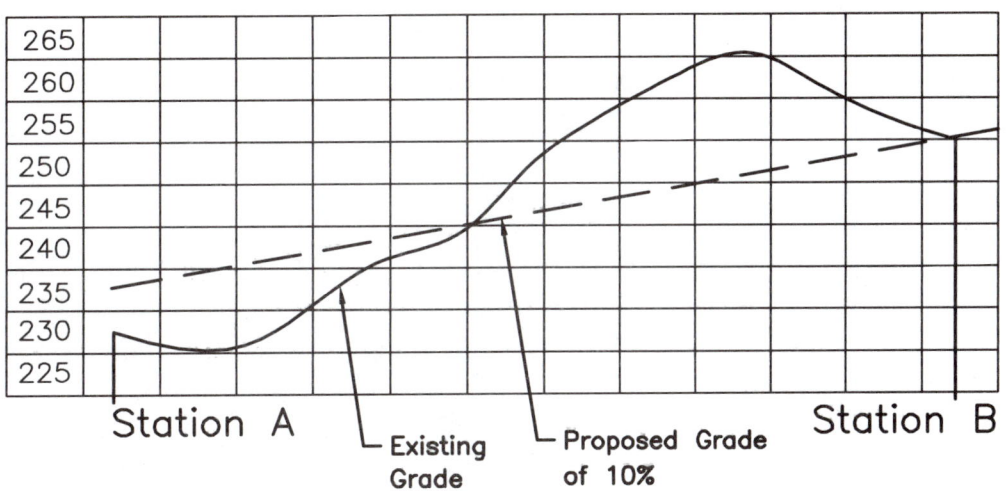

The polylines are contours showing the existing ground elevations for a proposed roadway. Use PEDIT to create smooth polylines. The original ground elevations are shown with contours at 5.00 m intervals.

The same drawing shows a profile route of the highway running from station A to station B at +10.00. On the profile, plot the highway elevations as well as the original road elevations.

STRETCH, TRIM, EXTEND, OFFSET, and ARRAY

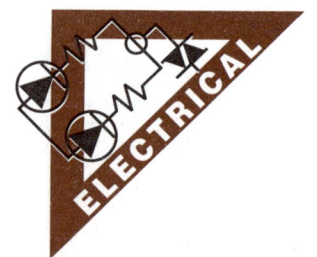

Exercise E5

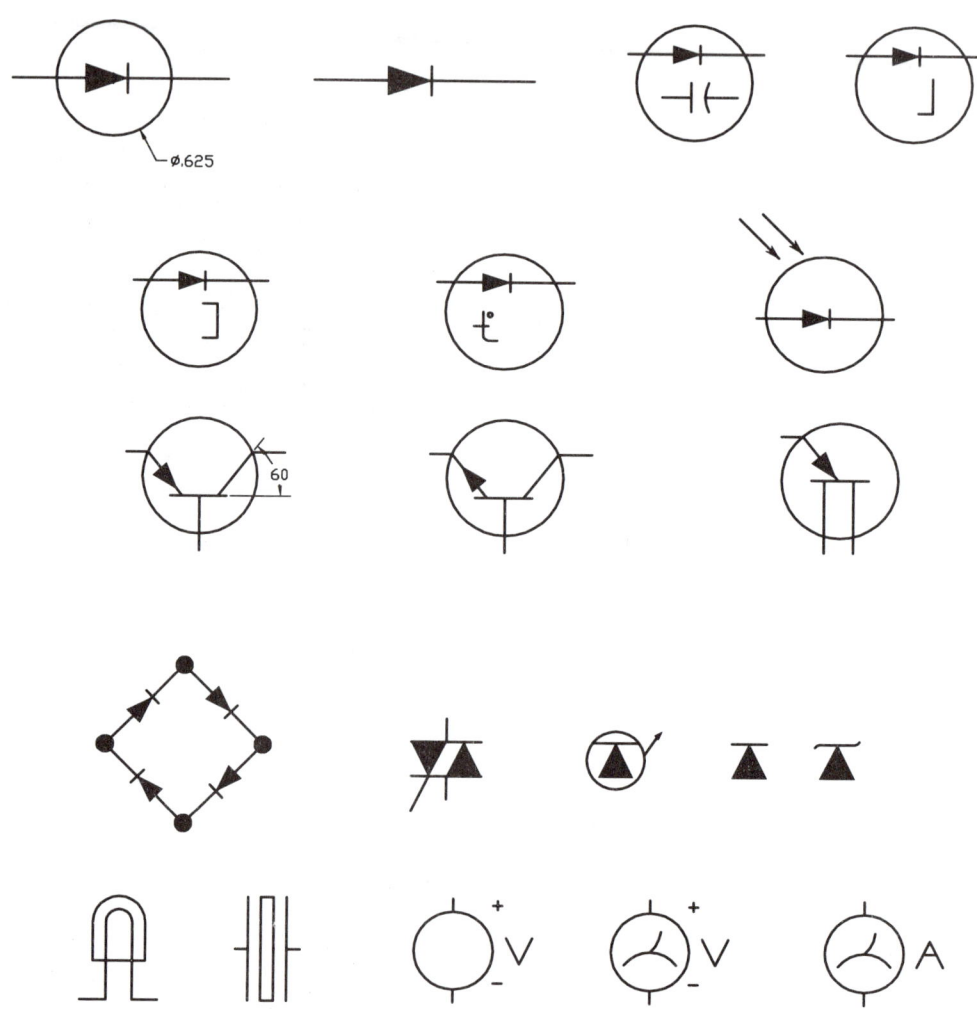

Create these symbols with the editing commands; use COPY, ROTATE, and MOVE to draw the parts more efficiently. Use the SCALE command to change the relative size of the PLINE when you enter it as an arrow.

When you are finished, use the symbols to create a logical schematic.

Exercise M5

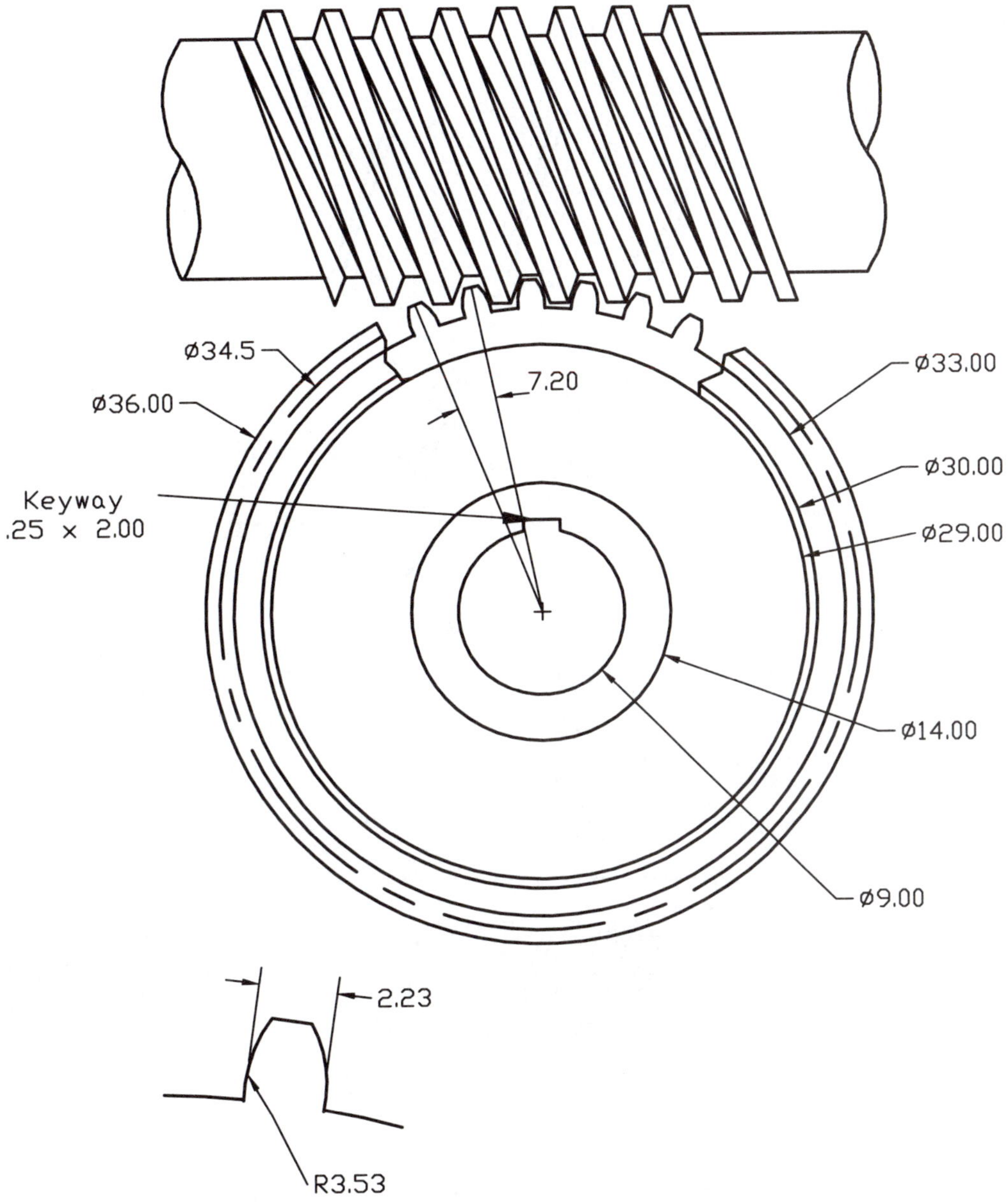

Use the commands on the following page to make this model.

Once you are finished, SAVE or END so you can add dimensions in a future exercise.

Hints on M5

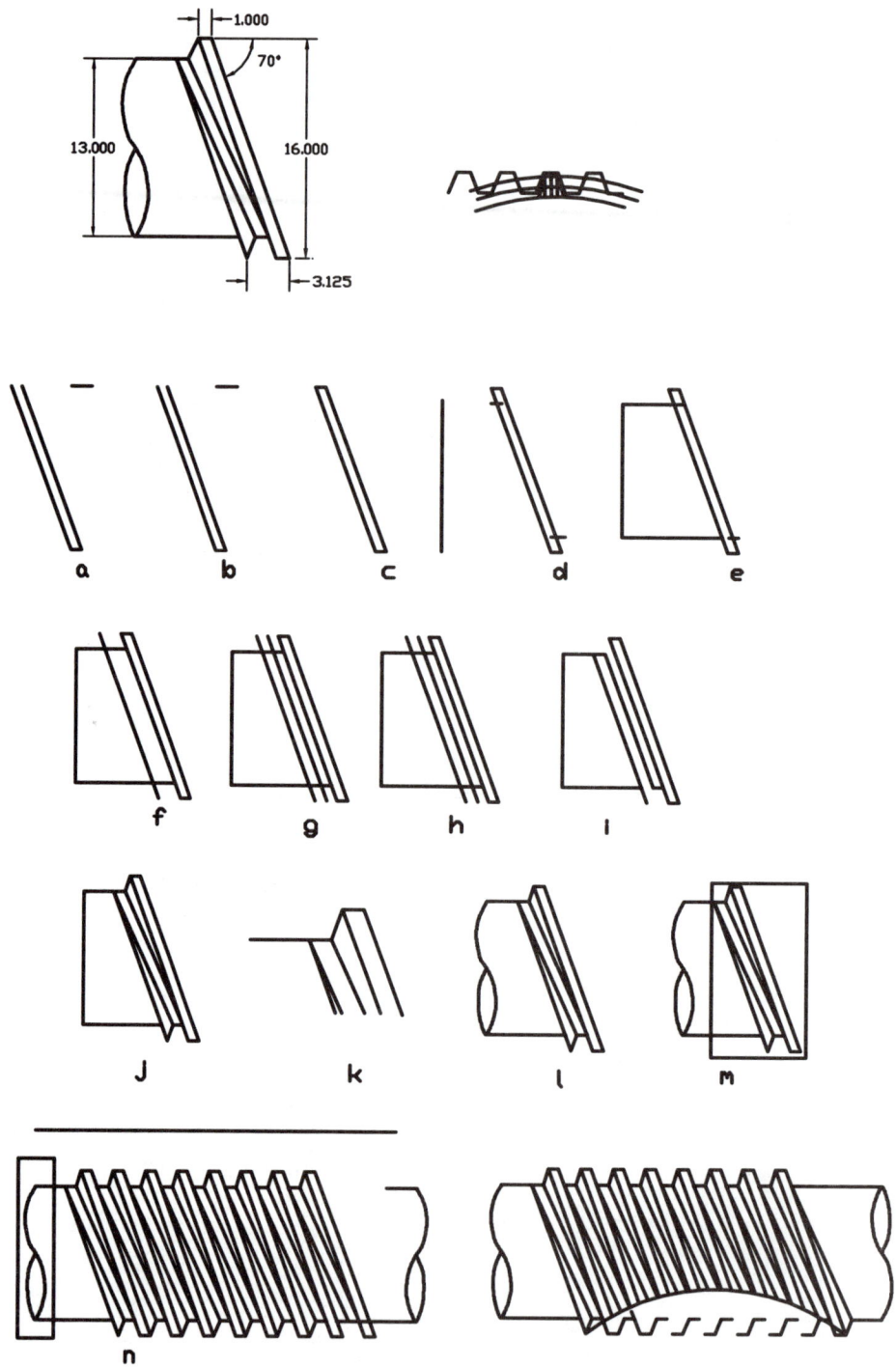

a. Draw the first line. b. OFFSET at the specified distance. c. Add the bottom line. d. OFFSET again to the top at 16 units, and draw in the line for the end. e. Use OFFSET to create the top and bottom, then TRIM. f. to i. OFFSET and TRIM. j. Draw the lines as shown. k. Draw in the small line section. l. and m. Use ARRAY to create a line of teeth. Then finish the part.

Challenger 5

Mapping and surveying are often used in conjunction with services. Try using COPY, ROTATE, and MOVE to create this transformer rack.

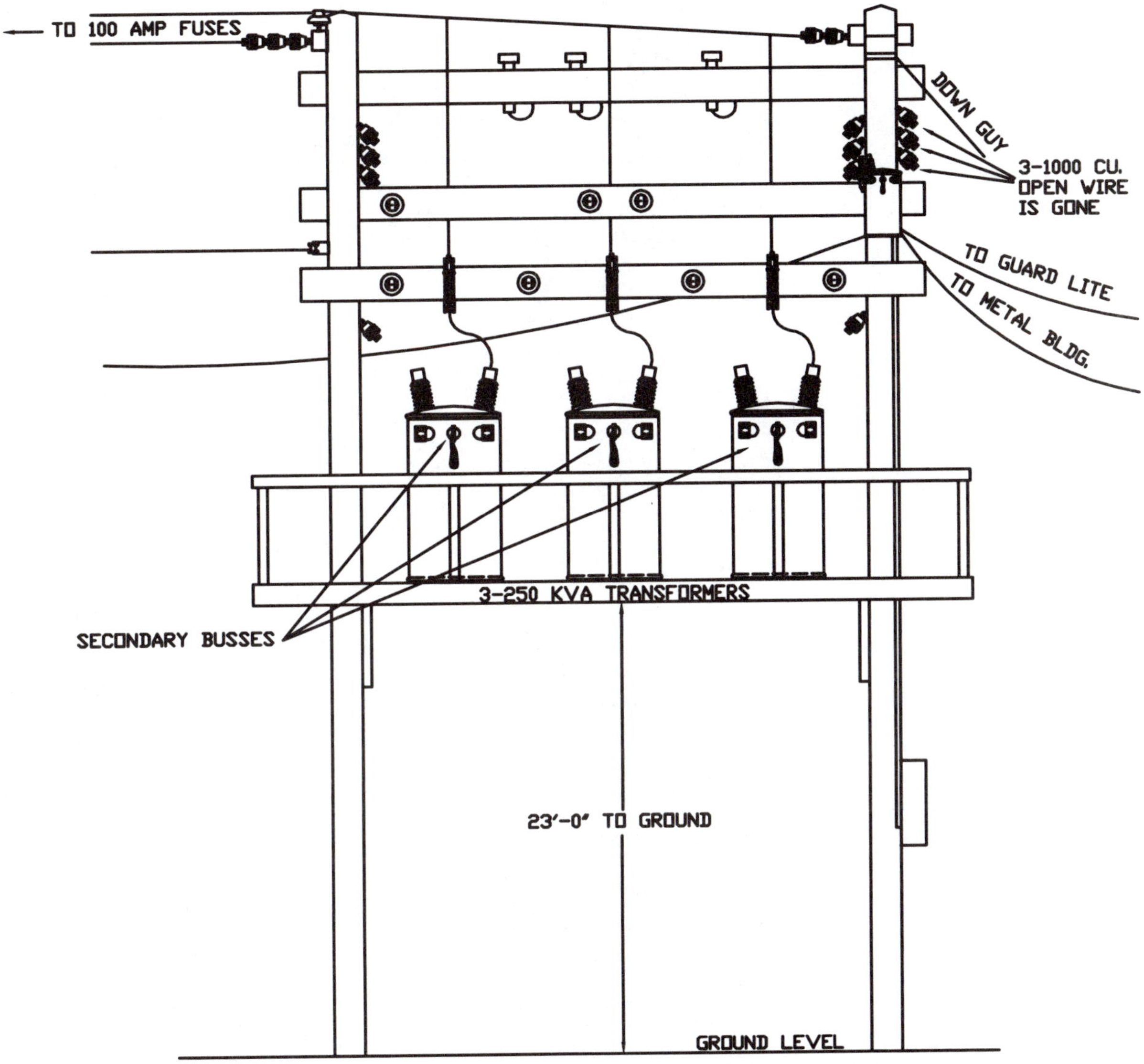

EXISTING TRANSFORMER RACK

The individual parts are actually quite simple to draw. Once drawn, they can be copied and rotated to fit the illustration. See the next page for overall measurements.

STRETCH, TRIM, EXTEND, OFFSET, and ARRAY

Use the overall dimensions above to create the individual items, then place them using MOVE, COPY, MIRROR, and ROTATE.

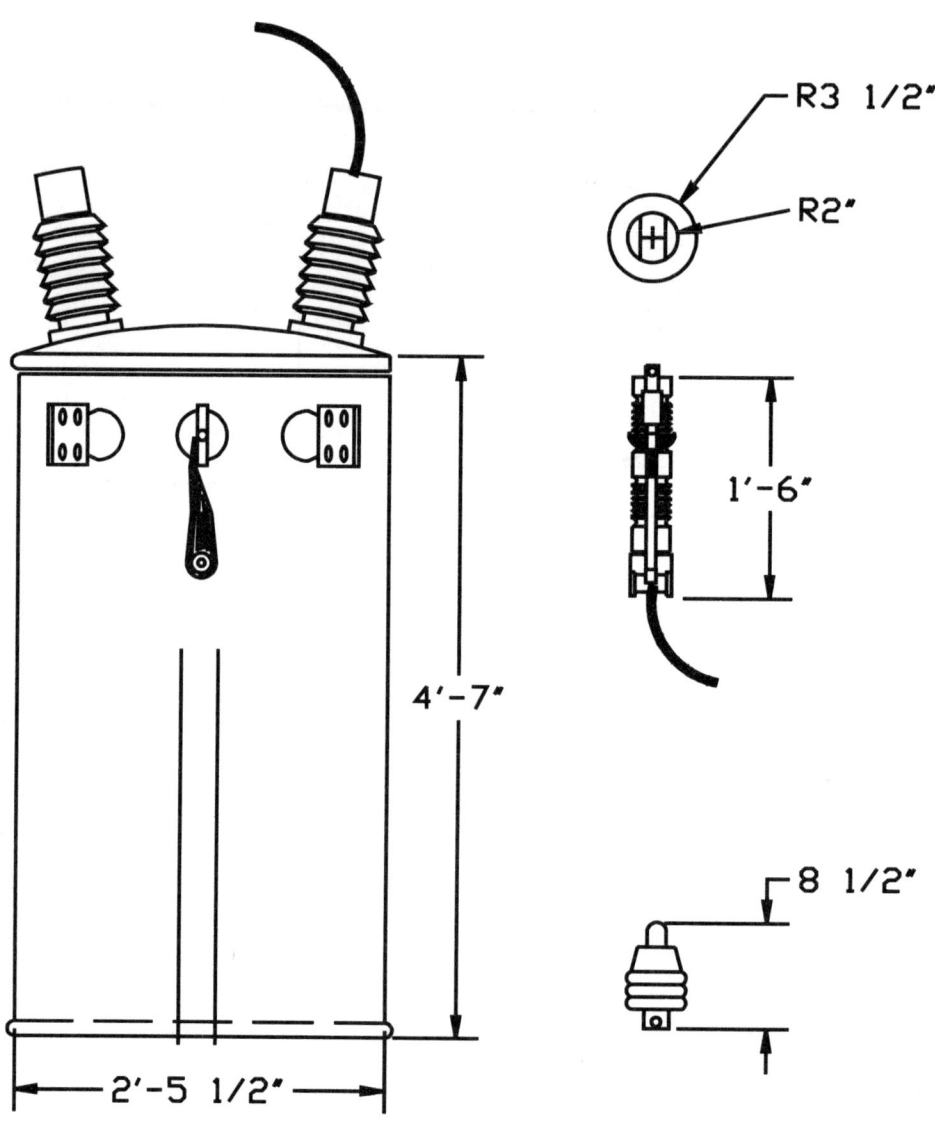

(Drawing: Compliments of Johnny Barton.)

Entity Properties: Layers, Colors, and Linetypes

Upon completion of this chapter, you should be able to:

1. Set up LAYERs and create geometry on them
2. Set colors to layers and use the COLOR command
3. Set LINETYPEs to layers and use the LINETYPE command
4. Change the properties of objects
5. List the properties of objects
6. Freeze/thaw and ON/OFF the LAYER
7. Lock and unlock layers

About LAYERs

The LAYER command allows you to control the drawing by means of visible entities. In a sense, it is like a set of transparencies or acetate overlays that contain different colors and linestyles which may be either visible or invisible. Layering is a powerful organizational tool and should be used on all drawings.

In AutoCAD, different colors and linetypes can be associated with different layers. To help complete the database, these layers can be either displayed or undisplayed, active or inactive, accessible or inaccessible. The number of layers you can use is unlimited.

Figure 6-1

Using LAYERs

There are two ways of accessing layering capabilities. The first and easiest is through the layer dialog box under the Data pull-down menu in DOS, or the Layer button in WINDOWS. This is the 'DDLMODES menu.

Windows Pick the Layer button:

DOS Under the Data menu, choose Layer.

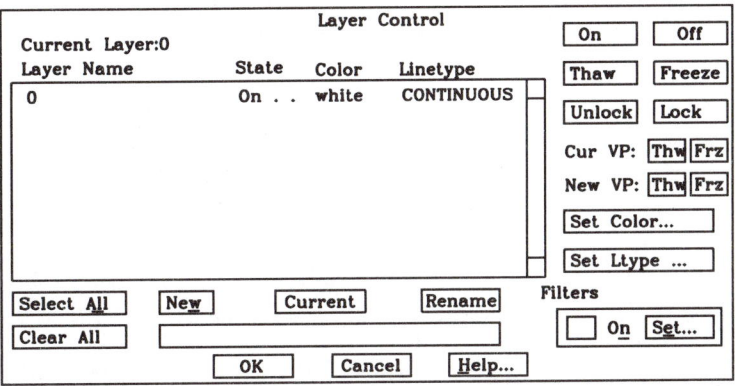

The dialog box will help you to view the layers, change the colors, turn layers off, freeze and thaw them, and lock or unlock them.

Current layer	The current layer name — the one you are working on.
Layer name	The name of the layer; no spaces or dots allowed.
State	Whether the layer is on, frozen, locked, or visible in only certain viewports.
Color	The color set to that layer.
Linetype	The linetype of the layer.
Select All	Selects or highlights all the layers listed.
Clear All	Clears or removes highlight on all selected layers.
On Off	Sets the selected layers on or off, which makes the layer invisible or visible.
Thaw Freeze	Freezes selected layers, making them invisible and not regenerated, or thaws (unfreezes) them.
Unlock Lock	Locks selected layers, making them visible but not accessible, or unlocks them.
Cur VP	Sets the state in the current paper space viewport.
New VP	Sets the state in a new paper space viewport.
Set Color ...	Sets the color.
Set Ltype ...	Sets the linetype.
Filters	Selects or filters through the layers.
Rename	Changes the name of a layer.
Cancel	Cancels the layer changes or additions you have made.
Help ...	Provides help files on the layer functions.

Creating a New Layer

There should be a blinking cursor in the blank area at the bottom of the screen, which is called the *layer name text box*; if there is not, pick this area, then enter the name of the layer that you would like to make. A layer name can have up to 31 characters. There are only spaces for ten layers in the display box provided, but you may enter as many layers as you like and the screen will scroll down.

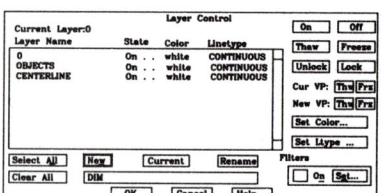

Figure 6-2

```
Command: (pick DDLMODES)
(type in a layer name)
(pick New)
```

Once the name is entered pick New and the layer will be added to the layer list at the bottom. The next time you enter the dialog box it will be listed alphabetically.

When typing layer names, keep the following rules in mind:

1. Use no spaces in the names. For example, to indicate the layer for a side bracket use SIDE-BR.
2. Use no slashes (/ or \) or periods (.), as the system will not accept them.
3. You may enter as many as 31 characters for a layer name, but you will only be able to read eight letters on the Status line. Try to keep the entry to that many characters.

You can enter more than one name at a time by separating them with a comma: **one,two,three,** etc.

Activating Layers for Changes

In order to change the name, state, color, linetype, or lock value of the layer, it must be identified or selected. Move the cursor to the layer name and pick it. The layer's line should turn blue.

Once a layer is selected, many of the options that were in grey-toned letters will become black-toned letters. These options are available only when a layer or series of layers is selected. Notice that the Current option is only highlighted in black when only one layer is selected. If more than one layer is selected, or no layers are, nothing is highlighted. You must be careful of having more than one layer highlighted, as the Set color and Set linetype options will affect all highlighted objects.

Changing a Layer Name

If you should type in the name incorrectly, pick the name from the list. The area now activated will turn blue. The name will appear in the Layer Name text box at the bottom of the screen. Now type in a new name or revise the current name and then pick Rename. The new name will replace the old.

```
Command: (pick DDLMODES)
(pick the layer from the list)
(type in a new name)
(pick Rename)
```

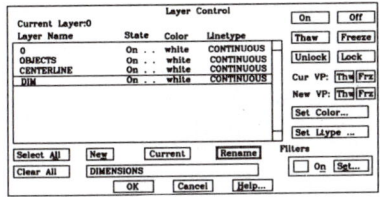

Figure 6-3

If you have created a layer that will never be used, delete it with PURGE. Save the file, use PURGE, then All, then save the file again (see page 262). A LISP routine called Dellayer is available that will remove all of the items on a layer.

Making a LAYER Current

In order to draw on a layer you must make it *current*. Activate the layer that you would like to have current, and pick the word Current.

```
Command: (pick DDLMODES)
(pick the layer from the list)
(pick Current)
```

The top line of the layer dialog box will show the current layer name. While working on a drawing or model, you can easily check to see what layer is current by reading the status line. All of your work from this point on will be in the current layer.

Changing Layer Color

First select the layer or layers that you would like to be in a particular color, then pick Set Color. This will invoke the Select Color dialog box. From the dialog box pick a color that you would like, or enter a color name or number in the Color text box at the bottom of the box.

```
Command: (pick DDLMODES)
(pick the layer from the list)
(pick Set Color ...)
(pick a color)
(pick OK)
```

When you pick the color that you want, the color will be shown in the Color text box on the bottom of the screen. Anyone who has trouble seeing colors on the screen can also type in the number of the color in the text box.

All computers use the same number code for the first seven colors.

1 = red

2 = yellow

3 = green

4 = cyan

5 = blue

6 = magenta

7 = black or white

8 = grey

Once you have chosen the desired color, select it by picking OK. You will then return to the Layer Control menu.

Setting Color Independent of Layer

You can set a color using the COLOR command. This will override your Layer Color setting. To do this, type in the word **COLOR** at the command prompt.

```
Command:COLOR
New object color<bylayer>:RED
```

All subsequent objects will be drawn in this color. Keep the COLOR command set to Bylayer and you should have no problem.

Danger

If your objects are being entered in a color different than your layer setting, you have set a color in the COLOR command. Set the COLOR setting back to Bylayer.

Loading LINETYPEs

In addition to colors, you can also have linetypes associated with each layer. The linetypes are not loaded with the default AutoCAD drawing, and thus may not be available under the listing of linetypes in the dialog box. To use different linetypes, they must be loaded into the file.

> **Windows** From the toolbar, pick the Linetype icon:
>
> **DOS** From the Data menu, pick Linetypes.
>
> OR:
>
> **Windows** and **DOS** From the Layer dialog box activate a layer, choose Set Ltype, then pick the Load button. Both dialog boxes will allow you to load the linetypes.

The command line equivalent is **LINETYPE**.

```
Command:LINETYPE
?/Create/Load/Set:
```

In order to add the hidden linetype on the layer dialog box, load the linetypes either through the dialog box or through the command line. In the dialog box choose:

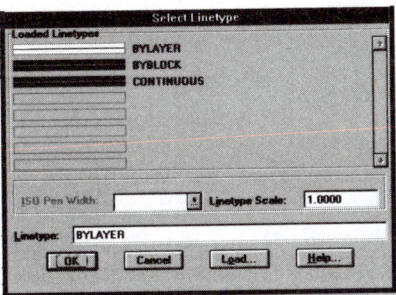
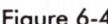
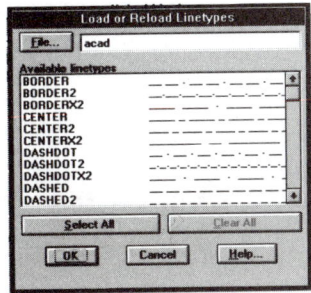

Figure 6-4   Figure 6-5

Pick Load, choose the linetypes that you require, then choose OK.

To load from the command string:

```
Command:LINETYPE
Set/Load/Create/?:L
Linetype(s) to load:HIDDEN
Select Linetype File dialog box:⏎
```

Where: **?**   = invokes a list of linetypes
       Create = creates a customized linetype
       Load   = loads the linetypes
       Set   = sets a current linetype that will override your layer setting

AutoCAD then displays the Select Linetype File dialog box. To accept the default linetypes, just pick OK or use ⏎.

```
File to search<ACAD>:⏎  (accepts the default file listing)
```

Be careful to type the linetype exactly as it is spelled in the listing; otherwise, it will not load. (As mentioned earlier, in Canada the spelling of "center" as "centre" is the most common stumbling-block.)

If you would like to load all of the available linetypes, use the DOS wildcard option *.

```
Command:LINETYPE
Set/Load/Create/?:L
Linetype(s) to load:*
↵
```

If you want all of the linetypes always loaded, open the ACAD.DWG file, load the linetypes, then SAVE the file as ACAD.DWG. This will update the AutoCAD prototype file and make the linetypes always available.

Once the linetype is loaded, you can go back to the layer control dialog box, 'DDLMODEs under the Settings pull-down menu, and choose the linetype in the same way you chose your color.

All future geometry added will use this linetype. If the linetype is the hidden type, all entities in that layer will be in hidden lines, unless they are overridden by the linetype command. Until you are sure of what you are doing, both the Color setting and the Linetype setting should be Bylayer.

Changing LTSCALE

Depending on the size of the object, the hidden lines may not show up as hidden. If this is the case, either the objects that you thought were on that layer are not, or the scale of the drawing is too large or too small for the linetype to show. You will need to change the linetype scale relative to the current drawing.

> **Windows and DOS** From the Options menu, choose Linetypes, then Global Linetype Scale.

The command line equivalent is **LTSCALE**.

```
Command:LTSCALE
New scale factor<1.00>:12
```

To determine the LTSCALE of the screen, take the furthest point in X and divide by 12. If you have defined the screen but have not as yet entered any points or geometry, take the furthest limit in positive X and divide by 12. If the limits are 200,160, the LTSCALE will be 200/12 or approximately 14. If the limits are 3, LTSCALE will be 3/12 or .25.

LTSCALE is often linked to the Hatch scale and text scale.

Using LTSCALE will cause your file to regenerate.

The LTYPEGEN Command

When LTYPEGEN is turned on, the linetype of the polyline is continuous regardless of the placement of the vertices in the PLINE. When LTYPEGEN is turned off, the linetype is displayed with a dash at each vertex.

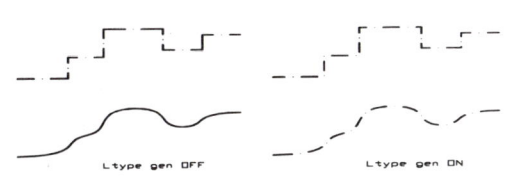

Figure 6-6

Windows and DOS Pick Options, Linetype, Linetype Generation.

What Can Go Wrong with LAYER, COLOR, and LINETYPE

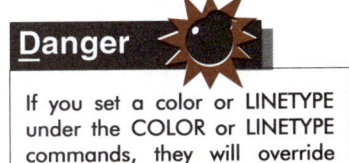

If you set a color or LINETYPE under the COLOR or LINETYPE commands, they will override your layer setting.

When working with LINETYPE and COLOR, keep in mind that both can be changed independent of LAYER. Both LINETYPE and COLOR can be changed under the Set option under the Options menus. This means that if COLOR, for example, is set to Red, all the geometry will show up as red even if the current layer is blue. Similarly, if the LINETYPE is set to Hidden, all of the lines will be hidden lines, even if the current layer asks for center lines.

When picking layer names for changing color, linetype, freezing, etc., make sure that *only* the layers you want to change are highlighted in blue. You must pick the layer again to deselect it.

LINETYPE and COLOR must be set to Bylayer to access the different linetypes and colors associated with each layer.

CHPROP and CHANGE with LAYERs

To change the properties of entities — the color, linetype, or layer — use CHPROP (CHange PROPerties), or the DDCHPROP dialog box, or the DDMODIFY dialog box.

Windows From the Object Properties toolbar, choose this button:
From the Edit menu, choose Properties.

DOS From the Modify menu, choose Properties.

The command line equivalent is **CHPROP**.

The dialog box will help you to change the properties of the selected object. If you select one entity, you will be shown the DDMODIFY dialog box. If you select more than one entity, the DDCHPROP dialog box will be shown. DDMODIFY will allow you to change other properties such as the size of the entity, the location, and the linetype scale.

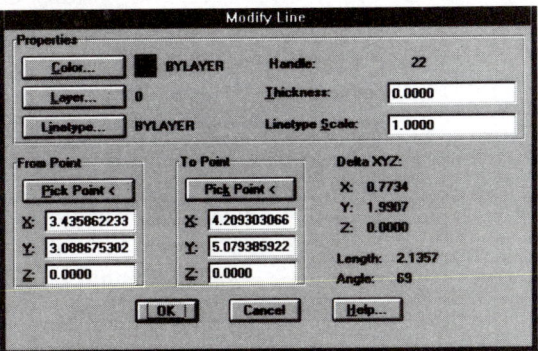

Figure 6-7

Entity Properties: Layers, Colors, and Linetypes

With both the DDCHPROP dialog box and the DDMODIFY dialog box, you are prompted to pick the objects you would like modified; then you can choose Layer, Color, Linetype, or Thickness on the dialog box. A secondary menu will appear that lists all available Layers, Colors, and Linetypes.

```
Command:CHPROP
Select objects:ALL
Select objects:↵
Change what property(Color/LAyer/LType/LtScale/Thickness)?:C
New color<varies>:BYLAYER
Change what property(Color/LAyer/LType/LtScale/Thickness)?:LT
New linetype<varies>:BYLAYER
Change what property(Color/LAyer/LType/LtScale/Thickness)?:↵
```

If everything changes to the correct color and linetype, check to see that COLOR and LINETYPE are set to Bylayer as well.

If objects still do not show the color or linetype expected, they are probably in the wrong layer. Again, use CHPROP.

```
Command:CHPROP
Select objects: (pick 1, 2) (pick the objects to change)
Select objects:↵
Change what property(Color/LAyer/LType/LtScale/Thickness)?:LA
New layer<varies>:HIDDENLINES (for example)
Change what property(Color/LAyer/LType/LtScale/Thickness)?:↵
```

The CHANGE command can also be used to change the properties of objects.

If things are still not as you want them, try REGEN. If this does not work, there is something wrong with the entity itself. Use LIST (see below) to show you the properties and position of the object or objects not reacting to the above commands. Often you will find the objects are not on the layer that you thought they were.

The LIST Command

The LIST command provides a list of the properties and position of an object.

```
Command:LIST
Select objects: (pick 1) (this selects the item on which you
  want to find information)
```

The screen will now offer you information on the item that you have chosen, including the layer of the object selected, and whether there is a color or linetype overriding those set in the layer.

Changing the State of the Layers

Once the layers are loaded, and objects have been placed in them, you can have them displayed or not displayed, displayed but not editable, and completely turned off. In Windows, you can access the state of the layers either through the dialog box or through the icons.

On/Off

Turning off a layer will make that specific layer invisible. To turn it off, use Off in the LAYER dialog box. The word On will be removed from the status area of the layer. Use to view it again.

Freeze/Thaw

Freezing a layer will tell AutoCAD to ignore any entities on the specified layers when regenerating the drawing, as well as making the layer invisible. When the drawing gets quite large, this option is used to save time in regenerating by removing items that will not be viewed. It is usually better to use Freeze than Off.

Frozen layers must be Thawed before they can be viewed or plotted.

To Freeze and Thaw, use the Layer dialog box. The state of the layer will be reflected by an F if it is frozen, and a dot (.) if it is not.

Lock/Unlock

This facility will help you to place objects relative to other objects without taking the chance of editing those objects. Locking prevents editing on visible objects.

To lock the layers choose Lock from the dialog box. This will be shown as an L on the State column.

Current VP and New VP will be dealt with in Chapter 11 when viewports are introduced.

The LAYER Command

In a perfect world, you would probably always use the dialog box. For many reasons, however, sometimes dialog boxes may not work. This could be a hardware problem, which is usually associated with the graphics card, or it could be because the LISP functions are disabled. You may also find it faster to key in the single letters to change the layer. In any case, if you cannot use the dialog box, or your FILEDIA is set to 0, use the LAYER command (type **LAYER** or **LA**).

```
Command:LAYER
?/Make/Set/New/On/Off/Color/Ltype/Freeze/Thaw/LOck/Unlock:N
New layer name(s):ELECTRIC
?/Make/Set/New/On/Off/Color/LType/Freeze/Thaw/LOck/Unlock:C
Color:1
Layer name(s) for color 1(red)<0>:ELECTRIC
?/Make/Set/New/On/Off/Color/LType/Freeze/Thaw/LOck/Unlock:L
Linetype or ?<continuous>:DASHED
Layer name for linetype DASHED<0>:ELECTRIC
?/Make/Set/New/On/Off/Color/LType/Freeze/Thaw/LOck/Unlock:
```

None of the LAYER options take effect until you have exited the command. Use to exit the command.

If you invoke a LTYPE using the Layer command as above, it loads that linetype automatically if it is not already loaded.

If you want to look at the list of layers, use the ? option. AutoCAD then prompts for the layers you want listed. The wildcard * can be used to list all of the layers, or to search for specific layers if you have more than one screen's worth of layers. The * wildcard and the ? wildcard can be used to turn on, turn off, freeze, and thaw a whole series of common-named layers.

Layer Filtering

Sometimes you may want only certain layers to be listed in the Layer Control dialog box. The Filter option allows you to limit which layers are listed. You can filter on the basis of:

- Layer names, colors, and linetypes
- Whether they are on or off
- Whether they are frozen or thawed
- Whether they are locked or unlocked

To Filter a Layer

Open the Layer Control dialog box through the icon, or in DOS under the Data pull-down menu.

In the Filters area of the Layer Control dialog box, choose Set ...

In the Set Layer Filters dialog box, select the settings. Then choose OK to close the Set Layer Filters dialog box.

Using this dialog box you can, for example, list all of the layers that are red.

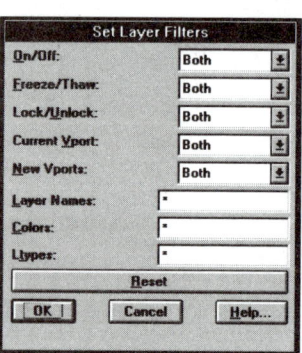

Figure 6-8

Layer Management

The default layer is 0. If you enter any dimensions, you will also create a layer called Defpoints. Neither of these layers is renamable. The Defpoints layer does not plot.

While creating a model in AutoCAD, keep in mind that other people may want to work with your file at some point. Layers should therefore be named in a logical, straightforward manner. Also, keep your LINETYPE and COLOR command set to Bylayer.

Many industries have developed layering standards so that there is no question about where objects will be located. If you are starting work with a company, make sure that you know what the layering standards for the company are, and find out whether there are any standards outside of the company that you should be aware of.

If you do not use the layer names suggested in the exercises, at least make sure your names are logical.

Freezing layers when you are not using them will save a great deal of time with larger models. Until they are regenerated, layers that have been frozen will not display when they are thawed.

If you are making a layer current, make sure that the layer is on as well. It is possible to have a layer current when it is also off, and you will be creating objects on the invisible layer without seeing what you are creating. If you have drawn objects that are not appearing, check to see that the layer is turned on.

Prelab 6 Layers, Colors, and Linetypes

In this example we will make an adjustable bearing with hidden lines and center lines.

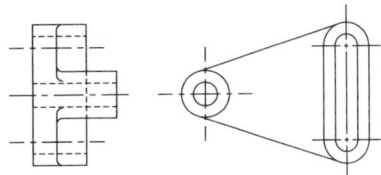

Step 1 First set up your layers from the pull-down menu.

> **Windows** Choose the Layer button:
>
> **DOS** Choose Data, then Layer.

In the Layer Name text box enter the following:

LINES

HIDDEN

CENTER

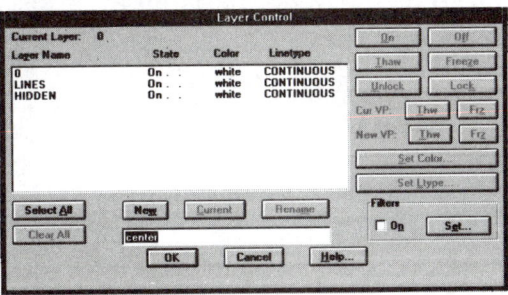

Step 2 Once the names are loaded, change the color of each layer by picking the layer name and then picking the Set color button. An overlay menu will offer you a selection of colors. Choose a color for that layer.

- LINES   White
- HIDDEN   Yellow
- CENTER   Red

Step 3 Now load and set the linetypes. Press OK on the layer menu, then:

> **Windows** Pick the Linetype button:
>
> **DOS** From the Data menu, pick Linetype.

In the Set Linetype dialog box, pick the Load button. Pick the Select all button, then pick OK.

Entity Properties: Layers, Colors, and Linetypes   **153**

Step 4 Now that the linetypes have been loaded, return to the Layer pull-down menu and load the linetypes onto the appropriate layer.

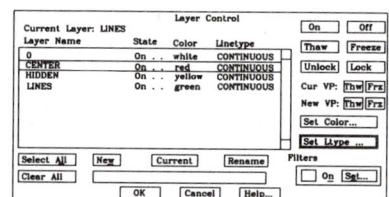

Windows Choose the Layer button:

DOS Choose Data, then Layer.

Activate the hidden layer, press LINETYPE, then choose the hidden linetype from the list. Press OK. Deactivate the hidden layer by picking it again, then activate the Center layer and set the center linetype to that layer. Use OK to exit from the Set Linetype dialog box, then OK again to exit from the Layer Control dialog box.

Step 5 Now that the layers are ready, use the same menu to make the LINEs layer current.

Your screen should look like this:

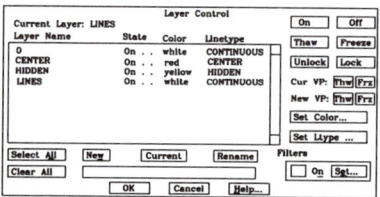

Step 6 Now with the LINEs layer current, draw in the front view of the adjustable bearing as shown. Do not draw in the dimensions.

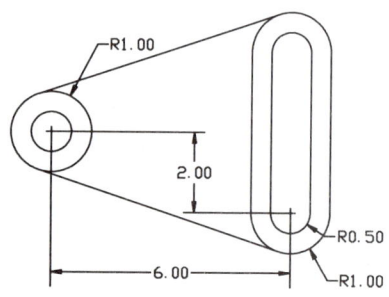

The limits and grid are not important in this drawing, but a snap of .25 would help to place the objects quickly. If you want the origin or 0,0 to be the center of the left circle, use UCS to place the origin about 2 inches up and 2 inches over from the bottom of the screen.

Use LINE and CIRCLE with OSNAPs to create the objects.

Step 7 Now return to the pull-down menu and make the CENTER layer current.

Draw in the lines as shown. They should be a different color as well as a different linetype.

Try changing the LTSCALE to see if there is any noticeable difference in the center lines.

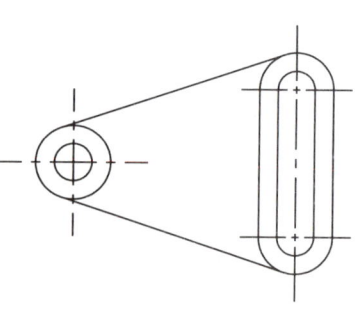

```
Command:LTSCALE

New scale factor<1.00>:1.5
```

Step 8 Return to the pull-down menu and make the LINEs layer current to draw in the side view as shown. You will need to use PAN to move the front view over. You can use LIMITS and pick the bottom left and upper right corners.

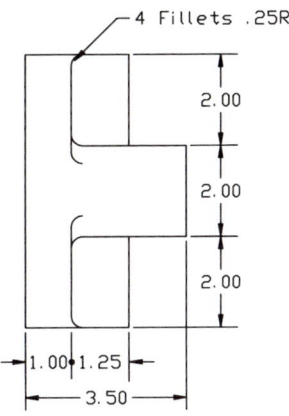

Step 9 Now return to your pull-down menu and make the HIDDEN layer current to place the hidden lines as shown. They should be a different color as well as a different linetype.

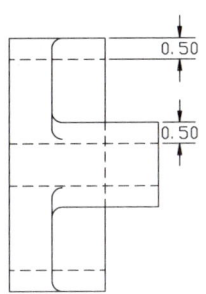

Step 10 Now make the CENTER layer current and add the center lines as shown. Again, a SNAP would be of use in this view. If the SNAP is too big, make it smaller.

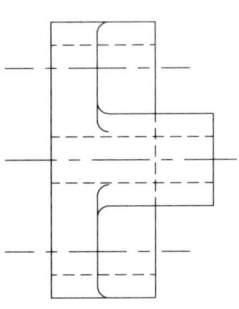

Step 11 Return to the layer pull-down menu and change the colors of the layers to see if you can make it look nicer. If the colors remain the same as you change the layer color, it is because the current color of an object is overriding the layer color. Set color back to Bylayer, then use CHPROP to change all of the objects to be colored Bylayer.

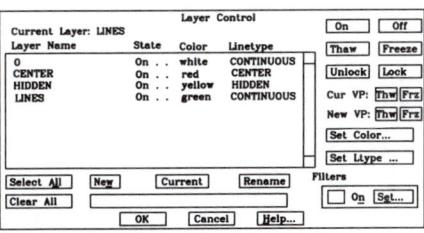

Step 12 We are now going to play with the Lock/Unlock facility to see how it works. It may be time to save the file — just in case.

```
Command: SAVE
(enter a name in the dialog box)
```

On the 'DDLMODE, select the layer for the hidden lines and the center lines, then pick Lock. In Windows, pick the Lock icon and then pick the center and hidden layers.

Entity Properties: Layers, Colors, and Linetypes

Once the layers are locked, you will be able to see them, but you will no longer be able to edit them. Make sure the LINEs layer is current before you start.

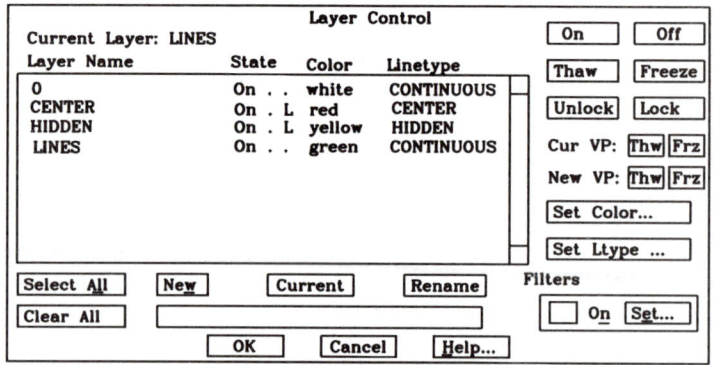

Step 13 Now back in the drawing or model, from the Modify menu pick Properties and change the color of the LINEs layer to blue.

> **Windows** From the Object Properties toolbar, choose this button:
> Or from the Edit menu choose Properties.
>
> **DOS** From the Modify menu, choose Properties.

Pick all of the objects on screen with either a window or a crossing, pick the color button, and pick the top blue square.

Notice that only the objects on the LINEs layer were changed, because the other layers were locked.

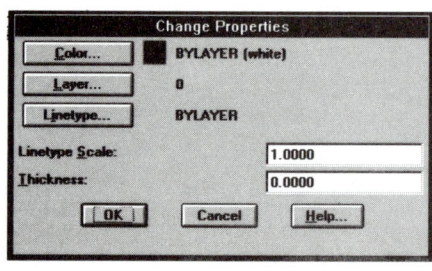

Step 14 Try turning Off and On, and Freezing and Thawing, the layers. The data for the layers is still on file, but the information is not displayed.

Note that you cannot change the current layer. If you switch the current layer Off, a warning box will appear.

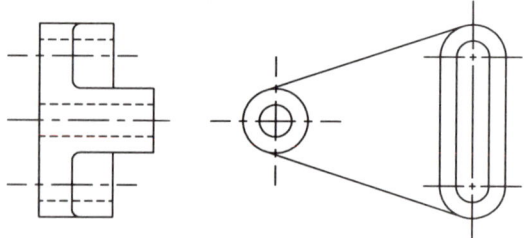

156 CHAPTER SIX

Command and Function Summary

CHANGE is an editing command that allows the user to change the properties and other aspects of existing objects.

CHPROP allows the user to change the properties of objects: Color, Layer, Linetype, Linetype Scale, and Thickness.

COLOR allows the user to set a new current color. This overrides the layer setting.

'DDCHPROP is the dialog box used to change the properties of an object.

'DDLMODE is the dialog box that allows manipulation of layers.

'DDMODIFY is the dialog box used to change the properties and other parts of objects.

Freeze/Thaw allows the user to deactivate and make invisible a layer.

LA or LAYER is the command that allows manipulation of layers. Layering within selected viewports will be covered in Chapter 10.

LINETYPE allows the user to set, create, and load linetypes.

Lock/Unlock allows the user to make selected layers visible (but not editable).

On/Off allows the user to make objects in layers either visible or invisible.

Practice Exercise 6

While drawing in these mechanical parts, have different layers for the center lines and for the hidden lines.

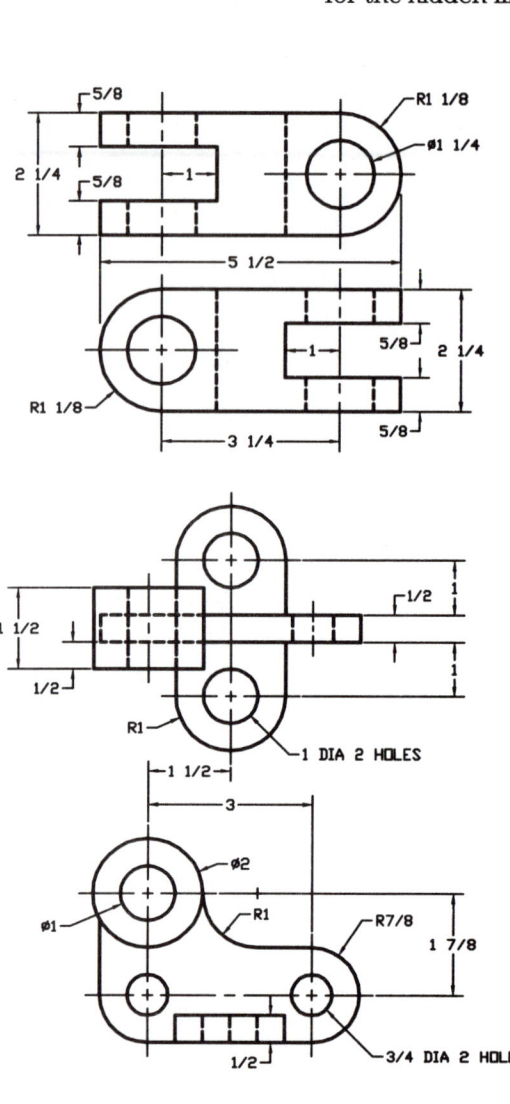
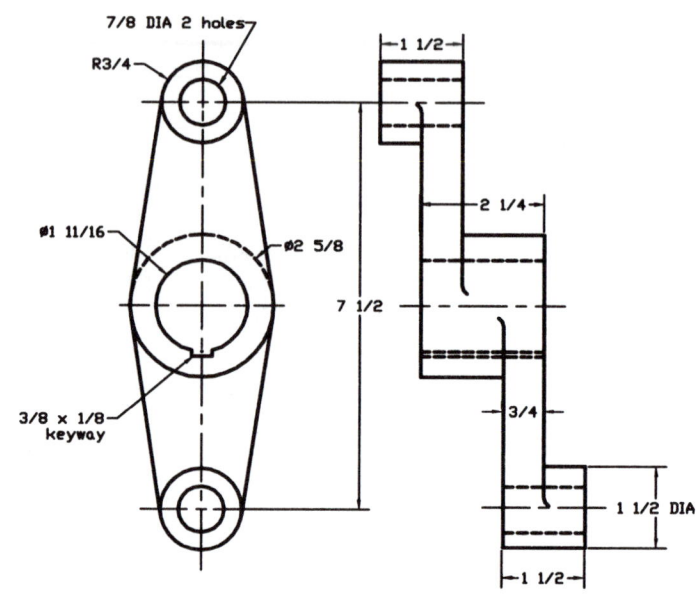
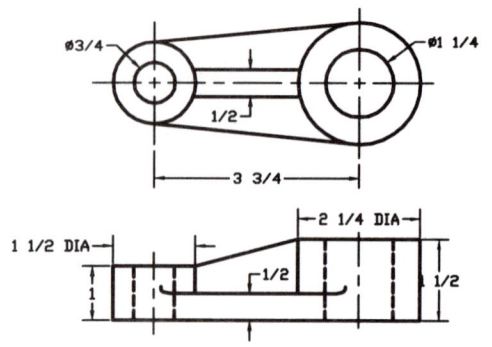
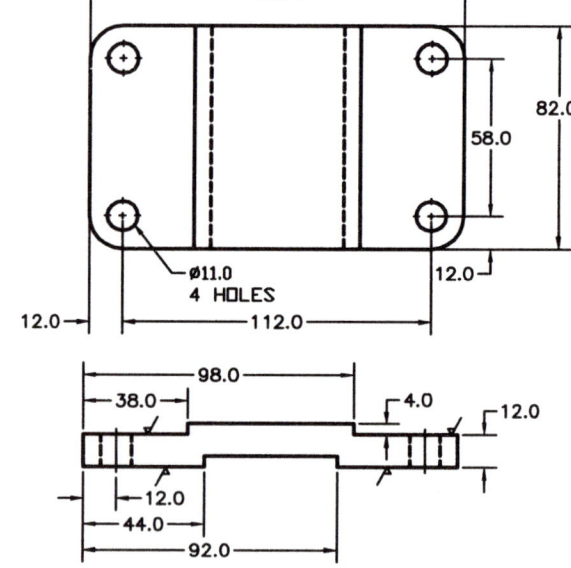

Top Plate

Exercise A6

Open FIRSTFL. Create new layers for Doors, Windows, Foundation, and Elevation. Use CHPROP to place the existing windows in the Windows layer. Use LINETYPE to load the linetype Hidden. In the LAYER pull-down menu, make the FOUNDATION layer current and draw in a foundation at 16″ around the first floor plan as shown. Change the LTSCALE to 48 to make the hidden lines visible.

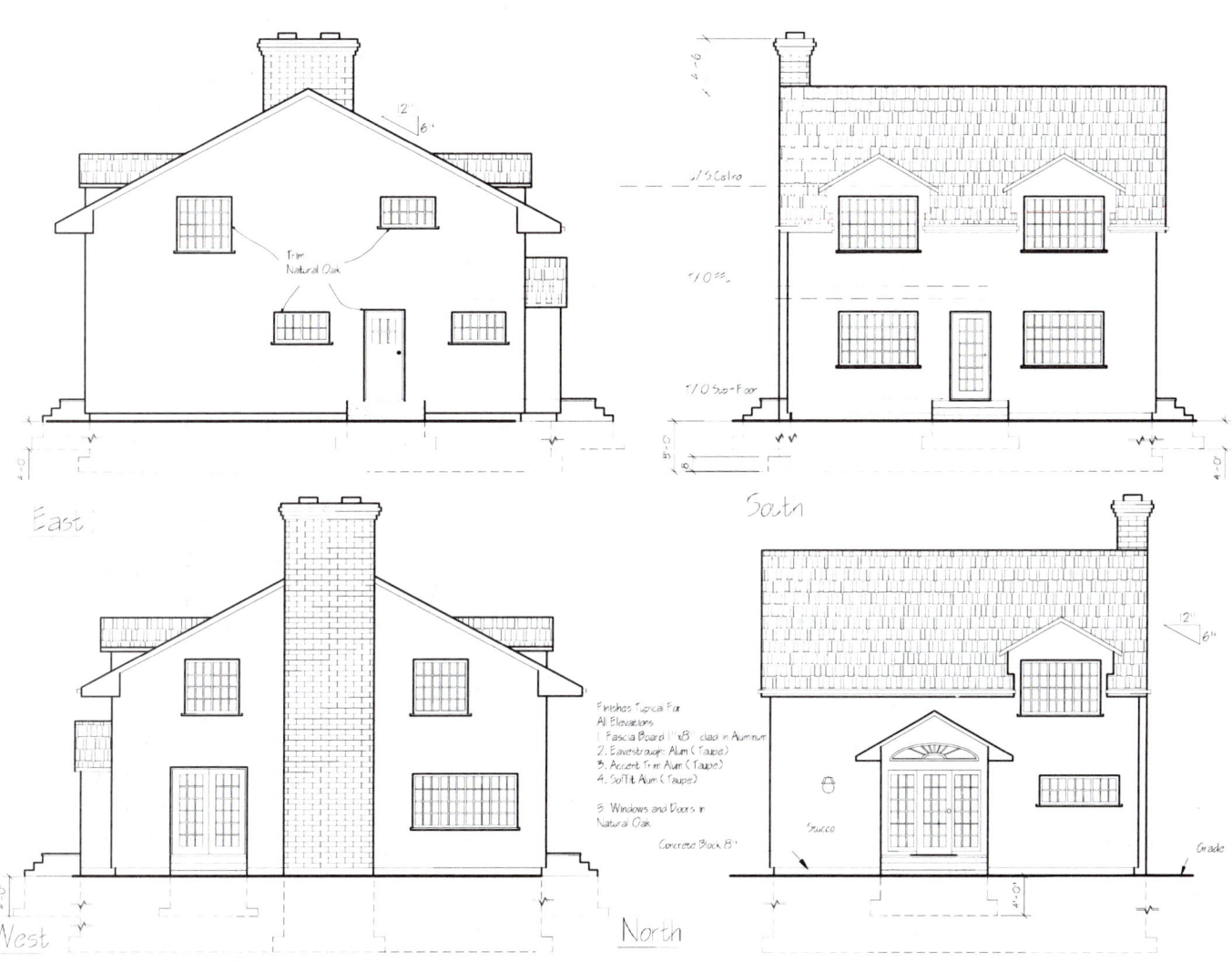

Entity Properties: Layers, Colors, and Linetypes

Exercise C6

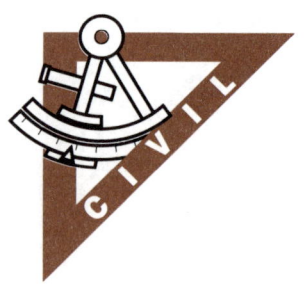

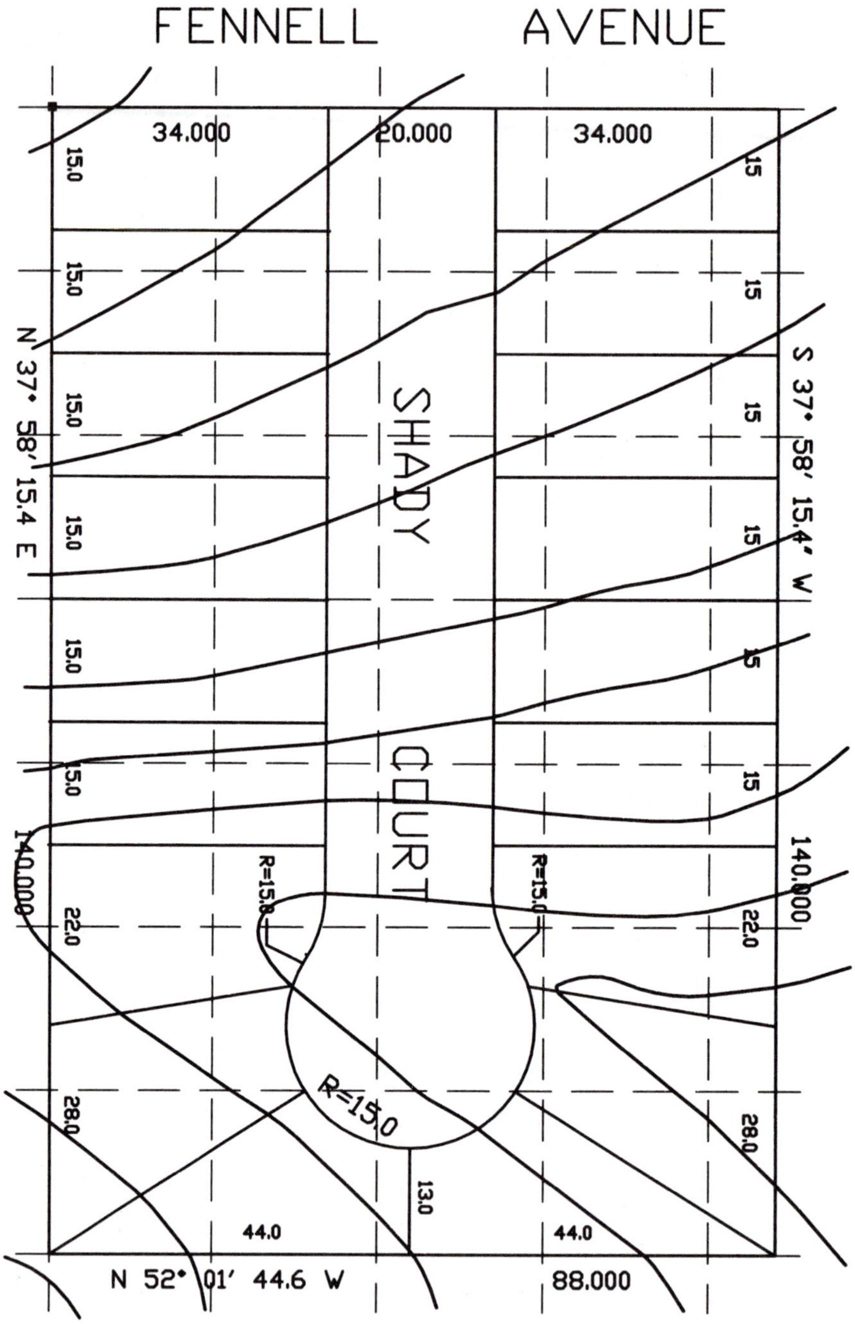

FENNELL AVENUE
34.000
20.000
34.000
SHADY COURT
N 37° 58' 15.4 E
S 37° 58' 15.4' W
140.000
140.000
R=15.0
R=15.0
R=15.0
13.0
44.0
44.0
N 52° 01' 44.6 W
88.000

Hints on C6

First create the geometry for this court. Change the units to surveying units.

Use OFFSET as much as possible. Do not forget the OSNAP commands.

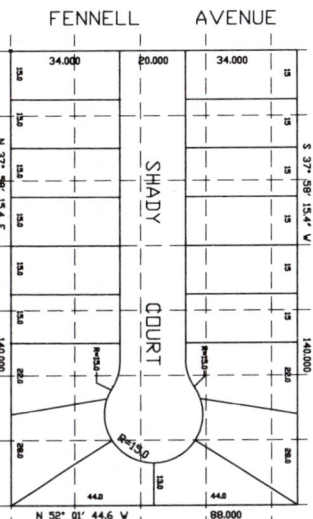

Now you can add the grid to the overall design. Add the linetype hidden under the LINETYPE command, then create a new layer for the grid. Use OFFSET to create the grid.

To generate the contours, use PLINE with PEDIT and the Spline function. You can also experiment with the OFFSET command on a spline in this case. Once it is in, it can be trimmed to the proper size.

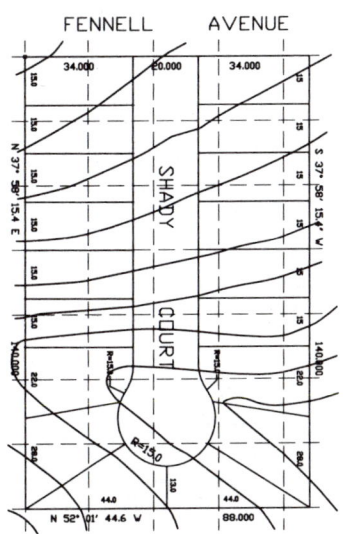

Going Further

Put each lot on a separate layer for use in a GIS package later.

Entity Properties: Layers, Colors, and Linetypes

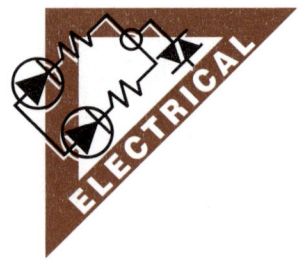

Exercise E6

This circuit diagram is created with LINE, CIRCLE, and PLINE. The text is placed by simply using DTEXT.

Create a new layer for the conductors. Put each conductor path on a different layer. Finally, enter the text.

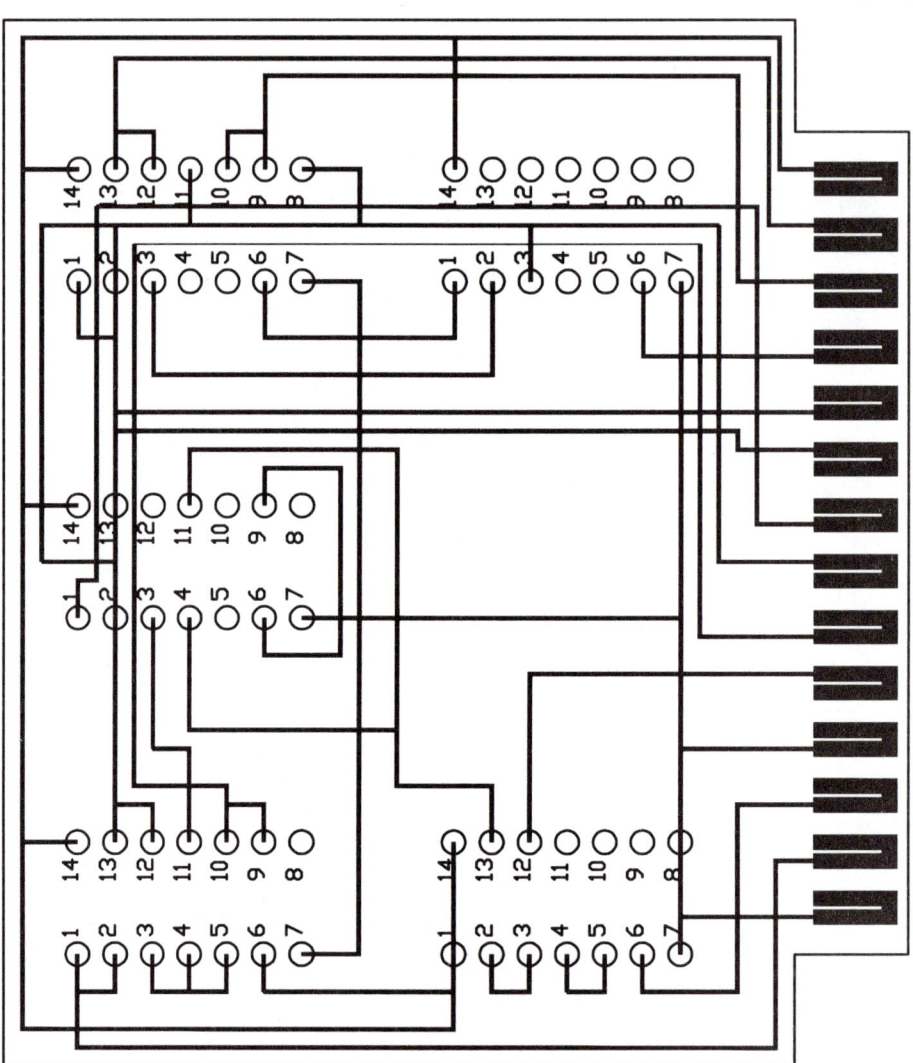

Hints on E6

To get the letters in exactly as you want them, create one letter, then use COPY with Multiple and SNAP to place the character in even rows. Use the CHANGE or DDEDIT commands to change the actual number once it has been placed.

Exercise M6

LIMITS 12,9
GRID .5
SNAP Various

Make different layers for each linetype and use different colors. The hatching is included as an aid for visualization.
Unless you are using LINE and ARRAY instead of HATCH, don't add it in at this point.

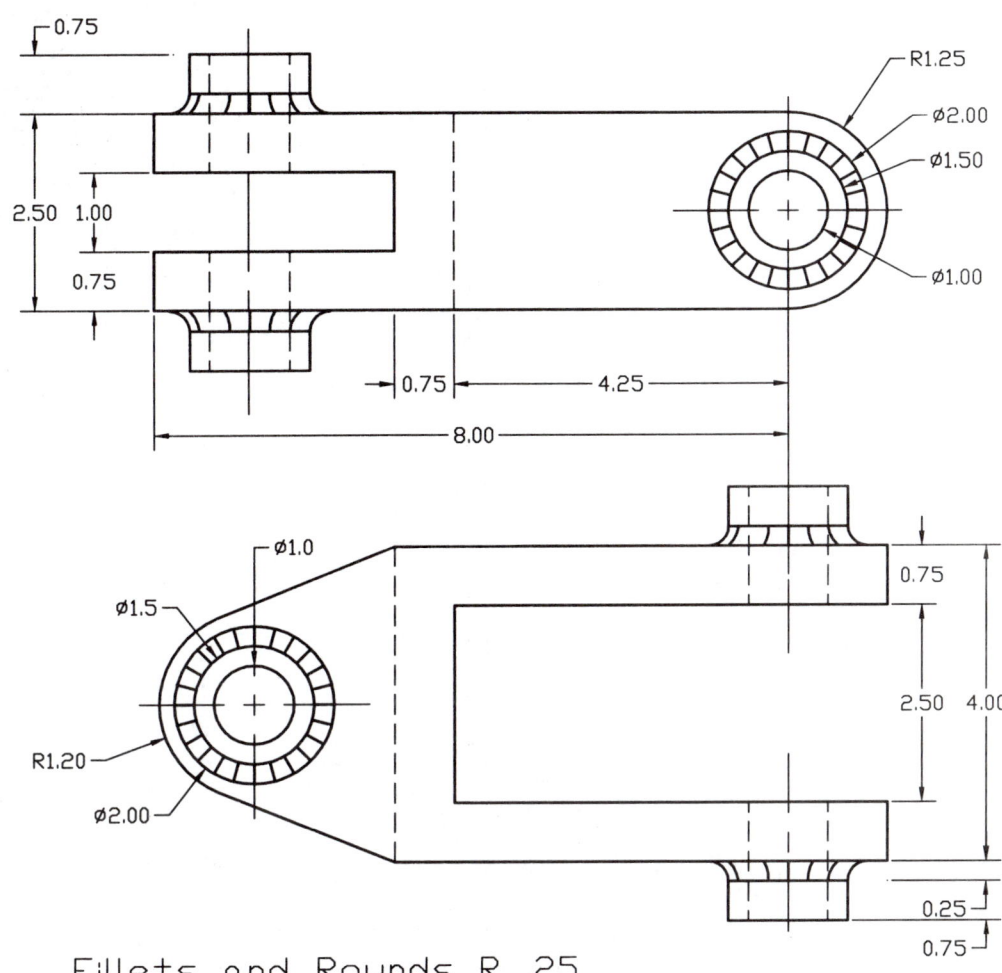

Fillets and Rounds R .25

Suggested Layer Schedule

Name	Color	Linetype
0	White	Continuous
One	Red	Continuous
Center	Blue	Center
Hidden	12	Hidden

Challenger 6

Use the editing commands to create the brickwork and block work on this gothic window. To create the hatch, freeze the layer that contains the brickwork and have just the boundary showing.

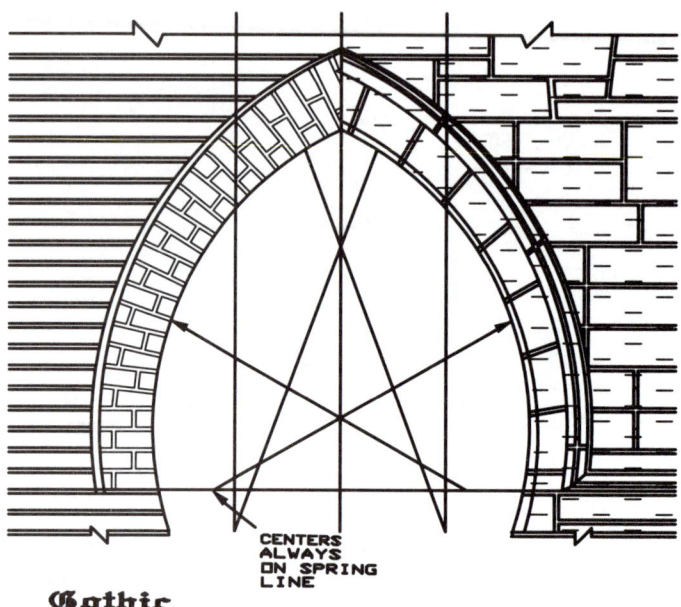

Gothic

NOTE : STONE JOINTS MAY BE HANDLED IN A VARIETY OF WAYS THIS IS ONE ILLUSTRATION

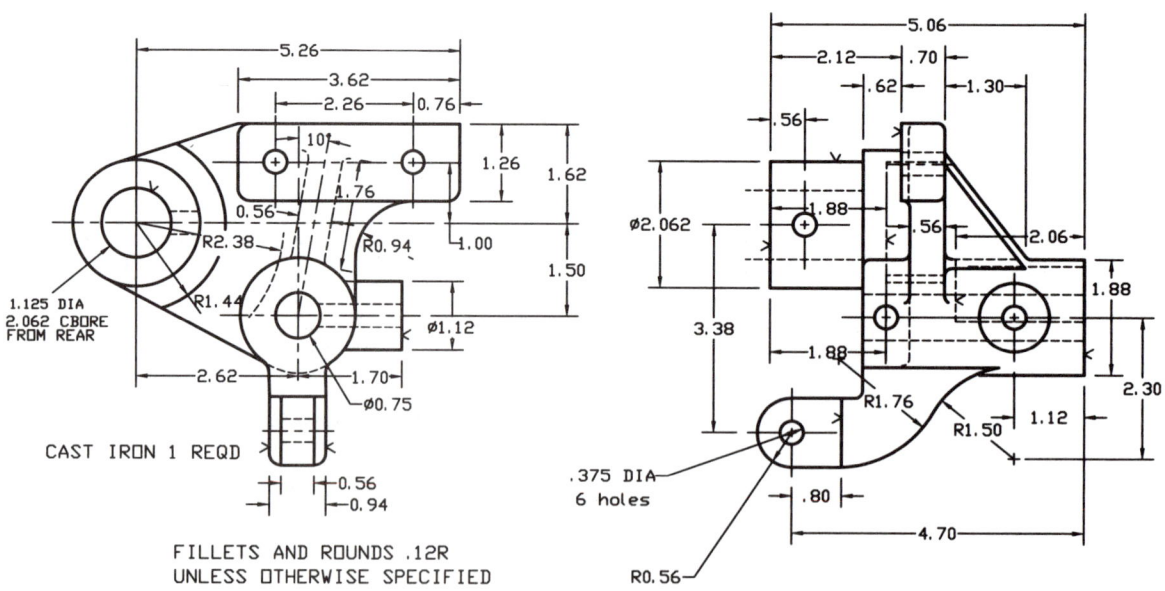

7 Dimensioning

OBJECTIVES

Upon completion of this chapter, you should be able to:

1. Set up a dimensioning style, using the DDIM dialog box
2. Add vertical, horizontal, and aligned dimensions
3. Add diameter and radius dimensions
4. Add baseline and continuous dimensions
5. Use DIMSTYLE, UPDate, and OVERRIDE to alter the dimensions

About Dimensioning

While the objects are being created, the size of the object is being programmed with the part's geometry. Lines, circles, arcs, etc. should be created perfectly every time. If you get into the habit of creating data in a lazy or slapdash manner, it will catch up with you later when dimensioning the drawing.

Dimensioning shows the measurements, the locations, and the angles of objects. The dimensioning commands are designed to extract the sizes that are already programmed with the part, and display them in accepted formats. Every discipline has a different set of drawing protocols. The dimensioning variables and dimension styles are used to set the dimensions to the required parameters.

AutoCAD offers a wide variety of ways to produce linear, baseline, radial, diameter, and angular dimensions. This illustration shows some of the basic dimension types.

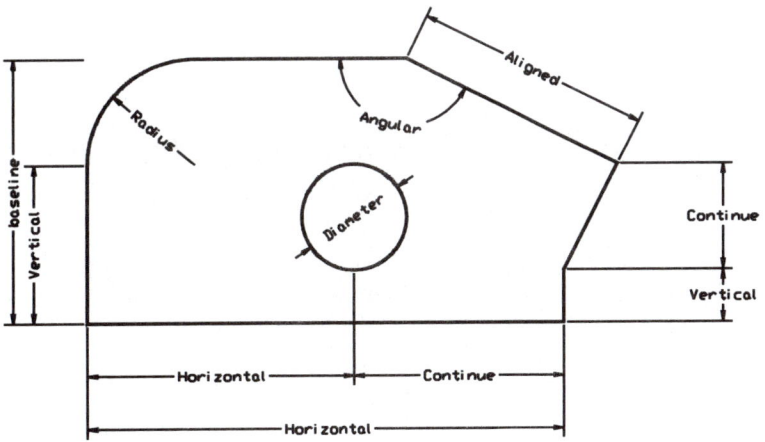

Figure 7-1

Dimensioning Components

Every dimension has several components. The ***dimension text*** states how big the object is. The ***dimension line*** holds the dimension text. The ***extension line*** extends from the object to the dimension line. The extension line is ***offset*** by .06 or 1/16 of an inch from the part itself and should extend .12 or 1/8 inch past the dimension line.

This illustration shows the components of a dimension.

The point at which you start your dimension is the reference point or **definition point**. AutoCAD automatically puts a gap between this point and the start of the extension line. AutoCAD also creates a layer for this point called *defpoints* that does not plot.

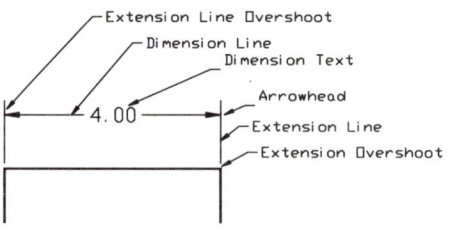

Figure 7-2

Dimension Mode

Prior to Release 13, you had to enter the dimension mode in order to create dimensions. This is no longer necessary. While you can use the command DIM to access the DIM mode commands, including UPDate, the dimensioning commands and variables are now available directly from the command line. As well, dimensions in Release 13 are run more from the pull-down menus and set up in dialog boxes than in earlier releases.

The dimension variables found in previous releases are still available through the DIM mode, but the Dimension style function has been developed to make dimensioning easier. To add dimensions and accept the default or standard dimension style, simply access the dimensions through Draw DIM on the screen menu, Dimensioning on the Draw pull-down menu, or dimensioning in Windows on the dimension toolbar.

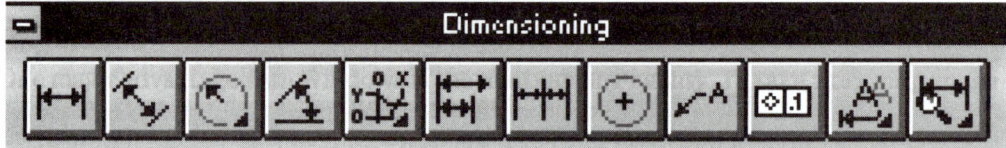

Figure 7-3

Accessing Linear Dimensions

To access the screen menus in DOS, pick Draw Dim.

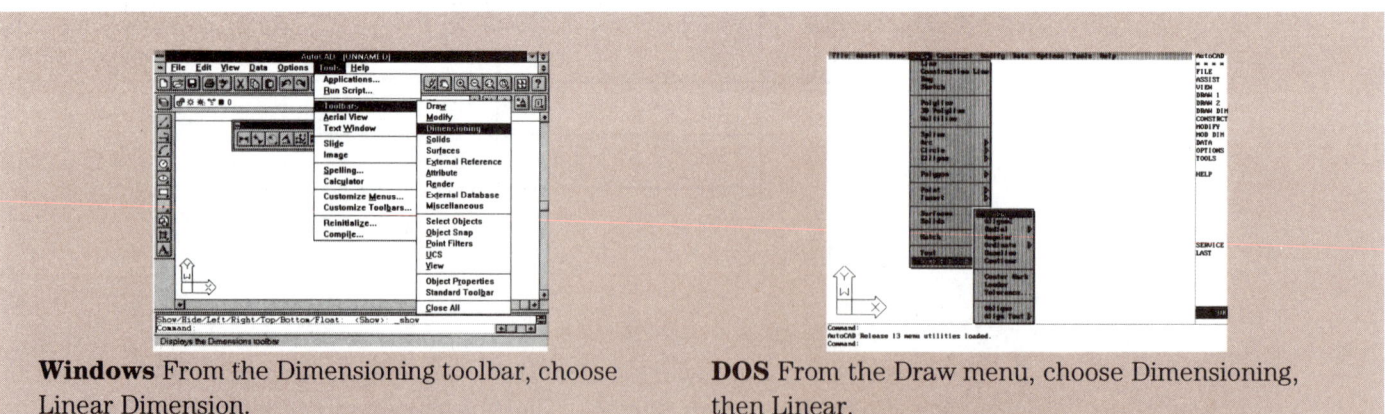

Windows From the Dimensioning toolbar, choose Linear Dimension.

DOS From the Draw menu, choose Dimensioning, then Linear.

Entering Dimensions

Horizontal and Vertical Dimensions

Like the Window and Crossing options, horizontal and vertical dimensions can be placed simply by using the LINEAR command, and they will show up at the desired spot depending on the extension line origins that you set or the point where you select an object. To override the horizontal or vertical default, choose **V** or **H**.

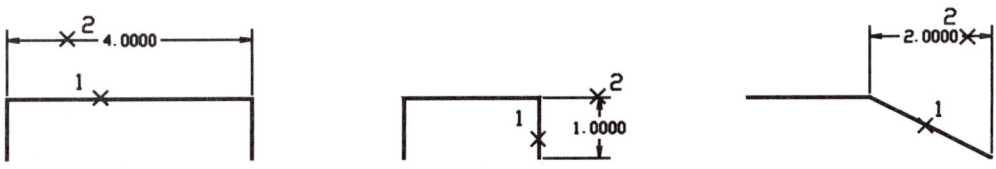

Figure 7-4

Once you have accessed the linear command, you will be prompted for the first extension line origin. Press ⏎ to select the first line rather than the two extension lines.

```
Command: (Draw pull-down menu, Dimension, Linear)
First extension line origin or RETURN to select:⏎
Select object to dimension: (pick 1)
Dimension line location<Text/Angle/Horizontal/Vertical/
   Rotated>: (pick 2)
```

In all three examples, the extension offshoot is shown because of the position of the adjoining lines. In the next examples, the first extension line is picked, then the second, and finally the placement of the dimension line is picked. In these, you would lose the extension line gap distance if you simply chose the object itself. Use SNAP and OSNAP for accuracy.

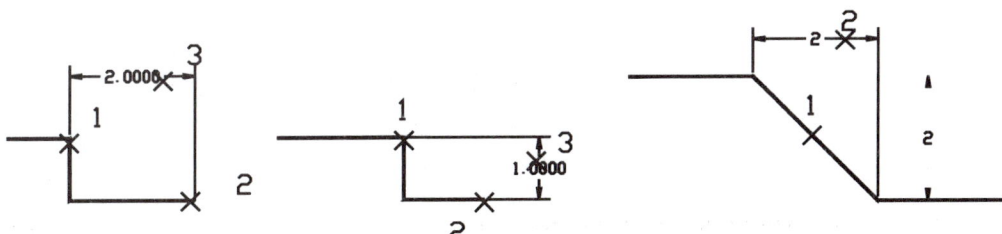

Figure 7-5

Before specifying the dimension line location, you are prompted to change the dimension text (Text), the text angle (Angle), the dimension direction (Horizontal or Vertical), or the dimension line angle (Rotated).

In the third illustration above, either the horizontal length or the vertical could be taken. Move your cursor to the position the dimension should be in, then pick the spot.

```
Command: (Draw pull-down menu, Dimension, Linear)
First extension line origin or RETURN to select:
Select object to dimension: (pick 1)
Dimension line location<Text/Angle/Horizontal/Vertical/
   Rotated>: (pick 2)
```
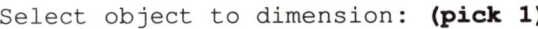

- To rotate the extension lines, type **R** for rotate.
- To override the horizontal or vertical default, type **H** or **V**.
- To override the text, type **T**.
- To rotate the text within the dimension line, type **A** for angle.

Aligned Dimensions

If the line is on an angle and you would like to rotate the dimension, use Aligned.

Windows From the Dimension toolbar, choose Aligned Dimension.

DOS From the Draw menu, choose Dimensioning, then choose Aligned.

```
Command: (Draw pull-down menu,
   Dimension, Aligned)
First extension line origin or
   RETURN to select: (pick 1)
Second extension line origin: (pick 2)
Dimension line location<Text/Angle>: (pick 3)
```
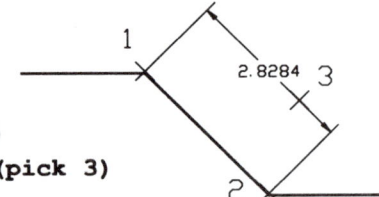

Figure 7-6

Like the vertical and horizontal, you can also simply choose the line itself to be dimensioned. To dimension circles, choose the circle and the end points of the diameter will be chosen for the end of the extension lines.

```
Command:_DIMALIGNED
First extension line origin or RETURN to select:
Select object to dimension: (pick 1)
Dimension line location<Text/Angle>: (pick 2)
```

To change the unit readout, change the dimension style (see page 171).

Continued Dimensions

Once you have either a horizontal, a vertical, or an aligned dimension, you can create baseline or continued dimensions.

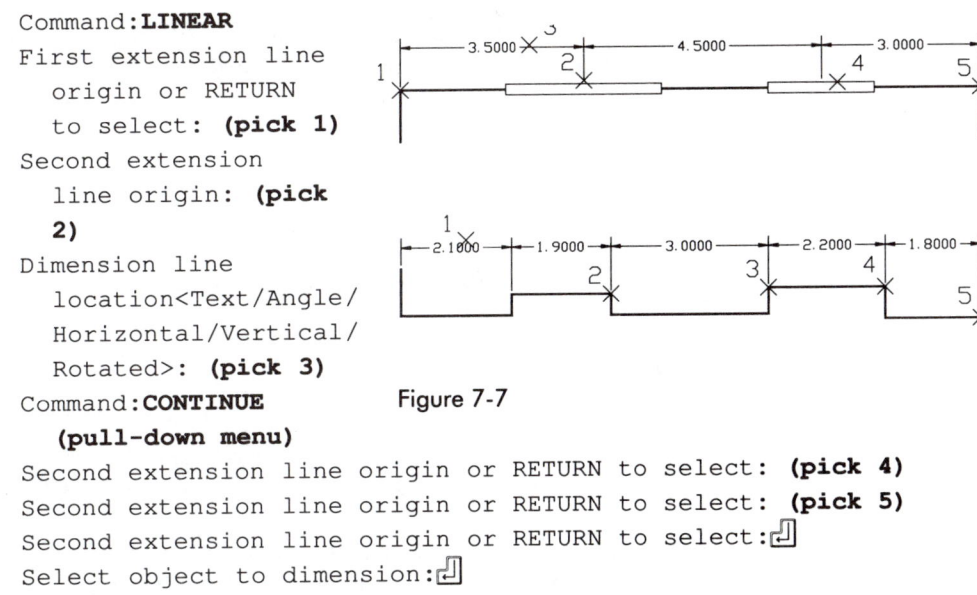

```
Command: LINEAR
First extension line
   origin or RETURN
   to select: (pick 1)
Second extension
   line origin: (pick
   2)
Dimension line
   location<Text/Angle/
   Horizontal/Vertical/
   Rotated>: (pick 3)
Command: CONTINUE
   (pull-down menu)
Second extension line origin or RETURN to select: (pick 4)
Second extension line origin or RETURN to select: (pick 5)
Second extension line origin or RETURN to select:⏎
Select object to dimension:⏎
```

Figure 7-7

Continued dimensions are multiple dimensions placed end to end. The first example continues from the dimension just put in. Choose a dimension that was entered earlier as the dimension object to have the successive dimensions line up with.

Notes

The distance between the dimension lines for both the continued dimensions and the baseline dimensions are set with the DDIM dialog box or the DIMension VARiables.

```
Command: _DIMCONTINUE
Second extension line origin or RETURN to select:⏎
Select object to dimension: (pick 1)
Second extension line origin or RETURN to select: (pick 2)
Second extension line origin or RETURN to select: (pick 3 and
   continue)
```

Baseline Dimensions

Similarly, for baseline dimensions, create the first dimension and continue from there.

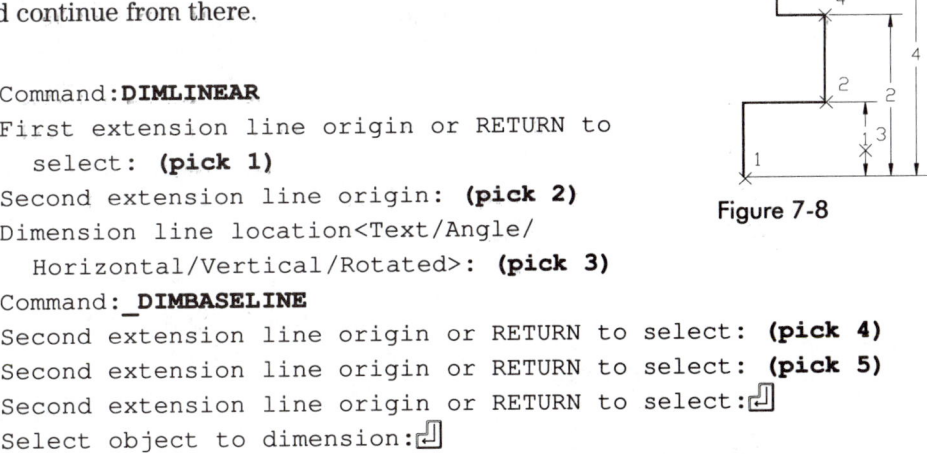

```
Command: DIMLINEAR
First extension line origin or RETURN to
   select: (pick 1)
Second extension line origin: (pick 2)
Dimension line location<Text/Angle/
   Horizontal/Vertical/Rotated>: (pick 3)
Command: _DIMBASELINE
Second extension line origin or RETURN to select: (pick 4)
Second extension line origin or RETURN to select: (pick 5)
Second extension line origin or RETURN to select:⏎
Select object to dimension:⏎
```

Figure 7-8

Dimensioning **169**

In both continued and baseline dimensions, if AutoCAD can't fit text and arrows between extension lines it will place the text on the side picked second. Use DIMTIX to override this.

Radial Dimensions

A ***radial dimension*** measures the diameter and radius of an arc or circle. The dimension will be placed according to the current dimension style or setup. Radial dimensions and diameter dimensions are very tricky.

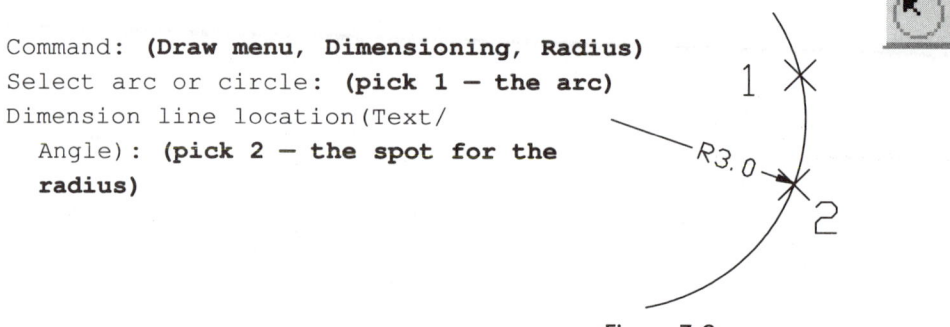

```
Command: (Draw menu, Dimensioning, Radius)
Select arc or circle: (pick 1 — the arc)
Dimension line location(Text/
   Angle): (pick 2 — the spot for the
   radius)
```

Figure 7-9

Dimradius can be used on both circles and arcs. Overriding text styles or Dimension Variables are often needed to place the text within the object where it belongs.

Use DIMTIX and DIMTOFL to place the text where needed, or to change the format and annotation options of the Dimension Styles dialog box.

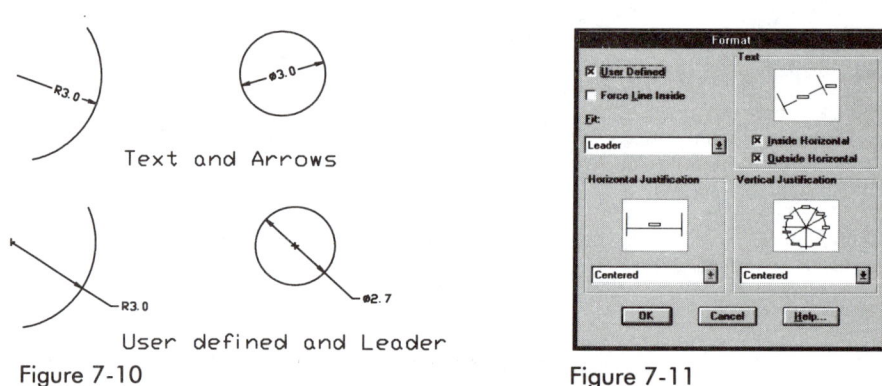

Figure 7-10

Figure 7-11

Diameter dimensions are affected by the same dimension variables as the radius dimensions. These are under the Format dialog box.

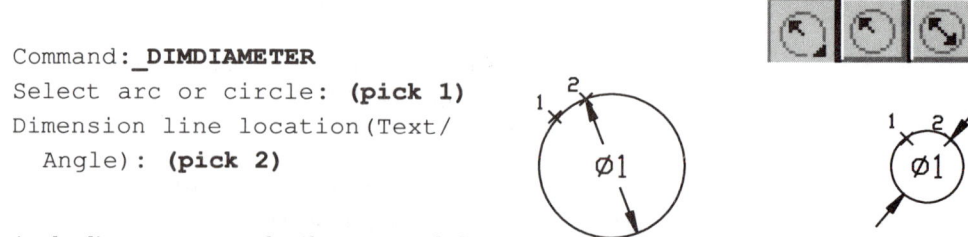

```
Command: _DIMDIAMETER
Select arc or circle: (pick 1)
Dimension line location(Text/
   Angle): (pick 2)
```

As in the linear commands, the text and the text angle can both be changed.

Figure 7-12

170 CHAPTER SEVEN

Angular Dimensions

The angular dimension command measures the angle between two nonparallel lines or three points. It can also measure the angle around a portion of a circle or the angle subtended by an arc.

The angle defined is determined by the placement of the dimension line.

```
Command: _DIMANGULAR
Select arc, circle, line, or
   RETURN: (pick 1)
Second line: (pick 2)
Dimension line location(Text/
   Angle): (pick 3)
```

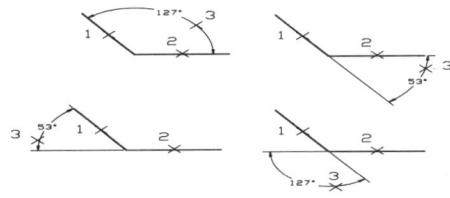

> **Notes**
> You can MIRROR a dimension to place it outside of the circle if you really get fed up.

The dimension line for the angular measurement is an arc that spans the measured angle and passes through the measured point.

Figure 7-13

Dimension Styles

Controlling Dimension Style

A ***dimension style*** is a named group of settings that determines the appearance of the dimension. Every dimension has an associated dimension style. If no style is applied before dimensioning, the Standard or default style is used. The style controls the unit readout, the text style, the color, the linetype scale, and many other factors.

In earlier releases of AutoCAD, the UNITS command controlled the dimension units; in Release 13 this command does not affect the units of the dimensions. They are determined by the dimension style.

Dimension styles are used to manage all variables. Once you have created a new style, it becomes the ***parent*** of a ***style family***.

To create a parent dimension style:

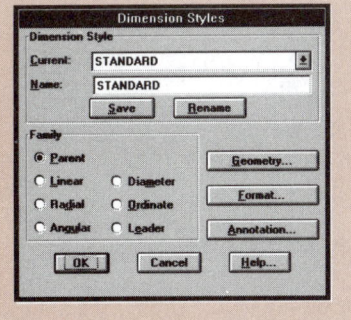

Windows From the Dimension style flyout on the dimensioning toolbar, choose Dimension Style.

DOS From the Data ... menu, choose Dimension Style.

The command line equivalent is **DDIM**.

```
Command: DDIM
```

Dimension styles are controlled through the DDIM dialog boxes. These take the place of many of the dimension variables.

Notes

Dimension variables can still be entered at the command prompt.

To create a new style, enter a style name in the Name box and pick Save. Changes that you make in the Geometry, Format, and Annotation boxes will be filed with the saved name.

The Geometry Dialog Box

The Geometry menu defines the appearance of the dimension line, extension lines, arrowheads and center marks for lines, and the scale of the dimension.

In the Dimension Styles dialog box, choose Geometry. This area helps to modify the dimension lines.

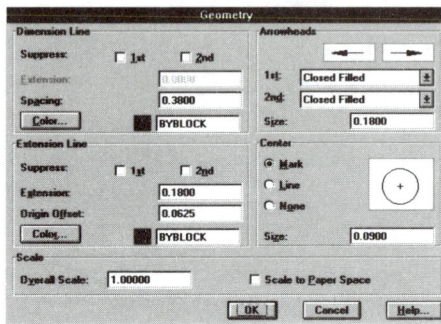

Figure 7-14

Scale or Dimscale

The overall size of the dimensions is determined within this submenu as well. If you change the Overall scale, you will change the arrowhead size, the extension line gap, the extension line overshoot, and the gap between the dimension and the dimension line. It is a much better idea to change the overall scale than to change each size individually.

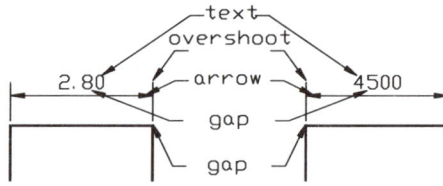

Figure 7-15

If you change the dimscale, all of these factors will change. If you change just the text size or just the arrowhead size, all of the other parameters will remain the same.

Paper space dimensions are dealt with in Chapter 11.

Arrowheads

First under Geometry are the arrowhead styles. The size of the Arrowhead is listed below. The styles can be changed by simply picking the current style, then choosing a new style from the available menu.

To enter a user-defined arrowhead style you must first define the objects, then block them using the BLOCK command. See Chapter 10.

```
Arrowheads
    None (DIMASZ 0)
←—2—→ Closed
●—2—● Dot
←—2—→ Closed Filled
╱—2—╱ Tick (DIMTSZ)
←—2—→ Open
○—2—○ Origin Indication
←—2—→ Right Angle
⋈—2—⋈ User Arrow (DIMBLK)
```
Figure 7-16

Center Marks

In order to change the center mark style, choose the appropriate style from the list.

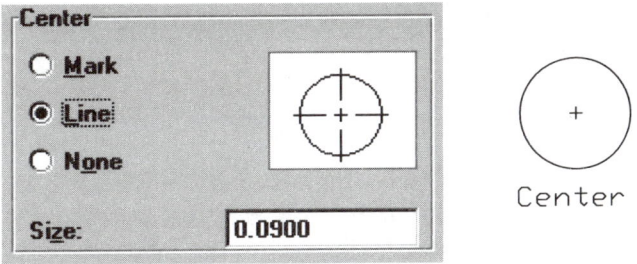

Figure 7-17

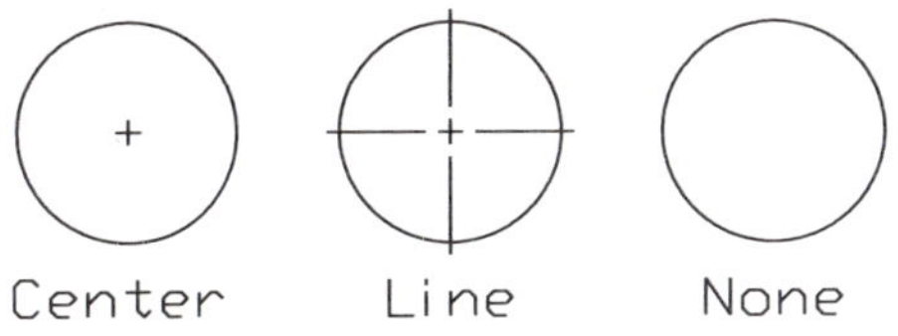

Figure 7-18

To add a center to an arc or circle, pick _DIMCENTER from the dimension list, then pick the circle. These are not real "center lines," and are placed on the current layer.

Suppressing Extension and Dimension Lines

In the geometry area, you can also cause either the extension line or the dimension line and arrowheads to be omitted. This is referred to as **suppressing** the extension or dimension lines. You can suppress both extension and dimension line in one command. Remember to return it to the previous setting.

To return to the main dimension style menu, choose OK. To cancel the changes that you have made, choose Cancel. To read the help files on this menu, choose Help.

Figure 7-19

Figure 7-20

You can change the color of the dimension lines and the text within this menu as well.

The Format Dialog Box

The Format dialog box will allow you to define the position of the dimension text, arrowheads, and leader lines relative to the dimension and extension lines.

To open the Format dialog box:

Windows From the Dimension style flyout, on the Dimensioning toolbar, choose Dimension style, then Format.

DOS From the Data menu, choose Dimension style, then Format.

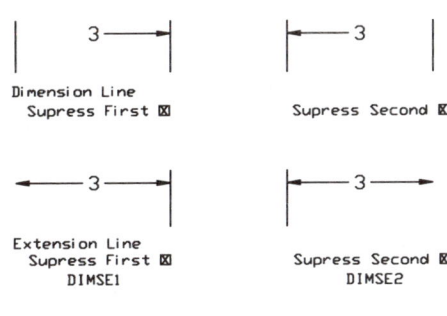

Dimensioning

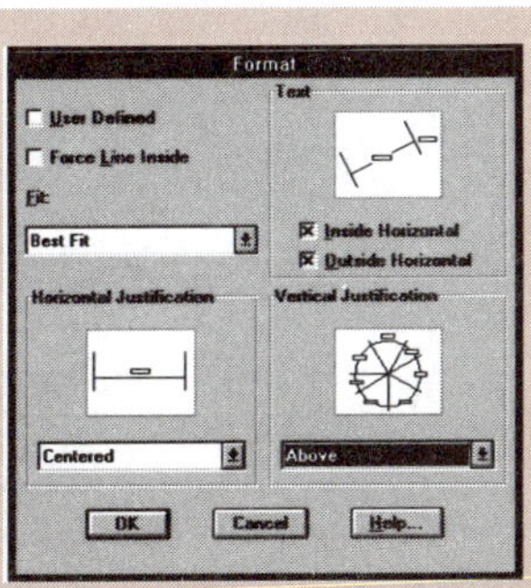

Best Fit

Many factors influence how arrowheads and dimension text fit within the extension lines. If possible, both text and arrowheads are accommodated between the extension lines no matter what fit option you choose, and AutoCAD will try to provide the "best fit." If you want to override the Best Fit, there are certain options.

If the best fit forces the text and arrowheads outside of the extension lines, you can have the line at least forced within the extension lines.

This was done in previous releases with the DIMTOFL DIM Var.

AutoCAD will follow the format set in the Format dialog box where possible, but once again this is determined by the sizes of the objects involved. If you set the text-fit options so that text only is meant to fit within the extensions, AutoCAD will include both text and arrows within the extension lines if there is room for both.

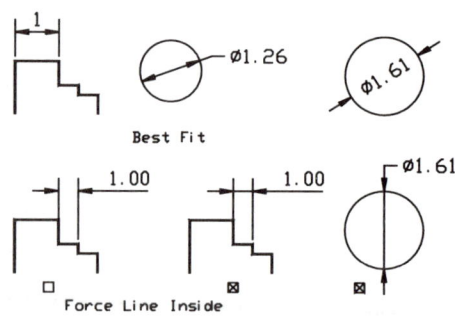

Figure 7-21

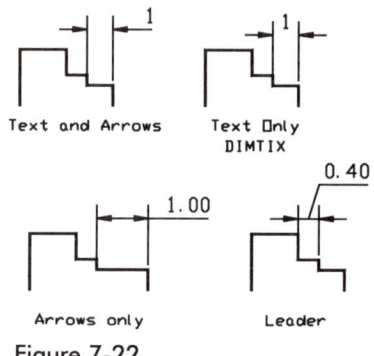

Figure 7-22

Usually the Arrows option is only used where there are more than three characters in the text string.

The Leader option places the text very much outside with a small leader pointing to it.

Text Orientation

Orientation of text is an important feature for dimensions. If you are doing civil or architectural drawings, the standard is to read vertical dimensions from the right with the text above the line as well.

Both the text inside and the text outside the extension lines can be changed. The results will follow both the dimension lines in vertical dimensions and in aligned dimensions.

The text is placed over the line with DIMTAD on or by using vertical justification.

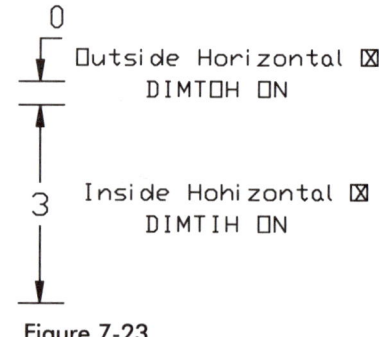

Figure 7-23

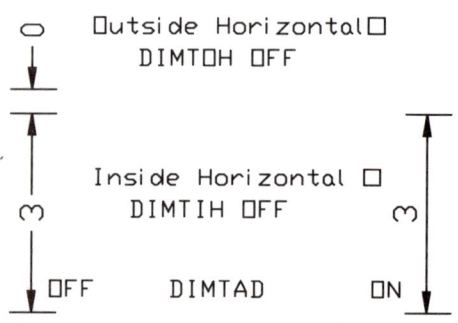

Figure 7-24

Vertical Justification

ANSI specifications generally have the text horizontal. There are a number of variables that can be used together to achieve the dimension required.

Both of the illustrations with circles show text not horizontal. The text can also be all outside the dimension lines.

The text can be fit above the dimension line or within the dimension line.

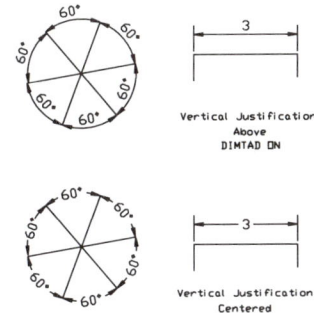

Figure 7-25

You will need to play around with the variables to see what combination suits your purpose.

Horizontal Justification

You can also change the horizontal justification of the text. Generally speaking, the text should be in the center of the dimension line. Use the options shown to change this.

If there is only one dimension that needs to be changed, you can use STRETCH to move it out of the way.

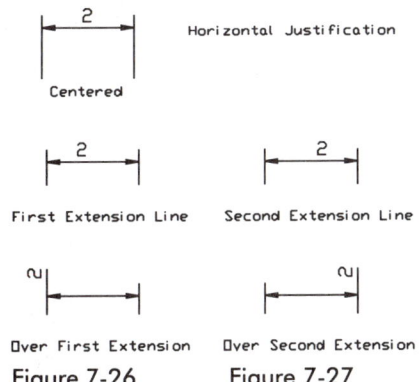

Figure 7-26   Figure 7-27

Annotation

The Annotation dialog box will allow you to control the dimension text with respect to the unit readout, alternate units, tolerances, text style, gap and color, and the rounding-off value.

To open the Annotation dialog box:

Windows From the Dimension style flyout on the Dimensioning toolbar, choose Dimension style, then Annotation.

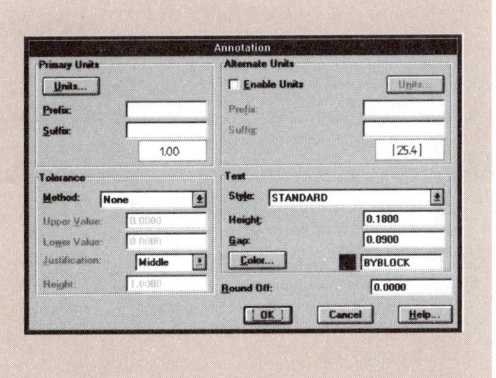

DOS From the Data menu, choose Dimension style, then Annotation.

In dimension text you must also consider prefixes and suffixes, alternate units, and lateral tolerances.

Dimensioning **175**

Alternate Units

Alternate units can be either imperial or metric, or they can represent another specific unit difference. The alternate unit will be shown in parentheses beside the primary unit readout. In the Annotation dialog box you can also control the degrees of accuracy of the alternate unit and the factor of the alternate unit.

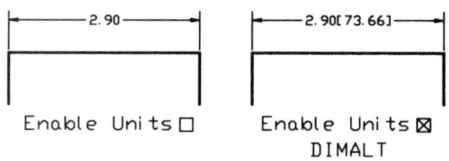

Figure 7-28

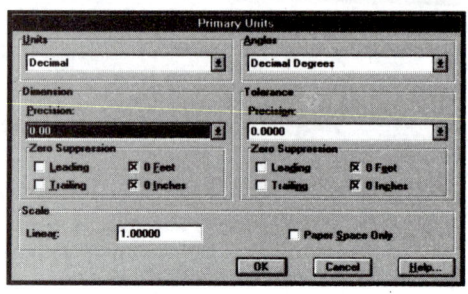

Figure 7-29

Primary Units

This will determine the main dimension values. With these units can be stored suffixes and prefixes such as diameter symbols, four holes, etc. As well, zero suppression is controlled and tolerances are set with this menu. Finally, you set the precision of the readout and the global measurement scale or a length scaling factor for dimensions created in paper space.

Select a unit type from the list for all dimension types except angular. For angular dimensions, select an angle type from the Angles list.

Under Dimension precision, you choose the precision value for the primary units.

Under Tolerance, select the precision value for the tolerance values.

Suppressing Zeros in Primary Units

You can suppress both leading and trailing zeros. With suppression of the leading zeros, 0.0400 becomes .0400. With suppression of the trailing zeros, 0.0400 becomes 0.04. If you suppress both, 0.0400 becomes .04. This option replaces DIMZIN.

Suppressing zeros is most important in dimensions using feet and inches. Here are some examples of how the options work.

Option	Effect	Examples		
0 feet and 0 inches	Suppresses zero feet and zero inches	1/2"	6"	1'-0 3/4"
No options	Includes zero feet and zero inches	0'-0 1/2"	0'-6"	1'-0 3/4"
0 inches	Suppresses zero inches (includes zero feet)	0'-0 1/2"	0'-6"	1'-0 3/4"
0 feet	Suppresses zero feet (includes zero inches)	1/2"	6"	1'-0 3/4"

Tolerances

In the manufacturing sector, tolerances are often used to control the degree of accuracy required for certain applications. The tolerances show the largest and smallest permissible deviation on the manufacturing.

To add tolerances, choose the tolerance area of the annotation menu and choose the method required. Upper and lower tolerances can be specified.

The types of tolerances can also be specified. In the illustration here, the first dimension is created with symmetrical or bilateral tolerancing and the second is created with the Limits option. Again, there are many variables that can be set in the tolerancing area to create the desired dimension style.

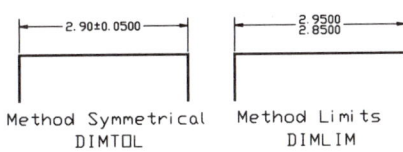

Figure 7-30

Dimension Text Style

In the Dimension text style area, you can modify both the font and the size of the text. Keep in mind that if you change the text size this will change *only* the size of the text; it will not change the size of the arrowheads or any of the other variables changed in the Scale option.

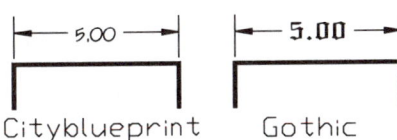

Figure 7-31

Before you select a style, it must be loaded using the STYLE command. See Chapter 8.

In choosing a text style for dimensions, keep in mind that certain fonts, such as Standard and Cityblueprint, are more readable than others, such as Gothic. The more intricate the font, the more difficult it will be to read and the more space it will take up on disk.

Rounding Off Dimensions

You can control the precision of dimensions by using the Round off area of the annotation menu. All dimension types can be rounded off, except for angular dimensions. The number of digits that follow the decimal point can be set under the Primary Units area or under Alternate units and Tolerancing units. The rounding-off value will be within this decimal value.

This option was performed in previous releases with the DIMRND DIM Var.

Using Dimension Style Families

Once you have made all of the changes necessary for a dimensioning style, you must save it as a particular style. Change the name in the Dimension Styles dialog box. This will be your parent style. Each style contains a font, just as layers contain a color and a linetype. Changes to the style alter all text within that style.

If there are certain options that you would like to change for some dimensions, you can make the changes and save them under the parent style. An example would be tolerancing on all of the radius or diameter dimensions but not on any of the linear dimensions. In this case, create a *family member style* that will allow tolerancing only on these dimension types.

To Create a Family Member Style

Open the Dimension Styles dialog box.

> **Windows** From the Dimension Style flyout on the Dimensioning Toolbar, choose Dimension Style.
>
> **DOS** and Windows From the Data menu, choose Dimension Style.

Under Family, select the Family button.

From the Current style list, select the parent style for which you want to create a family member style.

Make the required modifications to the parent style. Then select Save, and choose OK.

Overriding Styles

DIMOVERRIDE overrides dimensioning system variable settings associated with a dimension object, but doesn't affect the current dimension style.

Some of the dimensioning variables, such as extension line suppression and forcing text within extension lines, apply to only certain dimensions on the drawing. In a case like this, a dimension override would be appropriate.

An override is equivalent to changing a certain dimension variable without changing the overall style of the dimensions. To override existing dimensions, you can use the DIMOVERRIDE command to change only dimensions that you select.

```
Command:DIMOVERRIDE
Current value<Current>New value: (enter a value or press ⏎)
Dimension variable to override: (enter a name or press ⏎)
```

If you enter a dimensioning system variable name, AutoCAD redisplays the "Current value, New value:" prompt. If you press ⏎, AutoCAD prompts:

```
Select objects:
```

AutoCAD applies the overrides to the selected dimension objects.

UPDATING Existing Dimensions

There are three basic categories of ways to update your dimensions.

Dim:UPDate   will update the selected dimensions relative to the changes that you have made either in the dimension variables or on the Dimension Styles dialog box.

Command:DIMSTYLE Apply   will do the same.

DIMOVERRIDE   will let you update the selected dimensions or clear all of the dimension variables that are not the Standard dimensions on the selected dimensions.

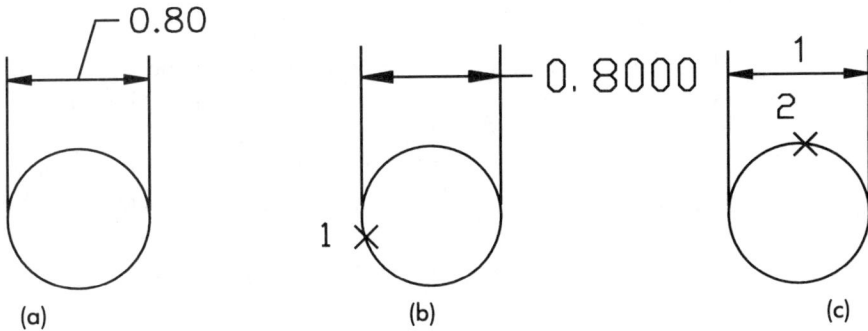

Figure 7-32

The aligned dimension on the left reflects some changes in the Dimension Styles dialog box. In Figure 7-32(b), DIMOVERRIDE was used; in Figure 7-32(c), DIMSTYLE Apply was used.

```
Command:DIMOVERRIDE
Dimension variable to override (or Clear to remove overrides):C
Select objects: (pick 1)
```

This has cleared all of the dimension variables that are currently set.

```
Command:DIMSTYLE
Dimension style:STANDARD
Dimension style overrides:
     DIMCLR 160
     DIMTAD 1
     DIMTIH 0       (these will only be listed if you
     DIMDEC 0          have used them as overrides)
     DIMSCALE 1
     DIMTIX 1
     DIMRND 1
Dimensions Style Edit (Save/Restore/STatus/Variables/Apply/?)
  <Restore>:A
Select objects: (pick 2)
Select objects:↵
```

Choosing the Variables option, then selecting a dimension, will list all the DIM variable settings for that dimension.

The Apply option of the DIMSTYLE command works like an update in overriding the current style on the dimensions you select.

```
Command:DIM
DIM:UPD
Select objects: (pick 2)
Select objects:↵
```

This should get you started on the way to good dimensions.

Dimensioning **179**

Editing Dimensions

Text

If you pick Properties from the Modify pull-down menu and select a dimension, you are given the Modify Dimension dialog box. Picking the Edit ... box puts you in the DOS editor or Windows notebook to change the dimension text. You can do this easier by typing **DIM**↵, then **NEW**. From the screen menu, choose Mod DIM, then DimEdit, then New. You will be prompted for the new text.

The DIMTEDIT Command

Moves and rotates dimension text.

Windows From the Dimensioning toolbar, choose the Align Dimension Text flyout.

DOS From the Draw menu, choose Dimensioning, then Align Text.

The command line equivalent is **DIMTEDIT**.

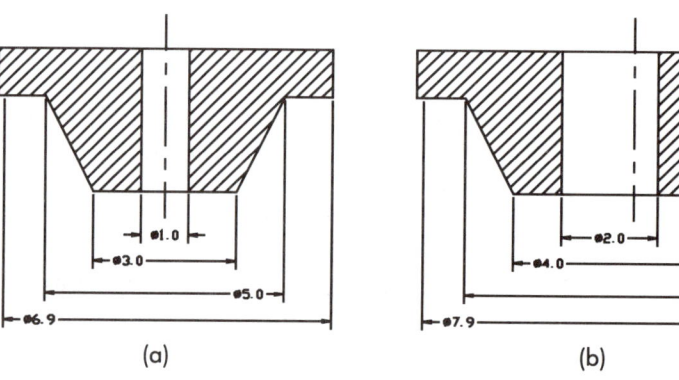

Figure 7-33

DIMTEDIT edits the position of the text on the dimension line.

Under the Annotation dialog box is the heading Prefix. You can add the prefix **%%c** for the diameter sign to add diameter signs to all of the dimensions shown.

Stretch

You can also stretch the object and have the dimensions automatically updated as in Figure 7-33(b). The STRETCH command can be used in the same way that DIMEDIT is used to move the text across the dimension line.

EXTEND and TRIM can also be used to have the dimensions show up exactly correctly.

When All Else Fails

Dimensioning can be quite annoying at first. If you have changed a series of dimension variables, and/or a series of options in the DDIM dialog box, you may find it just too cumbersome to try to figure out what is happening.

In this case, save the file, and open a new one. Use the INSERT command to insert the file into a fresh drawing environment. Use EXPLODE to revert the block into individual parts, and start again with your dimensioning variables.

See Chapter 10 for more details.

Prelab 7 Dimensioning

Step 1 Quickly draw the part as shown. The dimensions are shown for your convenience in this illustration, but do not try to add them yet. It will help with your placement later if the center of the lower left arc is drawn on a SNAP point. SNAP could be set to .25.

Notes

Systems may have different default factors. If you are *not* getting what is shown, just continue.

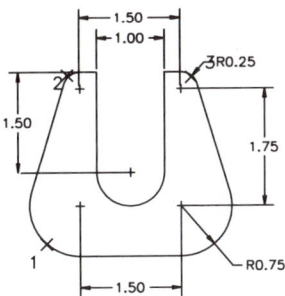

Step 2 Once the part is drawn in, let's take a look at what a dimension would look like using the standard dimensioning format.

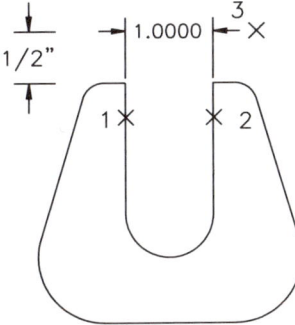

> **Windows** From the Dimensioning toolbar, choose Linear Dimension.
>
> **DOS** From the Draw menu, choose Dimensioning, then Linear.

```
Command:DIMLINEAR
First extension line origin or RETURN to select:END of (pick 1)
Second extension line origin:END of (pick 2)
Dimension line location<Text/Angle/Horizontal/Vertical/
   Rotated>: (pick 3)
```

The dimension line should be 1/2 inch away from the object line. With SNAP set at .25, this should be easy to place.

Step 3 Four decimal points of accuracy are not needed, and the scale of the dimensions is larger than needed. The next step is to make a dimension style that will incorporate the changes. First, access the Annotation area of the Dimension Style dialog box to change the precision.

> **Windows** From the Dimension style flyout on the Dimensioning toolbar, choose Dimension Style, then Annotation.
>
> **DOS** From the Data menu, choose Dimension Style, then Annotation.

Dimensioning   **181**

From the Annotation dialog box, pick Units under Primary Units, then 0.0000 under Precision. Pick the 0.00 option.

To exit the dialog box, pick OK three times in order to access the Dimension Styles dialog box. Now change the overall scale under the Geometry dialog box.

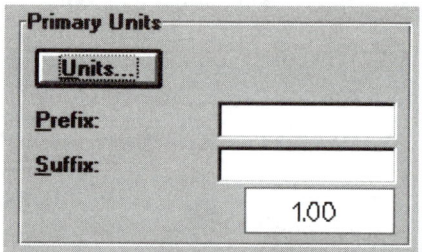

Pick .75 for Overall scale. Pick OK to return to the Dimension Style Dialog box.

Now name the dimension style that you have chosen so that the changes will be recorded and can be used on the dimensions made.

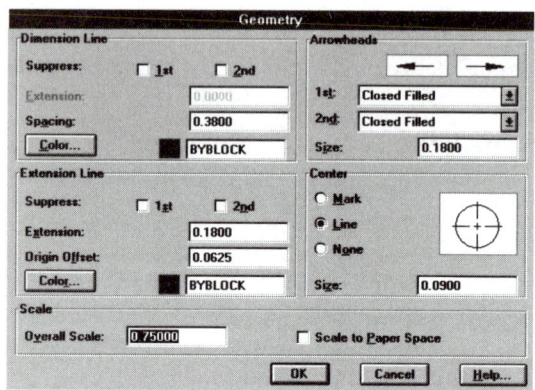
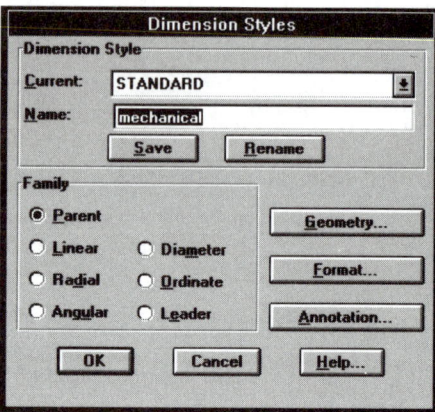

Beside "Name:", change the name to Mechanical, then Save the new dimension style. Finally choose OK to exit the menu.

Step 4 The dimension now needs to be updated to the new style.

At the command prompt, type in **DIMSTYLE**.

```
Command:DIMSTYLE
Dimensions Style Edit (Save/Restore/
   STatus/Variables/Apply/?)<Restore>:A
Select objects: (pick 1)
Select objects:
```

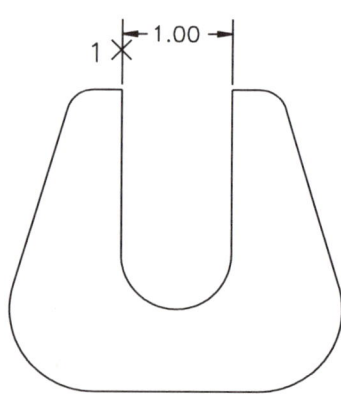

The dimension should update to the current style.

Step 5 Now that the style has been set, add the linear dimensions. First create the horizontal dimension between the centers of the two arcs on the top, then create the vertical dimension on the right.

Windows From the Dimensioning toolbar, choose Linear Dimension.

DOS From the Draw menu, choose Dimensioning, then Linear.

```
Command:DIMLINEAR
First extension line origin or
   RETURN to select:CEN of (pick 1)
Second extension line origin:CEN
   of (pick 2)
Dimension line
   location<Text/Angle/Horizontal/
   Vertical/Rotated>:(pick 3)
Command:⏎
First extension line origin or
   RETURN to select:CEN of (pick 4)
Second extension line origin:CEN
   of (pick 5)
Dimension line location<Text/Angle/Horizontal/Vertical/
   Rotated>: (pick 6)
```

Add the other vertical and horizontal dimensions in the same manner.

```
Command:DIMLINEAR
First extension line origin or
   RETURN to select:END of (pick 1)
Second extension line origin:CEN
   of (pick 2)
Dimension line location<Text/Angle/
   Horizontal/Vertical/Rotated>:
   (pick 3)
Command:⏎
First extension line origin or
   RETURN to select:CEN of (pick 4)
Second extension line origin:CEN
   of (pick 5)
Dimension line
   location<Text/Angle/Horizontal/Vertical/Rotated>: (pick 6)
```

Step 6   Now add the radial dimensions.

> **Windows** From the Dimension toolbar, choose Radial, then Radius.
>
> **DOS** From the Draw menu, choose Dimensioning, then choose Radial, then Radius.

```
Command:_DIMRADIUS
Select arc or circle: (pick 1)
Dimension line location (Text/Angle): (pick 2)
Command:_DIMRADIUS
Select arc or circle: (pick 3)
Dimension line location (Text/Angle): (pick 4)
```

Dimensioning

Notes

Again, some systems may have defaults set to avoid this problem. If your radius was entered correctly the first time, continue to Step 9.

Notice that the upper radius looks fine, because the leader was forced outside of the object lines. The lower radius looks messy, however, because the dimension has been forced inside the lines. We will erase the radius, then change one of the options on the dimension style and simply override the style for that one dimension.

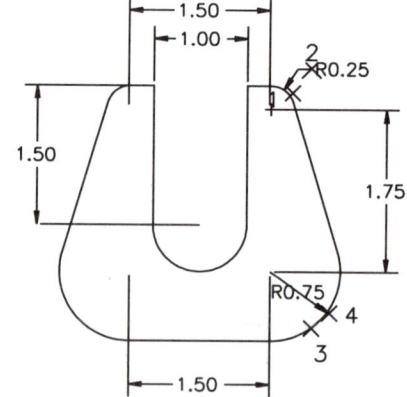

```
Command: E
Select objects: (pick the radius
   dimension at the lower right)
```

Step 7 Change the Dimension Style dialog box.

> **Windows** From the Dimension style flyout on the Dimensioning toolbar, choose Dimension Style, then Format.
>
> **DOS** From the Data menu, choose Dimension Style, then Format.

From the Format dialog box, pick User Defined, then Leader.

To exit the dialog box, pick OK to return to the Dimension Styles dialog box.

Exit the Dimension Style Dialog box without saving the style.

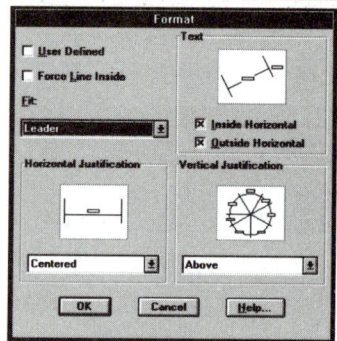

Step 8 Now add the radius dimension in place of the one that was erased. Notice that the user-defined style will allow you to enter the dimension on the outside of the part.

```
Command: _DIMRADIUS
Select arc or circle: (pick 3)
Dimension line location (Text/
   Angle): (pick 4)
```

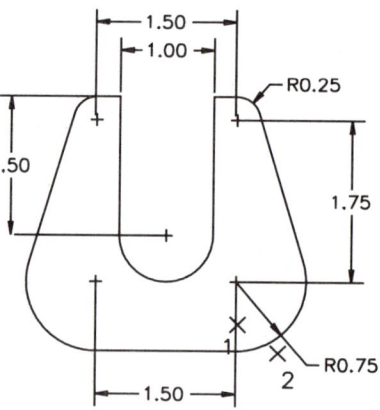

Once this is placed, return to the Format dialog box and remove the cross from the User Designed box.

Step 9 The dimensions for this part are now in place, but center marks should be added to all the center points. Use the Center Mark command to add the center marks to the arcs that don't have them.

> **Windows** From the Dimension toolbar, choose Center Mark.
>
>
>
> **DOS** From the Draw menu, choose Dimensioning, then choose Center Mark.

184 CHAPTER SEVEN

```
Command: _DIMCENTER
Select arc or circle: (pick 1, 2, 3)
```

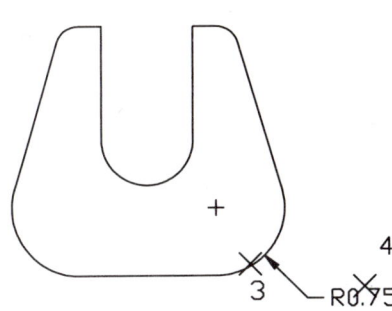

Step 10   The dimensions added are associative dimensions. This means that, if you have entered all the dimensions and then find that they are the wrong scale or the wrong units, that the arrowheads should be ticks, or some other problem, you can edit the dimensions singly or in groups with the aid of the UPDATE command.

The dimensions are entered relative to the points that you have identified. These are called defpoints, and there is a special layer created for them under your layer menu. This layer does not plot, so be careful not to make it current.

This means that you can also STRETCH the part and the dimensions will automatically update. Use STRETCH to edit the part and see if the dimensions update automatically.

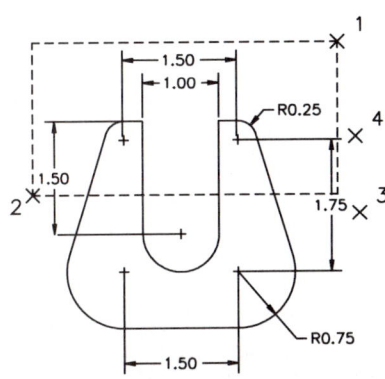

Windows From the Resize flyout on the Modify toolbar, choose STRETCH.

DOS From the Modify menu, choose STRETCH.

```
Command: STRETCH
Select objects to stretch by window ...
Select objects: C
First corner: (pick 1)
Other corner: (pick 2)
Select objects: ↵
Base point: (pick 3)
New point: @1<90 (or pick 4)
```

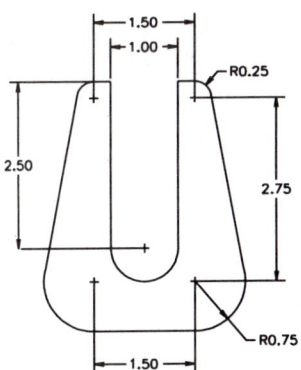

Your objects should stretch and the vertical dimension should change.

Dimensioning   **185**

Step 11 Finally, what if the dimensions should really be in fractional rather than decimal? You can change the dimension style and apply it to the current drawing.

First, save the drawing so that you have two copies, one with fractions and one with decimals.

From the File menu, choose SAVE_AS. Name the file PRELAB7.

The next step is to make a new dimension style that incorporates the necessary changes. First, access the Annotation area of the Dimension Style dialog box to change the dimension style.

> **Windows** From the Dimension style flyout on the Dimensioning toolbar, choose Dimension Style, then Annotation.
>
> **DOS** From the Data menu, choose Dimension Style, then Annotation.

From the Annotation dialog box, pick Units under Primary units, then the word Decimal. Once this is chosen, you will have a list of possible unit types. Pick Fractional.

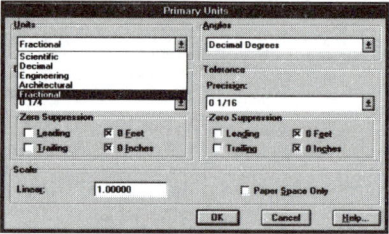

To exit the dialog box, pick OK until you are at the Dimension Styles dialog box. You can also change the overall scale under the Geometry dialog box.

Name the dimension style that you have created so that the changes will be recorded and can be used to update the current dimensions. Beside "Name:" change the name to Fractional, then Save. Finally choose OK to exit the menu.

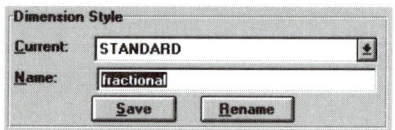

```
Command:DIMSTYLE
Dimensions Style Edit (Save/Restore/STatus/Variables/Apply/?)
  <Restore>:A
Select objects: (pick 1)
Select objects:↵
```

The dimension should update to the current style.

The final drawing should look like this.

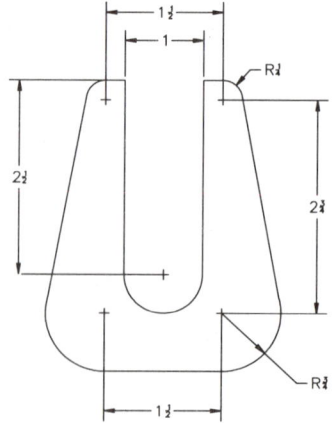

Command and Function Summary

Dimension Style dialog box allows the user to create a dimension style that will be used on all subsequent dimensions.

Geometry dialog box allows the user to control the appearance of the dimension line, extension lines, arrowheads and center marks for lines, and the scale of the dimension.

Scale or **Dimscale** controls the overall size of the dimensions.

Arrowheads controls the size and style of the arrowheads.

Center Marks controls the size and style of the center marks.

Extension and **Dimension Lines** controls the display of the relevant lines.

Format dialog box allows you to define the position of the dimension text, arrowheads, and leader lines relative to the dimension and extension lines.

Best Fit controls the placement of the text relative to the arrowheads and extension lines.

Text Orientation controls both the text inside and the text outside the extension lines.

Vertical Justification fits the dimension text above the dimension line or within the dimension line.

Horizontal Justification controls the horizontal justification of the text.

Annotation dialog box allows the user to control the unit readout, alternate units, tolerances, text style, gap and color, and the rounding-off value.

Alternate Units controls alternate units.

Primary Units determines the main dimension values.

Suppressing Zeros controls the number of zeros in both the primary units and the alternate units readouts.

Tolerances allows and controls reading of tolerances.

Dimension Text Style controls the style of the text.

Rounding off allows you to round the dimensions off to the desired unit readout.

DIMANGULAR allows you to add aligned dimensions.

DIMBASELINE allows you to add baseline dimensions.

DIMCONTINUE allows you to add continued dimensions.

DIMDIAMETER allows you to add diameter dimensions.

DIMLINEAR allows you to enter dimensions either vertically or horizontally.

DIMRADIUS allows you to add radial dimensions.

DIMSTYLE allows you to edit or update the current style.

UPD in the DIM mode allows you to update the dimensions to the current style.

Practice Exercise 7

The upper section is a roof assembly for fire-rating. The lower are simple mechanical parts that should help with your dimensioning skills.

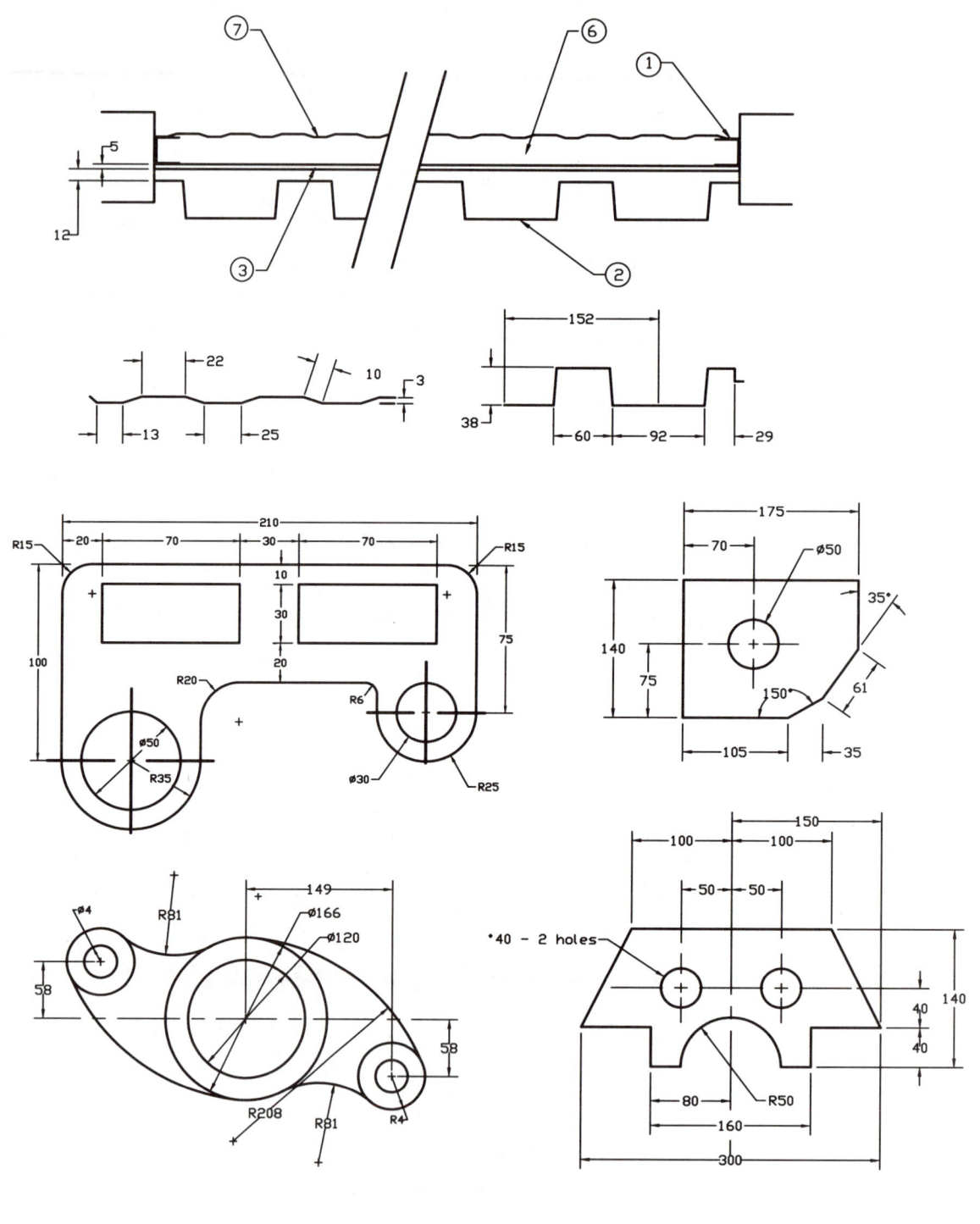

Exercise A7

Open the file SECONDFL. Create a layer called DIM2 and make it current; be sure to change the color so that it is easier to distinguish the extension lines from the object lines. Create the dimensions as shown, dimensioning to the MIDpoint of the windows and doors on the second floor. The first line of dimensions should be to the center of the wall partitions.

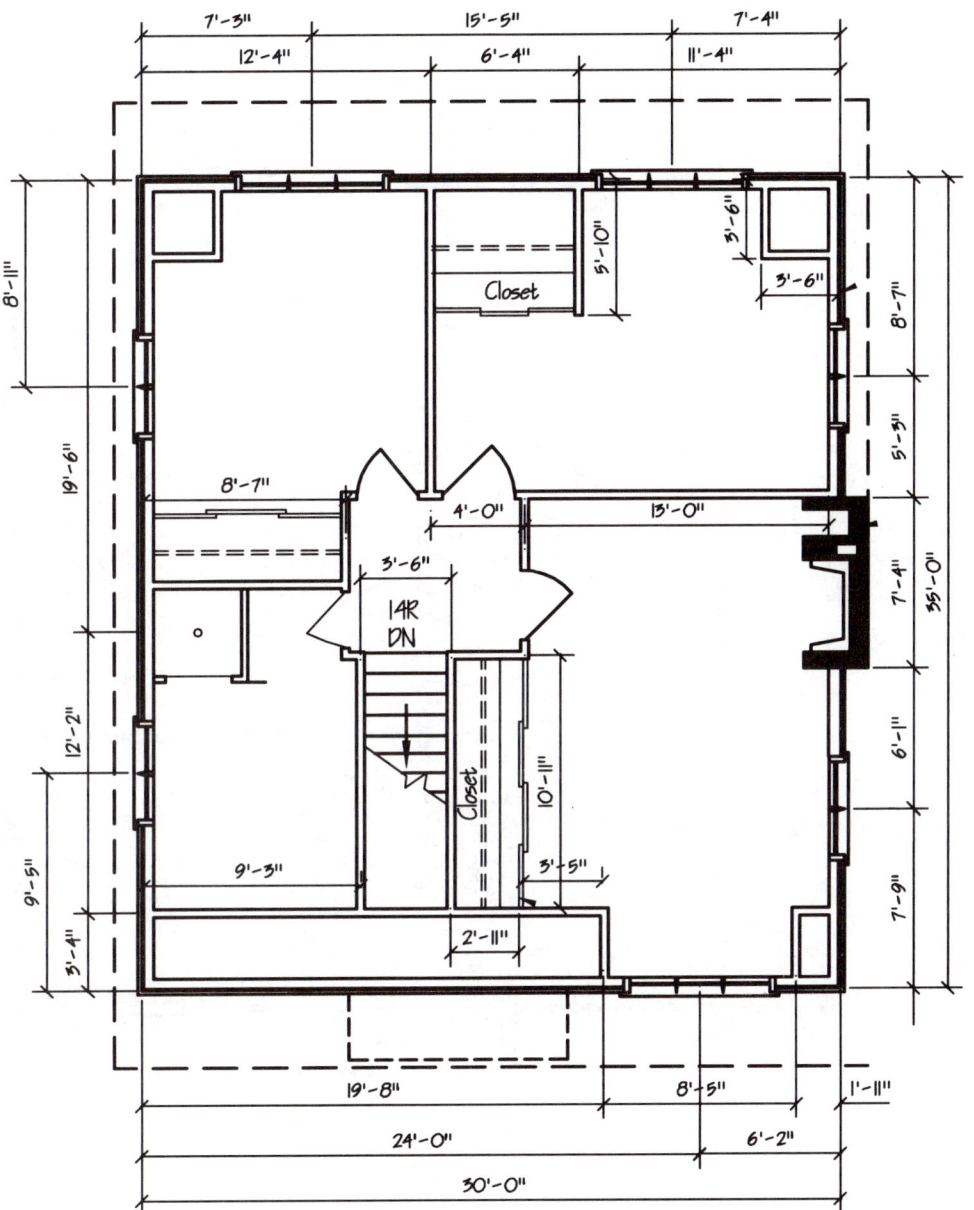

Second Floor

Dimensioning

Exercise A7 (cont.)

Now open FIRSTFL. Create a layer called DIM1 and make it both current and a different color. Place dimensions on the first floor.

Be sure to save both files when the dimensioning is complete.

Always use object snaps to place the dimensions accurately on the feature being dimensioned. Set the OSNAP mode to ENDpoint if most of your horizontal and vertical dimensions are from the ends of objects.

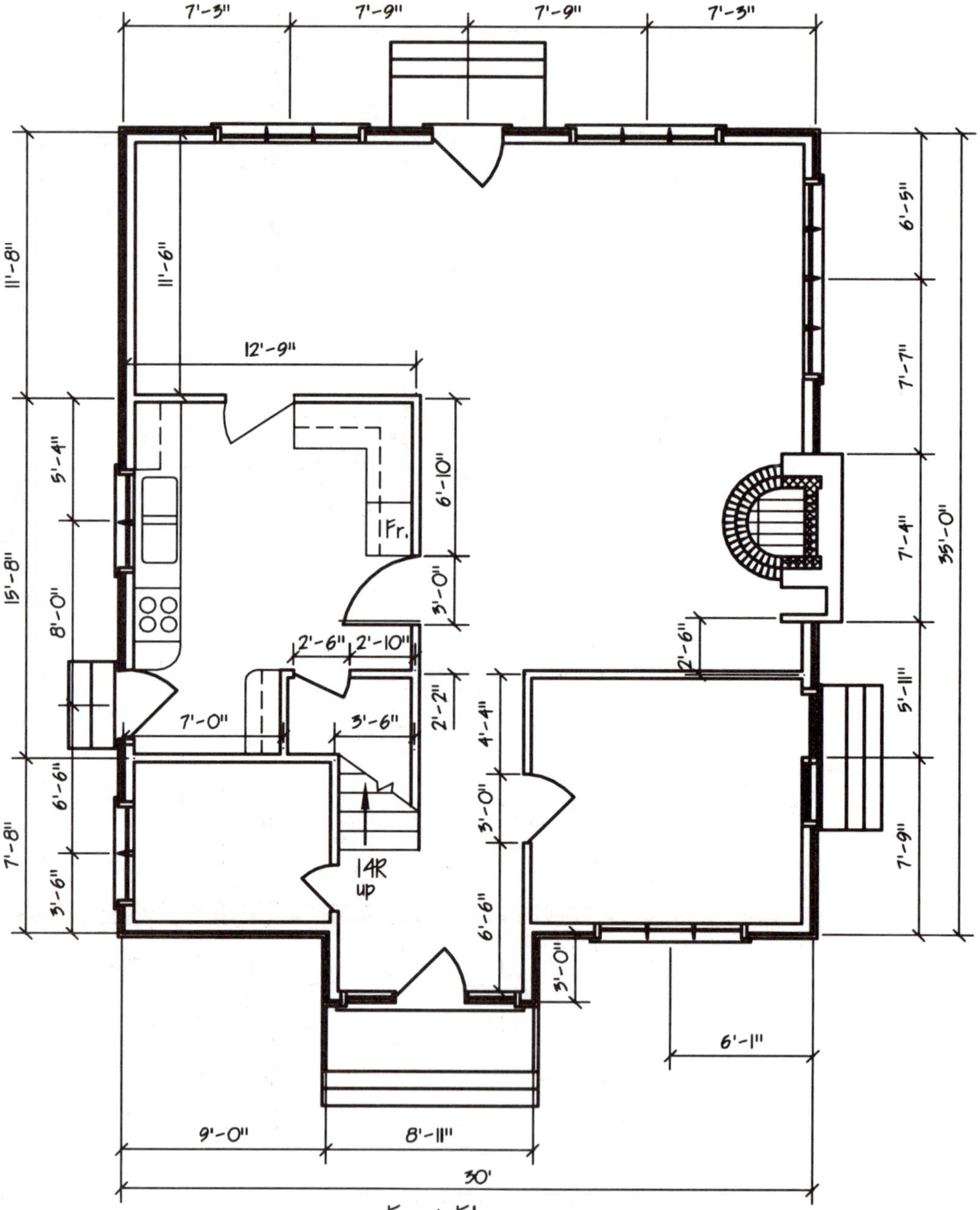

First Floor

Exercise C7

Create this footing drawing using different layers for the footings.

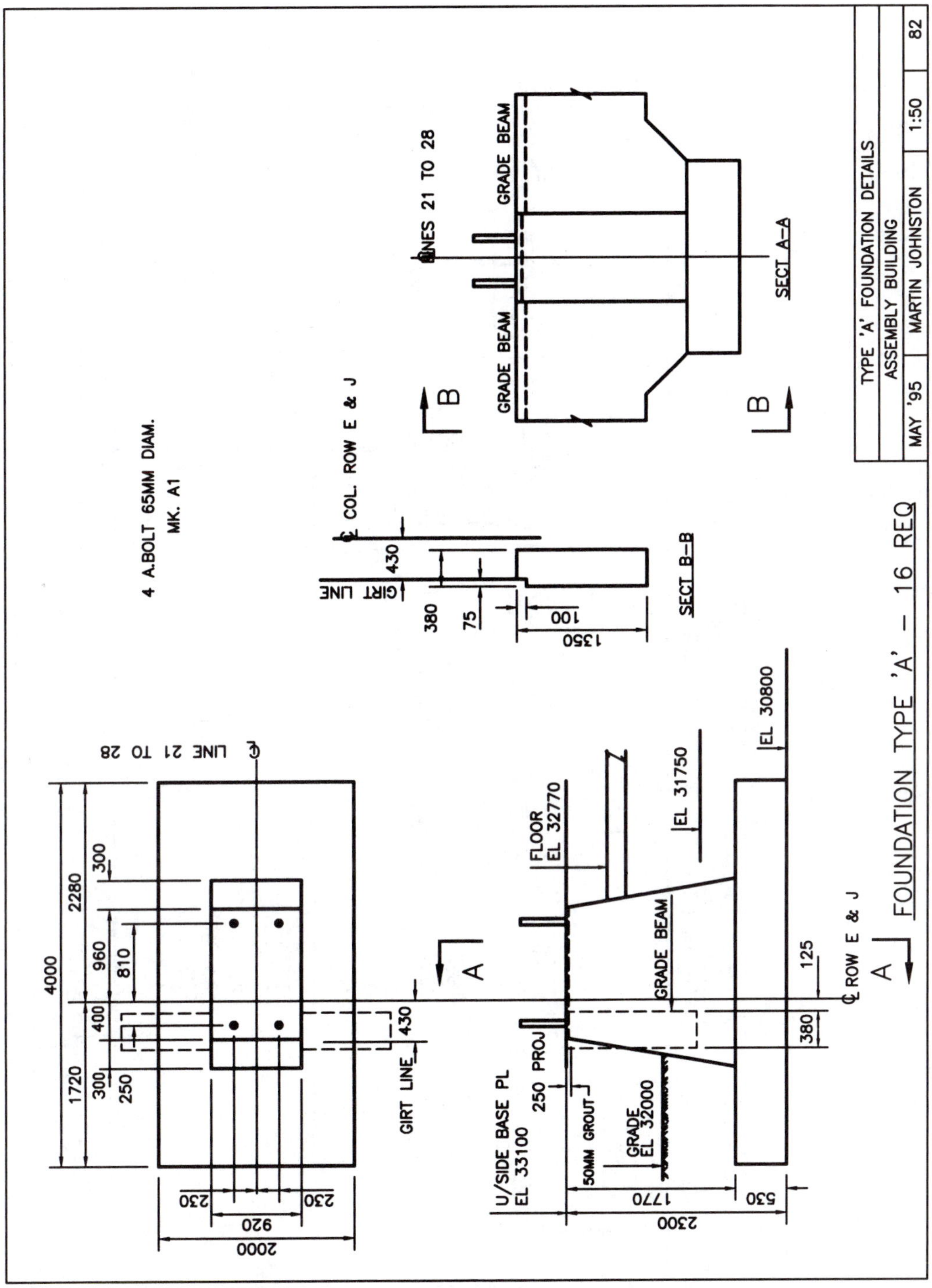

Exercise E7

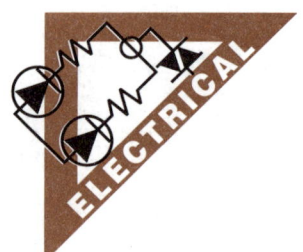

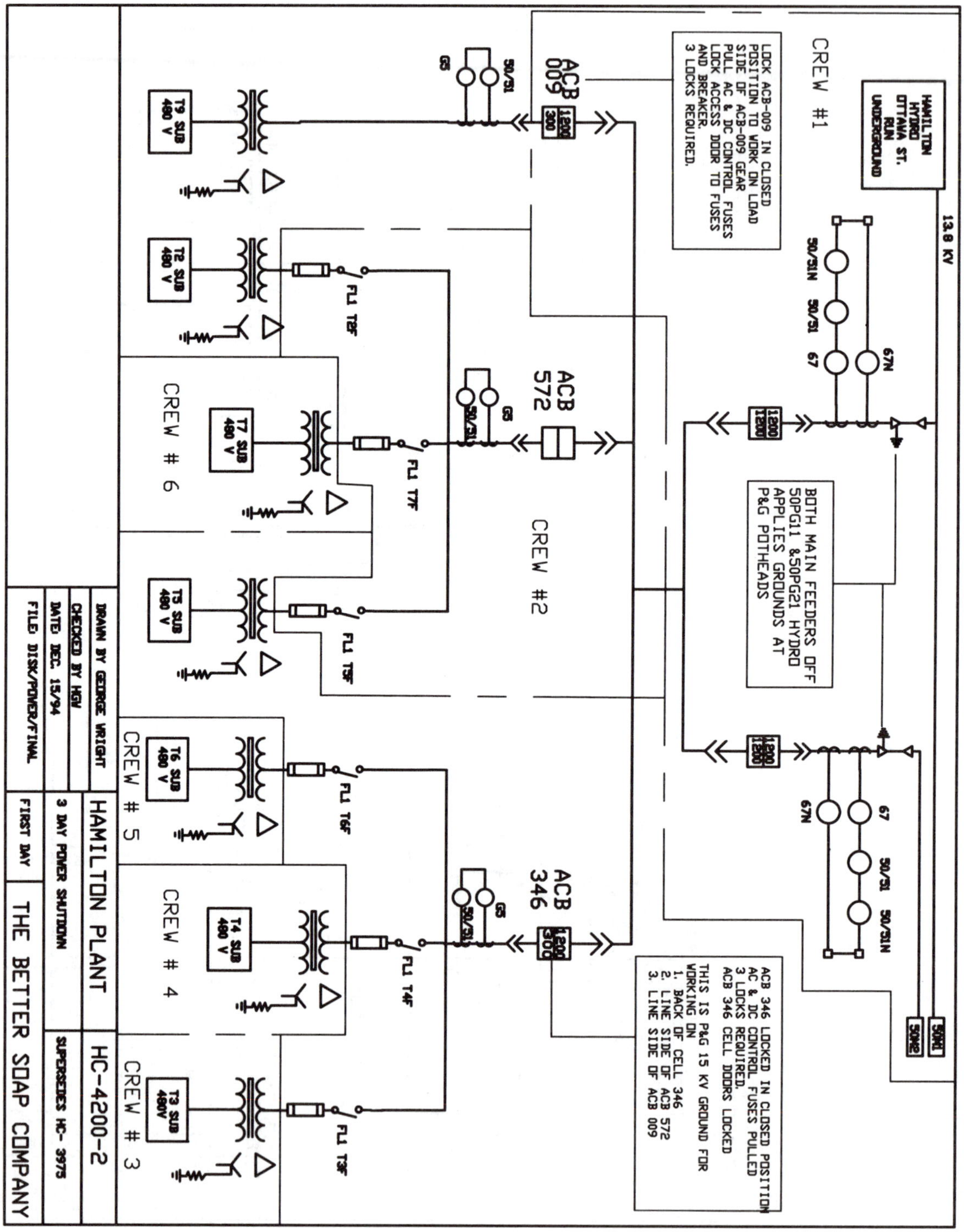
HAMILTON HYDRO OTTAWA ST. RUN UNDERGROUND
13.8 KV
CREW #1
LOCK ACB-009 IN CLOSED POSITION TO WORK ON LOAD SIDE OF ACB-009 GEAR PULL AC & DC CONTROL FUSES LOCK ACCESS DOOR TO FUSES AND BREAKER. 3 LOCKS REQUIRED.
ACB 009
50/51N 50/51 67 67N
BOTH MAIN FEEDERS OFF 50PG11 &50PG21 HYDRO APPLIES GROUNDS AT P&G POTHEADS
ACB 572
CREW #2
ACB 346
ACB 346 LOCKED IN CLOSED POSITION AC & DC CONTROL FUSES PULLED 3 LOCKS REQUIRED. ACB 346 CELL DOORS LOCKED
THIS IS P&G 15 KV GROUND FOR WORKING ON
1. BACK OF CELL 346
2. LINE SIDE OF ACB 572
3. LINE SIDE OF ACB 009
50M1 50M2
T9 SUB 480 V
T2 SUB 480 V
CREW # 6
T7 SUB 480 V
T5 SUB 480 V
CREW # 5
T6 SUB 480 V
CREW # 4
T4 SUB 480 V
CREW # 3
T3 SUB 480V
FL1 T2F
FL1 T7F
FL1 T5F
FL1 T6F
FL1 T4F
FL1 T3F
DRAWN BY GEORGE WRIGHT
CHECKED BY HGW
DATE DEC. 15/94
FILE DISK/POWER/FINAL
HAMILTON PLANT
3 DAY POWER SHUTDOWN
FIRST DAY
HC-4200-2
SUPERSEDES HC-3975
THE BETTER SOAP COMPANY

Exercise M7

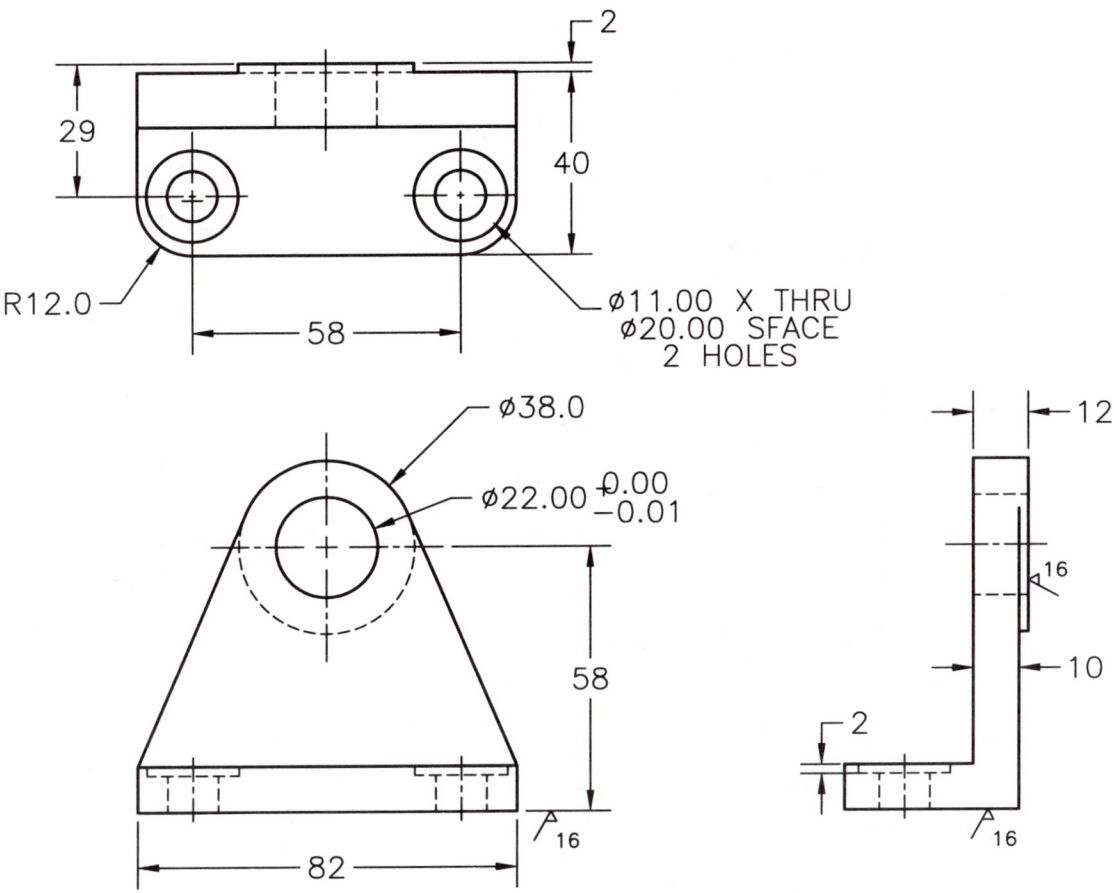

Axle Support

Set up the file with several different layers: Object, Hidden, Center, and Dimensions. Draw the objects, making sure that all lines line up with the corresponding views. Add center lines, hidden lines, and object lines on the appropriate views. Then add dimensions, making sure that the dimension lines also line up with the corresponding views.

Challenger 7

Be sure to use different layers.

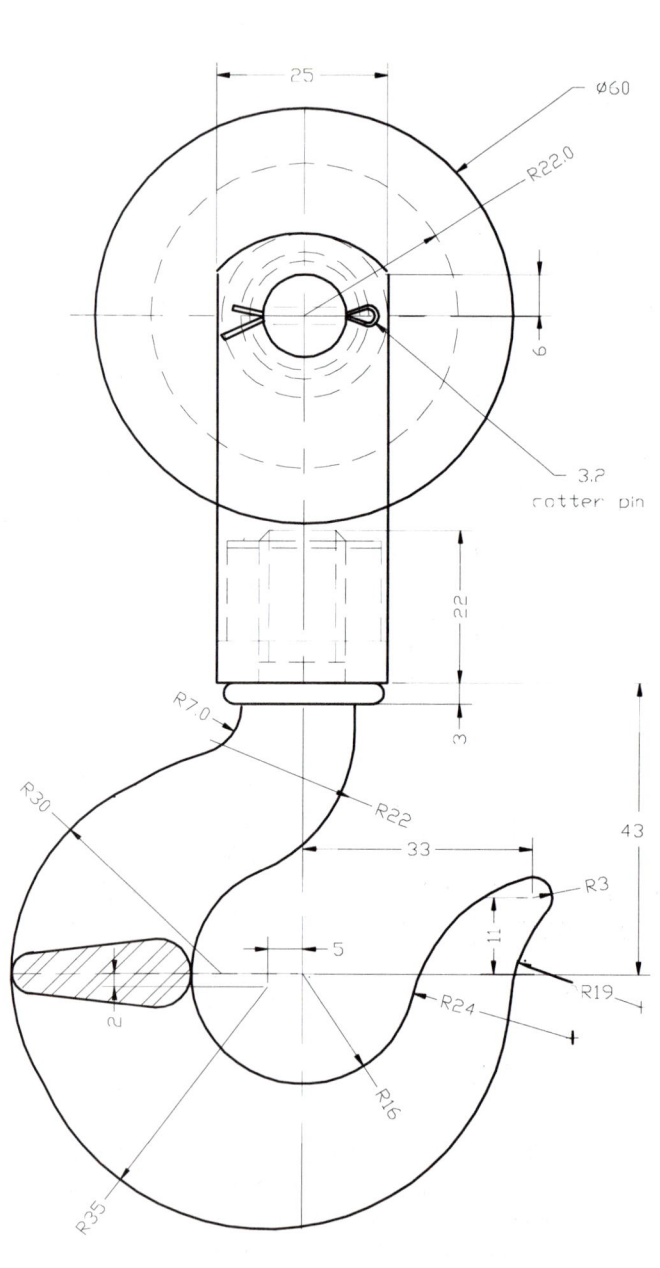
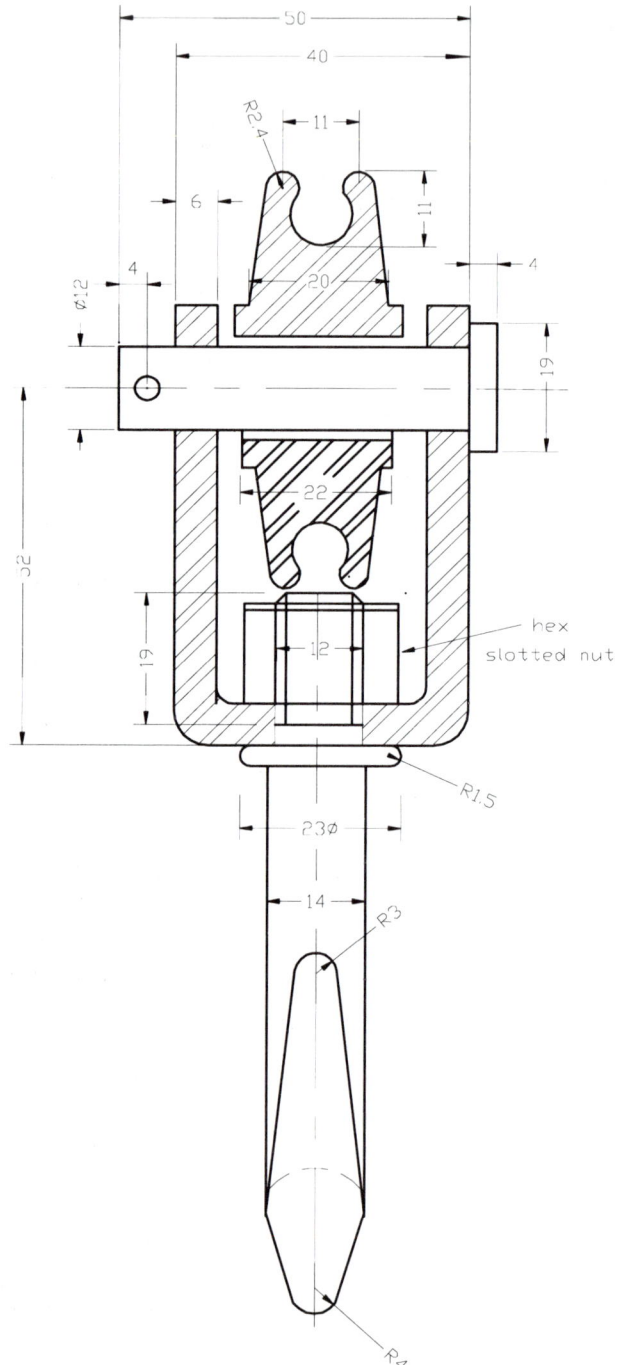

8 Text

Upon completion of this chapter, you should be able to:

1. Place linear text in any size at any rotation angle
2. Place paragraph text
3. Create and change collections of formats to make a text style
4. Change existing text
5. Set up an isometric text

AutoCAD provides two basic ways to create text. ***Linear text*** places simple entries of one or two lines. For longer entries ***paragraph text*** is used. Text is entered in the current text style, which incorporates the current format and font settings. With Release 13, however, there are several methods of saving and controlling alternate text styles.

Linear Text

As stated in Chapter 3, the commands TEXT and DTEXT will place strings of characters on your drawing. When entering text, AutoCAD will prompt you to choose a height for each character, a rotation angle for the string, and a point at which to place the text string on the model or drawing. Many people prefer DTEXT, as with it you can see the style and placement of the text as it is entered.

The TEXT Command

Windows From the Draw toolbar, choose this button:

DOS From the Draw toolbar, choose Text, then Single-line text.

The command line equivalent is **TEXT**.

The TEXT command is as follows:

```
Command:TEXT
Justify/Style/<Start point>:J
Align/Center/Fit/Middle/Right/TL/TC/TR/ML/MC/MR/BL/BC/BR:
```

Text 195

Where: **Justify** = the placement of the text
Style = controls the style of the letters; the styles must be loaded in AutoCAD to be accessible
Align = an alignment by the end points of the baseline; the aspect ratio (X vs. Y) will correspond to the preset distance
Center = the center point of the baseline; this option will fit the text through the center point indicated
Fit = an adjustment of width only of the characters that are to be fit or "stretched" between the indicated points
Middle = a placement of the text around the point, i.e. the top and bottom of the text are centered as well as the sides
Right = an alignment with the right side of the text

The examples in Figure 8-1 demonstrate the standard justifications. The default is left justification at the baseline of the text string.

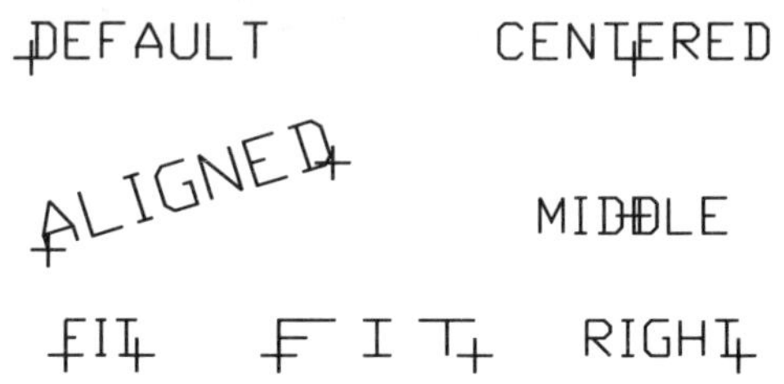

Height can be chosen by picking a point to indicate the height, or by typing in a number.

The double-initialled justification options are as follows:

  TL = top left
  TC = top center
  TR = top right
  ML = middle left
  MC = middle center
  MR = middle right
  BL = bottom left
  BC = bottom center
  BR = bottom right

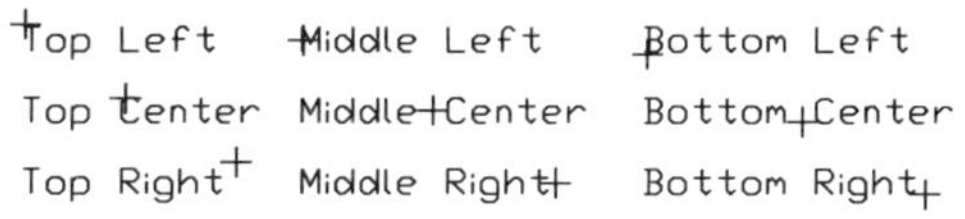

Figure 8-1

Figure 8-2

Once you have chosen a point at which to place your text, the command will prompt you for the height of the letters, the rotation angle, and the text or string of characters itself.

A text *string* is one line of text.

```
Command:TEXT
Justify/Style/<Start point>:C
Center point: (pick 1)
Height<.2000>:.5
Rotation angle<0>:⏎ (to accept the default)
Text:Front Elevation
```

Figure 8-3

In this example, the justify option Center was chosen, so the other options for placement were bypassed. If Justify had been chosen, the following line would have been offered:

```
Align/Center/Fit/Middle/Right/TL/TC/TR/ML/MC/MR/BL/BC/BR:
```

If TEXT was the last command entered, pressing ⏎ at the "Justify/Style/<Start point>:" prompt skips the prompts for height and rotation angle and immediately displays the "Text:" prompt. The text is placed directly beneath the previous line of text.

The DTEXT Command

When using the TEXT command, you will see the text string only at the bottom of the screen in the "Command:" prompt area. The DTEXT or Dynamic TEXT command allows the text to be displayed on the screen as you enter it. Many people find this a more useful format. DTEXT always displays text as left-justified on the screen (default) regardless of the format chosen. The justification will be corrected when the command is finished.

As in the ZOOM Dynamic command, there are those who will argue that the D stands for Difficult instead of Dynamic, because, once entered, it is more difficult to get out of. The advantage is that it allows you to enter text more quickly. The DTEXT command automatically offers you a second string for text. Press ⏎ to exit from the command. To find DTEXT:

Windows From the Draw toolbar, choose Text, then this button:

DOS From the Draw menu, choose Text and then Dynamic Text.

The command line equivalent is **DTEXT**.

```
Command:DTEXT
Justify/Style/<Start point>: (pick 1)
Height<.5000>:.25
Rotation angle<0>:⏎(to accept the default)
Text:Scale
Text: (pick 2)
Text:Date
Text: (pick 3)
(etc.)
```

To exit from DTEXT, use:

```
Text:⏎
```

Figure 8-4

Danger

If you use the Cancel button or **Ctrl-C** in DOS or **ESC** in Windows within the DTEXT command, all of the string of text that you have entered in that command will be lost. Always use ⏎ to exit from the DTEXT command.

You can also reposition your cursor and start a string of text in another area of the screen at any point within the command, by simply choosing another point. This lets you position text of a similar height in various places of your drawing with one DTEXT command. The pull-down menus are disabled throughout this command.

Special Character Fonts

You can underscore, overscore, or include a special character by including control information in the text string.

%%u = underscore
%%o = overscore
%%d = degree symbol
%%p = plus-minus tolerance symbol
%%c = diameter symbol
%%nnn = ASCII characters — example: **%%123 %%125** = { }

When using special character fonts with the DTEXT command, the special characters will be displayed as you type, i.e. **%%uFront Elevation%%u**. The entry will be updated to the desired text once the command is finished.

```
Command:DTEXT
Justify/Style/<Start point>:R
Right side: (pick 1)
Height<.2500>:↵
Rotation angle<0>:↵
Text:%%u%%c25 4 holes
Text:↵
```

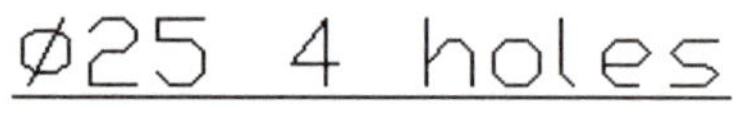

Ø25 4 holes

Figure 8-5

Multiline Text

A new text string will line up with the previously entered text string if there are no changes in the base point or justification options in both the TEXT and the DTEXT command. If Center is chosen, all of the text will be centered (see Figure 8-6(b)); if no option is chosen, all strings will be left-justified (see Figure 8-6(a)). Your last string of text will be highlighted to show where the next line will be lined up.

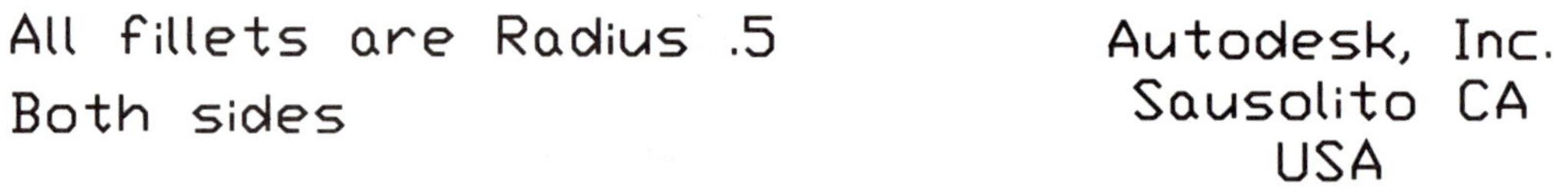

All fillets are Radius .5
Both sides

Autodesk, Inc.
Sausolito CA
USA

(a) (b)

Figure 8-6

If you do not want your text to line up with the last string entered, simply identify a new start point.

The DTEXT command allows multiline text in one command. In TEXT use ↵ to reenter the command after the first text string, then ↵ to accept the default position, size, and rotation.

Once your text has been entered, it is accepted as one item and can be edited using any of the edit commands, such as COPY, MOVE, ERASE, ROTATE, AND ARRAY. To edit the text itself, use DDEDIT, DDMODIFY, or CHANGE.

Paragraph Text

Paragraph text or MTEXT is for long, complex entries that have many lines of text. Any number of text lines or paragraphs can be entered to fit within a specified width. The paragraphs form a single object that can be moved, rotated, copied, erased, mirrored, stretched, or scaled. This is the default text command under the Draw toolbar in Windows Release 13.

You can apply overscoring or underlining, fonts, color, and text height to any individual character, word, or phrase of the paragraph.

Creating paragraph text is very different in Windows and in DOS. If you use Windows, you can create text in the Edit Mtext dialog box. If you use DOS, you must create the text in a third-party text editor. This editor can be changed using the MTEXTED system variable. In Windows, use Preferences to set up a different editor. You can also enter paragraph text on the command line in both Windows and DOS.

Paragraph Text for Windows

The Edit MText dialog box is a very efficient way to set properties that effect the entire paragraph or selected text. As in a word processor, you should set the width before you create the text. The paragraph will be displayed in a dialog box within the specified width. The text will wrap or spill in the direction defined by the current attachment setting. The text boundary can be realigned. To create paragraph text:

From the Text flyout on the Draw toolbar, choose Text. This will invoke the MTEXT command and the Edit MText dialog box.

Windows From the Draw toolbar, choose this button:

DOS From the Draw menu, choose Text.

The command line equivalent is **MTEXT**.

Specify the insertion base point for the text as follows:

```
Command:MTEXT
Attach/Rotation/Style/Height/Direction/Width/2Points/<Other
   corner>: (pick a point where the text will start)
```

Next you specify the width of the text by using one of the following methods:

- To define a diagonally opposite corner of a rectangular text boundary, specify a point.
- To define only the width of the text boundary, enter **W** and specify a width value. Entering **0** causes the text to extend horizontally until you press ⏎.
- To define the width by specifying two points, enter **2P**. Then specify the points.

The width of the text that you have chosen will be reflected on the dialog box. Type in the text you would like placed at the specified location on your file, then choose OK to write it to the file.

The text can be typed in as in a word processor; it will wrap according to the width chosen.

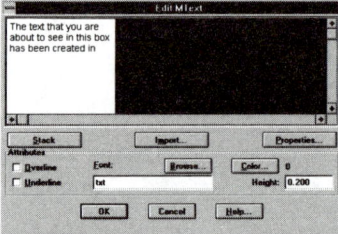

Figure 8-7

To edit the color of one word or phrase, or to have it underlined or overscored, select the text by "wiping" the mouse over it, then choose Overscore, Underline, or Color.

Some fonts cannot be displayed in the Edit MText dialog box. If text isn't shown, select a substitute font to represent the original font, then choose OK. When you are finished editing, the original font selected appears in the graphics area.

Paragraph Text for DOS

Paragraph text in DOS will switch you automatically to the DOS editor. Once you have specified the width of the text, the text will wrap according to the parameters chosen.

To place paragraph text:

> **DOS** From the Draw menu, choose Text, and Text again. This will invoke the MTEXT command.

The command line equivalent is **MTEXT**.

```
Command:MTEXT
Attach/Rotation/Style/Height/Direction/Width/2Points/<Corner>:
   (pick 1)
Other corner: (pick 2)
```

The screen will turn automatically to the text editor. Type the required text.

To return to the graphics screen, choose File from the pull-down menus on the text editor.

First Save, then Exit.

The text will be placed on your file as shown in the diagram.

Figure 8-8

See Paragraph Text for Windows for further placement options on the paragraph text placement.

MTEXT Options

Other options of the MTEXT command are as follows:

Attach controls the boundary alignment of your text. The option you select determines both text justification and text spill in relation to the text boundary. According to the option chosen, the text will justify to the right, the left, or the center.

TL/TC/TR/ML/MC/MR/BL/BC/BR:

Rotation	specifies the rotation angle for the text boundary.
Style	specifies the text style for the paragraph text.
Height	specifies the height of the uppercase text.
Direction	specifies the direction of the paragraph text object. This will create vertical text, not rotated text.
Width	specifies the width of the text boundary.
2P	allows you to pick two points for placement of the paragraph of text. See the earlier section "Paragraph Text for Windows."

Specifying a Text Editor

Any ASCII text editor can be used to create text for AutoCAD. The default command editor is Edit in DOS. In Windows you can specify a specific text editor in the Preferences dialog box.

A new text editor may permit some formatting options not offered with the standard Edit MText dialog box. If you use a text editor to create text, it is a good idea to keep the same editor for subsequent edits.

To specify a text editor: from the Options menu, choose Preferences.

In the Preferences dialog box, choose Misc.

In the Text Editor box, enter the name of the executable file for the ASCII text editor you want to use to create paragraph text. Enter **INTERNAL** to specify the AutoCAD MText editor.

Choose OK.

Alternatively, use the MTEXTED system variable to specify the name of a third-party ASCII text editor.

Text Styles and Fonts

Text styles are what the user names the style of the lettering chosen. *Fonts* are the style or design of the letters and numbers use to create the text string. The fonts can be supplied by AutoCAD or a third-party developer. A text style is stored with not only the lettering style but also a group of characteristic settings.

Setting	*Default*	*Description*
Style name	Romans	Name of up to 31 characters
Font file	ROMANS.TXT	File associated with font
Height	0	Character height
Width	1	Expansion or compression of characters (aspect ratio)
Obliquing factor	0	Slant of individual characters
Backward	No	Orientation of text (for mirrored images)
Upside down	No	Orientation of text
Vertical	No	Vertical, not rotated

The *obliquing factor* is the angle of the text characters themselves, i.e. the slant.

Here are some examples of what your text will look like.

```
                 ROMANS

     Width factor .5      AutoCAD      v A
     Width factor 1       AutoCAD      e u
     Width factor 1.5     AutoCAD      r t
                                       t o
                                       i C
     Oblique angle 0      AutoCAD      c A
     Oblique angle 30     AutoCAD      a D
     Oblique angle -30    AutoCAD      l

     Upside down          AutoCAD
     Backwards            AutoCAD
```

Figure 8-9

Note that the options Height and Width are in fact the **aspect ratio**, or relative *X-Y* value, of the text. If you change the Height of the text with the STYLE command, you forfeit the flexibility of changing it in the TEXT command. It is recommended that, while you are learning AutoCAD, you *never change the height of the text in the STYLE command*, because there are many instances — such as with dimensions and annotations — where this will cause you great inconvenience.

The Vertical option places text vertically. To rotate a text string, *do not* change Vertical in the STYLE command; rather, change the rotation angle in the TEXT command.

Using Text Styles

Each text style takes a font file from the AutoCAD list. The style is saved with the font style. If you change the font or vertical property of an existing style, all text using that style is regenerated using the new font or orientation. For MText objects, other properties change as well. When you change other properties of the text style such as height or obliquing angle, the text already created in that style using TEXT and DTEXT does not update, but all text created subsequently will reflect those changes.

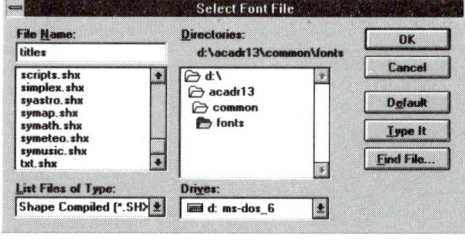

The default style is Standard, using the TXT.SHX font. To use any other type of lettering you must load or create that style.

Figure 8-10

To create a text style:

> **Windows and DOS** From the Data menu, choose Text Style.

The command line equivalent is **STYLE**.

The first prompt is for the text style name. Typing **?** will list all the text styles you have identified. The style name is the name you give the font. Enter the name of the style you would like to create at the prompt. The name must be less than 31 characters in length; AutoCAD will convert the name to uppercase letters.

Notes

When prompted for the style name, enter the name of the style you would like to create.

This will invoke the Select Font File dialog box. Select the font you want to use.

Change other settings as needed.

This will now be the current text font and will remain so until you change it.

Making Text Styles Current

Once the text style is created, you can use it as often as you like, then change the style or create a new one. To make the text style current use the following:

Windows From the Object Properties toolbar, choose Object Creation, then Text Style.

DOS From the Data menu, choose Object Creation, then Text Style.

Select the name of the style you want to use, then choose OK to exit the dialog box.

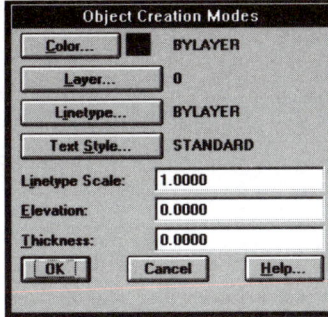

Figure 8-11

Other Text Fonts

A font is the particular style or design of the text characters. AutoCAD comes with a list of available fonts as seen in Figure 8-10. In addition to AutoCAD's standard list, you can also load TrueType or Adobe Type 1 PostScript fonts from third-party vendors.

First install the PostScript or TrueType fonts according to the vendor's instructions. Then add the directory to the AutoCAD library search path specified by the AutoCAD environment variable.

TEXTFILL and FFLIMIT variables are used to manage and control these font files.

The Unicode character encoding standard is also available in Release 13. Unicode formats can contain up to 65,535 characters. The Big fonts option on the Select Font File menu is invoked for this purpose.

If you use a third-party font, be aware that other AutoCAD users may not have it installed. This makes transferring of files complex. The font files have an extension of .shp, .shx, .pfb, or .pfm. Add these files to your disk when transferring.

Using Text Fonts

Now that you have been working with AutoCAD for some weeks, you should be becoming aware that the amount of disk space the file for a drawing occupies is directly related to the number of lines on the drawing. Take a look at your floppy disk files with the DOS command DIR A:, and you will see that the exercise file for Chapter 7 is much larger than the file for Chapter 1.

AutoCAD's text fonts are made up of a series of small line segments. Each style uses a different number of line segments. The more complicated the style, the more line segments there will be, and the more disk space the text uses. While the Gothic letters are attractive, they take up a lot of space and are not appropriate for many jobs. In addition, regeneration time is much longer, particularly in paper space. Therefore, to avoid overloading the disk, before choosing a font try to determine which is best suited to your job. Begin with a simple font; then, when you are finished, change the style's font to a more complex font.

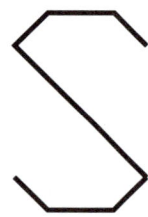  

Monotext
7 line segments

Roman Simplex
19 line segments

English Gothic
70 line segments

Figure 8-12

In Figure 8-12, you can see that the Gothic letter has ten times the amount of lines as the Monotext; it may be appropriate for a title, but not for a notation or a dimension.

Since text takes up memory space, to save time during regenerations it is a good idea to have a separate layer for text and freeze it.

The QTEXT Command

If you want to save time, but do not want to freeze the layer for text because you need to position objects in reference to the text, try QTEXT. This command can be accessed both in the Drawing Aids dialog box and at the command line with QTEXT.

QTEXT places a rectangle of the approximate size of the text in the space provided for the text. Text characters are made up of vectors, and thus occupy a lot of memory; when text is replaced with a simple rectangle, only four lines need to be regenerated per text string. This will save a lot of time during editing commands, zooms, and regenerations.

Remember to turn QTEXT off before plotting, or only the rectangles will plot.

Editing TEXT and DTEXT

A text string is considered an object and can therefore be moved, copied, changed into different layers, and created in different colors. You can also alter the text string itself as well as the height, the rotation angle, and the style of the characters.

Text objects also provide grips for stretching, scaling, and rotating. A paragraph text has grips at the four corners of the text boundary and, in some cases, at the attachment point. Line text provides a grip at the lower left corner and another at the alignment point. (Review Chapter 4 for more information on grips.)

Osnap INSERT is the insertion point to grab your text by the point you selected to create the text.

Editing Line Text

You can change both the text and the text content with the CHANGE command. This is useful for changing many lines of text, particularly when making charts. DDEDIT and DDMODIFY make editing single strings of text much easier.

```
Command: CHANGE
Select objects: (pick 1)
Select objects:↵
Properties/<Change point>:↵
Enter TEXT insertion point:↵
```

```
Text style:STANDARD
New style or RETURN for no change:ROMANT (part 1 of Figure
   8-13)
New height<5.0000>:10 (part 2)
New rotation angle<0>:30 (part 3)
New text<AutoCLAD>:AutoCAD (part 4)
```

This command gives you the options to change all of the variables in the TEXT command as shown in Figure 8-13.

If you pick more than one string, the command will repeat the prompts for each string.

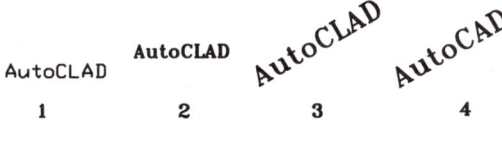

Figure 8-13

The DDEDIT Command

In long strings of text, DDEDIT makes the editing process much easier. It will change only the text, not the formatting or properties of the text.

To edit the text string content:

> **Windows** From the Special Edit flyout of the Modify toolbar, choose Edit Text.
>
> **DOS** From the Modify menu, choose Edit Text.

Select the line text object that you would like to edit. Pick the text where you want it to be changed, or type over or reenter the text, then choose OK to have it updated. Pressing Backspace will delete the highlighted text.

Select another line of text, or press ⏎ to exit the command.

Figure 8-14

The DDMODIFY Command

If you want to change more than the text line content, use DDMODIFY. This works on only one string of text at a time, but offers you a variety of things to change.

To edit the text string:

> **Windows** From the Edit menu, choose Properties.
>
> **DOS** From the Modify menu, choose Properties.

You can change the content of the text by editing or typing over the text in the appropriate box.

You can also override the text style parameters in the boxes illustrated.

Choose OK when you are ready to exit the dialog box.

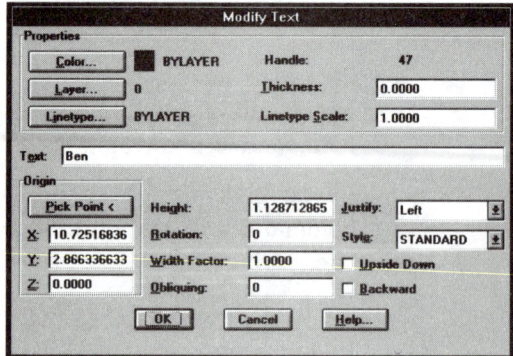

Figure 8-15

Editing Paragraph Text

You can use both DDEDIT and DDMODIFY to edit paragraph text. If you want to change the paragraph text in a text editor, use the following:

> **Windows** From the Special Edit flyout on the Modify toolbar, choose Edit Text.
>
> **DOS** From the Modify menu, choose Edit Text.

You will be prompted to select the paragraph text to edit. Type over or reenter the text. When your entry is complete, save the changes and exit the text editor.

Making Isometric Lettering

As mentioned earlier, SNAP sets a spacing for point entries, and Grid places a dot grid on the screen.

SNAP allows you to indicate points or positions on the screen at preset regular integers. SNAP also allows a rotated or isometric drawing to be entered.

```
Command:SNAP
On/Off/Value/Aspect/Rotate/Style:R
Base point<0'-0.00",0'-0.00">:
Rotation angle<0.00>:45    (will rotate at 45 degrees)
```

The grid size will follow the snap size unless changed by the GRID command.

The GRID and SNAP can be changed at any time during a drawing. This is particularly important in creating text, because text is often close to, but not on, an existing item or line.

To create isometric lettering, change the SNAP to Isometric. Then change the obliquing angle in the STYLE or FONT command to be either 30 or -30. This will adjust the slant. Then enter your letters using Fit.

Isometric Lettering — An Exercise

Use your own initials to fill in the spaces on the cube.

Step 1 Change the SNAP style to Isometric, and draw in a cube.

Step 2 Pick a font from the pull-down menu (Roman, Standard, or Monotext are the best for this purpose), and change the obliquing angle to 30. This is calculated by measuring the angle from the horizontal. Use Fit to place the text.

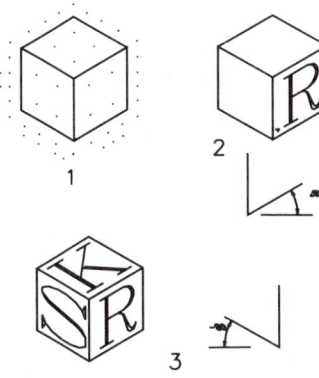

Figure 8-16

Step 3 Pick the same font, and change the obliquing angle to -30. Use Fit to place the text. Use your own judgment to create the last letter.

Using LEADER to Create Notations

The LEADER command in the Dim: menu gives you the facility of creating text with a leader line or series of leader lines and an arrowhead. The leader arrowhead emanates from the point picked. The command is as follows:

```
Command:DIM:
DIM:LEADER
From point: (pick 1)
To point: (pick 2)
To point (Format/Annotation/Undo)<Annotation>: (pick 3)
To point (Format/Annotation/Undo)<Annotation>:↵
Annotation (or RETURN for options):24.00%%c %%p0.02
MText:↵
```

You can enter as many points on the leader line as are necessary.

All special text characters can be used in this dimension as well.

The arrowhead size is set in the dimension style or with DIMSCALE or DIMASZ.

To enter multiple lines of text, either keep typing at the "MText:" prompt, or

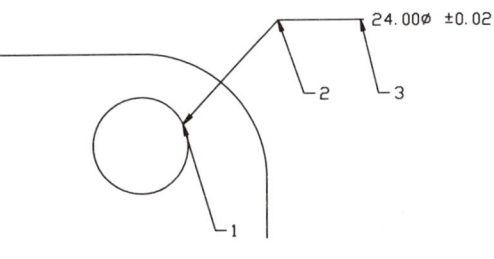

Figure 8-17

Text **207**

enter at the "Annotation (or RETURN for options):" prompt. If you choose the latter, you will enter the text editor and can continue entering text using it.

The options of Leader include:

Tolerance   offers a control frame containing geometric tolerances using the Geometric Tolerances dialog box.

Copy   copies text, a text paragraph, a block, or a feature control frame to the leader line.

Block   inserts a block at the end of the leader line.

Format   controls the way the leader is drawn and whether it has an arrowhead. Options include Spline, Straight, and Arrow.

To enter a leader line without related text, enter a single blank space when you are prompted for the dimension text. This can be done with all dimension entries.

As in the LINE command, you can use **U** to undo the previous point entry without exiting the command.

Prelab 8 Using Text and Text Styles

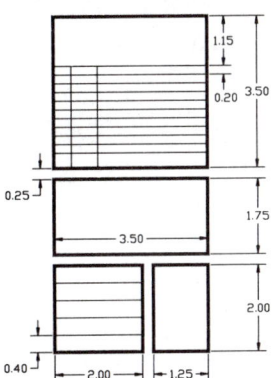

In this lab we will create a title block.

Step 1 Use PLINE and LINE to create this title block. If you set the SNAP to .25 to start, it will make drawing easier. Change the SNAP as needed.

The lower left area will be for Scale, Date, etc., the central area will be for the company title, which is "3D Design Studio," and the top area will be for revisions.

Step 2 Set up a text style.

> **Windows** From the Data menu, choose Text Style.
>
> **DOS** From the Data menu, choose Text Style.

The command line equivalent is **STYLE**.

At the "Command:" prompt, enter the text name.

```
Command:STYLE
Text style name (or
    ?)<Standard>:TITLES
```

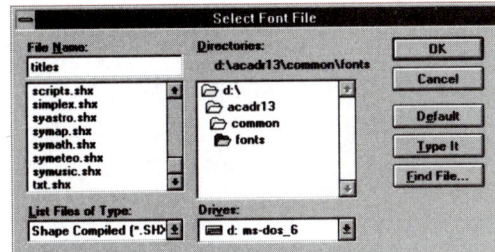

This will invoke the dialog box.

Slide the bar up until you find SAS.PFX. Choose this font, or another if this is not available; then press until you return to the command prompt.

Step 3 Set up DTEXT.

> **Windows** From the Draw toolbar, choose Text, then DTEXT.
>
> **DOS** From the Draw menu, choose Text, then Dynamic text.

The command line equivalent is **DTEXT**.

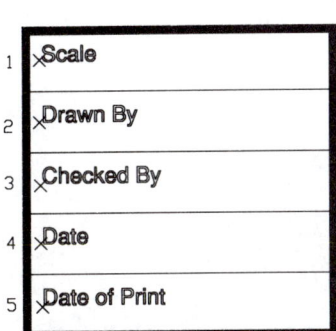

```
Command:DTEXT
Justify/Style/<Start point>:(pick 1)
Height<.5000>:.1
Rotation angle<0>:
Text:Scale (pick 2)
Text:Drawn By (pick 3)
Text:Checked By (pick 4)
Text:Date (pick 5)
Text:Date of Print
Text:
```

Text **209**

Step 4 The lettering is a bit large, so use the CHANGE command to make the letters a bit smaller. You can change them with DDMODIFY, but for this operation CHANGE is quicker.

```
Command:CHANGE
Select objects: (pick 1, 2, 3, 4, 5)
Select objects:⏎
Properties/<Change point>:⏎
Enter TEXT insertion point:⏎
Text style:TITLE
New style or RETURN for no change:⏎
New height<0.1000>:.08
New rotation angle<0>:⏎
New text<Scale>: (keep pressing ⏎ until you have changed all
    the sizes and exited from CHANGE)
```

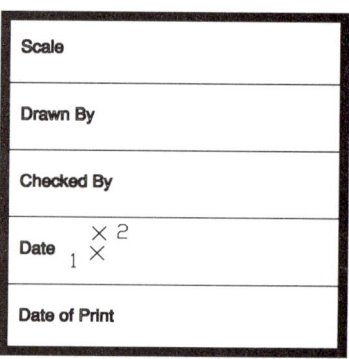

Step 5 Now use the MOVE command to move the text up a bit.

```
Command:MOVE
Select objects:P
Select objects:⏎
Base point: (pick 1)
Displacement: (pick 2)
```

Step 6 For the other text in this box, create a new text style and add the text as shown. The text font illustrated is Cityblueprint.

> **Windows** From the Data menu, choose Text Style.
>
> **DOS** From the Data menu, choose Text Style.

The command line equivalent is **STYLE**.

Then add the text as either DTEXT or single-line text.

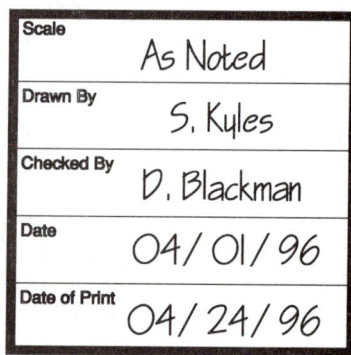

210 CHAPTER EIGHT

Step 7   Now make a cube with isometric lettering. Change SNAP and GRID to draw in the cube.

From the Data menu, pick Drawing Aids. Change the rotation angle of the SNAP to 30.

```
Command:SNAP
On/Off/Value/Aspect/Rotate/Style<1.00>:S
Isometric/Standard:I
Increment<1.00>:↵
```

The grid size will follow the snap size unless changed by the GRID command.

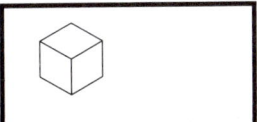

Use the line command to draw in the cube.

Step 8   Now use the STYLE command to load the ROMANT font and change the obliquing angle in order to load letters onto an isometric plane.

From the Draw menu, choose Text and then Text style.

```
Command:STYLE
Text Style name<Monotxt>:FANCY
Font File:ROMANT
Height<0.00>:↵
Width<1.00>:↵
Obliquing factor<0>:30
Backward<N>:↵
Upside down<N>:↵
Vertical<N>:↵
Command:TEXT
Justify/Style/<Start point>:F
First point: (pick 1)
Second point: (pick 2)
Height<.2000>:.35
Text:D
```

Now change the obliquing angle to -30 and add the second line.

```
Command:STYLE
Text Style name<Monotxt>:FANCY
Font File:ROMANT
Height<0.00>:↵
Width<1.00>:↵
Obliquing factor<30>:-30
Backward<N>:↵
Command:TEXT
Justify/Style/<Start point>:F
First point: (pick 3)
Second point: (pick 4)
Height<.3500>:↵
Text:3
```

Step 9 Change the obliquing angle of the ROMANT font to 0 and add the title and address using the Center option. Set SNAP back to normal before you start.

```
Command:SNAP
On/Off/Value/Aspect/Rotate/Style<1.00>:S
Isometric/Standard:S
Increment<1.00>:↵
```

```
Command:STYLE
Text Style name<Monotxt>:FANCY
Font File:ROMANT
Height<0.00>:↵
Width<1.00>:↵
Obliquing factor<0>:0
Backward<N>:↵
Upside down<N>:↵
Vertical<N>:↵
```

```
Command:DTEXT
Justify/Style/<Start point>:J
Align/Center/Fit/Middle/Right/TL/TC/TR/ML/MC/MR/BL/BC/BR:C
Pick center point: (pick 1)
Height<.3500>:.15
Rotation angle<0>:↵
Text:Design Studio
Text:85 Glen Fern Road
Text:Hamilton
Text:Ontario
Text:↵
```

Step 10 Use PAN to move the screen down so that you can change the STYLE to Italic and add the notes regarding the revisions. Use the dialog boxes to change the style.

```
Command: (Draw, DTEXT, Text Style)
Text Style name<Monotxt>:HIGHLIGHT
Font File:ITALICS
```

Do not change any options.

```
Command:DTEXT
Justify/Style/<Start point>:(pick the first point)
Height<.1500>:↵
Rotation angle<0>:↵
Text:Revisions (pick again)
Text:Date (pick again)
Text:Description
Text:↵
```

Step 11 Now create another box above the Revisions box, and add a paragraph of text regarding the date of tender for the drawing. Use PLINE to quickly draw in a rectangle. Then change the style to Romans. Finally add a paragraph of text using MTEXT.

Change style or font first.

Windows From the Text flyout on the Draw Toolbar, choose MTEXT.

DOS From the Draw menu, choose TEXT, and TEXT again.

The command line equivalent is **MTEXT**.

```
Command:MTEXT
Attach/Rotation/Style/Height/Direction/Width/2Points/<Corner>:
   (pick 1)
Other corner: (pick 2)
```
Now type:

Note: Drawings will be sent out for pricing by owner on June 6 1996.

Note: Drawings will be sent out for pricing by owner on June 6 1996.

Step 12 Use ZOOM All to view your title block.

Change some of your text using DDEDIT and file it for future use under the name PRELAB8.

Command and Function Summary

DDEDIT edits multiple lines of text and attribute definitions.

DDMODIFY edits text and properties of single lines of text.

DTEXT or Dynamic text creates text and displays it on screen while it is being created.

MTEXT creates paragraph text.

QTEXT allows the text to be displayed as a box to save regeneration time.

STYLE controls the style of the characters.

TEXT creates text in single-line format.

Practice Exercise 8

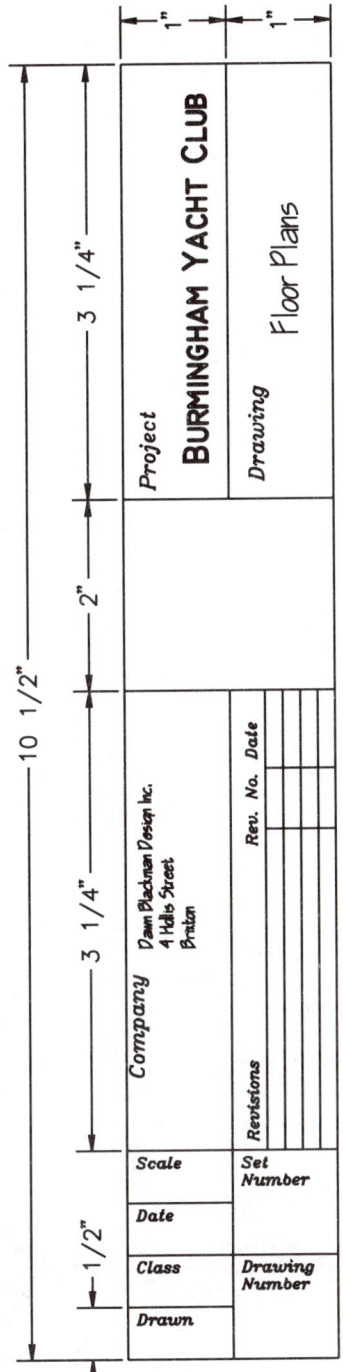
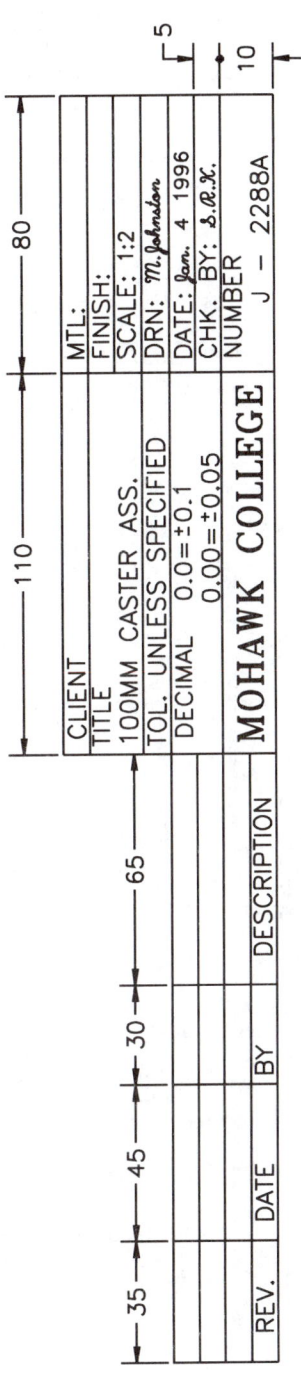

Use a PLINE to make the outer edges. Remember to use Close for the corners. Use at least three different lettering styles. Do not forget to change SNAP to line up the lettering.

If you prefer another style of title block, that is fine, as long as it is acceptable by your department.

Exercise A8

Retrieve the model called A1SHEET. Using the TEXT and DTEXT commands plus the Set style option, enter the text in the title block as shown. Retrieve the files FIRSTFL and SECONDFL and add the view titles at a size of 1'6". Then add the necessary notations for things like closets, stair risers, and room titles.

Since this is relatively easy, use the remainder of the lab time to create the interior finish. Offset at .5 and FILLET Rad 0 is the easiest way.

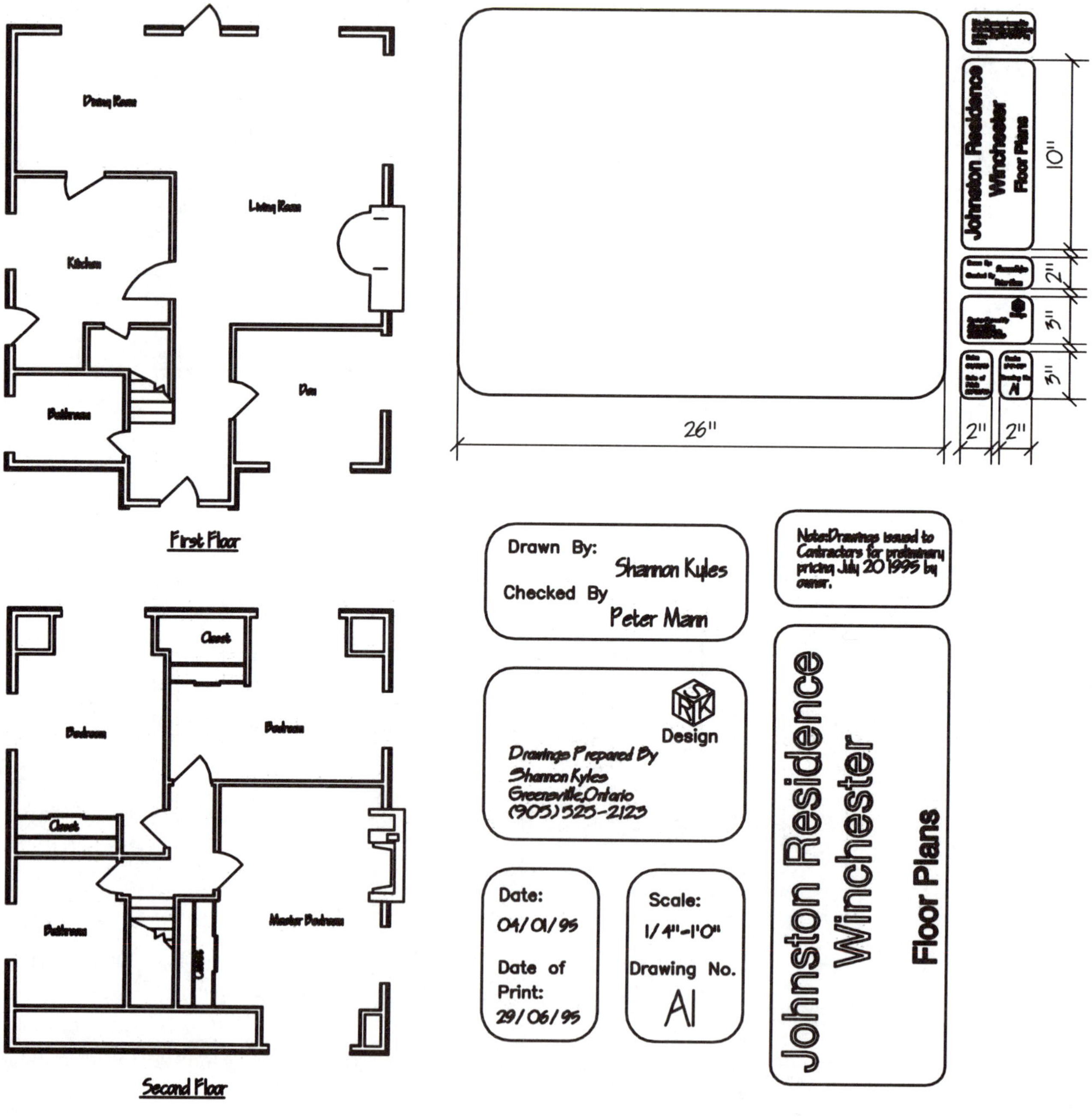

Exercise C8

Draw the title block in first, then draw in the foundations.
Sizes for footings are in Exercises noted.

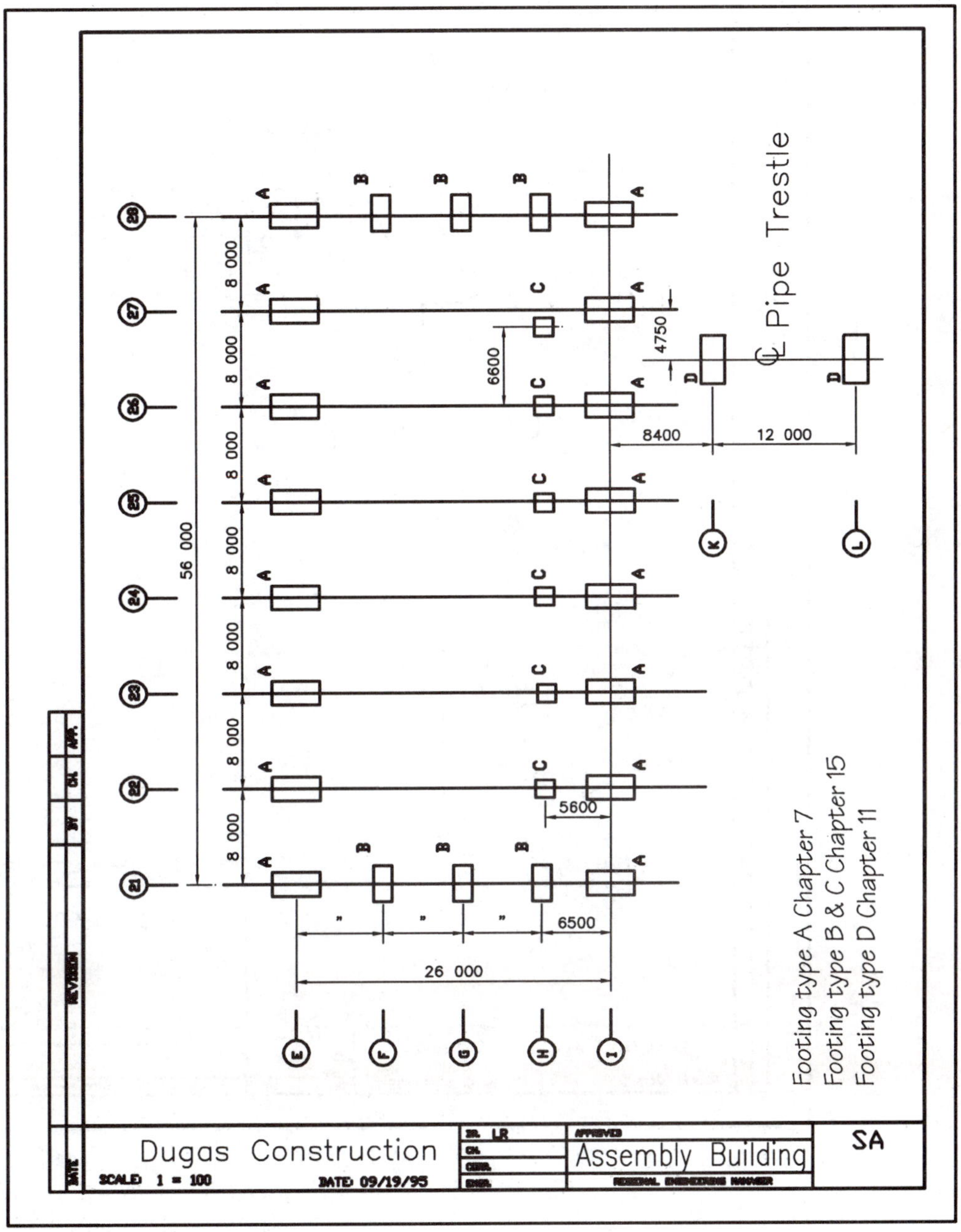

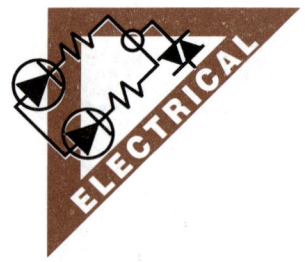

Exercise E8

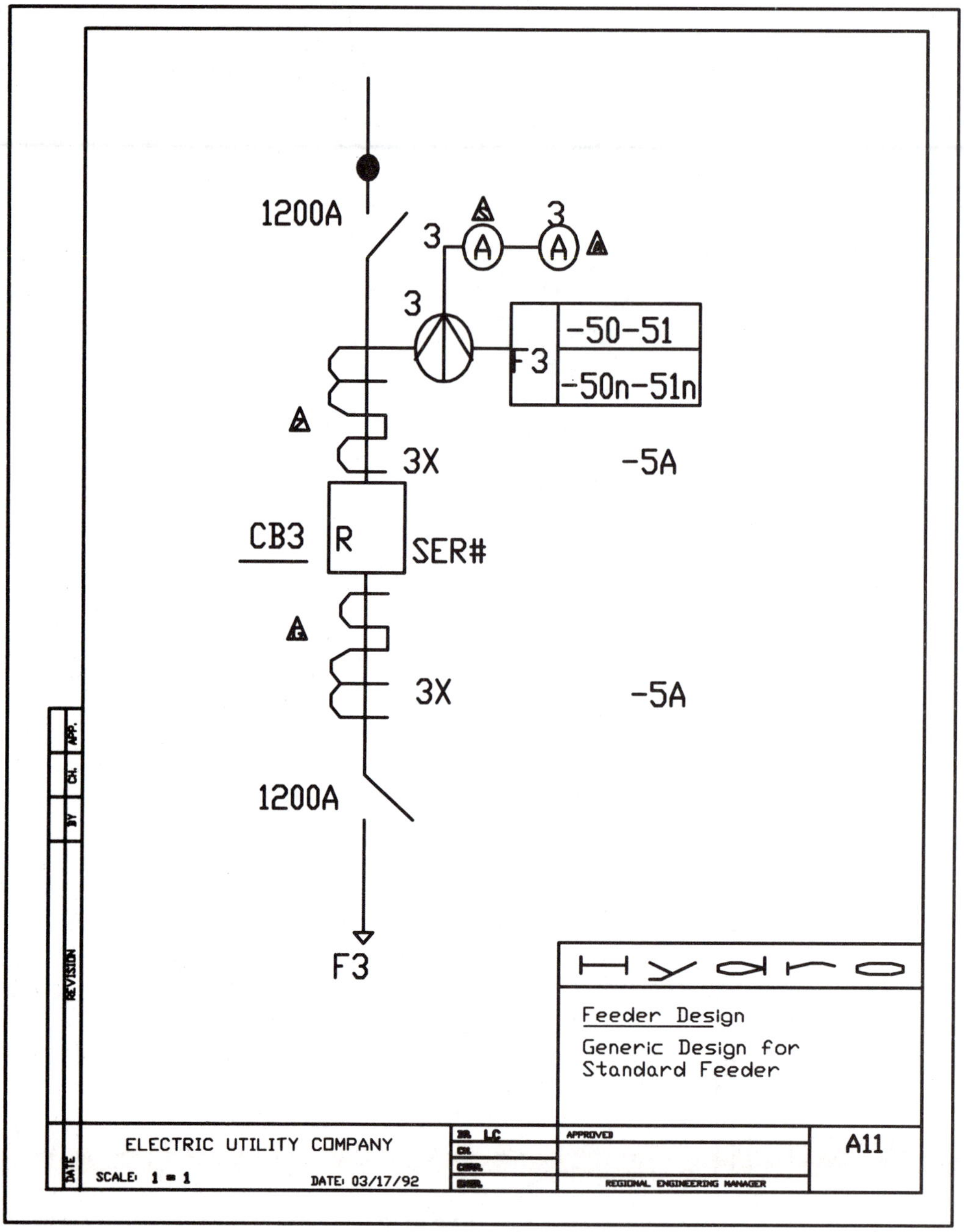

Draw the title block in first, then add the text. Once this is completed, you can start on the drawing and add the notations.

Exercise M8

Draw the title block in first, and add the text. Once this is completed, you can start on the drawing and add the notations.

Change the SNAP to get the isometric in easily.

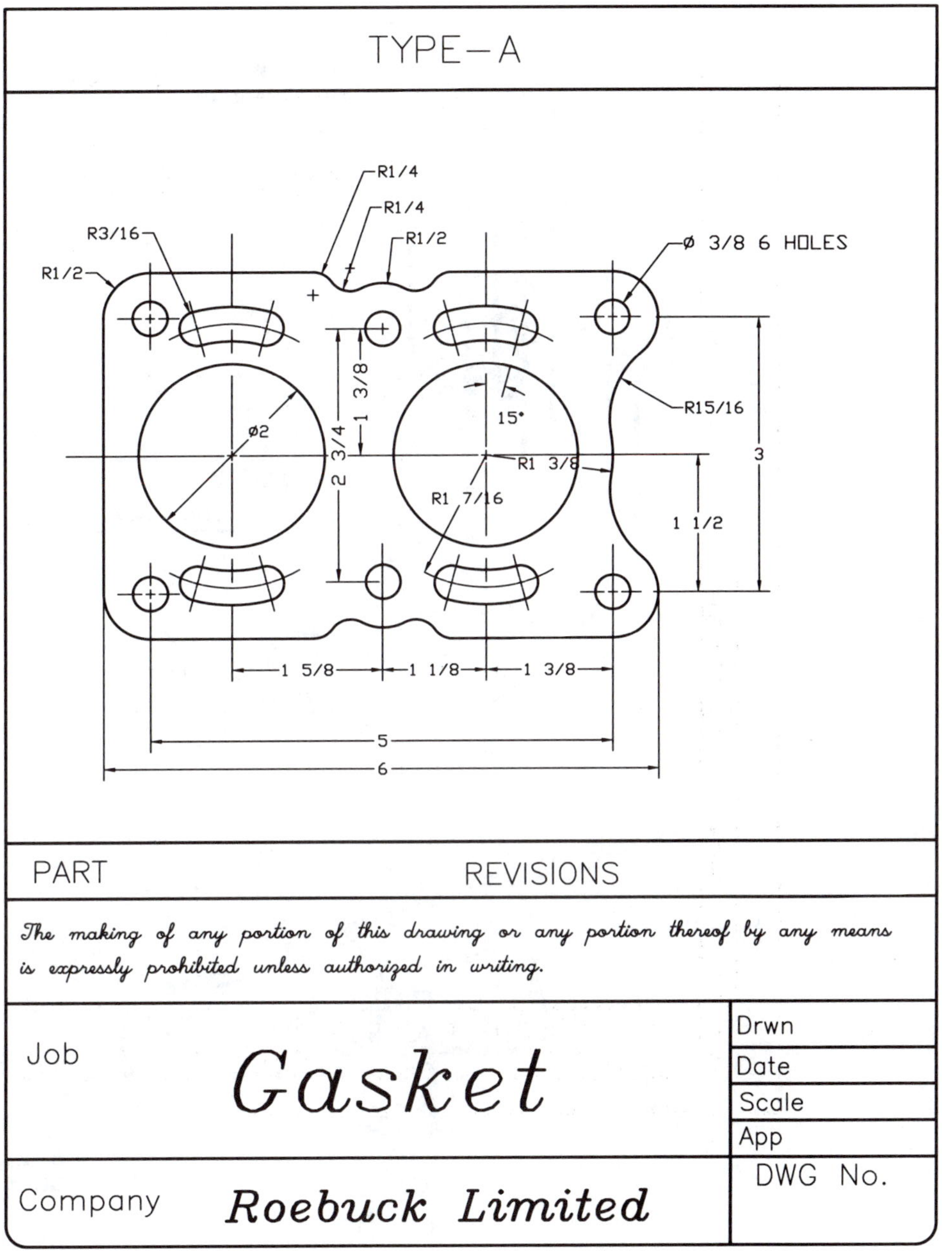

Challenger 8

These architectural details will be useful later.

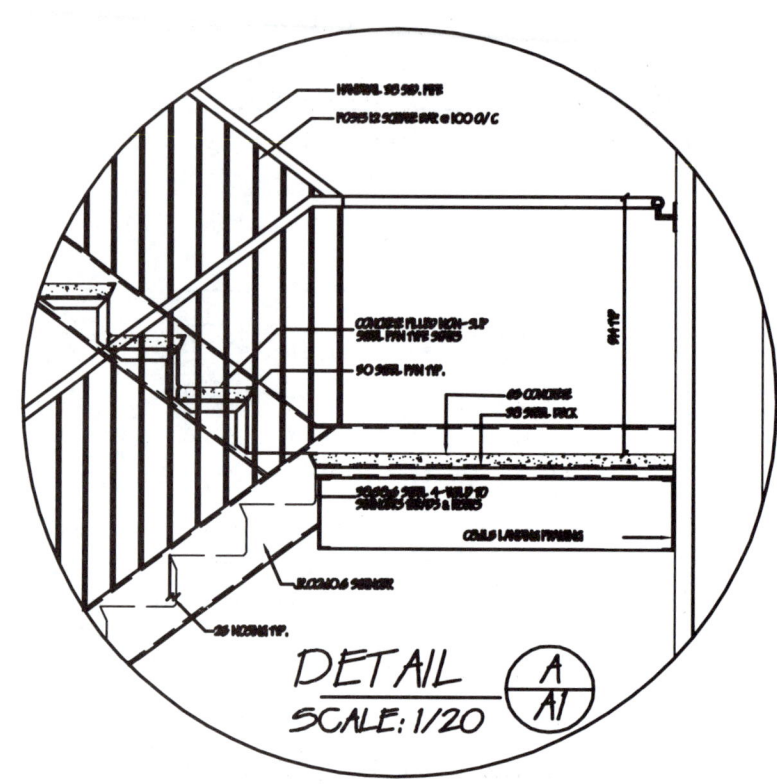

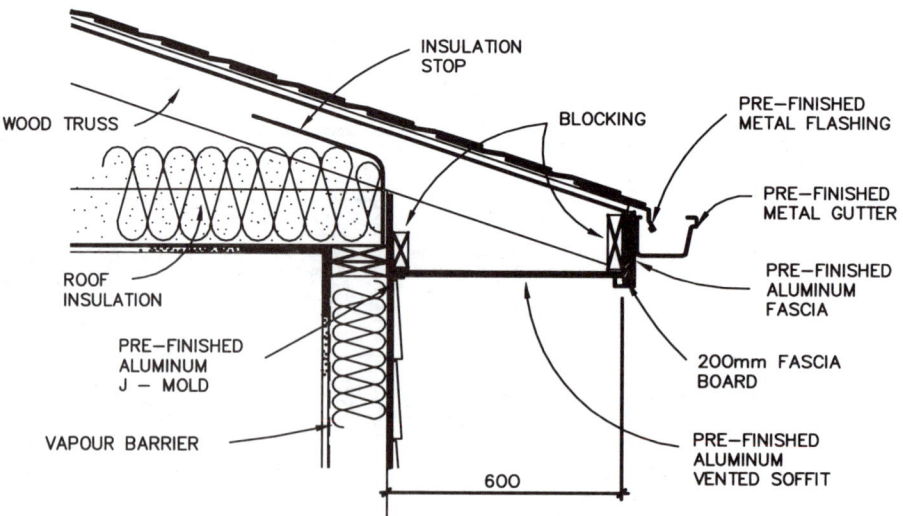

(Many thanks to UMA for this drawing.)

9 HATCH and SKETCH

Upon completion of this chapter, you should be able to:

1. Use the HATCH command
2. Use BHATCH with the dialog box
3. Edit existing hatches
4. Use the SKETCH command
5. Use point filters

The HATCH Command

The HATCH command fills an area with a specified pattern. It creates a nonassociative hatch pattern which will not update when the boundaries are updated. (The BHATCH command, described in the next section, creates a hatch pattern that will be updated along with the associated boundary.) Unless otherwise specified, HATCH combines the lines that make up the hatch into a block.

The HATCH command is accessed through the command prompt.

To start, you must create a boundary for the hatch. The geometry must be perfect: the corners must all meet, and there must be no gaps or overlapping items. Closed polylines, circles, and ellipses can be selected as single objects for hatch generation. The hatch command is found only by typing it in.

The default hatch pattern is a 45 degree angle.

Enter **HATCH** at the command line.

```
Command: HATCH
Pattern (? or name/U,style)<ANSI31>:
Scale for pattern<1>:
Rotation angle<0>:
Select hatch boundaries or RETURN for
   direct hatch option
Select objects: (pick 1)
Select objects:
```

Figure 9-1

In order to create a HATCH, you need four basic parameters: the pattern, the size, the rotation angle, and the items which are to be hatched.

Objects can be selected by individual selection, Window, Crossing, or any of the other object selection methods.

Direct Hatch

This option allows you to define the boundary of the hatch within the HATCH command. You can either retain the boundary or not. AutoCAD prompts for points until the boundary is complete. Both line and arc segments can be added to the boundary. Pressing ⏎ ends the command and creates the hatch.

If you opt not to retain the boundary, only the hatch pattern is drawn. This is useful for areas where only a hatch, not a boundary, is required.

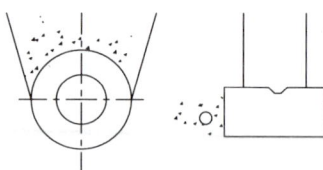

Figure 9-2

```
Command:HATCH
Pattern (? or name/U,style)<ANSI31>:⏎
Scale for pattern<1>:⏎
Rotation angle<0>:⏎
Select hatch boundaries or RETURN for direct hatch option
Select objects:⏎
Retain polyline?<n>:⏎
From point: (pick a point)
Arc/Close/Length/Undo/<Next point>:
   (pick the remaining points)
Arc/Close/Length/Undo/<Next point>:⏎
From point or RETURN to apply hatch:
```

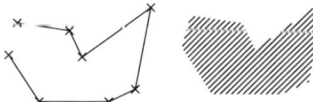

Figure 9-3

You can either choose one of the existing hatch patterns, or create a series of straight, parallel lines to form your own hatch pattern by selecting the User option. List the HATCH patterns in the HATCH command with the **?**.

```
Command:HATCH
Pattern (? or name/U,style)<ANSI31>:?
```

Scale, Patterns, Rotation angle, and Style for HATCH are the same as for BHATCH.

The BHATCH Command

BHATCH (Boundary Hatch) fills an enclosed area with an associative hatch pattern that will update when the boundaries are modified. In addition, BHATCH allows you to preview the hatch pattern and adjust the definition or options of the hatch such as scale and angle.

BHATCH is accessed through a dialog box.

The BHatch Dialog Box

> **Windows** From the Draw toolbar, choose Hatch, then this button:
>
> **DOS** From the Draw menu, choose Hatch, then Hatch.

The command line equivalent is **BHATCH**.

The BHatch dialog box controls every aspect of the hatch pattern, and under the Advanced option you can create a new boundary.

Pattern Type   lists the current pattern.
Pattern Properties   controls the ACAD standard hatch patterns. Hatch patterns are listed by name in the dialog box.
Boundary   controls the boundaries chosen.

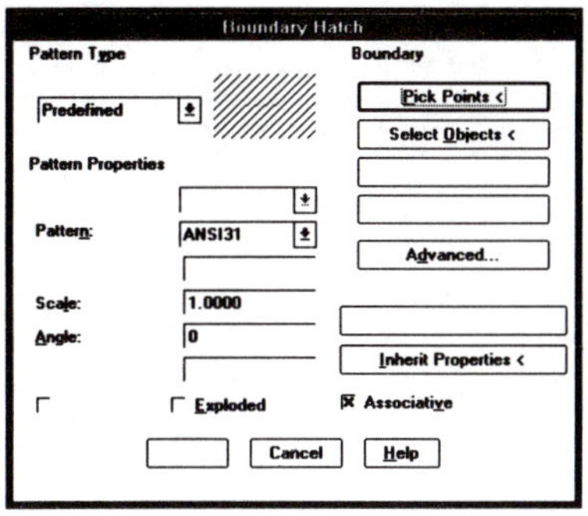

Pattern Type

Figure 9-4

Sets the pattern type, whether predefined or taken from the standard AutoCAD pattern list; user-defined or defined by a series of parallel lines; or a custom pattern from a customized file or third-party vendor.

Pick on the displayed pattern to see the next pattern in the predefined list.

Pattern Properties

Sets the properties specific to the chosen pattern type.

ISO Pen Width   controls the pen width if an ISO pattern is chosen from the Pattern list.

Pattern   allows the user to choose a pattern from the standard list. Pick the down arrow to view the available patterns. This area lists only the names of the patterns.

Patterns   are set up in groups. The ANSI patterns comply with the American National Standards Institute formats. The ISO patterns comply with the International Organization of Standardization formats. Architectural patterns preceded by AR usually have a scale factor 12 times that of the other patterns.

Custom Pattern   allows you to access any custom patterns that you may have.

Scale   allows you to control the scale of the hatch patterns. The default scale factor is 1. This means that the hatch is calculated to be displayed at the default screen size or an 11″ × 8 1/2″ sheet.

Scale

Each hatch pattern has a specific number of lines per inch. For example, the ANSI31 pattern generates three lines per inch at a scale of 1:1. If you have changed the limits and the model is larger than 11 units in X, you *must* change the scale factor.

When working in inch units, the scale factor of the hatch should be the maximum X value of the screen divided by 12, the same as the scale factor for LTSCALE. Example, limits at 0,0 and 24,18 would be a scale factor of 2.

> **Danger**
>
> If you do not adjust your scale factor, AutoCAD will attempt to create the hatch at the default scale. There is a good chance you will run out of room on your floppy disk and be forced to exit from the file. Save your file before hatching, and preview the hatch just in case.

HATCH and SKETCH

Other scale sizes are shown in Figure 9-5. Remember, if you are using architectural units or feet and inches, each unit is an inch, not a foot.

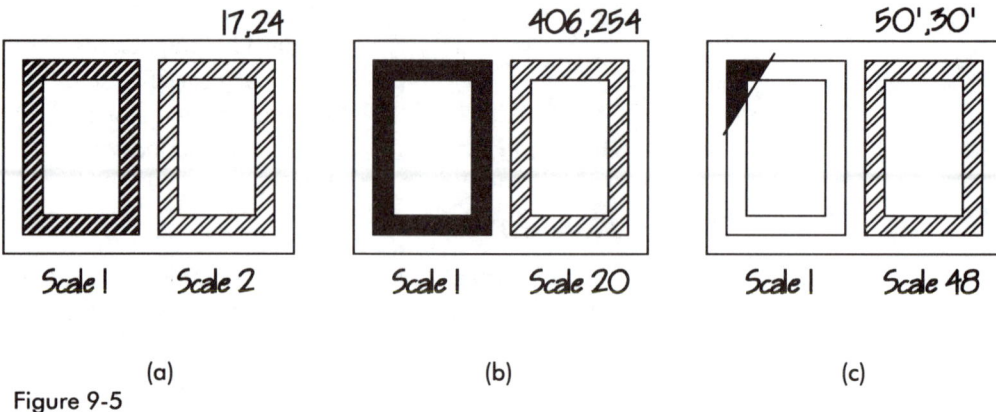

Figure 9-5

If you do not change the scale factor, there will be far too many lines to view, and the lines from the patterns will quickly fill up space on your disk. If you have not changed the scale factor, and the pattern is much too small, then it is possible you will have a system crash or be forced to exit your file. AutoCAD will make an attempt to save everything on your file up to the point where the hatch started, but there is no guarantee. You can be sure that, if this is going to happen, it will happen 20 minutes before the drawing is due.

Always use the Preview Hatch option just in case or create a small test pattern on the screen, about 1″ × 1″ relative to the real size or actual vectored inches of the screen, i.e. not relative to the limits but to the screen. You can tell by the test pattern if the scale is appropriate. If, on the other hand, the test area is totally filled in or remains blank, the scale needs to be changed.

Usually, people add hatches because they are going to be plotting onto paper. The hatch scale should be determined by the size of the final drawing, not the size of the object being hatched. If you have to change the scale of a hatch, keep in mind the final size of the plot as well as the size of the drawing relative to the screen.

In Chapter 11, there is a discussion on scale with regard to the size of the hatch and linetype scales and the size of the drawing. Before producing the final drawings, check that chapter to see that the hatch is accurate.

Angle If you are picking a hatch from the pull-down menu, the rotation angle will be exactly as shown. For example, Figure 9-6(a) uses ANSI33, which is displayed at an angle of 45 degrees. To pick this pattern as it appears on the screen, do not change the rotation angle; leave it at zero, because that is the angle of the pattern itself.

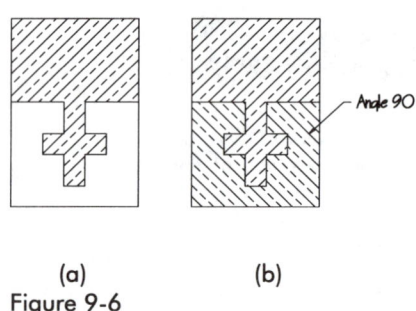

Figure 9-6

Figure 9-6(b) has ANSI33 at the default rotation and at a rotation angle of 90 degrees.

When using user-defined hatch patterns, the rotation angle is calculated at a horizontal, rotating counterclockwise.

Spacing is for user-defined hatch patterns; it refers to the distance between the first line and the second. This option is only available if User-defined is selected in the Pattern Type box.

Exploded specifies that the hatch pattern is to be created as individual line segments rather than as a block. EXPLODE can also be used once the hatch is in place to separate the hatch into individual line segments.

Boundary

For many students, this is where the difficulties begin, because objects must form a perfect boundary.

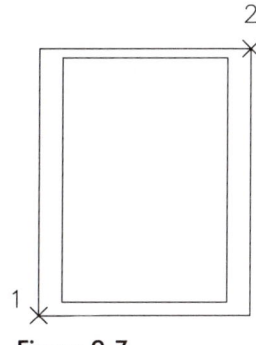

The problem many students encounter is that lines, circles, and other objects are sometimes not entered accurately. SNAP and OSNAP are not always used effectively and consequently the lines are crooked, do not have tidy intersections with adjacent items, or are otherwise defective. This means that these objects do not provide an adequate boundary for hatching.

Poor entry of objects will lead to other problems as well. If you have what appears to be four

Figure 9-7

objects on the screen, there should be only four objects (or one polyline) on the screen. If you list the objects using Window and have more than four objects, you may have trouble.

```
Command:LIST
Select objects:(pick 1) Other Corner: (pick 2) 7 found
```

If your hatch doesn't work because the point picked is outside the boundary or the boundary can't be identified, clean up your geometry and be more careful next time.

There are two basic ways of identifying the boundaries: by picking points and by selecting objects.

Pick Points identifies a boundary from existing objects that form an enclosed area. AutoCAD defines the boundary by analyzing all the closed objects in the area. Once you have chosen the Pick Points option, the dialog box disappears and AutoCAD prompts for point specification.

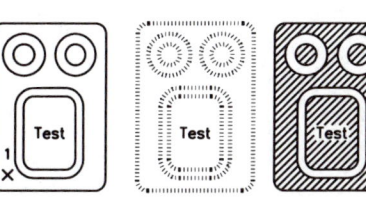

```
Command:BHATCH
(choose Pick Points)
Select internal point: (pick 1)
Select internal point:⏎
(choose Preview, then Apply)
```

This will provide the hatch as shown using the default pattern and scale.

Figure 9-8

Select Objects allows the user to identify the boundary by object selection rather than as an internal point. Sometimes both Select objects and Pick Points can be used successfully to identify one hatch boundary.

```
Command:BHATCH
(choose Select Objects)
Select internal point: (pick 1)
Other corner: (pick 2)
Select internal point:⏎
(choose Preview, then Apply)
```

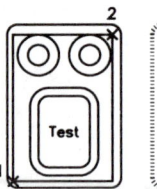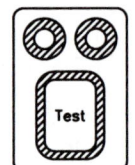

Figure 9-9

In both cases, the geometry must form a continuous boundary.

Preview Hatch displays the hatching before it is applied. This can save a lot of time and a lot of grief if you have not set the scale properly.

Inherent Properties allows the user to apply the properties of an existing BHATCH to a new boundary. The hatch style, rotation angle, and scale will all be applied to the new area. Once the Inherent Properties box is chosen, the dialog box disappears and the user is prompted to indicate the hatch pattern.

```
Select hatch object:
```

When the hatch has been identified, the dialog box returns so that Pick Points or Select Objects can be chosen.

Remove Islands removes from the boundary set objects defined as islands by the Pick Points option.

```
Command:BHATCH
(choose Pick Points)
Select internal point: (pick 1)
Select internal point:⏎
(choose Select Objects)
Select internal point: (pick 2,3)
Select internal point:⏎
(Preview — get the upper hatch)
(choose Remove Islands)
Select internal point: (pick 4)
(Preview — get the lower hatch)
```

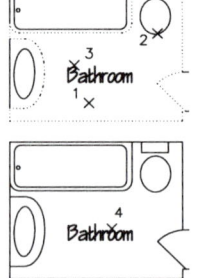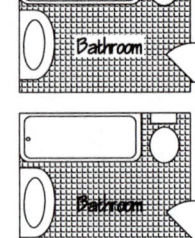

Figure 9-10

Remove Islands	selection sets act similarly to Remove Objects selection sets.
View Selections	displays the currently defined boundary set. This option is not available when no selection or boundary has been made.
Default Properties	resets the pattern properties to the current values in the hatching system variables.
Associative	is the default type of hatch pattern in BHATCH, meaning that the hatch will update if the boundaries are modified.
Apply	creates the hatch. Once the hatch is placed and previewed, choose Apply to have it become part of your drawing.

Advanced Options Dialog Box

This dialog box controls the definition of the boundary set.

Object Type	AutoCAD's hatches are created either in regions or in polylines.
Define Boundary Set	In large drawings this option allows the user to define certain objects or areas as the boundary set rather than taking all of the objects on screen. This saves time in producing boundaries.
Style	Once your data is correct, Window the objects you want to hatch, and the hatch pattern you have chosen will fill the boundary starting from the outer boundary. If there are closed boundaries within the outer boundary, the default is to have the hatch show up on alternating boundaries.

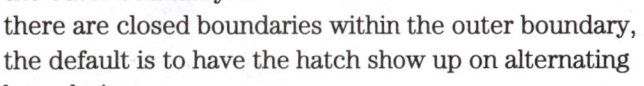

Figure 9-11

On rare occasions AutoCAD doesn't accept the boundary identified. In this case, use Ignore and TRIM to get the boundary where you would like it.

Hatching concave curves and multiple circles can cause hatching discrepancies.

Ray Casting	controls the way AutoCAD defines a hatch boundary. It helps to define boundaries when difficulties occur. If you encounter the prompt "Outside boundaries:" try the Ray Casting options.

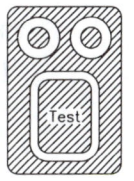 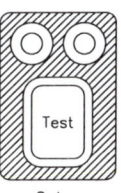 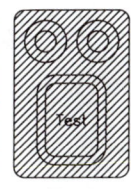

Normal   Outer   Ignore

Figure 9-12

HATCH and SKETCH   **227**

Island Detection specifies whether objects within the outermost boundary are used as boundary objects.

Retain Boundaries specifies whether your boundary will remain in the drawing after hatching is completed.

Editing Hatches

There are two ways of editing associative hatching. The first is to edit the boundaries; the second is to edit the actual hatch itself.

Editing Boundaries

With associative hatching, the hatch will update when the boundaries are changed.

```
Command: STRETCH
Select objects: (pick 1)
Other corner: (pick 2)
Select objects:↵
Base point: (pick 3)
Other point: (pick 4)
Analyzing associative hatch ...
```

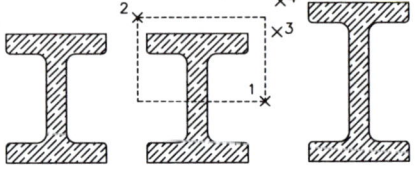

Figure 9-13

The hatch has updated along with the boundaries.

Editing the Hatch Properties

The Hatchedit dialog box and command allow for the modification of the hatch itself.

> **Windows** From the Modify toolbar, choose the Special Edit flyout, then the Hatch button:
>
> **DOS** From the Modify menu, choose Edit Hatch.

The command line equivalent is **HATCHEDIT**.

AutoCAD first prompts you to select the hatch object to be edited. Once this is done, the Hatchedit dialog box is invoked and you can choose the properties to be edited.

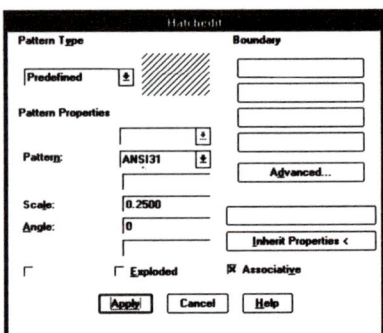

Figure 9-14

Figure 9-15(a) shows the base hatch. In (b), the pattern has been changed from ANSI35 to ANSI38. In (c), the scale of the hatch in (a) has been changed.

Most of the options are the same as those in the BHATCH command and dialog box.

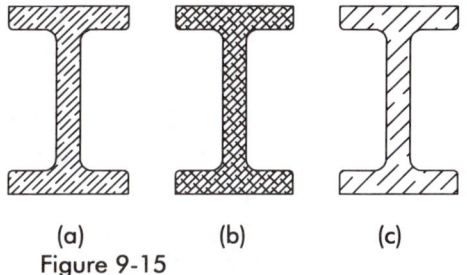

Figure 9-15

The SKETCH Command

The SKETCH command is included in this chapter because, like HATCH, it is often used to make final notations on drawings before they are plotted. In addition, just as the HATCH command has a tendency to take up a lot of room on a disk, the SKETCH command can also take up more room than expected on a disk because it can contain so many vectors.

SKETCH provides freehand sketching capabilities on a drawing. The motion of the mouse or digitizer determines the position of the sketch segments; the accuracy of the sketch segments is determined within the command.

The SKETCH command will prompt the user for the increment, or the distance between segments; for smoother curves, use smaller increments. The command is as follows:

Windows From the Miscellaneous toolbar, choose this button:

DOS From the screen menu choose Draw 1, then SKETCH.

The command line equivalent is **SKETCH**.

```
Command:SKETCH
Record increment<0.01>:.1
Sketch, Pen eXit Quit Record
  Erase Connect:P
Sketch:
```

Figure 9-16

Where: **Pen** = the pen being lowered. The system prompts to put your "pen down"; whatever move you make with your cursor will be recorded. **P** a second time will lift the pen up.

eXit = the end of the command. **X** will exit from the Sketch command, retaining all segments created up to that point.

Quit = a cancellation of the segments to date. **Q** will leave SKETCH without saving the segments you have drawn.

Record = a save of all the lines drawn so far. **R** will make a permanent record of your lines without exiting from the command.

Erase = an erasing of some of the line segments drawn. **E** will allow you to erase lines from a certain point; it acts as a "backspace" over segments created.

Connect = a continuation of a previous sketch. **C** lets you pick up sketch again at the last entered end point after it has been ended.

Segments are not recorded on the disk until you use the Record option. Because each segment is added as a separate object and can take up a lot of room on a disk, the computer will offer a series of warning beeps to let you know that you are moving too quickly or that the disk is full. Raise your pen, press pick, and record your lines to date before continuing.

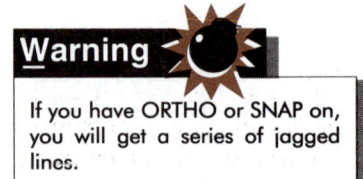

Warning
If you have ORTHO or SNAP on, you will get a series of jagged lines.

The SKPOLY Variable

AutoCAD captures sketching as a series of independent lines. Setting the SKPOLY system variable to a nonzero value produces a polyline for each contiguous sequence of sketched lines, rather than multiple line objects. This will save space on your disk.

Point Filters

X, Y, and Z *point filters* are used to make the entry of geometry easier.

With OSNAPs, the user can extract both the X and the Y value of points that are part of existing geometry; when you access the CENter of a circle, you are extracting the X and Y coordinates of that point. With point filters, you can extract just the X or just the Y value.

Just as with the OSNAP entries, a space must be entered after your filter option if you are typing it in.

```
Command:LINE
From point:.X (space) of MIDdle of (space)
```

In Figure 9-17, assume that you have the circle and the line. Now you want to place a doughnut at the X value of the center of the circle and the Y value of the top point on the line.

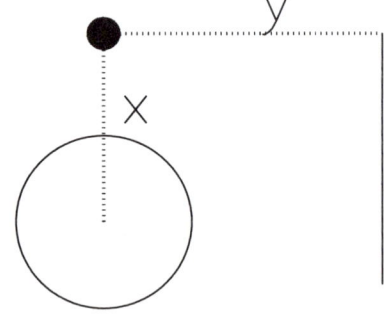

Figure 9-17

```
Command:DONUT
Inside diameter<0.5000>:0
Outside diameter<1.0000>:1
Center of doughnut:.X (space) of CENter of (pick the circle)
  (this gets the X value)
(Need YZ):.Y (space) of END of (pick the line) (this gets the
  Y value)
(Need Z):0 (positions the circle in the middle of the Z0 plane)
```

In Figure 9-18, assume that you have the vertical and diagonal LINEs and are trying to construct the PLINE starting at the same *X* value as the middle of the diagonal line, and the same *Y* value as the top of the vertical line.

```
Command:PLINE
From point:.X of MIDdle of (pick the
   diagonal)
(Need YZ):.y of END of (pick the
   vertical)
(Need Z):0
(complete the PLINE with options
   desired)
```

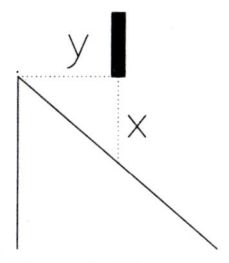

Figure 9-18

For MIDdle of and END of, the OSNAPs were used.

Rather than using filters, you could have created a horizontal line from the top of the vertical line and a vertical line from the middle of the diagonal line, and then started the PLINE from the INTersection of the two construction lines. But filters are much easier when you get used to them, and you don't need to either create or erase construction lines.

Using X, Y, and Z Filters

Filters "copy" coordinates of existing geometry.

With the construction lines, the draftsman lines up one view with another using his/her straight edge.

In English, you might say "Let's draw a horizontal line here, taking this other line and lining the straight edge up with it."

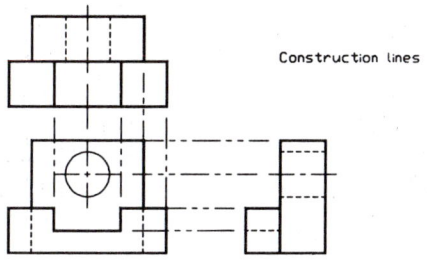

With the CAD drawing, the user picks points on the other views to have the computer calculate the perfect spot on the new view.

An English translation might be "Let's enter a line here, taking the *X* value of the end of this object and adding a *Y* of 1."

Or, even better, "Let's enter a line here, taking the *X* value of this line and the *Y* value of that."

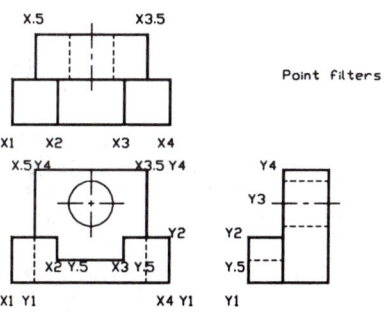

Now you have determined the first point of the line and you simply need to put in the second point.

Figure 9-19

On a part where the views are all on regular increments, the SNAP might be just as easy to use. But on a part where there are areas not on SNAP points, this offers two great advantages:

1. The drawing will be made much quicker, because you don't have to go to LIST or ID to pick up the point needed.
2. The accuracy will be far greater, because you can trust the system to enter the correct number instead of you figuring it out, writing it down so you won't forget,

and then typing it in. In the three steps needed to determine and enter a coordinate, there may be a high rate of errors.

Point filters work much the same way as OSNAP, in that you are asking for a point or parameter on something that already exists. The difference is that you are asking for only the X or the Y or the Z value instead of all of the coordinates attached to a particular object.

When used with OSNAP, filters filter out only the coordinate that you need.

Filters Without OSNAPs

If you are using a point in space to reference your filter, the value requested — .X or .Y — is taken from the point that you actually pick. If your SNAP is on, you will pick the coordinate with a preset integer; if not, you will pick exactly the point that you hit on the Z0 plane of the UCS.

If you are referencing an object with the use of one of the OSNAPS — CENter, ENDpoint, MIDdle, etc. — you must request first the filter desired, .X or .Y.

Entering Filters

A space must be entered after your filter option if you are typing it in. When you hit the space bar or ⏎ after this entry, your screen will show .X of or .Y of. Then you can pick the OSNAP that suits your purpose.

```
Command: LINE
From point:.X of ENDpoint of (pick 1) (this gets the X value)
```

Instead of ENDpoint or MIDdle of an object, you are asking for only the X value or the Y value of the specific portion of the object you have chosen.

To call up the filter, pick .X or .Y from the LINE menu, or type in .X or .Y. (Don't forget the period.) AutoCAD will then pick up only the X or Y value of the point indicated. Once you have picked up the value you require, you are prompted to fill in the other coordinates needed to complete the point.

The following is an easy exercise illustrating two-dimensional point filters.

X, Y Filters — An Exercise

We will just create a cross section of a simple bushing.

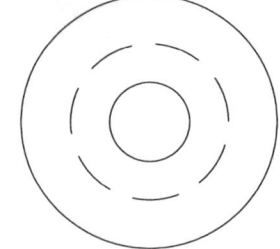

Step 1 First create three concentric circles on the left side of the screen. Keep the SNAP off so that you will not be tempted to use SNAPs rather than filters.

Step 2 Now extract the Y value of the top circle to create the first line of the section.

```
Command: LINE
From point:.Y (space) of QUADRANT of (pick 1) (this gets the Y
   value)
(Need XZ): (pick 2) (positions the XZ of the first point of
   the LINE)
```

CHAPTER NINE

```
To point:.Y of QUADrant of (pick 3)
(Need XZ): (press F8 = ORTHO on)
   (pick 4) (positions the first LINE)
To point:⏎
```

Make sure that ORTHO is on when placing the lower point on the line.

Step 3 Pick a spot to the right on your screen, then extract the Y value of the central hidden line circle. If you knew the length of the bushing, you could enter that, but for this exercise simply pick a point in space.

```
Command:LINE
From point:⏎ (this will pick the last point entered)
To point: (pick 5)
To point:.Y of QUADrant of (pick 6)
   (this gets the Y value)
(Need XZ): (pick 7)
To point:⏎
```

Step 4 Now add the horizontal line at any distance between the two verticals.

```
Command:LINE
From point:⏎ (this will pick the last point entered)
To point: (pick 8)
To point:.Y of ENDpoint of (pick 9)
   (this gets the Y value)
(Need XZ): (pick 10)
To point:ENDpoint of (pick 9 again)
To point:⏎
```

Step 5 Now that this section is in, you could of course simply mirror it through the center point. But you already know how to do that, so try to create the bottom half using both *X* and *Y* filters.

```
Command:LINE
From point:.X of ENDpoint of (pick 11)
(Need YZ):.Y of QUADrant of (pick 12)
(Need Z):0
To point:.Y of QUADrant of (pick 13)
(Need XZ): (pick 14) (ORTHO should
   still be on)

To point:⏎
```

Step 6 Using both *X* and *Y* filters finish the rest of the cross section.

Each point is going to need an *X* and a *Y* value.

HATCH and SKETCH **233**

Prelab 9A Point Filters

Looking at the drawing on the right, you can see that the side view would be easy to create once the front view is done, simply by lining up the horizontal lines with existing lines on the front view. Instead of using construction lines, try point filters.

For a really accurate model, *X,Y* filters are the easiest method.

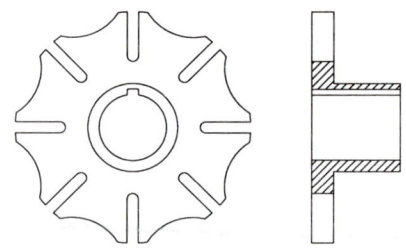

Step 1   Generate the geometry on the right using LINEs, ARCs, EXTEND, CIRCLE, and ARRAY. Make the center of the circle at 0,0.

The easiest way to approach it would be to **draw the two inner CIRCLEs plus the keyway, then draw one of the exterior scallops completely and ARRAY it.**

Once completed, you can start entering the cross section using the existing geometry as a guide.

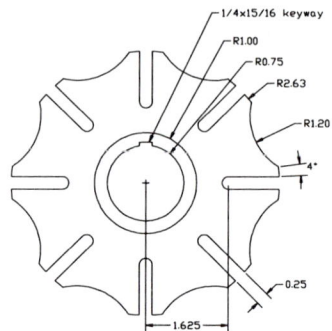

Step 2   Now add the vertical line from the top of the section to the center.

```
Command:F8 (= ORTHO on)
Command:LINE
From point:.Y of ENDpoint (pick 1)
(Need XZ):4,0
To point:.Y of QUADrant of (pick 2)
(Need XZ): (pick 3)
To point:@2,0 (gives the length of
  the line at 2 in X)
To point:⏎
```

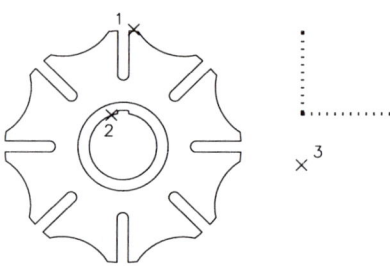

Step 3   Continue adding the horizontal lines across the bottom.

```
Command:LINE
From point:ENDpoint of (pick 4)
To point:.Y of QUADrant of (pick 5)
(Need XZ): (pick 6)
To point:@-1.5,0
To point:⏎
```

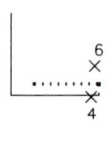

ORTHO should still be on.

234   CHAPTER NINE

Step 4   Continue adding lines.

```
Command:LINE
From point:ENDpoint of (pick 7)
To point:.Y of ENDpoint of (pick 8)
(Need XZ): (pick 9)
To point:@-.5,0
To point:↵
```

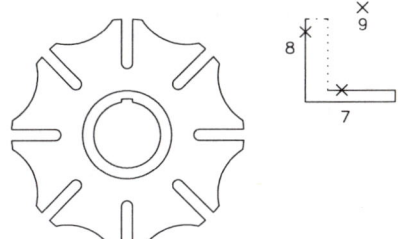

Step 5   Add the final two lines.

```
Command:LINE
From point:.Y of ENDpoint of (pick 10)
(Need XZ):NEAR (pick 11)
To point:PERpendicular to (pick 12)
To point:.Y of QUADrant of (pick 13)
(Need XZ):NEAR (pick 14)
To point:PERpendicular to (pick 15)
To point:↵
```

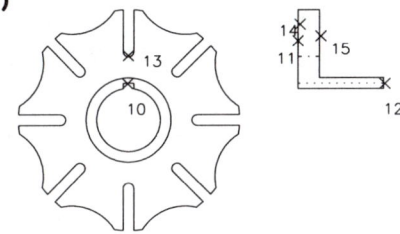

Remember that the default for a filter is to pick the point in space that you have picked. The filter won't pick up a portion of an existing object without the help of an OSNAP.

Step 6   Now that all of the lines are in, mirror the part through the center of the circle.

Erase the keyway line on the lower section. Use BHATCH to hatch the sectioned area and then add any lines needed to make it look like the illustration above Step 1.

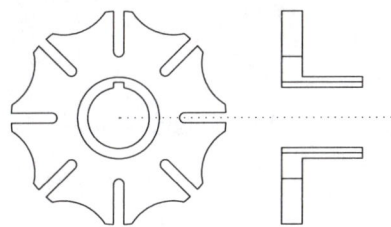

HATCH and SKETCH

Prelab 9B Hatch Using BHATCH

Step 1   In order to make this half-section assembly drawing, quickly draw up the part as shown. Notice that the increments are all in .25 units, so a SNAP of .25 may be of use.

Use Arc SER or Circles and Trim.

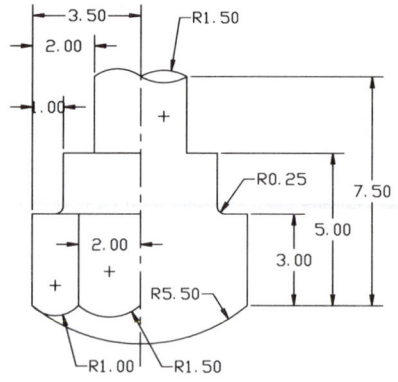

Step 2   Once this is completed, draw in the second half as shown.

You'll have to draw the full 180 degree arc and trim the dotted lines.

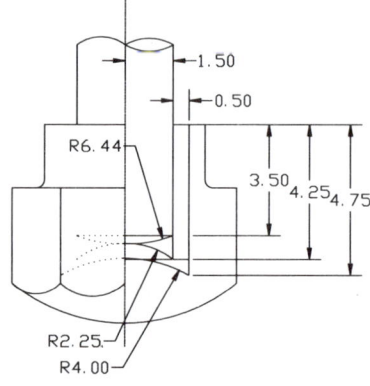

Step 3   Now access the BHATCH dialog box.

> **Windows** From the Draw toolbar, choose Hatch, then this button:
>
> **DOS** From the Draw menu, choose Hatch, then Hatch.

The command line equivalent is **BHATCH**.

This is the Boundary Hatch menu. From it, choose ANSI31.

```
Command: (pick ANSI31)
```

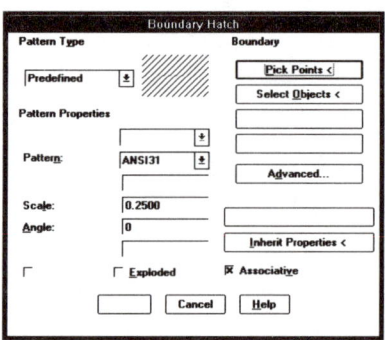

Now choose the ANSI32 pattern from the list.

```
Command: (pick ANSI32)
```

Step 4   Once you have chosen this pattern, you can choose the Pick Points option and choose the appropriate area.

    Command: **(pick Pick Points)**

    The dialog boxes have disappeared, and the part is visible. On your command line you will see the prompt:

    Select internal point: **(pick 1, 2)**

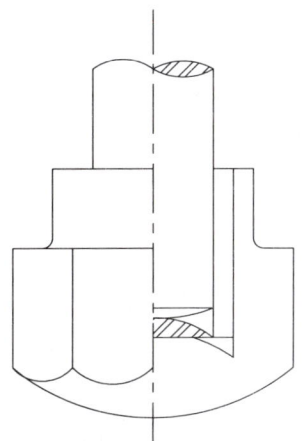

    Indicate by picking inside the areas that you would like hatched. The area boundary will be highlighted, showing where the hatch will go.

    Make sure the same areas as shown are highlighted; if not, recheck your geometry.

Step 5   When you have pressed  to indicate the end of the selection set, you will be returned to the Boundary Hatch menu. Choose Preview Hatch to make sure that it is OK. If satisfied, press ⏎ and return to the menu, then press Apply and the hatch will appear in the areas indicated.

    Command: **(pick Preview Hatch)**
    Command:⏎
    Command: **(pick Apply)**

Step 6   On the next hatch, follow the same menus with the following responses.

    Command: **(pick Draw, Hatch ...)**
    (pick ANSI32)
    (choose ANSI35)
    (pick Points, pick 3)
    (pick Apply)

    You will get the hatch pattern in the area shown.

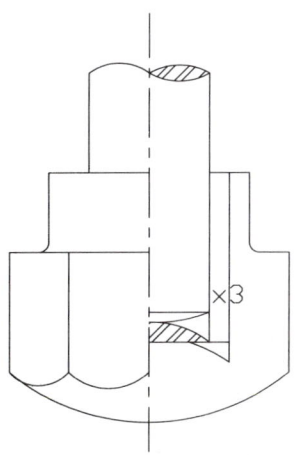

Step 7 For the final hatch, use the user-defined hatch patterns. Notice that the highlighted menu choices will differ according to the options you choose.

```
Command: (pick Draw, Hatch ...)
(user-defined hatch pattern)
Angle 135
Spacing .25
(OK)
(pick Points)
```

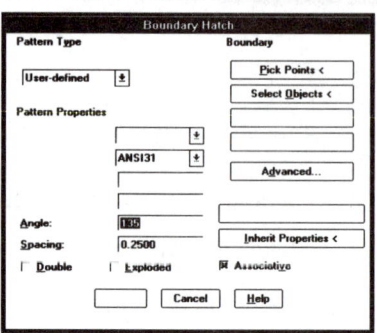

The ANSI31 hatch is the same, but try this one just for the practice.

Your final part should look like this.

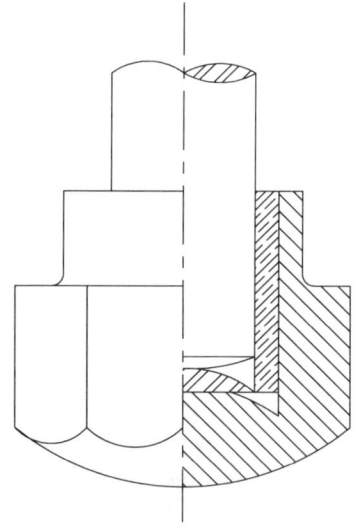

Command and Function Summary

BHATCH fills an enclosed area with an associative hatch pattern.

Point Filters filter an *X*, *Y*, or *Z* value from an existing point on an object for defining a point on a new object.

HATCH fills an area with a pattern.

SKETCH creates a series of freehand line segments.

Practice Exercise 9

Use the HATCH command to complete these examples. Change the angle of the user pattern to 55 degrees if there is an angle on the object line that is 45 degrees.

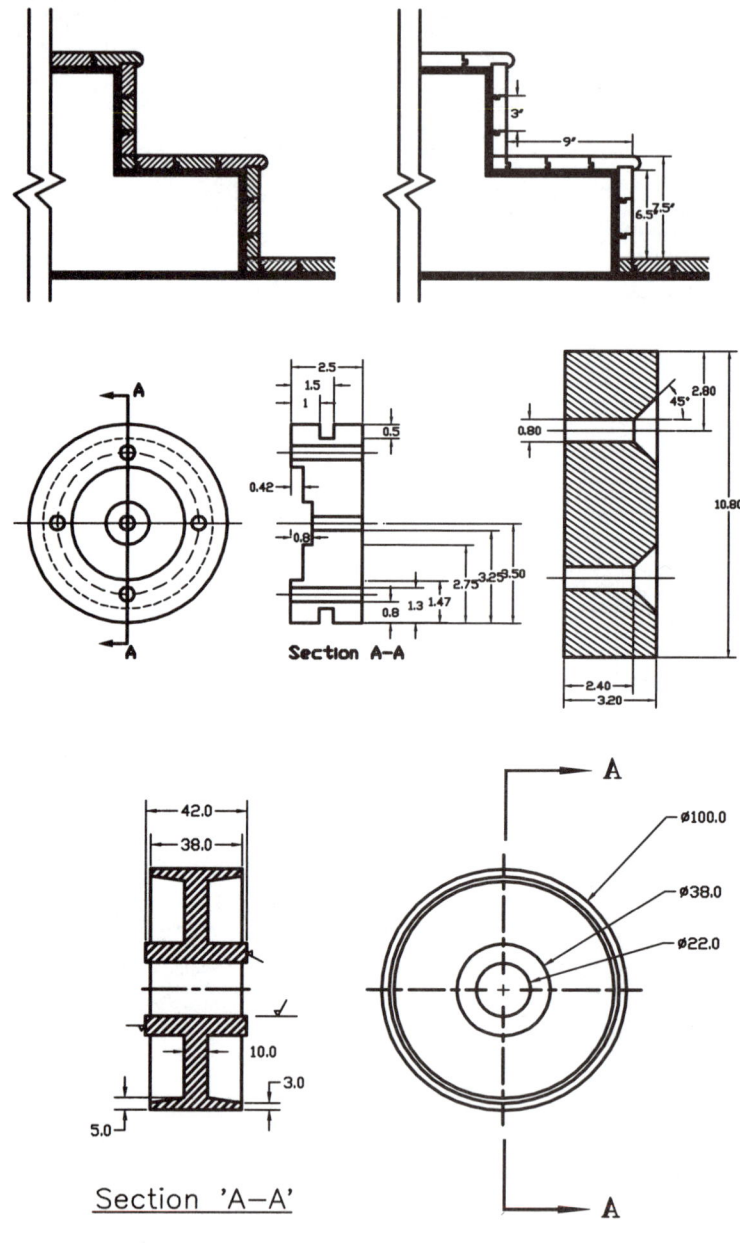

Section 'A–A'

Wheel

Exercise A9

Open FIRSTFL and create layers called Veneer, Fireplace, and Firebrick. Make veneer current and draw in the outline for the veneer. Use HATCH with ANSI31 at a scale of 48 to fill it in. On the fireplace, draw in the outlines with the appropriate layer and fill in the areas as shown. Do the same for the second floor.

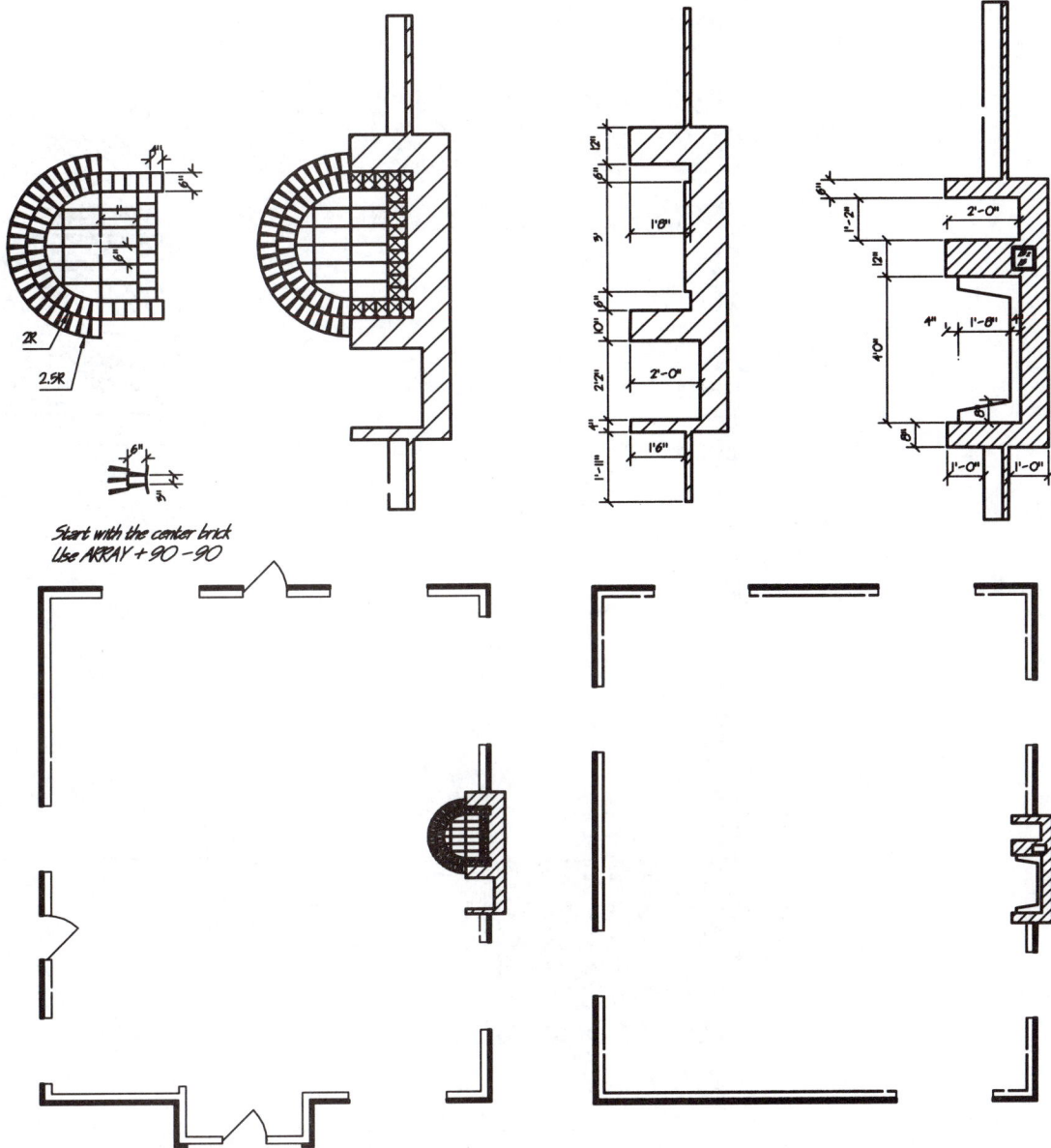

Exercise C9

Cavity Wall at Corner Parapet

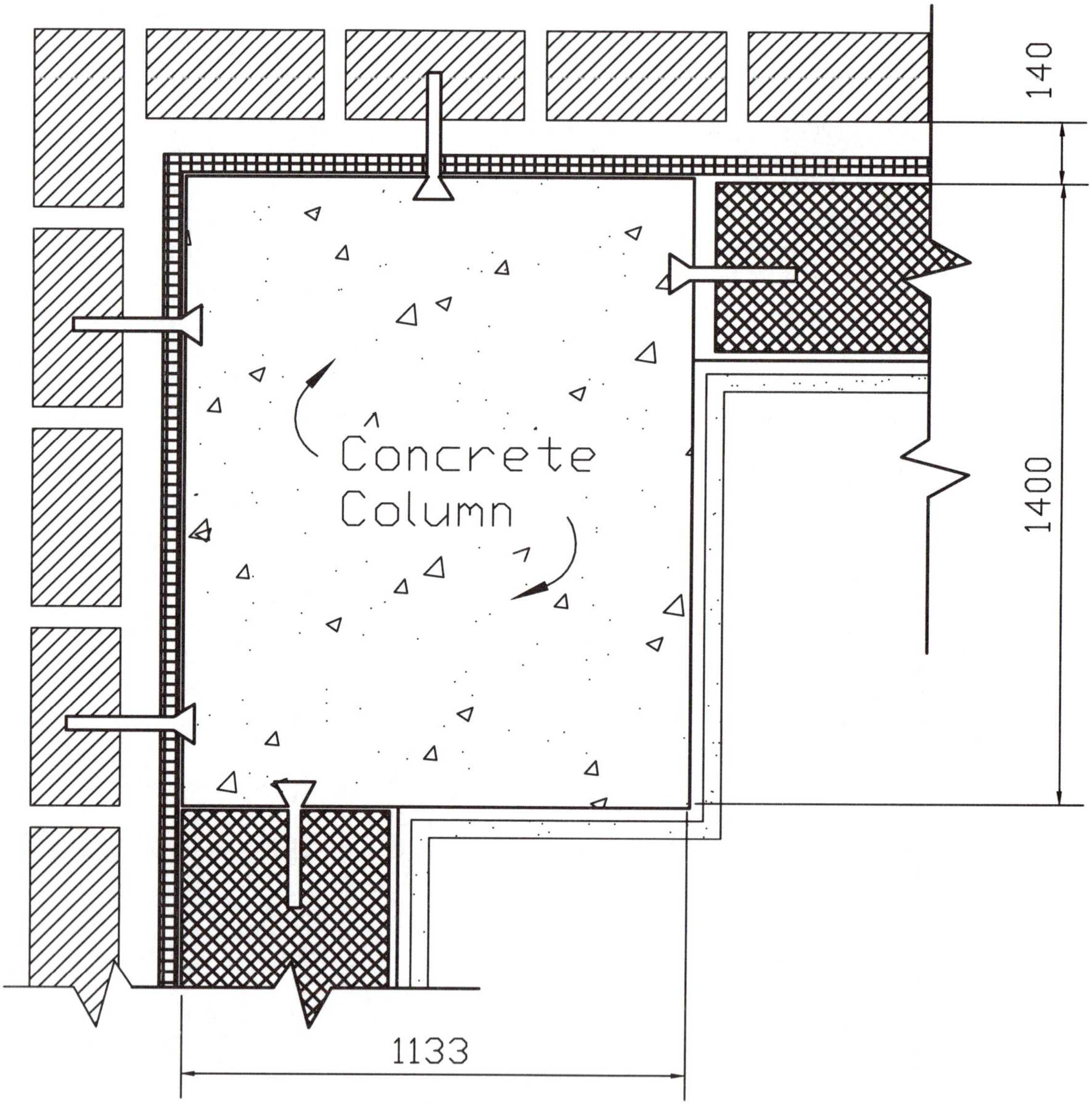

Hints on Exercise C9

Turn off the layer of sections you are not using when creating hatches. Remember you will need to change the hatch scale factor because the area is large.

Keep the text within the boundary when hatching, and it will be accepted as a separate boundary. If you are finished early, add a layer for the mortar, the pins, and the notations.

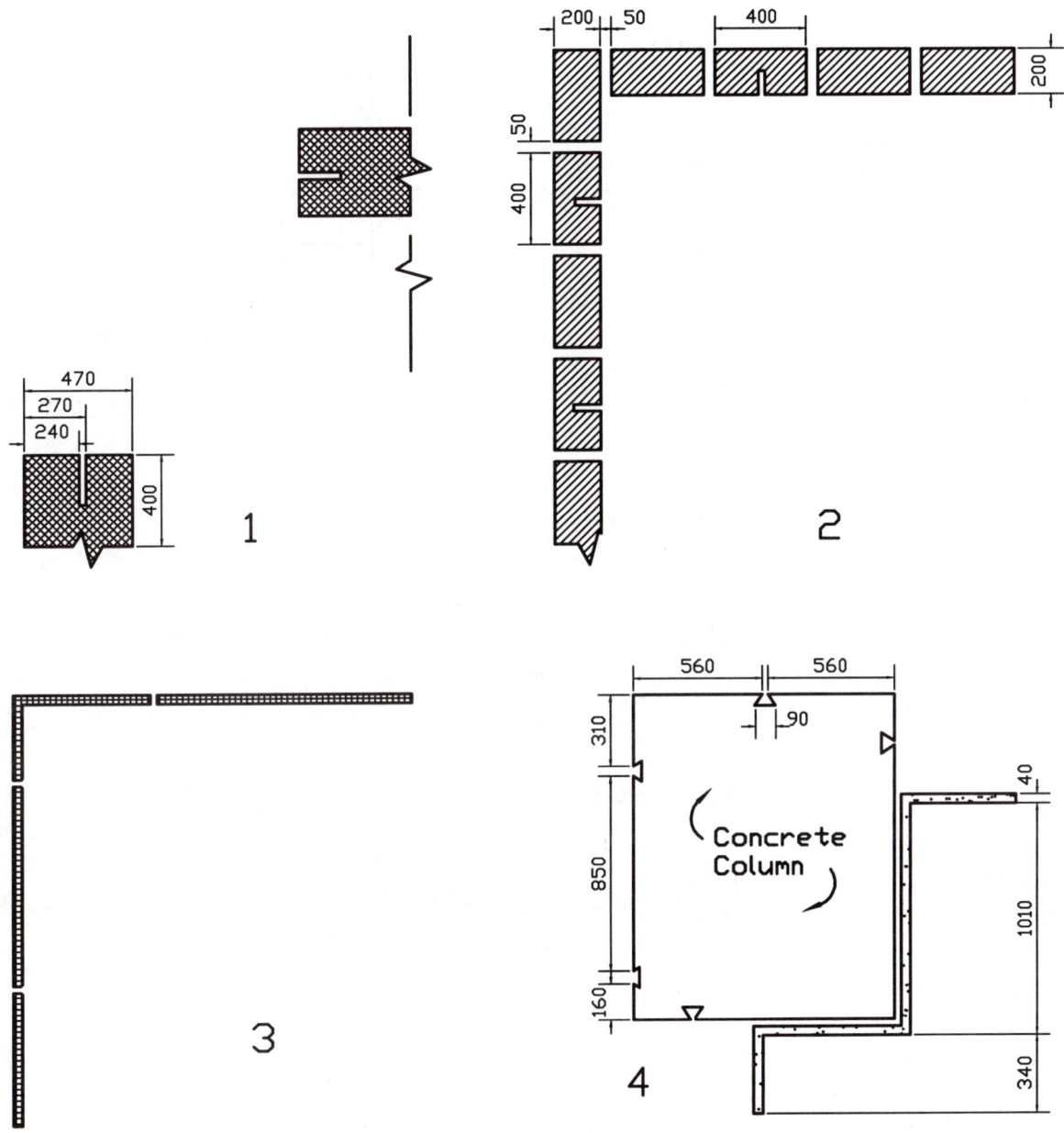

Exercise E9

This layout should fit on an 8.5″×11″ sheet.

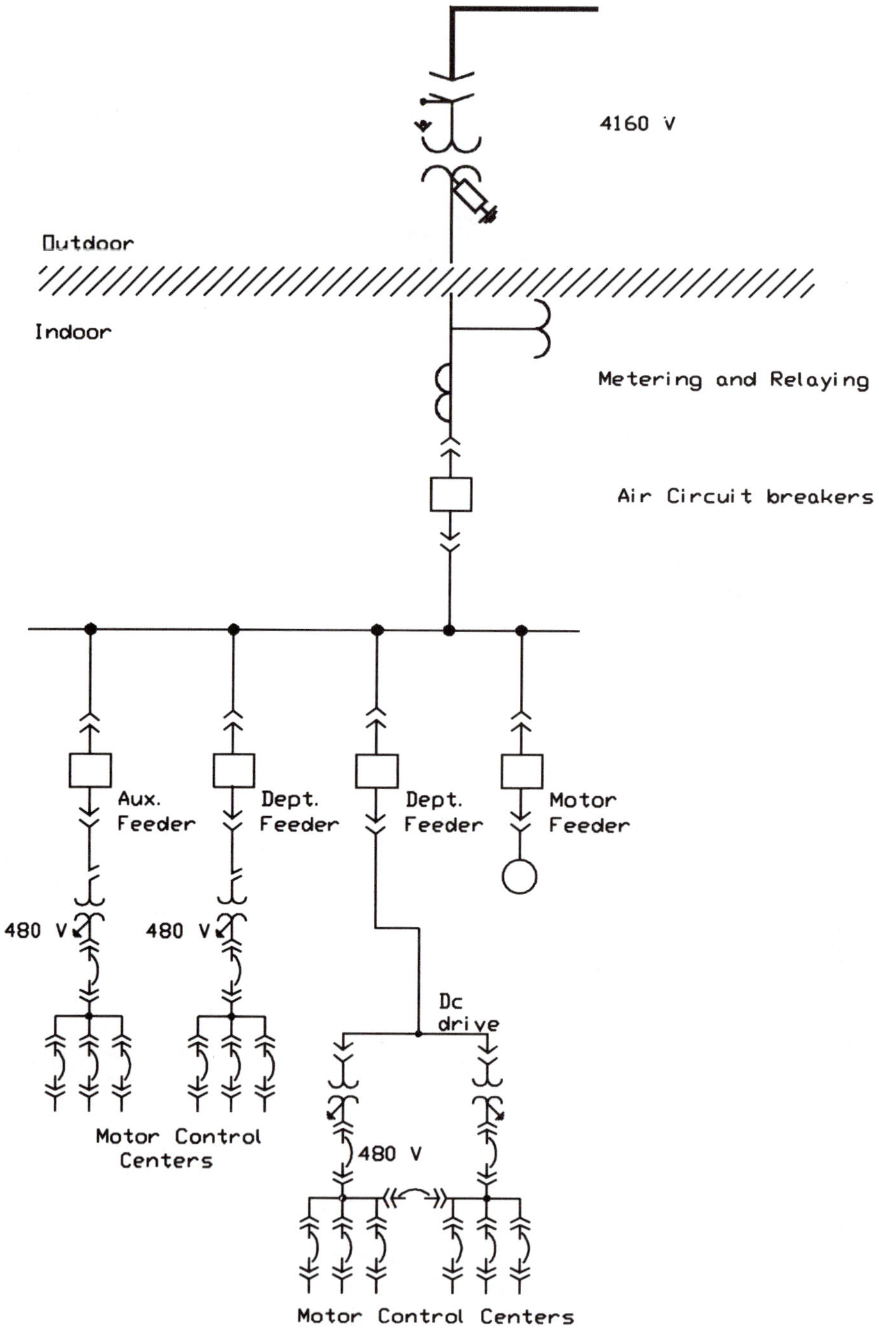

Exercise M9

Complete the drawing as in "Hints on Exercise M9" (page 246; use the dimensions listed there). Separate layers can be useful.

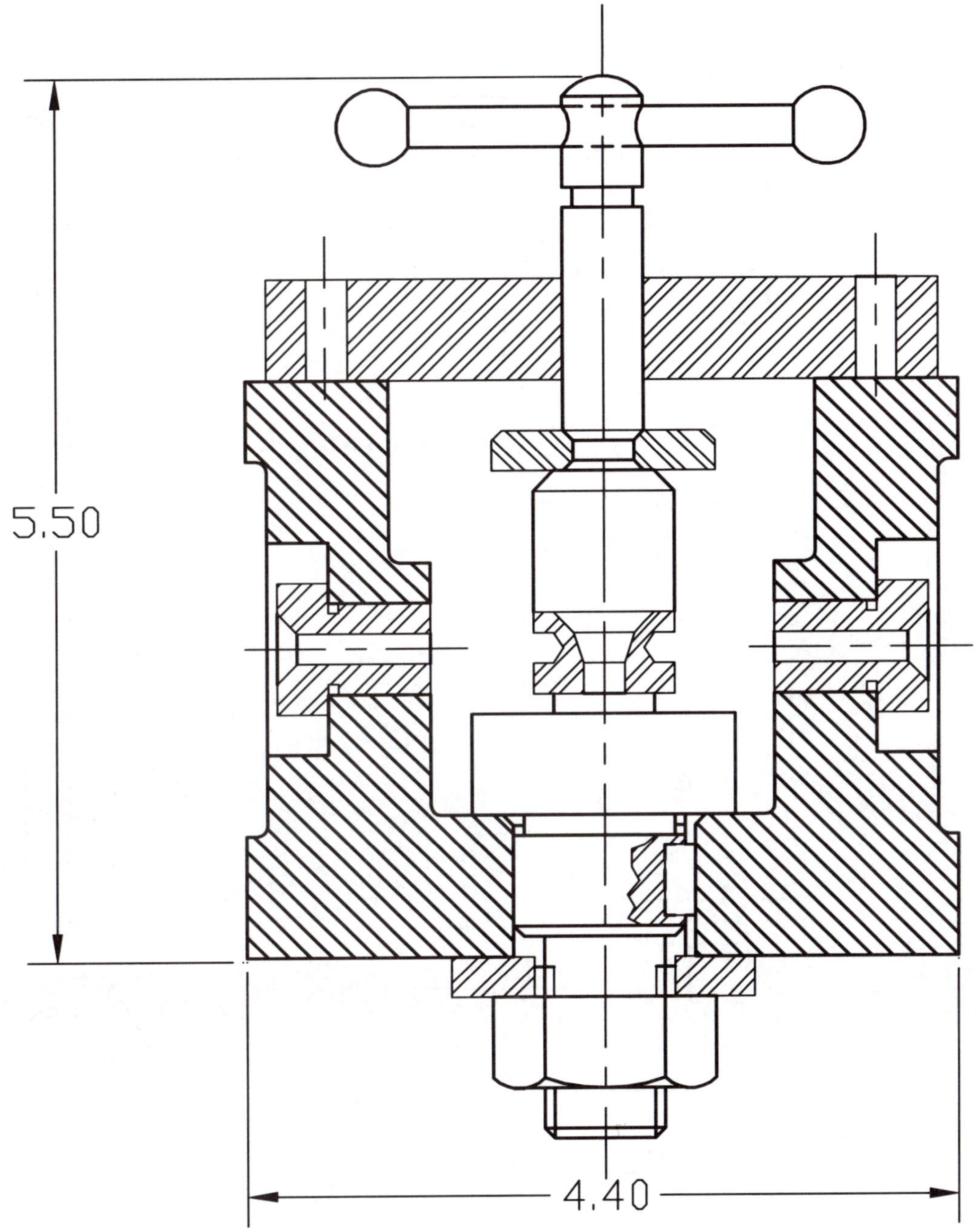

Hints on Exercise M9

Use the Pick Points option in the HATCH command to fill in the various boundaries. You will need to rotate one hatch by 90 degrees. Please note that the dimensions indicated are for you to create the part only.

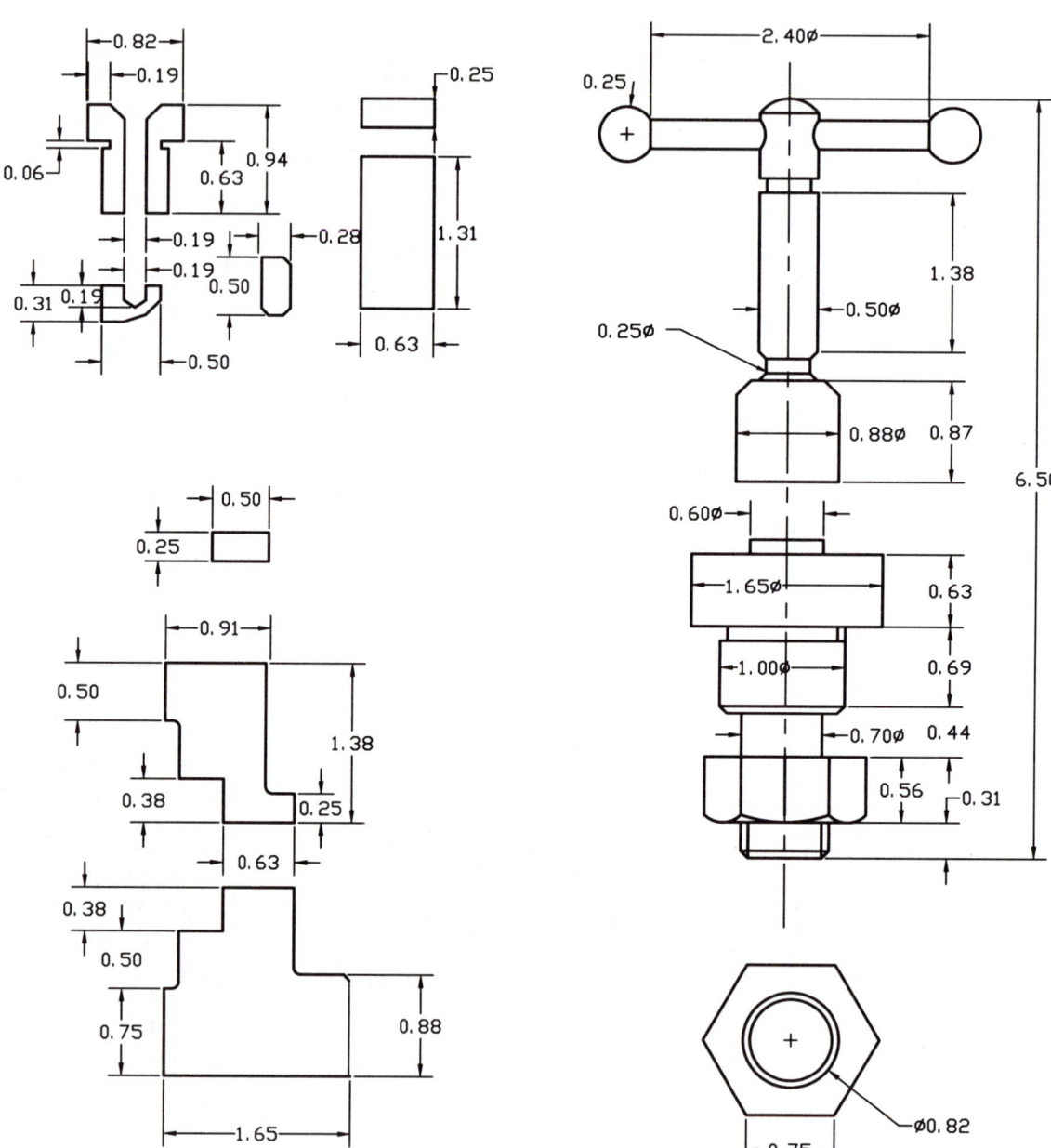

Challenger 9A

LIMITS: 20',15'
SNAP: 6''
GRID: 1'
ZOOM: All

Dimensions are on the following page.

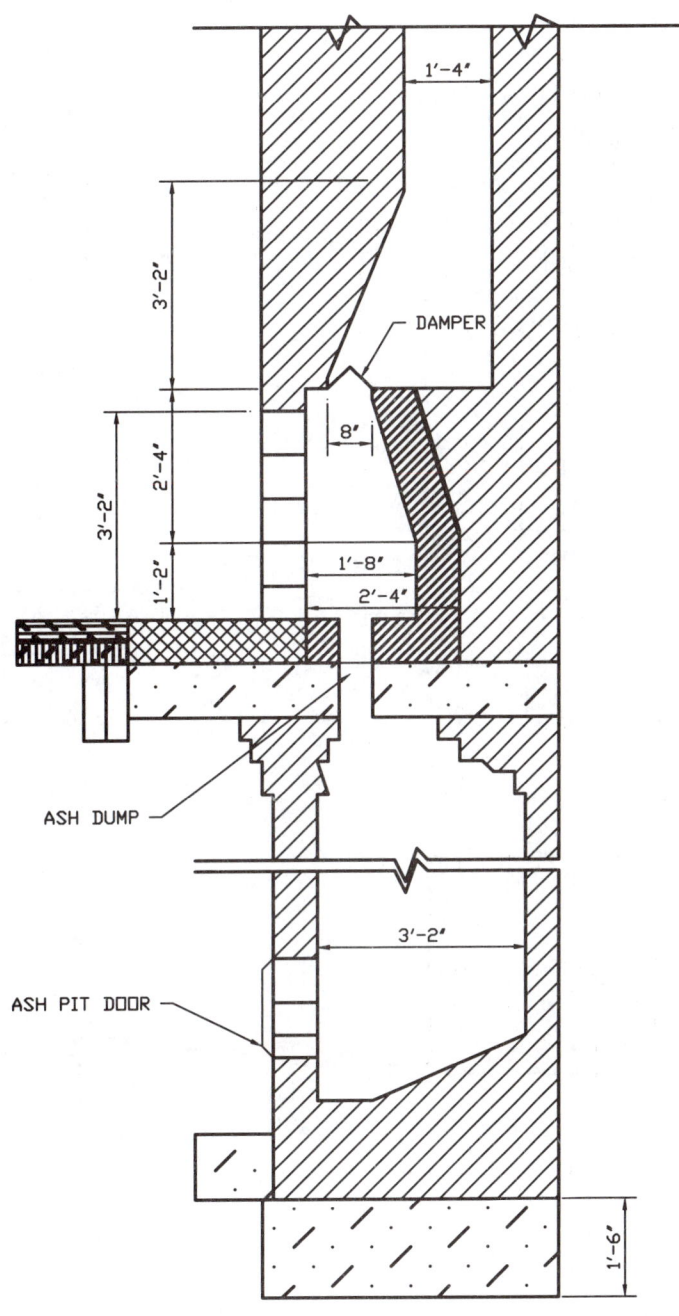

Hints on Challenger 9A

These sections do not need to be on separate layers after Release 11. It may be a good idea, however, to create different layers for insul-brick for estimating purposes later.

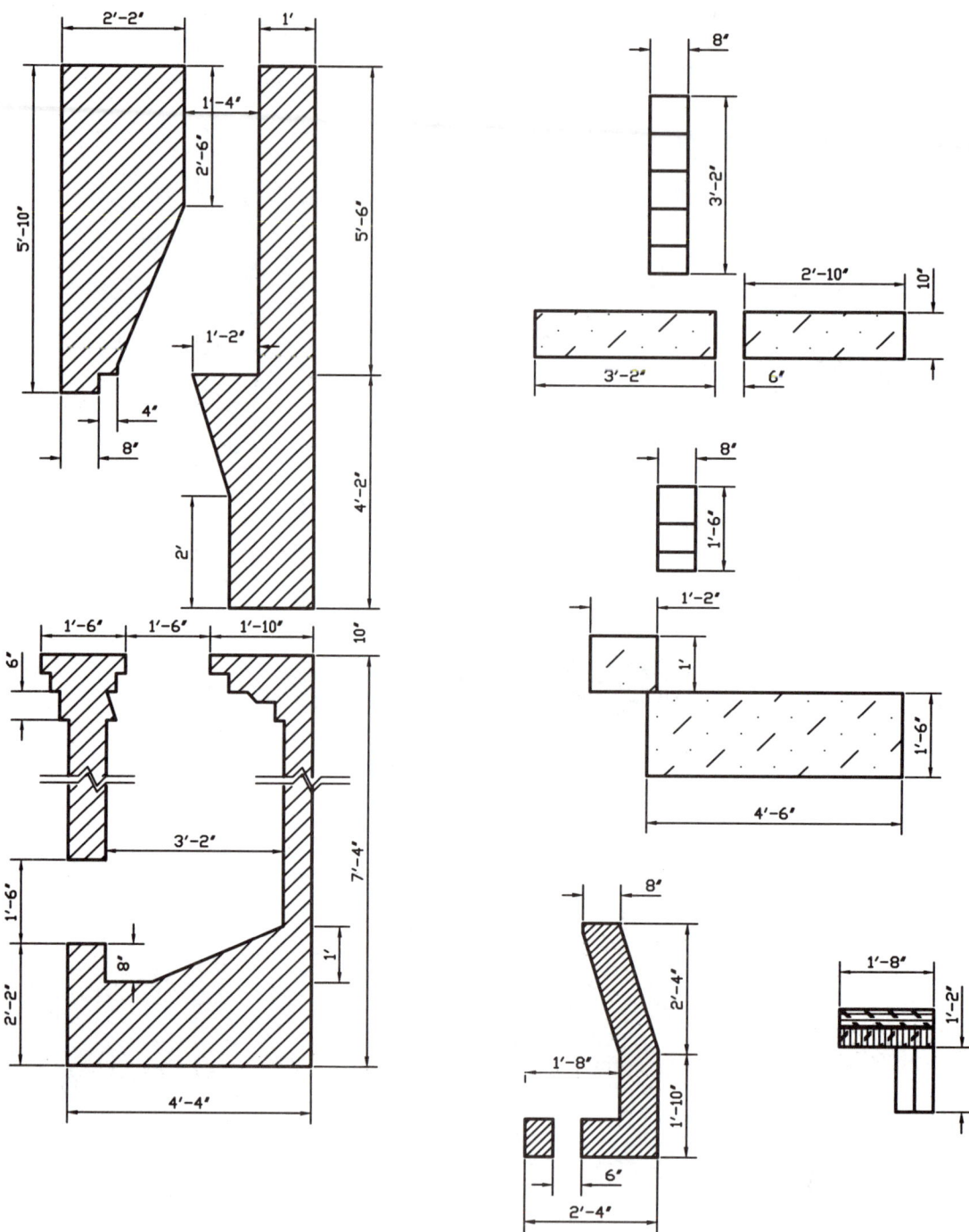

Use the Pick Points option of the HATCH command to access the different areas to hatch. You will need to change the rotation on at least one pattern. If you are using architectural units, make sure that you change your hatch scale factor to at least 12 (1″ = 1′0″).

Challenger 9B

This second example will give you experience with multiple hatches.

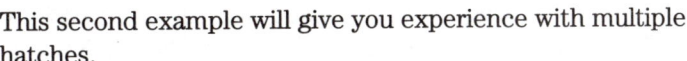

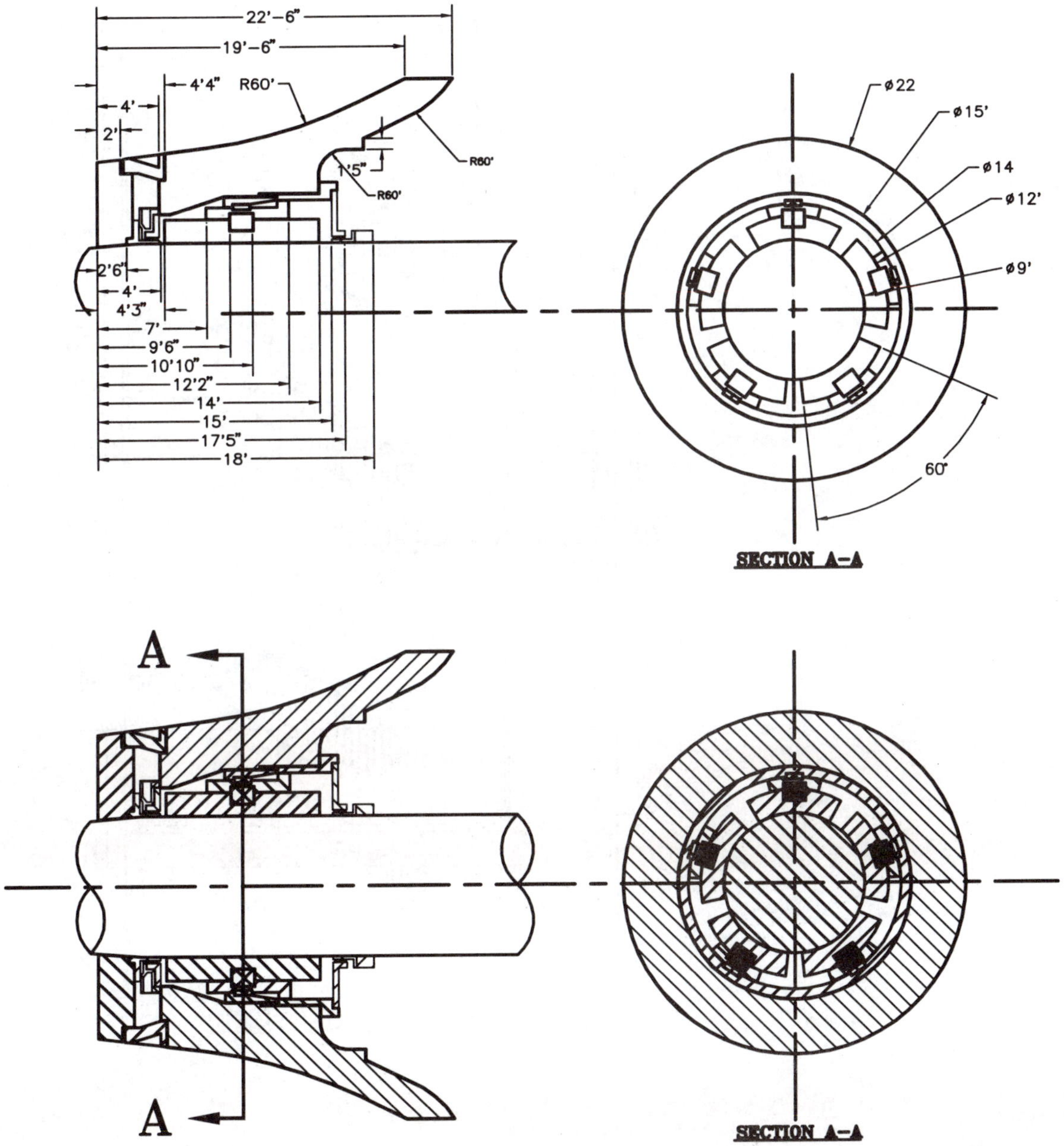

HATCH and SKETCH

Challenger 9C

This drawing can be compiled with the illustrations on page 250, 358, and the top detail on 220 to make a complete commercial stair design.

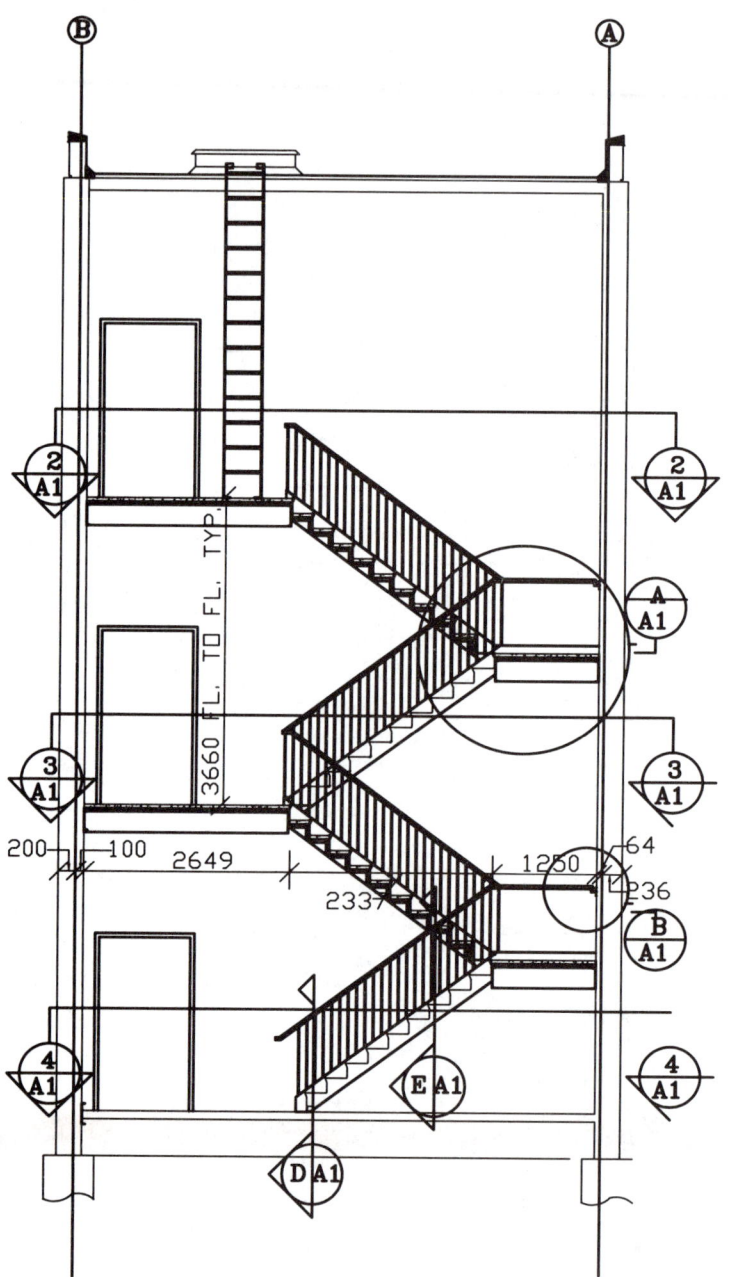

STAIR LAYOUT—SECTION

250 CHAPTER NINE

10 Blocks and Wblocks

Upon completion of this chapter, you should be able to:

1. Create a block
2. Create a wblock
3. INSERT both blocks and wblocks
4. EXPLODE blocks and wblocks
5. Update existing blocks
6. Use COLOR, LAYERS, and other data with blocks

Introduction

For the past ten chapters, you have been working on AutoCAD models and drawings as individual sets of data. The geometry was entered relative to a specific origin, and consisted of a variety of geometric and text information.

You have also seen that Layers are important as a management tool for keeping portions of the file separate for viewing purposes.

Another important management tool is **blocking**. With BLOCKs, the CAD operator can create a library of parts that can be used repeatedly. The operator can also have a library of drawing conventions such as title blocks, section markers, typical details, and other views that might be found on many drawings.

By using BLOCKS on large projects, you can cut down design time by creating portions of the drawing separately and then assembling them on a final drawing.

In a manufacturing environment, models or drawings of bolts, plugs, fasteners, and other standard parts are kept in a central database, and you would insert the object you needed rather than drawing it onto a new drawing. In architecture, a firm would have a library of staircase details, plumbing fixtures, windows and doors, and other often-used objects. In City Hall, there would be road signs, traffic lights, major service components, and other common symbols. In electrical applications, all the symbols for electrical components would be on file, so that the drawings would in fact be compiled more than drawn.

There are two types of blocks used in AutoCAD. An **internal block** created with the BLOCK command is part of the base drawing and cannot be accessed except within the drawing. An **external block** or **wblock** is a drawing file.

Any drawing file can be inserted onto another drawing file at any time.

Dimensions and Hatches are also considered blocks by AutoCAD, but these are not made with the BLOCK commands.

The five commands that are connected with blocks are:

BLOCK   creates an internal block.

WBLOCK   creates a .DWG file or external block.

INSERT   inserts either a block or a wblock (external drawing file).

MINSERT   inserts blocks in rectangular or polar arrays like the ARRAY command.

EXPLODE   reverts the blocked data back to individual objects.

The BLOCK Command

The BLOCK command is used to create a grouping or set of objects that are identified by a given name. Once blocked, an object can be inserted into your drawing many times. Blocks created with the BLOCK command are internal, i.e. available only within the current file. To make them accessible in other files you must use WBLOCK.

Objects must be drawn in before they can be blocked.

Internal blocks are used on drawings where an object or group of objects needs to be accessed a number of times. Instead of ZOOM All and COPY, the BLOCK command allows you to store a set of objects under a given name for insertion at any time.

An example could be chairs for an office layout. Let us take a variation of the layout from Challenger 4. Assume that the layout is complete except for the chairs.

> **Windows** From the Draw toolbar, choose Block, then this button:
>
> **DOS** From the Construct menu, choose Block.

The command line equivalent is **BLOCK**.

```
Command:BLOCK
Block name (or ?):STENO
Insertion base point:MID of (pick 1)
Select objects: (pick 2)
Other corner: (pick 3)
Select objects:⏎
```

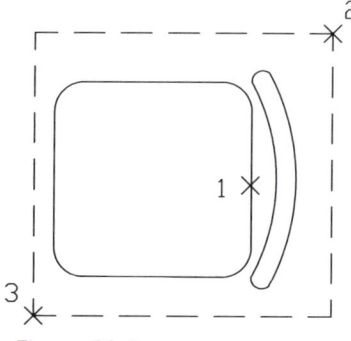

Figure 10-1

The block of objects will disappear. These objects will reside in memory until you are ready to insert them onto the drawing. You can return the objects to the screen with the OOPS command. Do not use Undo or the block will be erased.

The "Insertion base point" on this object is particularly important, because it is the point of reference for inserting, much like the base point in the COPY or MOVE command. If this point is placed logically on the object, the insertion of the block should be perfect every time.

Once the first command is completed, you can create another object and block it under a different name.

```
Command:BLOCK
Block name (or ?):ARMSTENO
Insertion base point:MID of (pick 4)
Select objects: (pick 5)
Other corner: (pick 6)
Select objects:⏎
```

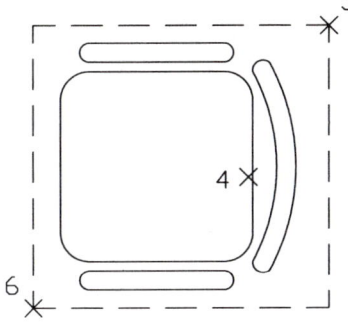

Figure 10-2

The blocks are now complete and you are ready to insert them into a drawing.

The INSERT Command

The INSERT command is used to place previously defined blocks into the current drawing.

> **Windows** From the Draw toolbar, choose Block, then this button:
>
> **DOS** From the Draw menu, choose Insert, then Block.

The command line equivalent is **INSERT** or **DDINSERT** for the dialog box.

```
Command:INSERT
Block name (or ?):Steno
Insertion point: (pick 1)
X scale factor<1>/Corner/XYZ:⏎
Y scale factor (default = X):⏎
Rotation angle<0>: (move your cursor
   around until the chair is properly
   positioned, then pick)
```

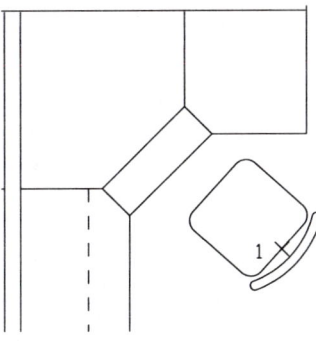

Figure 10-3

The *X* scale factor defaults to 1. Any number smaller than 1 will make the block smaller than its original size; any larger number will make it larger than the original.

The *Y* scale factor defaults to *X*. If you make the *X* value 2, the *Y* value will also be 2, unless you change it. Changing the *Y* value relative to the *X* value will distort the image of the original block.

The rotation angle is a counterclockwise rotation around the insertion base point.

Inserting blocks is much like using COPY, except that with COPY the original objects must be on screen. With the INSERT command, you are identifying a stored group of objects by name rather than identifying a selection set.

Use DDINSERT to enter the other chairs.

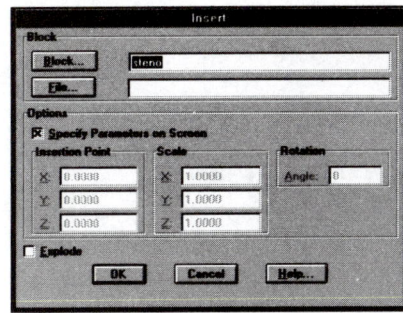

Figure 10-4

Blocks and Wblocks

```
Command: INSERT
Block name (or ?)<STENO>:⏎
Insertion point: (pick 2)
X scale factor<1>/Corner/XYZ:⏎
Y scale factor (default = X):⏎
Rotation angle<0>: (move and pick)
Command: INSERT
Block name (or ?)<STENO>:⏎
Insertion point: (pick 3)
X scale factor<1>/Corner/XYZ:⏎
Y scale factor (default = X):⏎
Rotation angle<0>: (move and pick)
```

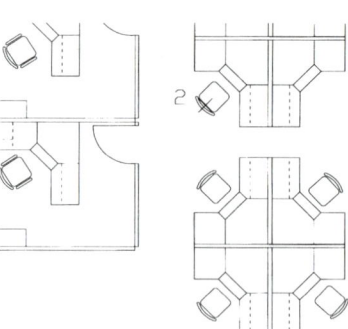

Figure 10-5

Now that the layout is complete, we could add a title block. This would be a drawing file already on disk.

Inserting External Blocks

Once the BLOCK or drawing has been INSERTed into the file, it is referred to as the ***block instance***. The original still exists in memory, but a copy of it has been put in the current drawing. This is similar to a rubber stamp and the stamp impression. On the office layout there are now two identified blocks and many block instances.

Let us take the title block from Chapter 8. First we must find out the name of the file.

Listing Files

Use the following to get a listing of files.

> **Windows** From the DDINSERT dialog box, pick File, then pick the directory and file that you need.
>
> DOS At the command prompt, type in DIR, then the directory name:
>   Command:DIR A:
>
> You will get a listing of your files with the dates. Pick F1 to return to the graphics screen.

Inserting the Block

Now insert the name of the title block onto the file.

Use DDINSERT to place the title block onto the file.

You can insert a drawing file onto your current drawing at any time.

```
Command: DDINSERT
Block name (or ?)<STENO>:A:TITLE
Insertion point: (pick 1)
X scale factor<1>/Corner/XYZ:⏎
Y scale factor(default = X):⏎
Rotation angle<0>:⏎
```

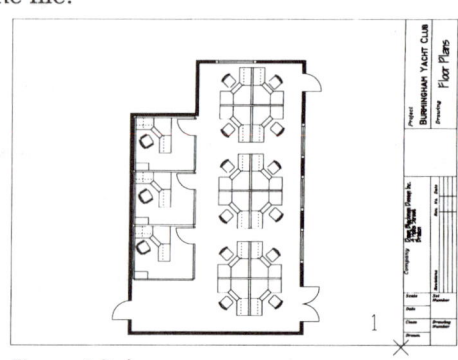

Figure 10-6

The WBLOCK Command

BLOCKs exist within the drawing and cannot be accessed except within the base drawing. This is why they are referred to as internal blocks.

WBLOCKs are separate files. They are stored separately as drawing files on your disk and have the extension .DWG. They do *not* need to be blocked.

Any file on disk can be inserted as a wblock into any drawing.

WBLOCK is used when you want to take only a portion of an existing file and use it on another file. Because the wblocks exist outside of the current file, they are referred to as external blocks.

An example of WBLOCK use is the extraction of a "north arrow" from a drawing for use on another drawing.

To start, you must open the file that contains the information that you would like to export as a block. Then access the WBLOCK command.

> **Windows** At the command prompt, enter WBLOCK.
>
> **DOS** From the File menu, choose EXPORT, then Block.

The command line equivalent is **WBLOCK**.

```
Command:WBLOCK
File name:A:NORTH
Block name (or ?):↵
Insertion base point: (pick 1)
Select objects: (pick 2)
Other corner: (pick 3)
Select objects:↵
```

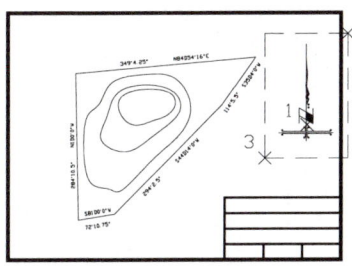

Figure 10-7

Once you have chosen the WBLOCK command, the Create Drawing File dialog box will be invoked. Enter the name of the file in the File Name box.

When you press ↵, the dialog box will disappear and the command will continue.

When you are identifying the file name, be sure to type in the drive or directory as well.

To check to see that the file is in the directory, list the files using the DIR command in DOS or File Utilities in WINDOWS.

```
Command:DIR
File specification:*.DWG
```

This will give you a listing of the drawing files on your current directory. Using DIR will give you the date plus the size. Among the files you should read the following:

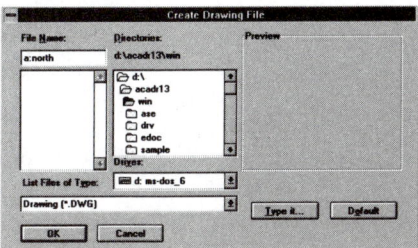

Figure 10-8

Blocks and Wblocks **255**

To place this file on another drawing, open the drawing and use INSERT.

```
Command:DDINSERT
Block name (or ?)<STENO>:A:North
Insertion point: (pick 1)
X scale factor<1>/Corner/XYZ:48
Y scale factor (default = X):⏎
Rotation angle<0>:36
```

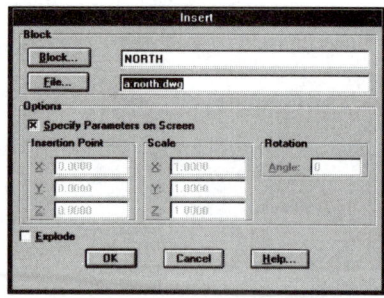

Figure 10-9

The north arrow is now part of the current file.

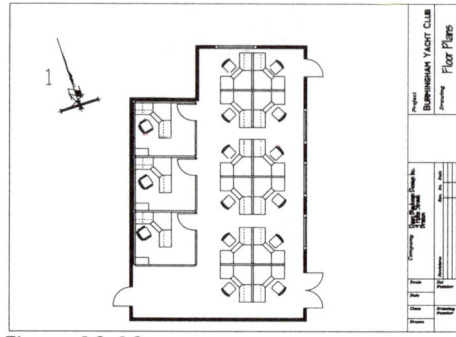

Figure 10-10

Inserting Existing Drawings as Blocks

The origin or 0,0 of the drawing is the default base point for drawings inserted into the current drawing. To change the default, open the original drawing and use BASE to specify a different insertion base point. AutoCAD uses the new base point the next time the drawing is inserted.

Using a BLOCK to Create a WBLOCK

If you have already created a block internally called N-ARROW and would like to have this block written to an external file saved as NORTH on your A: drive, use the following.

```
Command:WBLOCK
File name:A:NORTH
Block name (or ?):N-ARROW
```

This will take the internal block N-ARROW and create an external file called A:NORTH.

The insertion base point from your original block will be identified as the insertion base point of the file or wblock as well.

The listing gives you the name of the drawing, the extension showing that it is a .DWG file or drawing file, the size of the file, and the date and time when it was created.

If you want to use a portion of a file in another file, repeat this process, but press ⏎ at the "Block name (or ?):" prompt. Now you are prompted for the base point and selected objects.

WBLOCK with =

After entering the name of the wblock, entering an equal sign specifies that the existing block and the output file shall have the same name. If no block of that name exists in the drawing, AutoCAD redisplays the "Block name (or ?):" prompt.

WBLOCK with *

Entering an asterisk (*) writes the entire drawing to the new output file, except for unreferenced symbols. AutoCAD writes model space objects to model space and paper space objects to paper space.

Scaling Wblocks

Title blocks, north arrows, and other drawing symbols should be created at a scale of 1:1. For example, a title block could be perhaps 12 × 4 inches; a north arrow could be 3 inches; a section marker could be 1 inch.

When these are inserted onto a file, the scale of the block should be expanded to fit the size of the drawing. In the floor plan above, the drawing might be plotted at a scale of 1/4″=1′0″ or 1/48 of the actual size of the floor plan.

The north arrow would then be inserted at a scale of 48.

```
Command: INSERT
Block name (or ?)<STENO>:A:NORTH
Insertion point: (pick 1)
X scale factor<1>/Corner/XYZ:48
Y scale factor (default = X):⏎
Rotation angle<0>:36
```

In mechanical, if the final drawing is to be 1:20 and you are drawing in millimeters, insert the drawing symbols at 20 so that they will be the correct size on the drawing when plotted at 1/20 of that size or 1:20.

Charts of appropriate sizes are given in Chapter 11. For the exercises today, simply adjust the size to fit.

The MINSERT Command

MINSERT is used to insert multiple block instances of a block in a rectangular array. The command melds the INSERT command with the ARRAY command. If used properly, this can be a very useful command.

Windows From the Miscellaneous toolbar, choose:

DOS From the Draw menu, choose Insert, then Minsert.

The command line equivalent is **MINSERT**.

```
Command:MINSERT
Block name (or ?):LOT
Insertion point: (pick 1)
X scale factor<1>/Corner/XYZ:.25
Y scale factor (default = X):⏎
Rotation angle:180
Number of rows(----)<1>:3
Number of columns (|||):6
Distance between rows:4'
Distance between columns:5'
```

Figure 10-11

Minserted blocks cannot be individually edited or exploded. If you erase one, you erase them all.

Editing Blocks

Once a block has been placed in a file, it is edited as a single entity. An entity select pick will pick the whole block. If you pick up one object on the block, the whole block will be selected.

Try moving a block instance to the left.

```
Command:MOVE
Select objects: (pick 1)
Select objects:⏎
Base point or displacement: (pick 2)
Displacement: (pick 3)
```

Notice that your block moves as a unit. If you erase a block, only one pick point is needed to identify the block. Move the block back to its original position by using **U** ⏎.

Figure 10-12

If you want to edit or change the block once it has been inserted, you will need to return the block to its original objects. The EXPLODE command is used for this purpose.

The EXPLODE Command

The EXPLODE command reduces an object to its original entities. A block can be exploded into its original parts, a polyline or polygon can be exploded into line or arc segments, and a dimension can be exploded into lines, arrowheads, and text.

Title blocks are generally saved with only the general information such as Scale, Date, etc., and not the information regarding the specific drawing. If you would like to change any of the information already on the drawing, use EXPLODE and then CHANGE or DDEDIT.

Windows From the Modify toolbar, choose the Explode flyout, then this button: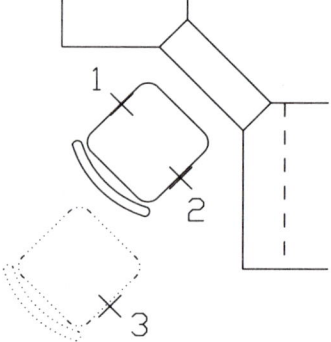

DOS From the Modify menu, choose Explode.

The command line equivalent is **EXPLODE**.

```
Command:EXPLODE
Select block reference, polyline,
    dimension, or mesh: (pick the
    block instance)
```

Now use CHANGE or DDEDIT to change the text as shown.

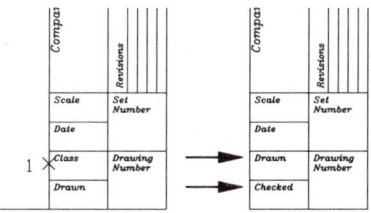

Figure 10-13

If you are inserting a block or wblock and know you want to change portions of it later, type * before the block name or pick the box Explode on the dialog box. This inserts the block as separate entities that can be edited. The * helps avoid the EXPLODE command. You are not given the option of changing the scale factor if you pre-explode the block.

```
Command:INSERT
Block name (or ?)<BUSHING>:*TITLE
Insertion point: (pick a point)
Rotation angle<0>:⏎
```

Now, if you try to edit anything on the inserted BLOCK, it will be identified as a single unit even though it is a block instance.

Updating Blocks

There can be only one definition of a block under each specific name within a file. Should you change or update the original block, the new block instances will be the new block and the old block instances will update. This can be very useful.

For example, if you have shown the drawing of the office layout to your client, and he or she has decided to specify a different chair, explode one chair. Make the changes and block it again under the same name, using the same insertion point to update the other chairs.

```
Command:EXPLODE
Select block reference, polyline,
    dimension, or mesh: (pick the block
    instance 1)
```

Figure 10-14

The block of the chair will now revert back to the original lines and arcs used to create it.

Now make the changes to the chair that you would like to see.

Because the chairs in the layout have all been rotated, make sure that you rotate the new block back to the original rotation angle before reblocking.

Once complete, use BLOCK to resave the chair under the original name.

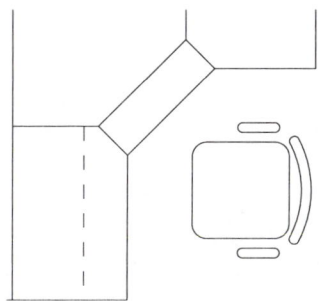

Figure 10-15

```
Command:BLOCK
Block name (or ?):STENO
Block STENO already exists. Redefine
   it?<N>:Y
Insertion base point:MID of (pick 1)
Select objects: (pick the objects)
```

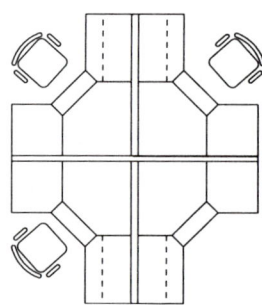

Figure 10-16

Compiling Drawings with BLOCK

Drawings are often compiled as opposed to being drawn. Generally you would have title blocks and symbols and details already on file. In an architectural firm you would find libraries of parts such as plumbing details, light fixtures, windows, doors, etc. already on file and you simply need to insert them rather than draw them in again.

It is a good idea always to compile drawings on the current file of the most important drawing. That way, if there are changes to the base plan, they can be made on the current drawing and then replotted rather than having to recompile the drawing many times. Chapter 11 deals more in detail with compiling of drawings.

Blocks, Wblocks, Color, and Layers

You may have noticed that the default layer is 0, and that this layer cannot be renamed or deleted. Layer 0, the universal layer, cannot be changed because it is used extensively with regard to BYBLOCK and BYLAYER settings for color and linetype. The layers blocks sit on and the color of the block can be set either bylayer or byblock.

Once you have created a number of blocks or wblocks, file management becomes more difficult, because you must remember the size of the blocks as well as the directory they are in and the layer they are on.

If items are created on a layer other than 0 in a wblock, these new layers will be added to the list of layers in the current file when the wblock file is inserted.

All objects on layer 0 will be automatically placed on the current layer of the drawing on screen, and will assume the current layer color and linetype when the block is inserted. These are not necessarily added bylayer. If color or linetype are not set bylayer, the current setting overrides the layer setting. To check to see what the current settings are, use the STATUS command or pick Object Creation (DDEMODES) from the Data menu.

If you want a particular wblock to always have a specific layer, linetype, and color, assign it explicitly; do not leave it on layer 0.

To illustrate this point, make sure that your current layer DIMS has a different color than white, then insert a drawing from one of the first tutorials. The drawing called TEMPLATE is a good choice, because it is quite small.

```
Command:INSERT
Block name (or ?):A:TEMPLATE
Insertion point: (pick a point)
X scale factor<1>/Corner/XYZ:↵
Y scale factor (default = X):↵
Rotation angle:↵
Command: (pick Zoom All from the Display pull-down menu)
```

Notice that your template file is in the current layer color. If you use LIST to find out the parameters of the model, you will see that the layer for the block is listed as the current layer.

Byblock

If color and linetype are defined by the Byblock option in the original file, the colors and linetypes can be changed once the file is inserted. If you want the color and linetype of the wblock to always assume those of the layer on which it is inserted, use bylayer rather than byblock as the setting for the color and linetype.

Once a block has been inserted, the objects will retain the original layers. Neither CHANGE nor CHPROP can be used to alter the layer of a portion of a block. You must first explode the block in order to change the layers on the block entities.

Naming Blocks, Wblocks, and Layers

From the previous discussion you can see that if five wblocks are inserted into a file, and each one has five separate and unique layer names, you will have a file with 25 different layer names. This can be a problem and a terrific waste of time. When creating drawings, try to use standard layer names to avoid problems. You may think layer names such as 1 and 2 or A and B are easier to enter and that you will remember what is in each layer as you are working.

You won't!

Furthermore, when you insert several drawings, you can bet on having problems with data management.

Using the date on the readout to locate files is useful, but every time you retrieve a file and then END or SAVE it, the date will change.

When working on large projects that have several different files and several different layers, it is a good idea to keep a project designation sheet containing the names of all of the files needed, the date they were last updated, the names of all layers, and what is contained within those layers. This will save you a lot of time later trying to figure out the database you are dealing with.

> **Danger**
>
> If you start your drawing from a floppy disk, remember not to remove your floppy disks while you are in the AutoCAD drawing editor. If you do, you may be creating bad clusters on your disk which will eventually cause a crash.

Accessing Files for Inserting

If you are using floppy disks, you should copy all of the files either onto the hard drive or onto your working floppy disk before inserting them onto a base drawing.

If you have two or three files on different disks, make sure you are writing to the C: drive, then copy the files to a clean new disk and insert them.

Removing Unwanted Blocks

Each time a block or drawing file (wblock) is inserted into a file, a copy of it is placed in the drawing memory or default area. If you want to clean your file, you must PURGE the blocks. The PURGE command will erase all unused blocks, layers, text styles, linetypes, etc. from a file. Any blocks, layers, etc. that have been brought into the file but never used can be purged. It erases the information from the file defaults and settings.

If you don't PURGE the memory of the original drawing file or wblock before you insert a new one, the old block file will be inserted, even if you have created a new file.

When a drawing file or wblock is inserted, then erased, the copy of the wblock remains in memory in case it is needed again. The original file *must* be purged in order to accept a new file.

The PURGE Command

PURGE removes unused named references, such as unused blocks or layers from the database.

If you are inserting an updated drawing file and you keep getting the original file, use PURGE. Unlike in previous releases, in Release 13 PURGE does not need to be the first command in the editing session.

```
Command:PURGE
Purge unused Blocks/Dimstyles/LAyers/LTypes/SHapes/STyles/
   Apids/Mlinestyles/All:A
Purge Block Title<n>:Y
```

Once the old block is purged, you can enter the new or updated one.

```
Command:DDINSERT
Block name (or ?):A:TITLE
Insertion point: (pick a point)
X scale factor<1>/Corner/XYZ:
Y scale factor (default=X):
Rotation angle:
```

This gives you the new title block.

Another way to insert an updated block is to use FILENAME=. This will indicate that the file has been updated, and the base file will search for the updated file. A message will be displayed stating that the block has been redefined.

```
Command:INSERT
Block name (or ?):TITLE=
Insertion point:0,0
X scale factor<1>/Corner/XYZ:
Y scale factor (default = X):
Rotation angle:
```

This command sequence will insert the new copy of the file called TITLE instead of the old one. Using PURGE, however, is a better use of memory.

Tips

Objects must be visible to be included in the block, but they can be on many layers. If you are trying to block information from only one layer, turn the others off for easy selecting.

When in doubt, use 0,0 as the insertion point. If all of the files are created at a 1:1 scale, the data will always fit.

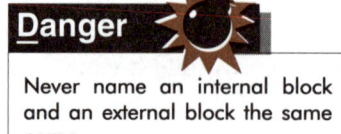

Danger
Never name an internal block and an external block the same name.

If you have inserted the wrong block, undo it, and the block reference on the file will be deleted as well. This is the same as undoing a HATCH rather than erasing it.

Paper space Objects

When you insert an entire object into the current drawing, any paper space objects will not be included in the current drawing's block definition. To use the paper space objects in another drawing, open the original drawing file and use BLOCK to define the paper space objects as a block. The block will then be included in the original drawing database when you insert the drawing into another drawing.

Creating a Symbol Library

There are various methods of creating a symbol library. This will help you to define a library of parts that can be accessed to compile your drawings.

1. Create a directory on your disk, e.g. SYM, and make individual files.
2. As above, except draw all the symbols on one sheet and WBLOCK them individually to the disk.
3. Draw all the symbols on one sheet, calling the drawing ELEC-SYM or MEC-SYM, and block them individually. Save the drawing. To access the symbols, simply insert the file ELEC-SYM into your current drawing at 0,0 and then use DDINSERT to access the block you want.

Also note that you can insert an old file if you want some blocks from it. Use **Ctrl-C** when asked for the insertion point. This loads the drawing but only into the database, saving much time on regenerations. Once you have extracted the blocks you want, you can purge the old file or WBLOCK*.

Prelab 10 BLOCK, WBLOCK, INSERT, and MINSERT

In this lab we will draw a standard lot, block it, insert it both as an individual lot and as a survey, and insert the title block from Chapter 8.

Step 1 Draw in a rectangle that is 40 × 30. Use ZOOM to get it on screen, and then ZOOM by .5X. Add the house as shown.

```
Command:L
From point:0,0
To point:30,0
To point:30,40
To point:0,40
To point:C
Command:Z
All/Center/Dynamic/Extents/Left/
   Previous/Vmax/Window/X/XP:A
Command:Z
All/Center/Dynamic/Extents/Left/
   Previous/Vmax/Window/X/XP:.5X
```

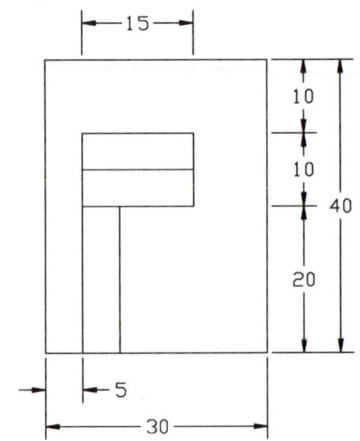

Step 2 Use BLOCK to create a block of this data.

> **Windows** From the Draw toolbar, choose Block, then this button:
>
> **DOS** From the Construct menu, choose Block.

The command line equivalent is **BLOCK**.

```
Command:BLOCK
Block name (or ?):LOT
Insertion base point:0,0
Select objects:W
Select objects: (pick 1)
Select objects: (pick 2)
Select objects:
```

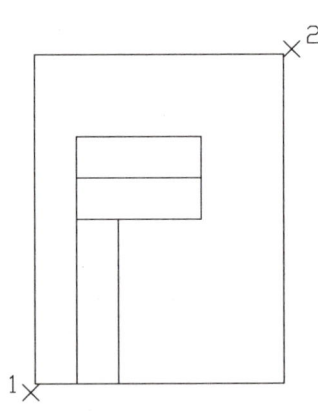

The object will disappear from the screen.

Step 3 Now use INSERT to place the block back on the screen.

> **Windows** From the Draw toolbar, choose Block, then this button:
>
> **DOS** From the Draw menu, pick Insert, then Block.

At the command prompt, type **INSERT**.

264 CHAPTER TEN

```
Command: INSERT
Block name (or ?): LOT
Insertion point: (pick the lower
   left of your screen)
X scale factor<1>/Corner/XYZ: ⏎
Y scale factor (default = X): ⏎
Rotation angle: ⏎
```

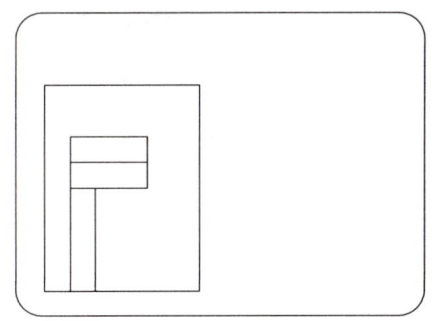

Step 4 Create a new layer called SURVEY with color set to Magenta. Make it current.

```
Command: DDLMODES CREATE LAYER SURVEY
```

Step 5 Use the MINSERT command to place a survey on the page.

> **Windows** From the Draw menu, choose Insert, then Multiple Blocks.

```
Command: MINSERT
Block name (or ?): LOT
Insertion point: (pick 1)
X scale factor<1>/Corner/XYZ: .25
Y scale factor (default = X): ⏎
Rotation angle: ⏎
Number of rows (----)<1>: ⏎
Number of columns (|||): 6
Distance between the columns
   (|||): 7.5
```

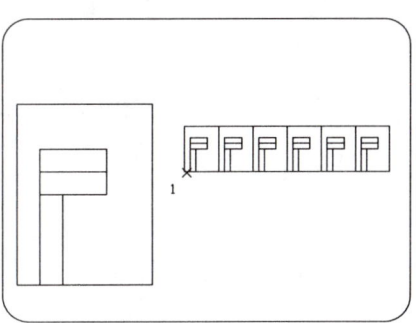

Step 6 Use MINSERT again to place the lots on the other side of the street.

```
Command: MINSERT
Block name (or ?): LOT
Insertion point: (pick 1)
X scale factor<1>/Corner/XYZ: .25
Y scale factor (default = X): ⏎
Rotation angle: 180
Number of rows (----)<1>: ⏎
Number of columns (|||): 6
Distance between the columns
   (|||): 7.5
```

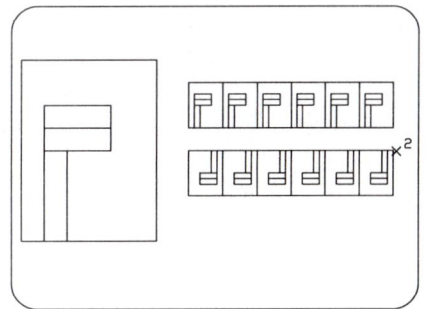

If the two surveys are not perfectly aligned, use MOVE to move one of them. Window is not needed; an object selection pick is all that is needed to pick up the parts.

Blocks and Wblocks **265**

Step 7 Now insert the title block from Prelab 8. Use DDINSERT to find the file.

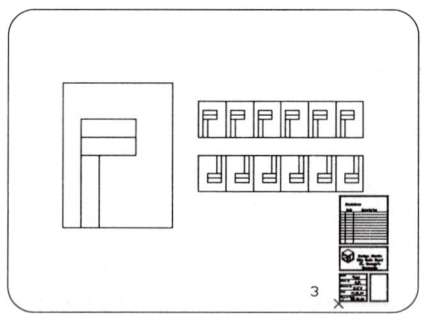

```
Command:DDINSERT
Block name (or ?):A:EXAMP8
Insertion point: (pick 1)
X scale factor<1>/Corner/XYZ:3
Y scale factor (default = X):⏎
Rotation angle:⏎
```

Notice that the date and names may be wrong on the inserted title block. If you try to use CHANGE to change the text, the entire entity will be chosen. You must first EXPLODE the block instance before you can change the text.

```
Command:EXPLODE
Select Block/Polyline/Dimension: (pick the block or L for last)
```

Now use to DDEDIT to update the data.

Notice that the title block originally inserted took on the current layer, which was SURVEY, but once exploded it went back to layer 0.

Step 8 Now that you have this drawing complete, what happens if you would like to use a block that is in this drawing on another file? Use WBLOCK to make a named internal block an external file on your disk.

> **Windows** Use File, Export, then Block.
>
> **DOS** Use File, Export, then Block.

```
Command:WBLOCK
File name:A:LOT
Block name (or ?):LOT
```

This will take the block LOT and create a file of it. Check to see that this has been written to a file by using Utilities or DIR A:.

```
Command: (pick Files, List)
```

OR:

```
Command:DIR
Name of directory:A:
```

Your listing should include a file called LOT.DWG.

Step 9 Now what happens if you would like to use a portion of this drawing on another file? Use WBLOCK to take a portion of this file and have it accessible on other files.

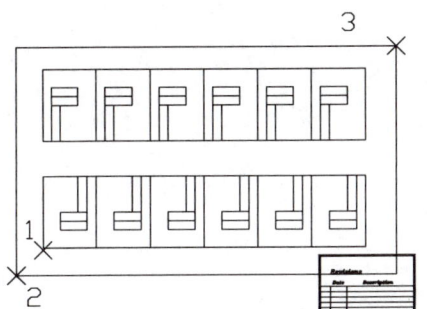

```
Command:WBLOCK
File name:A:SURVEY
Block name (or ?):⏎
Insertion base point: (pick 1)
Select objects: (pick 2)
Other corner: (pick 3)
Select objects:⏎
Command: (pick Files, List)
```

Now save your file. Start a new file called SURVEY2. Use DDINSERT to insert the file SURVEY.

In DDLMODES note the addition of the layer SURVEY.

Remember: *A file is not a block until it is inserted into another file.* You do not need to block a drawing to insert it into another file.

Command and Function Summary

BLOCK creates an identified group of objects for insertion within the current file.

EXPLODE breaks a compound object such as a block, a dimension, a hatch, or a pline into its component objects.

INSERT allows you to insert predefined blocks and drawing files (wblocks).

MINSERT inserts multiple instances of a block in a rectangular array.

PURGE allows you to remove any layers, blocks, linetypes, text files, or shape files that you have entered into your file and not used.

WBLOCK creates an external block or .DWG file from specified objects on the screen.

Practice Exercise 10

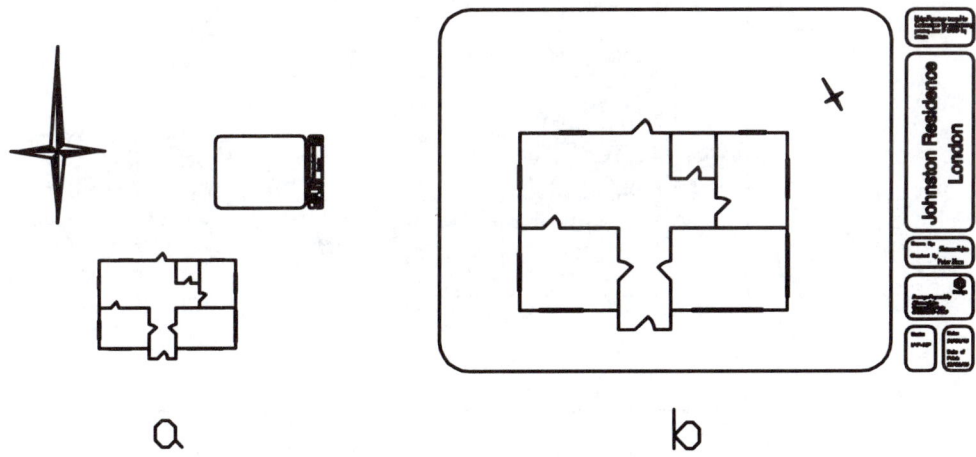

a. b.

1. Draw the first house. BLOCK it. INSERT it at different scale factors and aspect ratios. Using the same block, create some row housing with MINSERT.
2. a. Draw a north arrow (or retrieve it from Chapter 3) and file it as a separate file. Draw a very simple house plan at 1:1. Draw a simple title block (Chapter 8). b. Open the floor plan. Insert the title block at a scale of 48. Insert the north arrow to fit.

Blocks and Wblocks **269**

Exercise A10

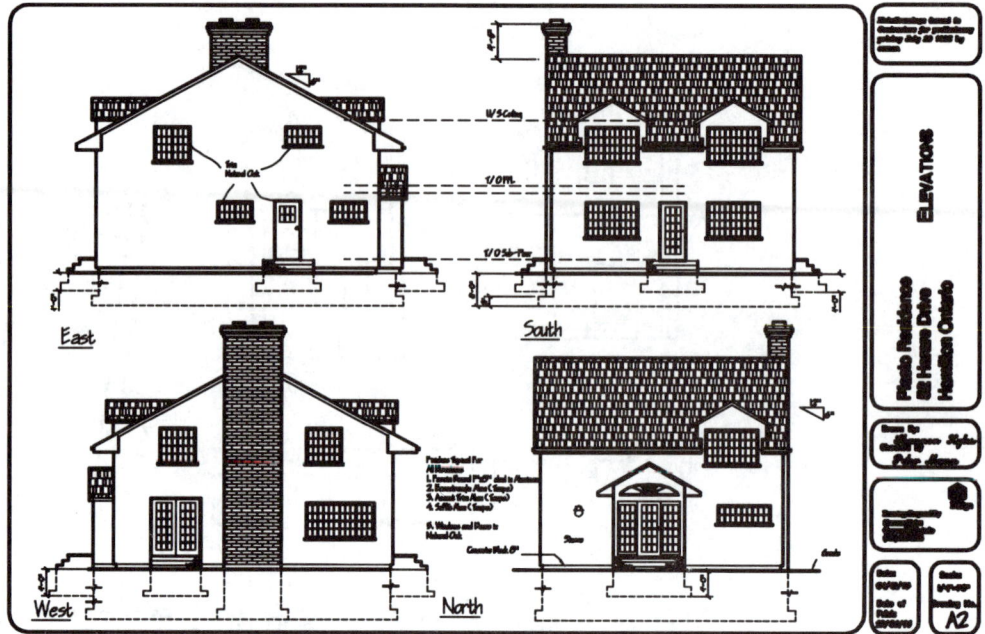

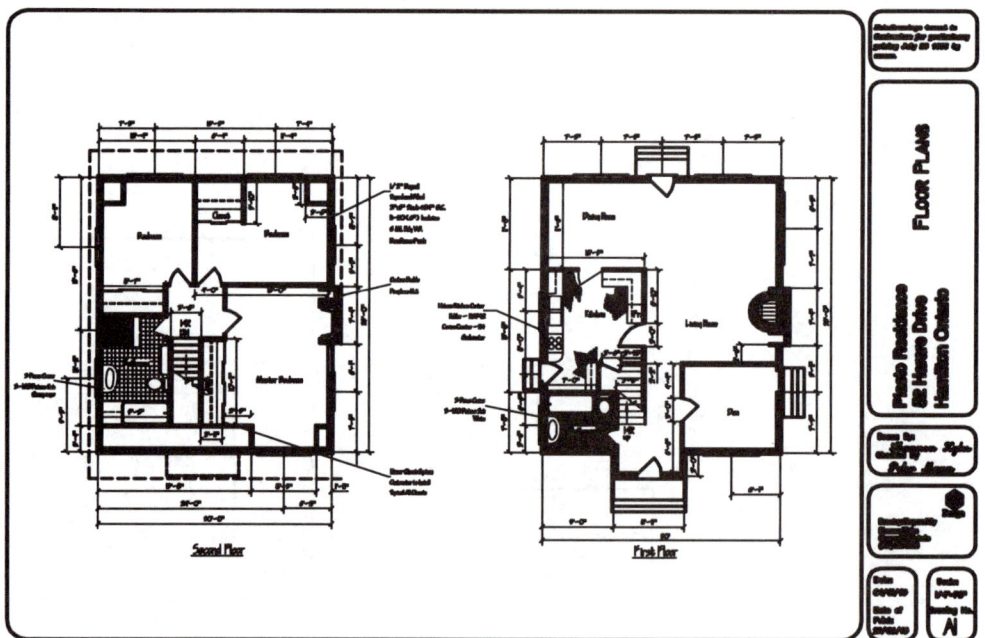

1. Retrieve your floor plan for the first floor.
2. Insert your floor plan for the second floor.
3. Insert the title block from Chapter 8 at a scale of 48.
4. Use SCALE to adjust the size of your title block to suit the paper size.
5. If there is room, insert the elevations — one or many — from Chapter 6. If not, insert the title block again using *NAME so that you can change the drawing number and the description.
6. Your drawing is now ready to plot.
7. Add any notations or titles at this point.

Exercise C10

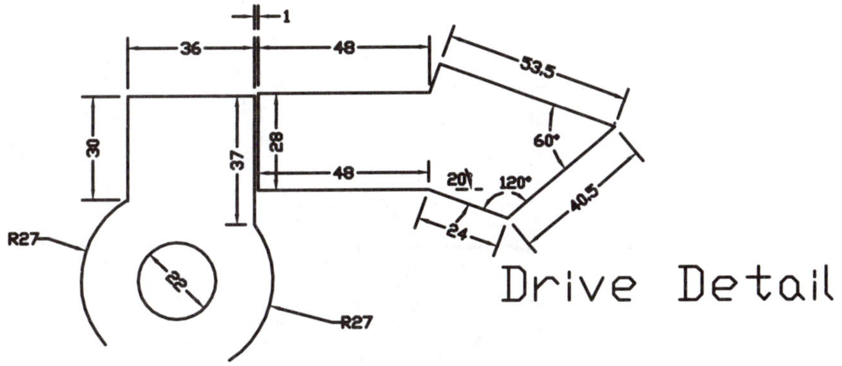

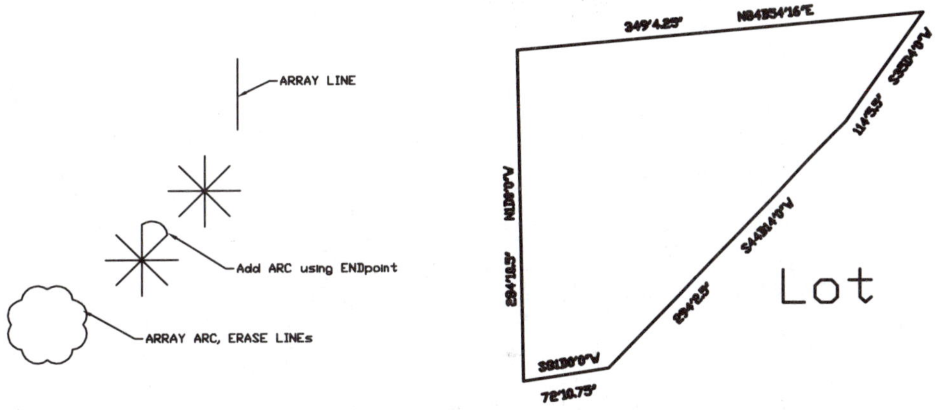

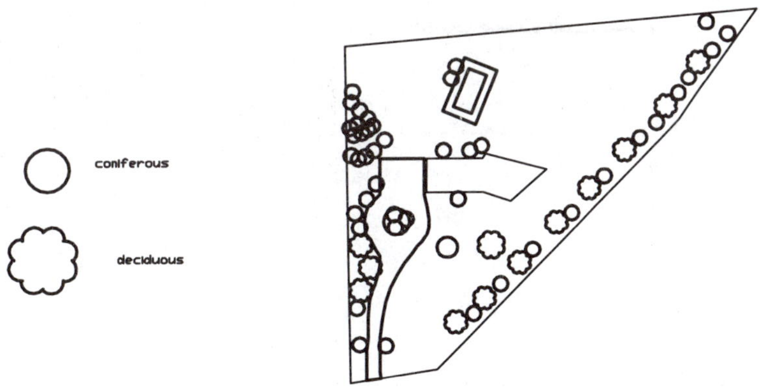

1. Draw the lot at the size suggested.
2. Place the house and driving circle on the property. Add a driveway.
3. Make a tree using ARRAY. Block it and name it DECID for deciduous. Make another for CONIFerous. INSERT the trees on the lot.

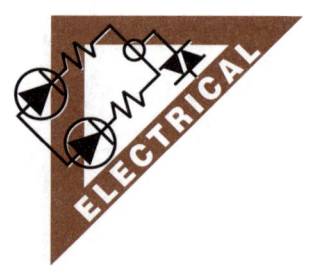

Exercise E10

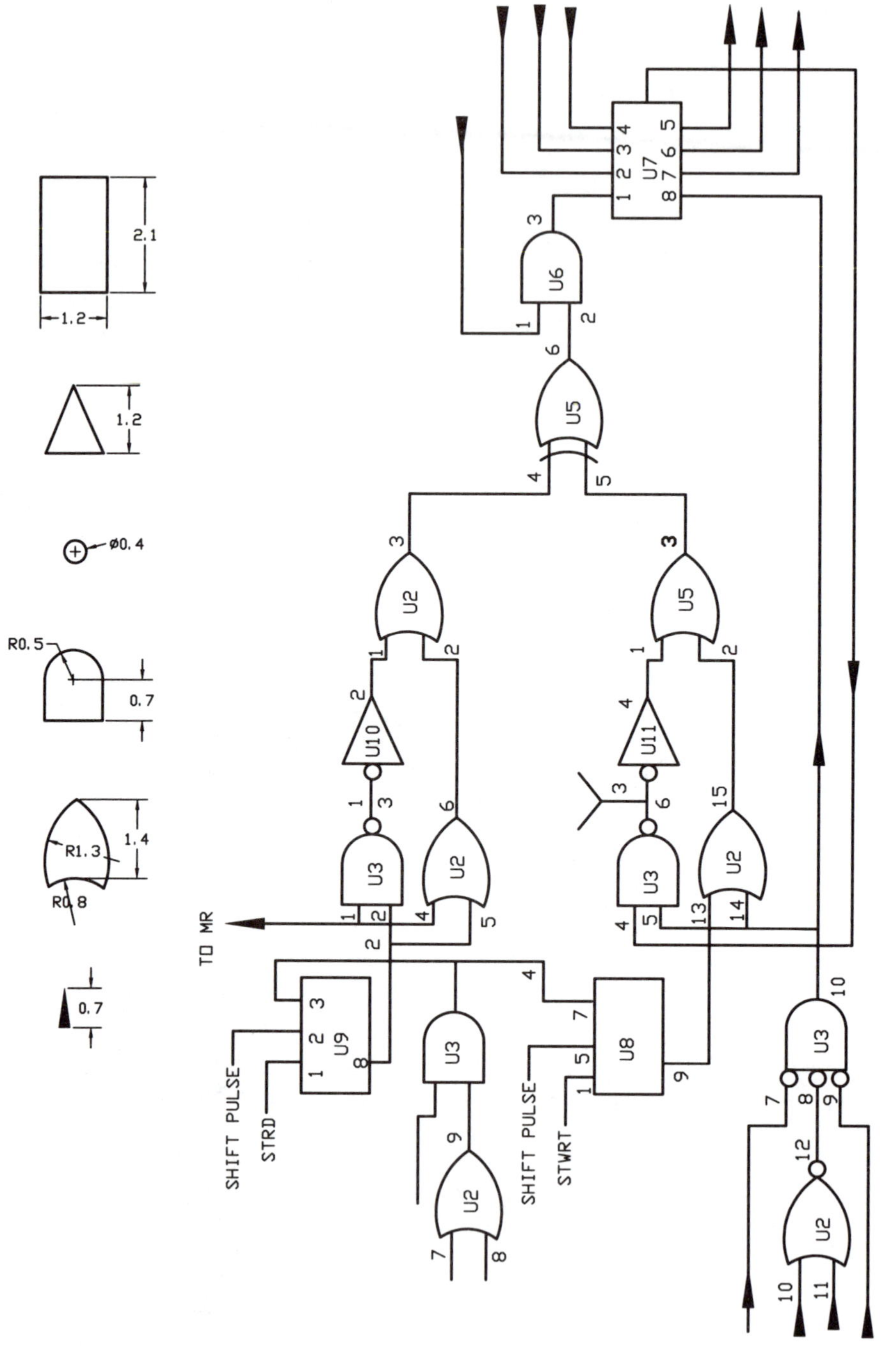

Create the blocks needed to make this computer logic diagram. Then insert the blocks and add the lines to create the layout.

If you have time, add a title block.

Exercise M10

Create the jig body, the jigleg, and the body on one file as shown.

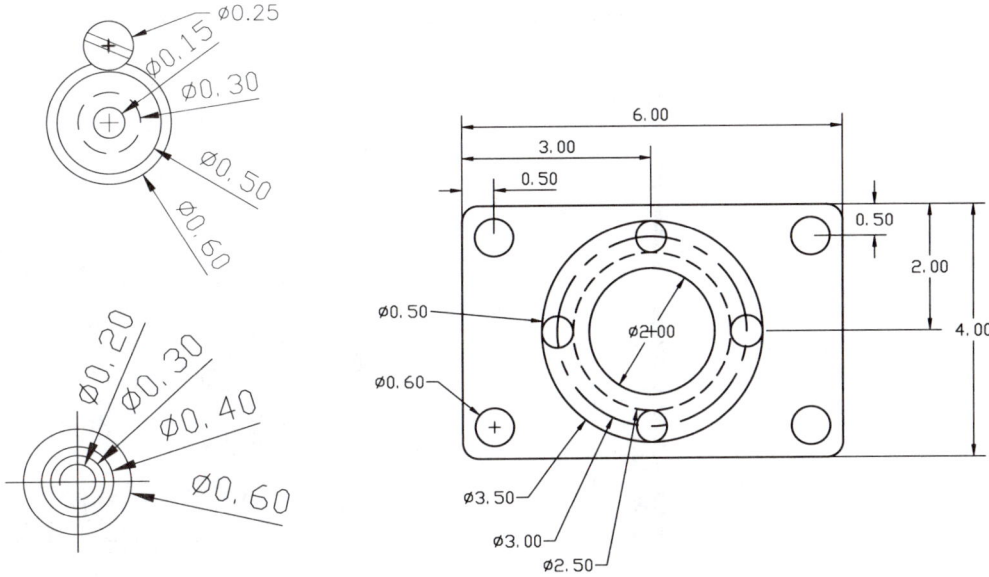

Block the jigleg as JIGLEG. Use the center of the circle as the insertion base point.

Block the bushing as BUSHING using the center of the large circle as the insertion base point.

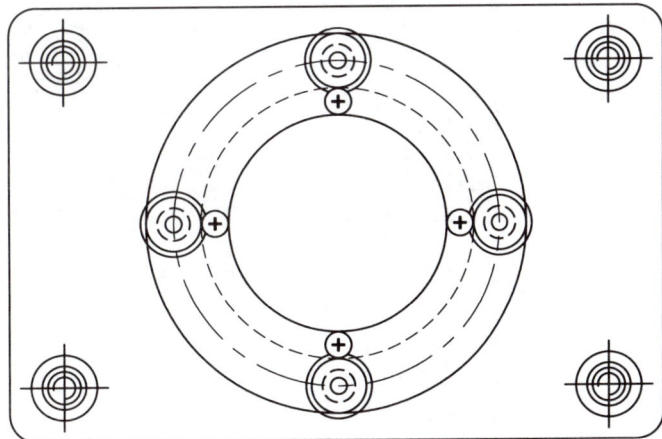

Once blocked, insert the jiglegs and bushings as shown in the above illustration.

Blocks and Wblocks **273**

Challenger 10

Use the dimensions from the Exercises listed to create files that can be merged into this final part. (Note: Pr-6 = Practice Exercise, Chapter 6; M-7 = Mechanical Exercise, Chapter 7.)

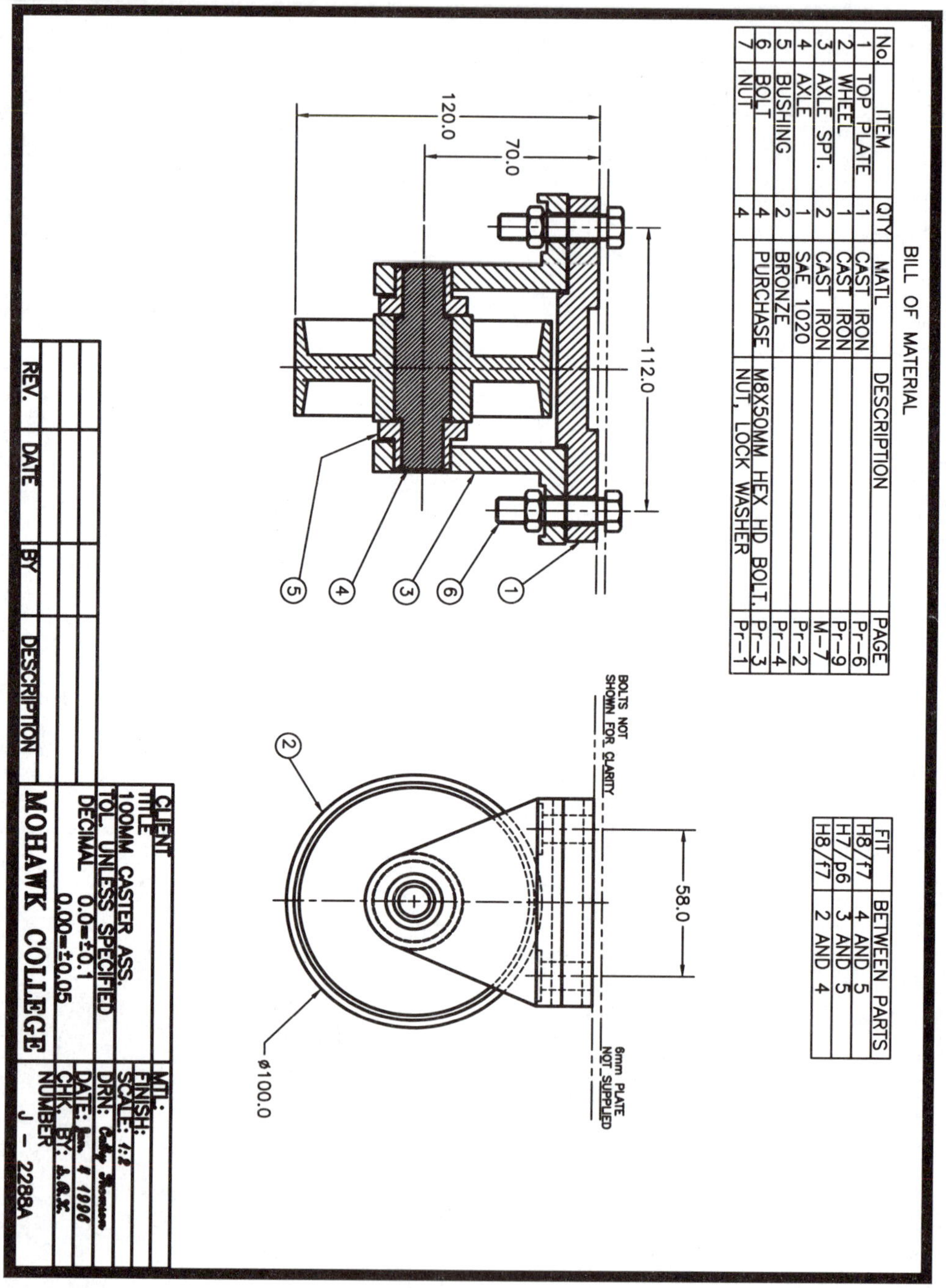

11 Setting Up Drawings and PSPACE

Upon completion of this chapter, you should be able to:

1. Create a plot of a single view
2. Set TEXT, HATCH, and LINETYPE sizes for any drawing
3. Use blocks to compile multi-view drawings
4. Use paper space and tilemode to access a paper environment
5. Use MVIEW to set up a drawing with four views on a standard title sheet
6. Use VPLAYER to have layers visible in selected viewports only
7. Use MVSETUP to lay out a drawing

Set Up and Scale for Simple 2D Drawings

For the past ten sessions, we have been working on AutoCAD models as individual sets of data. The geometry is being entered relative to the origin (0,0). The data, once entered to perfection, can then be used for a variety of purposes, including:

1. Computer-aided manufacturing (CAM)
2. Analysis
3. Downloading to a computer graphics program for marketing purposes
4. Downloading to a stereolithography system
5. Estimating and costing
6. Downloading as a drawing for communication to others

If the model is initially scaled to fit a paper, many of the above applications are complicated. With the model scaled at 1:1, all we need do is extract from the model the views we want to see on a drawing and then scale them appropriately.

There are two ways of creating a scaled plot. The first is to draw the model at 1:1, then insert the title block at an appropriate scale factor and plot the file at a specific scale.

For example, say you have a layout for a house that is 40′ × 30′ and drawn in real units at a scale of 1:1. You want to have this plotted at 1/4″=1′0″. As you can see by Chart 1 (page 276), 1/4″=1′0″ is the same as 1/48.

Title Blocks

Your title block should be drawn at 1:1 as well. It must also be large enough to contain the floor plan. In this example it is set up for a 24 × 36 sheet. In order to have this visible at the correct scale on the layout, you can insert the title block at a scale of 48. Then, when plotting the model, use a plot scale of 1/48.

This is by far the best method of plotting and scaling, as it allows you to maintain the model at a scale factor of 1:1. For all of the information that requires scaling such as Overall Scale in the dimension style, HATCH, LINETYPE, TEXT Size, etc., 48 can be used.

Title Block, Dimension, Hatch, and Linetype Scale

The ANSI standard for the text in dimensions is 3/16″ or 5 mm. This is also the default for the dimension text in AutoCAD. This will need to be changed, however, if you are plotting the part or view at a scale different than 1:1. You want the part to be scaled, but you still want your dimension text to show up at 3/16″ on the paper.

To do this, you must determine the text size needed for each particular scale. Remember, by changing the DIMSCALE or resetting the Scale in the Dimension Style dialog box and then using the UPD command in the DIM mode, you can automatically change the size of the dimensions if they are not the correct size for the plot of your drawing.

Chart 1: Determining Scale for Drawings

Scale Factor (Architectural or Imperial)		Decimal Value	Fraction
3″=1′0″	3:12	.25	1/4
1″=1′0″	1:12	.0833333	1/12
1/2″=1′0″	.5:12	.0416666	1/24
1/4″=1′0″	.25:12	.0208333	1/48
3/16″=1′0″	.1875:12	.0156246	1/64
1/8″=1′0″	.125:12	.0104166	1/96
1/16″=1′0″	.0625:12	.0052083	1/192
Mechanical			
3/4″=1″	.75:1	.75	
1/2″=1″	.50:1	.5	
1/4″=1″	.25:1	.25	
Metric			
1:10		.1	
1:50		.02	
1:100		.01	
1:1000		.001	

The first step is to *determine* the scale factor for the plot, and change the overall dimension scale factor or the DIMSCALE and other notations to fit. You can also change your DIMTXT, but this is not advisable, because the arrowheads, overshoot, and other variables will not change with the text size.

On a simple drawing with only one scale factor, Chart 1 can be used to determine the plotting scale factor and the block insertion factor for title blocks and other blocks.

To use this chart, first determine the final scale for the drawing, then work back from there.

For example, if you are creating a plot or drawing that will be 3/16″ =1′0″, draw the floor plan at a scale of 1:1. Add the dimensions at a DIMSCALE of 64. Add the hatch at a scale of 64 (unless it is an architectural hatch), then insert the title block at a scale of 64. Finally, plot the drawing at a scale of 1/64.

Chart 1 will help you to scale the model correctly. The dimensions, text, hatch, and linetype must be scaled to fit the drawing.

Scaling the Text and Annotations

You will have noticed that, when creating text and dimensions, you often had to change the scale factors in order to see them correctly on the screen. This is even more of a difficulty when they are merged with a group of other files or views in a drawing at various scale factors.

If, for example, you want to make sure the text on the above floor plan is at 1/8″ on the final drawing, you would need to find the relationship between 1/8″ on the paper and the size of the text on the floor plan. If 1/4″ on the paper will be equivalent to 12 inches on the model, then 1/8″ or half of that value will be 6″ on the model. If you create your notations and dimensions at 6″ when creating the part, the final drawing at 1/4″=1′0″ will show text that is 1/8″ in height.

Text, dimensions, and related notations should be scaled to attain the appropriate size on the paper. Chart 2 (see next page) shows you at what sizes to scale your text to obtain the proper size on the final drawing. Find the size that you want your text to be, the scale you intend to use, and make the text the suggested size.

To use Chart 2, determine the final size of text that you require and set DIMSCALE and text size accordingly. Again, do not forget you can change existing dimensions with UPD or create a new style and have it Applied (page 178).

For dimensions on a drawing that will be 1/4″=1′0″, change the DIMSCALE to:

```
DIM:DIMSCALE
Current value<.1800>:48
```

If you take your final plot scale and multiply it by the default sizes and scale factors, you will arrive at workable scales and heights.

LTSCALE and HATCH Scale

While working with views to be placed on a drawing, you must also consider HATCH and LINETYPE. In creating these drawing aids for display on a screen, we determined that the scale for these functions should be the furthest value of X on your screen divided by 12. This provided a working area that was visible on the screen and easy enough to work with, without risking the possibility of filling the disk.

To place views onto a sheet of paper with the linetype and hatch at the proper paper format, use the same setting that you would for dimension scale or final plot scale.

Using Blocks to Compile Drawings

For plotting different drawings at different scale factors, you can insert drawings as blocks on a title block or another accepted drawing sheet. But though this will work, it is not recommended, because any changes will have to be made in the original file.

In the following drawing there are two separate files inserted onto a drawing sheet that is 36″ × 24″.

Chart 2: Dimension and Text Size

Plotted Text Size (Architectural or Imperial)	Scale	Text on the Model	DIMSCALE
1/8"	1/16"=1'0"	24"	192
(0.1800 units)	1/8"=1'0"	12"	96
	3/16"=1'0"	8"	64
	1/4"=1'0"	6"	48
	1/2"=1'0"	3"	24
	1"=1'0"	1.5"	12
1/4"	1/16"=1'0"	48"	
	1/8"=1'0"	24"	
	3/16"=1'0"	16"	
	1/4"=1'0"	12"	
	1/2"=1'0"	6"	
	1"=1'0"	3"	
3/16"	1/16"=1'0"	36"	
	1/8"=1'0"	18"	
	3/16"=1'0"	12"	
	1/4"=1'0"	8"	
	1/2"=1'0"	4.5"	
	1"=1'0"	2.25"	
Mechanical			
.25"	1:2	.5"	
	1:10	25.0"	
	2:1	.125"	
.125"	1:2	25"	2
	1:10	12.50"	10
	2:1	.0625"	0.5
.1875"	1:2	.3875"	
	1:10	18.75"	
	2:1	.09375"	
Metric			
3 mm	1:10	30	
	1:100	300	
	1:500	1500	

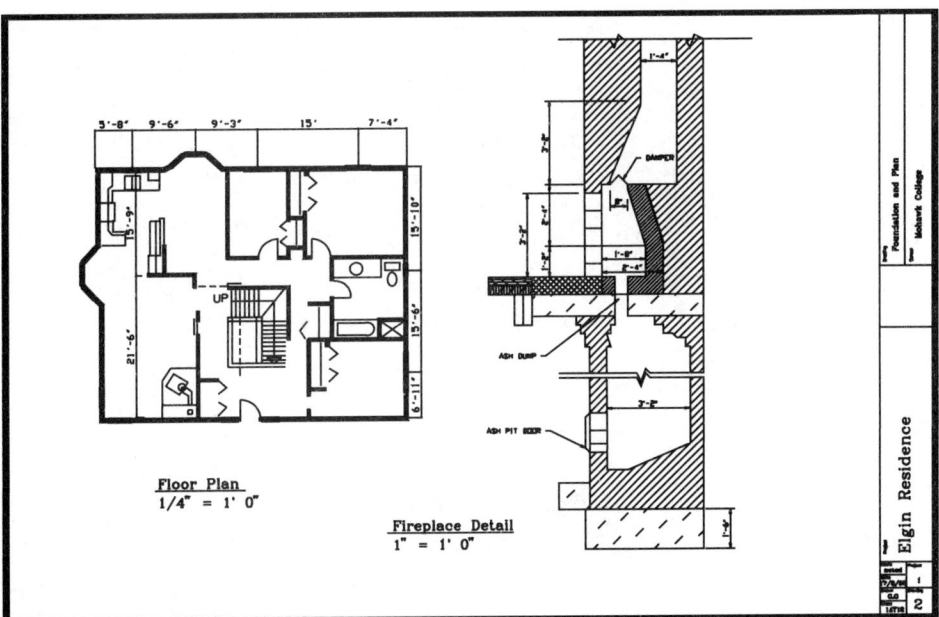

Figure 11-1

The drawing shown in Figure 11-1 consists of three separate stored files:

1. A title block at 36" × 24"
2. A fireplace detail at 16' vertically
3. A floor plan 35' × 30' (EXA6)

Since the floor plan is the most important view and the one most likely to have changes made, it is the base file. The other two files are inserted in. The floor plan was inserted at a scale of 48 and the fireplace, being a detail, was inserted at a scale factor of 4 times its original size, which will be 1"=1'-0".

```
Command: INSERT
Block name (or ?): TITLE
Insertion point: 0,0
X scale factor<1>/Corner/XYZ: 48
Y scale factor (default = X): ↵
Rotation angle: ↵
```

This method of creating drawings is useful for many reasons. Often, you have details that can be used on various drawings. If the details were merged with each file plotted, this would take up a lot of room on your disk. Time and space are saved when details are kept as separate files and simply merged when a drawing is ready to be plotted. The drawing can be compiled with many views, plotted, and then erased.

If a drawing is to be plotted at many different scale factors, create dimension styles at those different sizes and have the correct dimension scale factor displayed on the drawing that is plotted.

You can insert drawings onto the title block and plot at a scale of 1:1, or you can insert the title block and details onto the main view and plot at a determined scale.

Reconfiguring for Extra Space

If you are working on extremely large files — in 3D this is particularly important — you can have temporary files and swap files located in a reserved directory of the hard drive to speed up REGENs, shading, etc. Use CFIGDHAR.EXE to place swap files on reserved directory if you have not already done so.

A batch file can be used to delete contents of these directories when AutoCAD is started.

Using Paper space to Compile Drawings

The most efficient way to compile a drawing that has details to be shown at different scale factors is through paper space.

Essentially, **paper space** is a 2D document layout facility. Once the views are completed with all the dimensioning required, paper space takes these views and places them on a "paper" much like a cut-and-paste routine so that they can be compiled as a drawing. With paper space, you will be creating viewports so that you will have the views of the object set up on a 2D format. paper space acts like a sheet of paper through which the object is seen; you can then annotate as if you were drawing on paper.

Paper space makes use of the multiple viewport facility shown in the 3D section of this book. Multiple viewports are used to place the various views of the objects onto different portions of paper layout both in 3D and in 2D.

With paper space, it is as if you have taken photographs of the model and now you are compiling them on a board or paper for use in communicating the information to someone else. The paper space limits are set to the size of the plotted sheet.

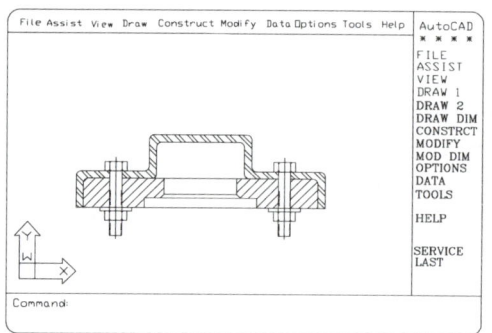

Model space   Paper space

Figure 11-2

In Figure 11-2, the screen on the left is in **model space** and the screen on the right is in paper space. Both are filled with the same drawing. One has one view, the other has three.

In paper space the crosshairs cover the screen. In the view on the left you have a regular UCS icon; on the right there is the UCS paper space icon. In the screen on the left is one view of the drawing; on the right you have one screen cut into three sections.

The borders for the views will not be plotted on the final drawing if they are made on a different layer.

The main view of the part seen in the drawing on the right will be produced at a scale of 1:1. The details of the views will be produced at a scale of 2:1.

While you can create the same type of drawing with blocks, paper space not only is much easier to use than blocks, but also saves a lot of space both on your disk and on the drawing, because there are many fewer overlapping lines. When using blocks to compile a drawing, there are four sets of lines in a four-view drawing. With paper space, the data within the file is only one model, and thus the storage is much smaller.

To get into paper space, use **TILEMODE**.

Paper space and Tilemode

Getting from model space to paper space is quite simple.

> **Windows** On the status bar, double-click Model. The toggle changes to PAPER. You can also choose Paper space from the View menu.
>
> **DOS** From the View menu, choose Paper space.

The command line equivalent is **TILEMODE**.

```
Command: TILEMODE
New value for TILEMODE<1>: 0
Entering paper space. Use MVIEW to insert model space
   viewports.
Regenerating drawing.
```

Where: **TILEMODE 1** gives you model space for modelling (= MSPACE)
TILEMODE 0 gives you paper space for drawing (= PSPACE)

Once this command is completed, you will not be able to see your drawing until you make some "holes in the paper," or viewports, using MVIEW.

When you have created your views, you can use either TILEMODE, which will toggle between the "tiled" views, or the commands MSPACE and PSPACE in paper space, which will toggle between the two modes.

Floating Viewports or MVIEW

The MVIEW or Floating Viewports command is used to identify views that will be used in paper space. The screen before MVIEW is the same as a blank piece of paper. The MVIEW command or Floating Viewports option allows you to add views of the model to this paper format.

> **Windows** From the View menu, choose Floating Viewports.
>
> **DOS** From the View menu, choose Floating Viewports.

The command line equivalent is **MVIEW**.

Once paper space has been accessed and the views have been made, you can toggle between model space and paper space quite readily. When you turn TILEMODE back on (returning to model space), AutoCAD restores the drawing as it was before paper space was entered.

When using the Floating Viewports or the MVIEW command, you can create various configurations of viewports; select viewports for hidden-line removal, and have different layers active in different viewports.

From the Floating Viewports menu, choose the number of viewports you want. If you only want one, pick the lower left and upper right corner for the viewport. The default setting for the command line is one viewport.

```
Command:MVIEW
ON/OFF/Hideplot/Fit/2/3/4/
   Restore/<First Point>: (pick 1)
Other Corner: (pick 2)
```

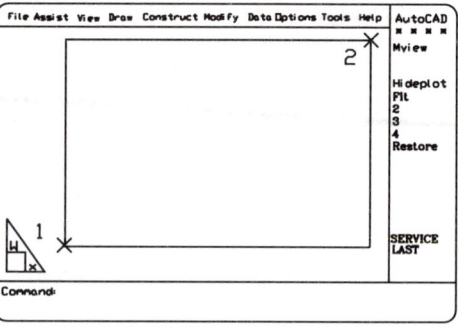

Figure 11-3

You can use this single viewport to create a single view of the model for plotting and drawing purposes. The single view on screen contains the view of the model that was current in model space. If you want more than one view, choose another option. Other options more relevant to 3D work are covered in detail in the 3D version of this book.

The 2/3/4 Option

If you would like to open a few viewports at once, pick 2, 3, or 4 viewports from the menu.

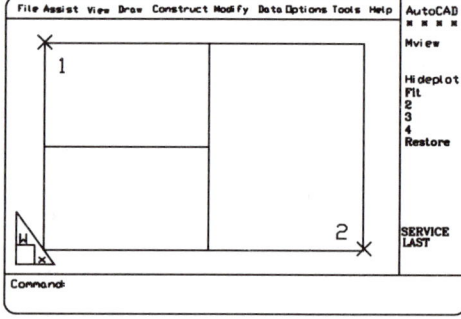

```
Command:MVIEW
ON/OFF/Hideplot/Fit/2/3/4/
   Restore/<First Point>:3
Fit/<First Point>:F
First Point: (pick 1)
Other Corner: (pick 2)
```

Figure 11-4

This will create a layout of three quadrants as shown. The view in each quadrant will be the view in the active viewport in model space. (If you only had one viewport, it will be that view.)

Using Model space and Paper space

TILEMODE On allows you to have one or more views of the model with the borders of the views side by side. You must be in model space.

TILEMODE Off allows you to move the views and place them on a page. (Going into TILEMODE Off will place you in paper space. Use MSPACE to toggle back.)

When TILEMODE is off, you can toggle from model space to paper space very easily using **MSPACE** or **MS** and **PSPACE** or **PS**.

```
Command:MSPACE (toggles back to model space)
Command:PSPACE (toggles back to paper space)
```

Paper space (PSPACE) only lets you access the paper view of the model. Model space (MSPACE) lets you access the model.

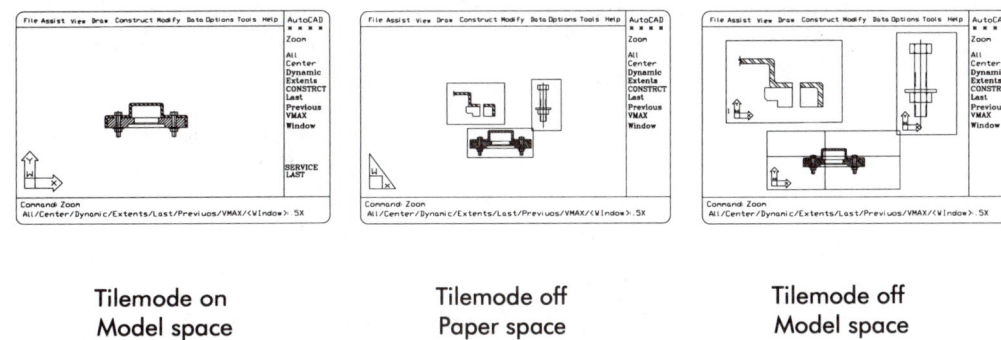

      Tilemode on   Tilemode off   Tilemode off
      Model space   Paper space   Model space

Figure 11-5

With TILEMODE off you can access the model through model space, or you can create the drawing in paper space. With TILEMODE on you can only access the model through model space.

When you use the ZOOM command in paper space, it affects the entire page or paper. A ZOOM .5X will result in the paper with the views intact at half the size that it was before.

When you use the ZOOM command in model space, the model or drawing within the view will be affected. Be careful not to ZOOM the views in model space after you have scaled them with ZOOM XP.

In model space you can PAN the objects within the viewports; in paper space the PAN command will affect the view of your paper environment.

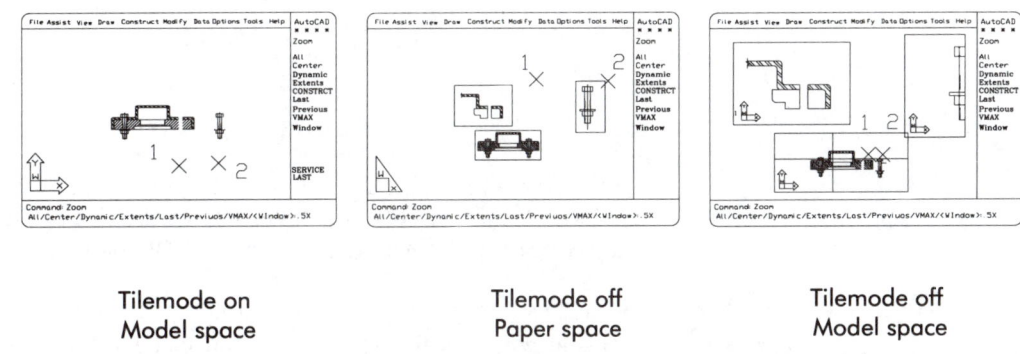

      Tilemode on   Tilemode off   Tilemode off
      Model space   Paper space   Model space

Figure 11-6

In model space the MOVE command will move the objects within the views. In paper space the MOVE command will move the views relative to the other views.

Manipulating the Views

You can use either construction lines (see hidden line) or point filters to move the information across the screen to line it up. The data contained within the views is accessible for reference either for filters or for OSNAPs, but is not accessible for editing purposes.

Any of the Modify commands will work on the viewports in paper space the same way that they work on the objects in model space. You can use MOVE, COPY, STRETCH, etc. to place the viewports where you want with the necessary information.

Scaling Views Within a Drawing

In model space, the command ZOOM All causes you to lose any relation to the actual scale of the part being designed; the image expands to fit the space provided. You now want the views to be the size that they will be plotted in, so use the ZOOM "times paper space" XP option to set the size.

While still in Tilemode Off, change your paper space environment back to model space, and set the zoom in the current viewport. To set the other viewports, simply pick those viewports, then use zoom again.

```
Command:MSPACE
Command:ZOOM
All/Center/Dynamic/Extents/Left/Previous/Vmax/Window/<Scale
   (X/XP)>:.5XP
(pick another view)
Command:ZOOM
All/Center/Dynamic/Extents/Left/Previous/Vmax/Window/<Scale
   (X/XP)>:.5XP
```

In paper space you want the views to be scaled to a relative size. Use the Zoom option XP to scale the object relative to the paper scale units. While the ZOOM option X scales the object relative to its current size within the viewport, the scale factor of XP gives you a zoom factor "times the paper scale" or relative to the paper and also relative to the actual part. Zoom XP scales can be seen in the following chart.

For a Scale Of	Use	Size Relative to Actual
1:1	ZOOM 1XP	1
1/2"=1"	ZOOM .5XP	1/2
1:50	ZOOM .02XP	1/50
1/4"=1'0"	ZOOM 1/48XP	1/(4 × 12)

A ZOOM of 2XP will result in a view twice the size of the original. You can use the scales on page 276 to determine the size of ZOOM needed.

The VPLAYER Command

Because you are using one model to create the drawing, and because each view of the drawing will contain dimensions and other information relative to that view and no others, you will need to be able to display certain layers in only one view of the drawing and not the others. VPLAYER allows you to do this.

The VPLAYER command allows you to perform LAYER freezing in selected viewports rather than as a global command. You can see the layers in DDLMODE and access layers by picking them to change the viewport visibility with Cur VP or New VP.

In model space, make one viewport current, then use DDLMODE to select which layers you would like to have current here. With Cur VP buttons, you can freeze and thaw layers in the current viewport. Freezing and thawing layers in this way overrides the global setting.

New VP freezes or thaws selected layers for all new objects you create.

As with the Floating Viewports or MVIEW command, this command only shows up in paper space, when the TILEMODE is off (or set to 0). You can use VPLAYER in either model space or paper space TILEMODE 0, but the system will automatically switch to paper space if you try to use it in model space.

> **Windows** or **DOS** From the Data menu, choose Viewport Layer Controls, then choose Freeze or Thaw.

The command line equivalent is **VPLAYER**.

```
Command:VPLAYER
?/Freeze/Thaw/Reset/Newfrz/Vpvisdflt:
```

Where: **?** = a request for a listing of the LAYERs that are frozen in any selected viewport

Freeze = freeze a LAYER or LAYERs in selected viewports

Thaw = the reverse of Freeze; turns on selected LAYERs in selected viewports

Newfrz = create a LAYER that is new in the VPLAYER command; this LAYER is created frozen in all viewports, and is primarily used for creating a LAYER to be viewed only in one viewport ever; you first create the LAYER, then you Thaw it in the desired viewport

```
Command:VPLAYER
?/Freeze/Thaw/Reset/Newfrz/Vpvisdflt:N (Newfrz)
New viewport frozen letter names:TITLE (name of the LAYER)
```

Where: **Reset** = the default display for layers created in the LAYER command is Thawed; in the VPLAYER command, the display defaults to Frozen; with the Reset option, the layers are returned to their original default setting

```
Command:VPLAYER
?/Freeze/Thaw/Reset/Newfrz/Vpvisdflt:R (Reset)
Layer(s) to Reset:0,DIM1,DIM2 (no space between names)
All/Selected/<current>:
```

Where: **Vpvisdflt** = set a default visibility for any layer in any viewport; this could be useful for Resetting the layers; determines the default visibility of all layers in existing viewports

For most of the above options, none of the changes will take place until you complete the command.

The prompts allow you to choose the LAYER to be modified, and the objects to be selected for freezing. You can enter multiple LAYERs by putting commas between the layer names.

Dimensioning in Paper space

To use different scales in different viewports on the same drawing, set the DIMSCALE value to 0. This is paper space scale, and will cause all dimensions to be shown in the default size regardless of the scale of the view. Once you have set the DIMSCALE to 0, you must update each viewport in MSPACE. Use the VPLAYER command to freeze dimension layers that appear in the wrong viewport.

The same is also true of LTSCALEs. To neutralize the effect of different scale factors in paper space, set the PSLTSCALE variable to ON and regenerate the file.

The MVSETUP Command

MVSETUP is a quick way to set up the specifications of a drawing.

In Model space

In model space, MVSETUP can be used to set the units, type, drawing scale factor, and paper size. Using the settings you provide, AutoCAD draws a rectangular border at the drawing limits. If you are just starting a drawing this is not terribly useful, as it somewhat cuts down on your ZOOM capabilities. This is an old drawing attitude. It is better CAD practice to draw the object and add the title block and notations later. This will make better use of the screen.

In Paper space

In paper space or with TILEMODE off (0), you insert one of several predefined title blocks into the drawing and create a set of floating viewports within this title block. You can specify a global scale as the ratio between the scale of the title block in paper space and the model geometry in model space.

The Floating Viewports menu is available only when TILEMODE is off or set to 0. The first prompt AutoCAD displays depends on the tilemode setting. In paper space find MVSETUP from the View menu.

> **Windows** and **DOS** From the View menu, choose Floating Viewports, then MV Setup.

The command line equivalent is **MVSETUP**.

```
Command:MVSETUP
Align/Create/Scale viewports/Options/Title block/Undo
```

Where: **Align** = (This option is explained below.)

Create = create viewports within the paper space environments

Scale viewports = adjust the scale factor of the objects displayed in the viewports; this functions like the ZOOM XP factor

Options = set the layer, limits, units, and Xrefs within each view

Titleblock = prepare paper space, orient the drawing by setting the origin, and create a drawing border and title block

Align will provide the following prompt:

```
Align/Horizontal/Vertical alignment/Rotate view/Undo:
```

Where: **Align** = pan the view in a viewport so that it aligns with a basepoint in another viewport
Horizontal = pan the view in one viewport until it aligns horizontally with a basepoint in another view
Vertical alignment = pan the view in one viewport until it aligns vertically with basepoint in another view
Rotate view = rotate the view in a viewport around a basepoint
Undo = undo the last option in the command

This method of preparing a drawing incorporates the VPLAYER, ZOOM XP, and MVIEW commands.

See Prelab 11B for a walk-through of the MVSETUP command.

Prelab 11A Architectural Example

In this example we will take a floor plan of a house that has a circular fireplace and a kitchen, and we will create one drawing sheet with a plan view and the two details described. This drawing will have three scales as follows:

Plan view of house	1/4″ = 1′0″
Detail of fireplace	1″ = 1′0″
Detail of kitchen	1/2″ = 1′0″

Step 1   Retrieving the File

Retrieve the file on which you want to create a drawing.

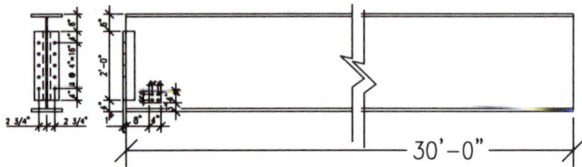
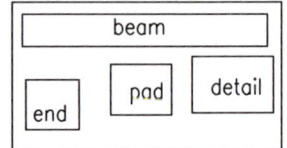

It may be a good idea to make a simple sketch of the final drawing noting the view scales and the DIMSCALEs. The final drawing will be plotted on a 36″ × 25″ sheet.

Step 2   Accessing Paper space

Assuming that the data is complete on the file, turn TILEMODE off and create a paper on which to place the views that you have decided on.

Paper space is a 2D document layout facility. You are now taking the file, laying out three rectangles on paper for the three desired views, and adding to the part geometry all of those things that relate to paper.

In order to create the viewports you must enter paper space.

> **Windows** On the status bar, double-click Model. The toggle changes to PAPER. Also, you can choose Paper space from the View menu.
>
> **DOS** From the View menu, choose Paper space.

The command line equivalent is **TILEMODE O**.

```
Command:TILEMODE
New value for TILEMODE<1>:0
Entering paper space. Use MVIEW to insert model space
   viewports.
Regenerating drawing.
```

Don't panic! Your screen should be blank and you should have the paper space icon in the lower corner.

Step 3   ***Setting Up Your File for Paper space***

First create the following layers, making BORDER current.

```
Command:'DDLMODES
Add layers with colors as shown.
Layer Name      Color          Color Number
DIM2            yellow              2
DIM3            green               3
BORDER          magenta             6
PSPACE          blue                5
```

The model may be on many different layers. Those listed above are added for the PSPACE information.

Views of the model will be added to the paper in paper space. These views will have borders or frames. As the borders for the views will be frozen before plotting, make your BORDER layer current before adding the views so that the view borders will be in that layer.

LIMITS, SNAP, and GRID can be used to help place the views properly. First, LIMITS will set the size of the paper in paper space, 34 inches by 23 inches. Again, this has nothing to do with the size of the model.

```
Command: (make BORDER layer current in 'DDLMODES)
Command:LIMITS
Lower left corner<0'-0",0'-0">:↵
Upper right corner<43'-9",35'-0">:34",23" (your default will
   be whatever size your original drawing is made in; make the
   LIMITS the actual size of the paper, in this case 34 × 23)
Command:SNAP
Snap spacing or ON/OFF/Aspect/Rotate/Style<1">:.5
Command:GRID
ON/OFF/Value/Aspect/<2">:1
Command:ZOOM
All/Center/Dynamic/Extents/Left/Previous/Vmax/Window/<Scale
   (X/XP):A
```

This will set the actual size of your paper. *There will still be nothing on your screen.*

Step 4   ***Creating the Views***

Now you can add your views relative to this paper using either the Floating Viewports or the command MVIEW. Account for 3 inches on the left of the paper for the plotter.

> **Windows** From the View menu, choose Floating Viewports, then 1 Viewport.
>
> **DOS** From the View menu, choose Floating Viewports, then 1 Viewport.

```
Command:MVIEW
ON/OFF/Hideplot/Fit/2/3/4/Restore/<First Point>:3.5,2
Other Corner:18,22
```

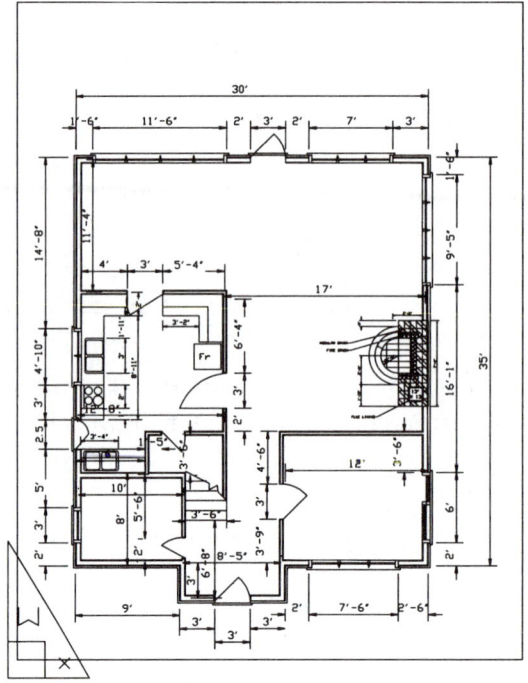
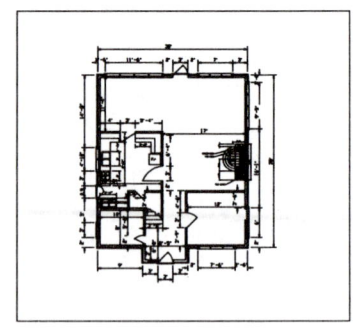
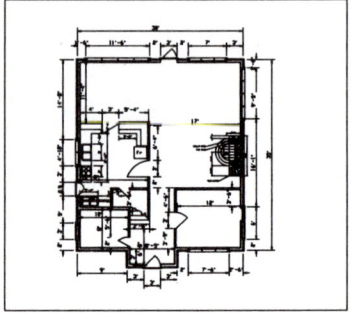

Add the two other views. Each view becomes an individual viewport entity.

```
Command:MVIEW
ON/OFF/Hideplot/Fit/2/3/4/Restore/<First Point>:20,2
Other Corner:31,11
Command:MVIEW
ON/OFF/Hideplot/Fit/2/3/4/Restore/<First Point>:20,12
Other Corner:31,22
```

Notice that your model space display is repeated in each viewport. The borders for the views are in layer BORDER, color magenta. If they are not, use CHPROP to put them in that layer.

Step 5 *Setting the Zoom with ZOOM XP in the Plan View*

We will decide on the scale of each view. Use ZOOM XP to scale each view relative to the viewport, and add dimensions and notations in each model space viewport.

Start with the plan view. Since this view is to be plotted at a scale of 1/4″=1′0″, the Overall scale and HATCH scale should be 48.

Toggle back to MSPACE so that you can access the information within the viewports.

```
Command:MSPACE
```

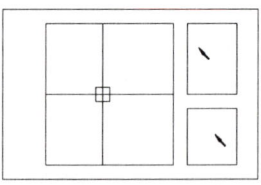

Notice that your crosshairs are in one view only. Pick the large view on the left to activate it for scaling and dimensioning.

Use ZOOM XP (times paper space) to create a view of the plan at the size in which it will appear on the drawing.

```
Command: (pick the plan view)
Command:ZOOM
All/Center/Dynamic/Extents/Left/Previous/Vmax/Window/<Scale
   (X/XP):W
(use Window and PAN to place the view within the viewport)
Command:Z
All/Center/Dynamic/Extents/Left/Previous/Vmax/Window/<Scale
   (X/XP):1/48XP
Command:PAN (pan the view onto the screen if necessary)
```

The plan view should be correct. The dimensions should be on in layer DIM1. If not, use CHPROP or add the dimensions and notations now.

Step 6   *Accessing the Fireplace Detail*

In model space you can only access the information within the viewports; you can't PAN or ZOOM the viewport configuration. Therefore, toggle back to paper space in order to get a detail of the fireplace viewport. This will make dimensioning easier.

```
Command:PSPACE
Command:ZOOM
All/Center/Dynamic/Extents/Left/Previous/Vmax/Window/<Scale
   (X/XP):W
First Corner: (pick 1) (pick 2)
```

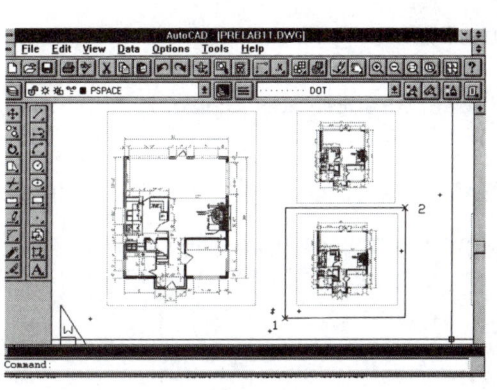
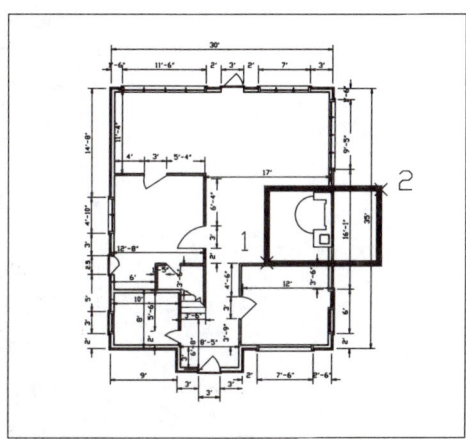

Now in MSPACE within the viewport, use ZOOM W to get a general area of the fireplace, then ZOOM XP to get the correct scale of 1″=1′-0″.

```
Command:MSPACE
Command: (pick the fireplace viewport)
Command:ZOOM
All/Center/Dynamic/Extents/Left/Previous/Vmax/Window/<Scale
   (X/XP):W
First Corner: (pick 1)
Other Corner: (pick 2)
```

```
Command:ZOOM
All/Center/Dynamic/Extents/Left/Prevs/Vmax/Window/<Scale
   (X/XP):1/12XP
Command:PAN (use PAN to place the view correctly)
```

Step 7 *Dimensioning the Views*

This view is going to be 1″=1′0″, so change your DIMSCALE to 12, make layer DIM2 current, and create dimensions on the fireplace. Use the layer dialog box and Cur VP to Freeze DIM1 to get the first set of dimensions off the screen.

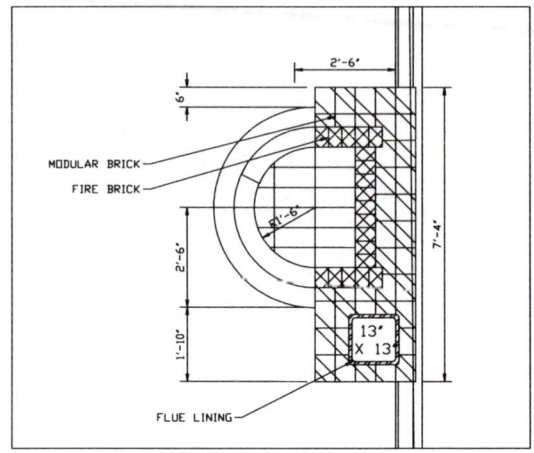

```
Command: (make DIM2 current, Freeze DIM1 in current VP)
Command:MSPACE (you should be in MSPACE already, but if not,
   use MSPACE now)
Command:DIM
Dim:DIMSCALE
Scale factor<48>:12
Dim: (continue to add dimensions)
```

Make the HATCHes on the same layer. The HATCH scale should be 12.

Once the view is complete, make the kitchen active and add dimensions there.

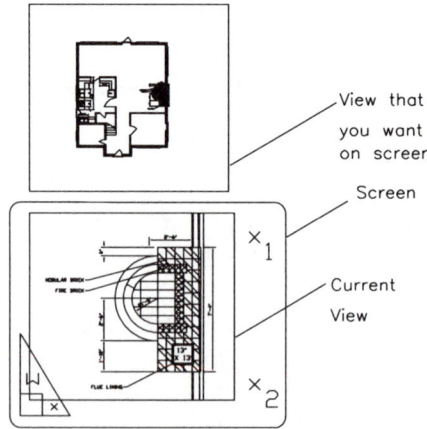

```
Command:PSPACE
Command:PAN
Displacement: (pick the top of the screen)
Second Point: (pick much lower on the screen)
```

Now use ZOOM XP to scale the view to 1/2"=1'0".

```
Command: MSPACE
Command: ZOOM
All/Center/Dynamic/Extents/Left/Previous/Vmax/Window/<Scale
   (X/XP): W
First Corner: (pick 1)
Other Corner: (pick 2)
Command: ZOOM
All/Center/Dynamic/Extents/Left/Prevs/Vmax/Window/<Scale
   (X/XP): 1/24XP
Command: PAN (use PAN to place the view correctly)
```

Make Layer DIM3 current and add the dimensions for the kitchen with a DIMSCALE of 24. Add any hatches or hidden lines at a scale of 24 as well.

You have dimensioned your views in the correct dimscale; now try DIMSCALE 0 and UPDate the views to see if there is any change. Set PSLTSCALE to 1.

Step 8   Adjusting the Size of Your View (Optional)

If the viewport is not large enough, return to paper space to STRETCH the size of the viewports to fit the views. Always adjust the size of the viewport rather than the scale of the view if possible. You may need to use ZOOM .7X in paper space to get the viewport configuration to fit more accessibly on the screen.

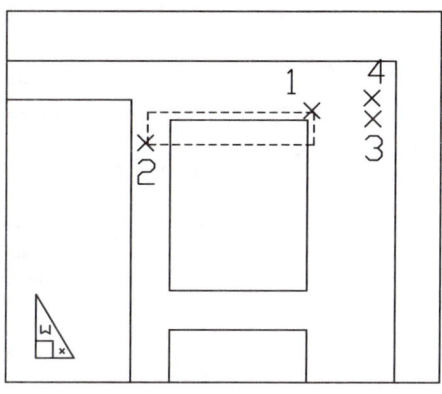

```
Command: PSPACE
Command: ZOOM
All/Center/Dynamic/Extents/Left/
   Prevs/Vmax/Window/<Scale
   (X/XP): .7X
Command: STRETCH
Select objects: C
First corner: (pick 1)
Other corner: (pick 2)
Select objects: ↵
Base point: (pick 3)
New point: (pick 4)
```

By picking up the outside edge of the viewport, you can stretch it as large or as small as you like. You can also use MOVE to place the view where you want it. Pick the border of the view at the "Select objects:" prompt.

Step 9   Using VPLAYER to Freeze Layers in Certain Viewports

The VPLAYER command performs LAYER freezing in selected viewports rather than as a global command. After ZOOM All, notice that the plan view has the dimensions for the kitchen and fireplace. The dimensions are on different layers at different scale factors.

Use VPLAYER to freeze DIM3 in the plan view and the fireplace view. First go to 'DDLMODES to thaw DIM1.

```
Command:'DDLMODES (thaw DIM1)
Command:VPLAYER
?/Freeze/Thaw/Reset/Newfrz/
   Vpvisdflt:F
Layer(s) to Freeze:DIM3
All/Selected/<current>:S
Select objects: (pick the plan and
   the fireplace)
?/Freeze/Thaw/Reset/Newfrz/
   Vpvisdflt:⏎
```

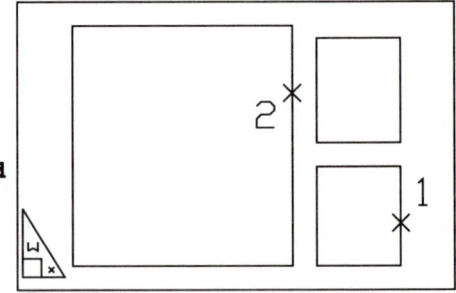

Once you become familiar with the command you can do all freezing and thawing in one command. Now freeze DIM2 in the kitchen and the plan, and DIM1 in the fireplace and kitchen.

```
Command:VPLAYER
?/Freeze/Thaw/Reset/Newfrz/Vpvisdflt:F
Layer(s) to Freeze:DIM2
All/Selected/<current>:S
Select objects: (pick the plan and kitchen) ?/Freeze/Thaw/
   Reset/Newfrz/Vpvisdflt:F
Layer(s) to Freeze:DIM1
All/Selected/<current>:S
Select objects: (pick the kitchen and fireplace)
?/Freeze/Thaw/Reset/Newfrz/Vpvisdflt:⏎
```

Having used the VPLAYER command, toggle back to MSPACE and use the layer dialog box to make the remainder of the layers frozen in the appropriate viewports.

Step 10   Adding the Title Block and Drawing Notations

Add a title block. You may have one from Tutorial 8; if not, draw in a title block in paper space. Make layer PSPACE current before you start. Use DIR A: to find the title block if you have one.

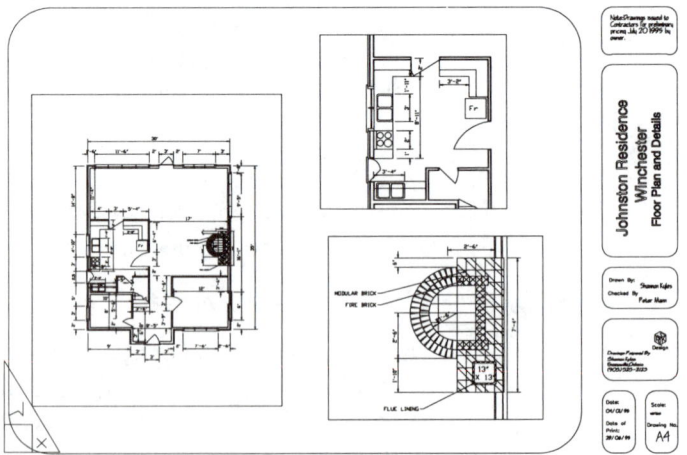

```
Command: (make PSPACE layer current in 'DDLMODES)
Command:INSERT
Block name (or ?):TB33X23 (or another title block if you have
   one)
```

```
Insertion point:3.5,.5
X scale factor<1>/Corner/XYZ:↵
Y scale factor (default=X):↵
Rotation angle<0>:↵
Command: (move the title block if you need to)
```

Make sure that your title block allows for the plotter's rollers; leave at least .5″ on three sides and 3″ on the left side.

If you want to reposition any of the views at this point, do so.

```
Command:MOVE
Select objects: (pick the border of the view)
Select objects:↵
Base point or displacement: (pick a point on the border)
Second point of displacement: (pick a point where you would
   like the view)
```

Step 11 Repositioning the Views with Point Filters (Optional)

Filters can be used to help line up the views.

```
Command:MOVE
Select objects: (pick 1)
Select objects:↵
Base point or displacement:END of
   (pick 2)
Second point of displacement:.X of
   END of (pick 3)
Needs YZ:F8 (pick 4)
```

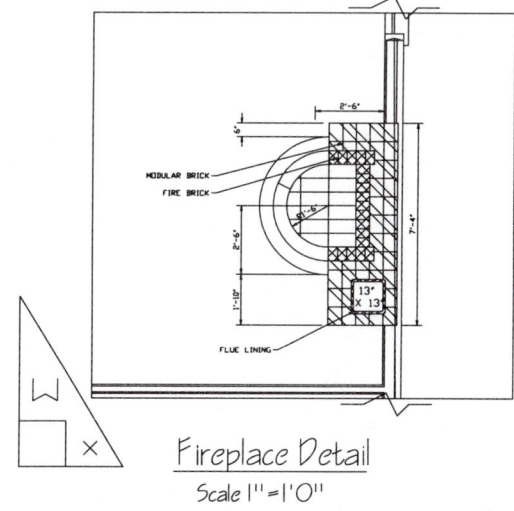

Points on objects within the view can be accessed as object snaps, but cannot be accessed for editing.

Remember that the borders will be turned off, so you will need break lines at the ends of the lines that will be cut as well as view titles and scales.

Setting Up Drawings and PSPACE **295**

If you have entered a title block with incorrect information, change it now. If it is a block, you may need to explode it.

Step 12   *Turning Off Paper space Frames or Borders*

The frames or borders for the tiles used to position the views are no longer necessary in the final drawing. Freeze layer borders so that the view borders will be removed.

Your drawing is now ready to plot at a scale of 1:1.

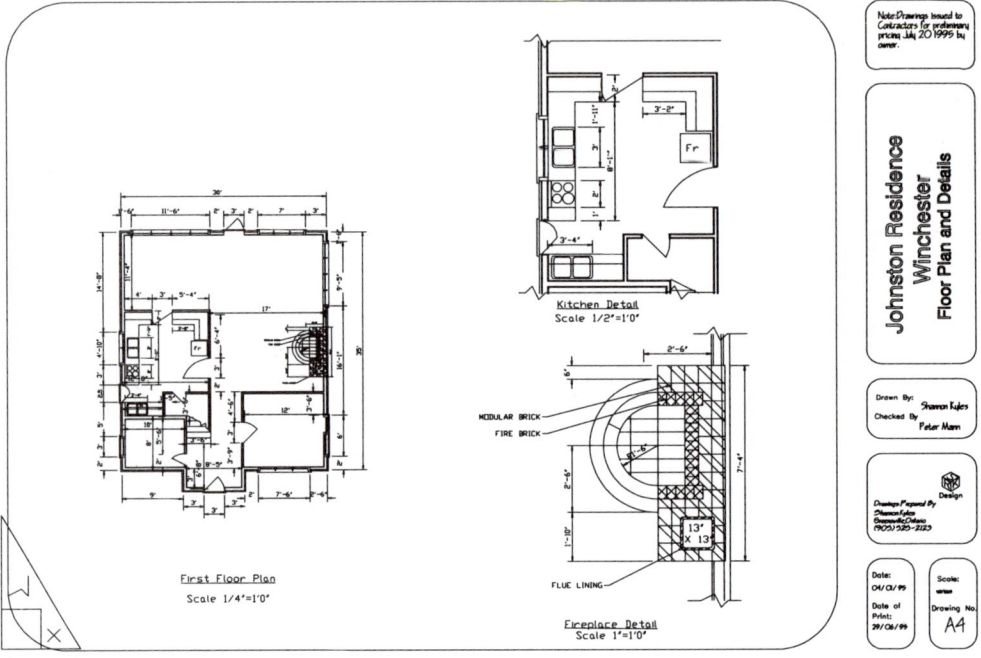

Prelab 11B Mechanical/Civil Drawing Using MVSETUP

In this example we will take a structural steel drawing that has an end detail, a PAD detail, and a section detail. We will create one drawing in model space, then create several views in paper space.

Beam elevation	1/4"=1'0"
End elevation	1 1/4"=1'0"
Pad detail	6"=1'0"
Detail A	1 1/4"=1'0"

Step 1 *Retrieving the File*

Retrieve the file for the beam or quickly draw it up.

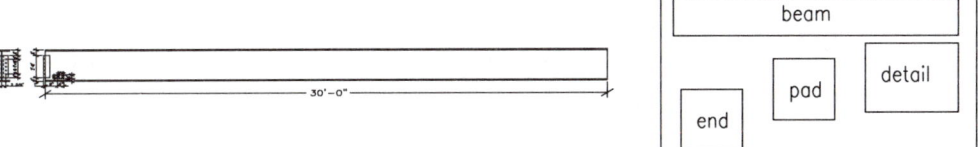

It may be a good idea to make a simple sketch of the final drawing noting the view scales and the DIMSCALEs. The final drawing will be plotted on an 8 1/2" × 11" sheet.

Step 2 *Accessing Paper space*

Assuming that the data is complete on the file, turn TILEMODE off and create a paper on which to place the views that you have decided on.

Paper space is a 2D document layout facility. You are now taking the file, laying out three rectangles on paper for the three desired views, and adding to the part geometry all of those things that relate to paper.

In order to create the viewports you must enter paper space.

> **Windows** On the status bar, double-click Model. The toggle changes to PAPER. Also, you can choose Paper space from the View menu.
>
> **DOS** From the View menu, choose Paper space.

```
Command:TILEMODE
New value for TILEMODE<1>:0
Entering paper space. Use MVIEW to insert model space
   viewports.
Regenerating drawing.
```

Don't panic! Your screen should be blank and you should have the paper space icon in the lower corner.

Step 3 Setting Up Your File for Paper space

First create the following layers, making PSPACE current.

```
Command:'DDLMODES
Add layers with colors as shown.
Layer Name      Color         Color Number
DIM2            yellow        2
DIM3            green         3
DIM4            cyan          4
PSPACE          blue          5
BORDER          magenta       6
```

The model may be on many different layers. Those listed above are added for the PSPACE information.

Views of the model will be added to the paper in paper space. These views will have borders or frames. The borders for the views will be frozen before plotting, so make your BORDER layer current before adding the views so that the view borders will be in that layer.

LIMITS can be set using MVSETUP, but GRID and SNAP are useful too, so use the normal commands to set up the drawing.

```
Command: (make PSPACE layer current in 'DDLMODES)
Command:LIMITS
Lower left corner<0'-0",0'-0">:⏎
Upper right corner<43'-9",35'-0">:9,12 (your default will be
   whatever size your original drawing is made in; make the
   LIMITS the actual size of the paper, in this case 8.5 × 11)
Command:SNAP
Snap spacing or ON/OFF/Aspect/Rotate/Style<1">:.25
Command:GRID
ON/OFF/Value/Aspect/<2">:.5
Command:ZOOM
All/Center/Dynamic/Extents/Left/Previous/Vmax/Window/<Scale
   (X/XP):A
```

If you want to set the units and limits in MVSETUP, use the following.

> **Windows** From the View menu, choose Floating Viewports, then MV Setup.
>
> **DOS** From the View menu, choose Floating Viewports, then MV Setup.

```
Command:MVSETUP
Align/Create/Scale viewports/Options/Title block/Undo
```

To change the options, use **O**, then:

To specify the current layer, use **L** for layer. Then enter the name **PSPACE**. If this layer is not already made, this command will make it, but you must go to the layer menu to make it current.

To reset the drawing limits, use **LI** for limits, then enter **Y**.

The size will adjust to slightly larger than the chosen title block.

To specify how drawing limits should be expressed, enter **U** for UNITS. Then enter **F** for feet, **I** for inches, **ME** for meters, and **MI** for millimeters.

Press ⏎ to return to the original prompt.

This will set the actual size of your paper. *There will still be nothing on your screen.*

Step 4   Adding the Title Block

Now you can add your title block and views relative to this paper using Floating Viewports and MVSETUP.

In the same command, enter **T** for Title block. Press ⏎ to display a list of standard paper sizes. Enter the number of the size you want to use, which would be:

```
Command:MVSETUP
Align/Create/Scale viewports/Options/Title block/Undo:T
Add/Delete/Redisplay/<Number of entry to load>:7
```

This will enter a preloaded title block.

Step 5   Adding the Viewports

From the same command, load the viewports.

```
Command:MVSETUP
Align/Create/Scale viewports/Options/Title block/Undo:C
Delete objects/Undo/<Create viewports>:
Available Mview viewport layout options

0: None
1: Single
2: Std. Engineering
3: Array of Viewports

Redisplay/<Number of entry to load>:1
```

Add the three other views. Each view becomes an individual viewport entity.

Notice that your model space display is repeated in each viewport.

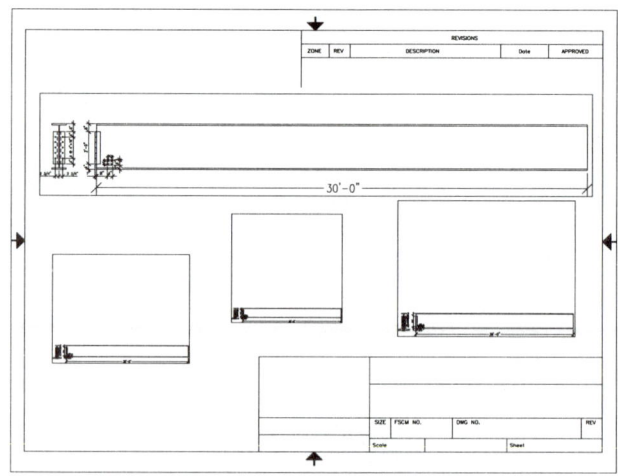

Step 6 Scaling the Viewports

You can scale the viewports using the MVSETUP command in PSPACE, but first you should go into MSPACE and ZOOM Window to get the approximate size that you want of each view.

Command:**MSPACE**

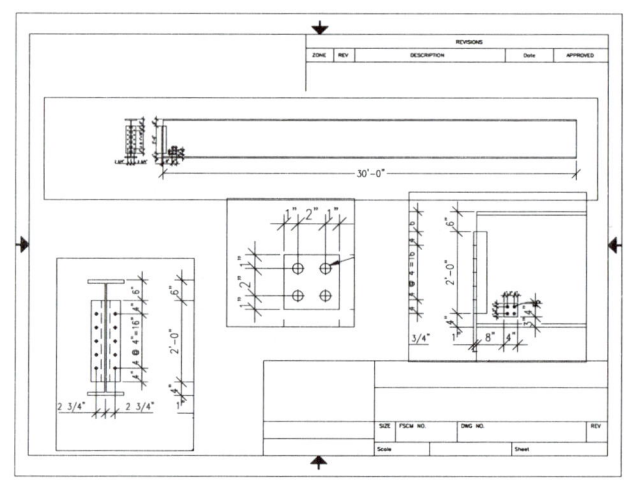

Notice that the crosshairs are only in one view. Zoom into the approximate area that you want on that view, and then pick another view and ZOOM Window on it.

Now toggle back to PSPACE and use MVSETUP to scale the views to the exact scale that you want. Start with the beam view.

```
Command:MVSETUP
Align/Create/Scale viewports/Options/Title block/Undo:S
Select objects: (pick the full beam view)
Enter the ratio of paper space units to model space units ...
Number of paper space units<1.0>:1/48
Number of model space units<1.0>:1
Align/Create/Scale viewports/Options/Title block/Undo:S
```

1/16 will be the fraction for 1'1/4'', and 1/6 will be the fraction for 6''.

You may need to PAN the information over on the screen. You must be in MSPACE to PAN a view within a viewport.

```
Command:MSPACE
Command:PAN
```

Step 7 Dimensioning Within the Views

The dimensions for the beam should have been done in model space. Now that the views are all assembled, use DIMSCALE 0 with UPDATE to make all of the dimensions the same size within each viewport.

Using the Dimension style mode under the Data menu, access the Geometry menu, then activate the Scale to Paper space button.

Step 8 Adjusting Layer Visibility

Use the DDLMODE Layer dialog box to turn off the layers not needed in each viewport.

```
Command:MS (MSPACE must be on to access each individual view)
Command: (pick the beam detail)
```

Select the DIM2, DIM3, and DIM4 layers, then pick FRZ beside Cur VP.

Do the same for the other views.

Make sure that all of the views contain the necessary information.

Step 9 Turning the BORDER Layer Off

The views should contain only the information needed. Use MOVE and STRETCH to make sure that the viewport boundaries are in the correct spot. You cannot change the size of the viewports once the BORDER layer is frozen.

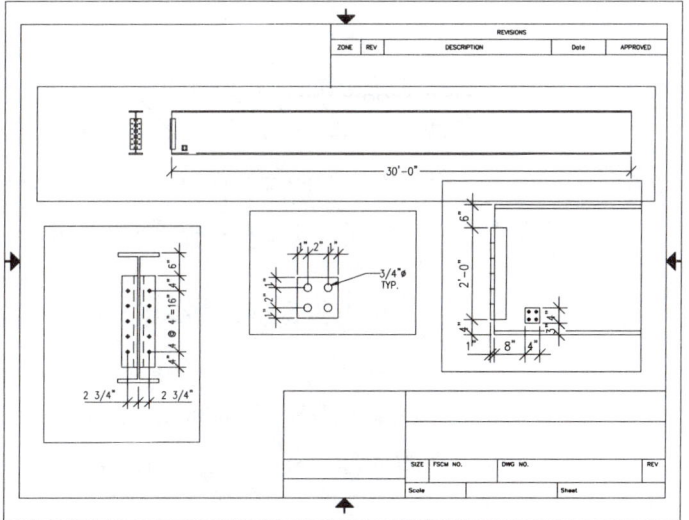

When everything is in correctly, freeze the BORDER layer.

Setting Up Drawings and PSPACE **301**

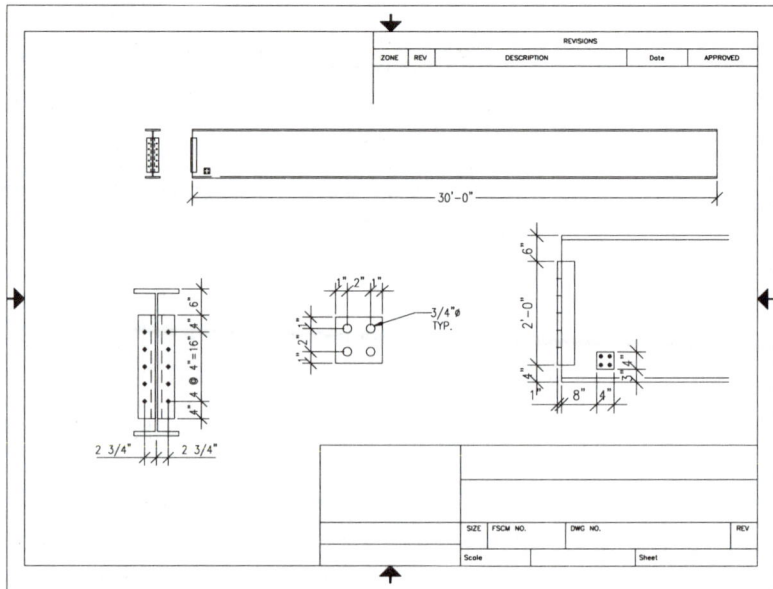

Step 10   Adding Notations

With the PSPACE layer current, add all of the information that you would need on a drawing such as view titles and scales. Fill in the title block.

Your drawing should now be ready to print.

For plotting parameters, see Appendix D, "Plotting and Printing" at the back of the book.

Command and Function Summary

Ctrl V switches viewports in model space.

Model space allows for the creation of a model or drawing in 2D.

MSPACE toggles from model space to paper space viewports.

MVSETUP in paper space is used to automate the process of setting up a drawing of a model.

Paper space allows for the layout of multiple views for drawing purposes.

PSPACE toggles from model space to paper space viewports.

TILEMODE is a system variable that controls paper space access.

VPLAYER allows layers to be visible in selected viewports.

Practice Exercise 11

Create this drawing in model space, then use paper space and MVSETUP to create the final drawing.

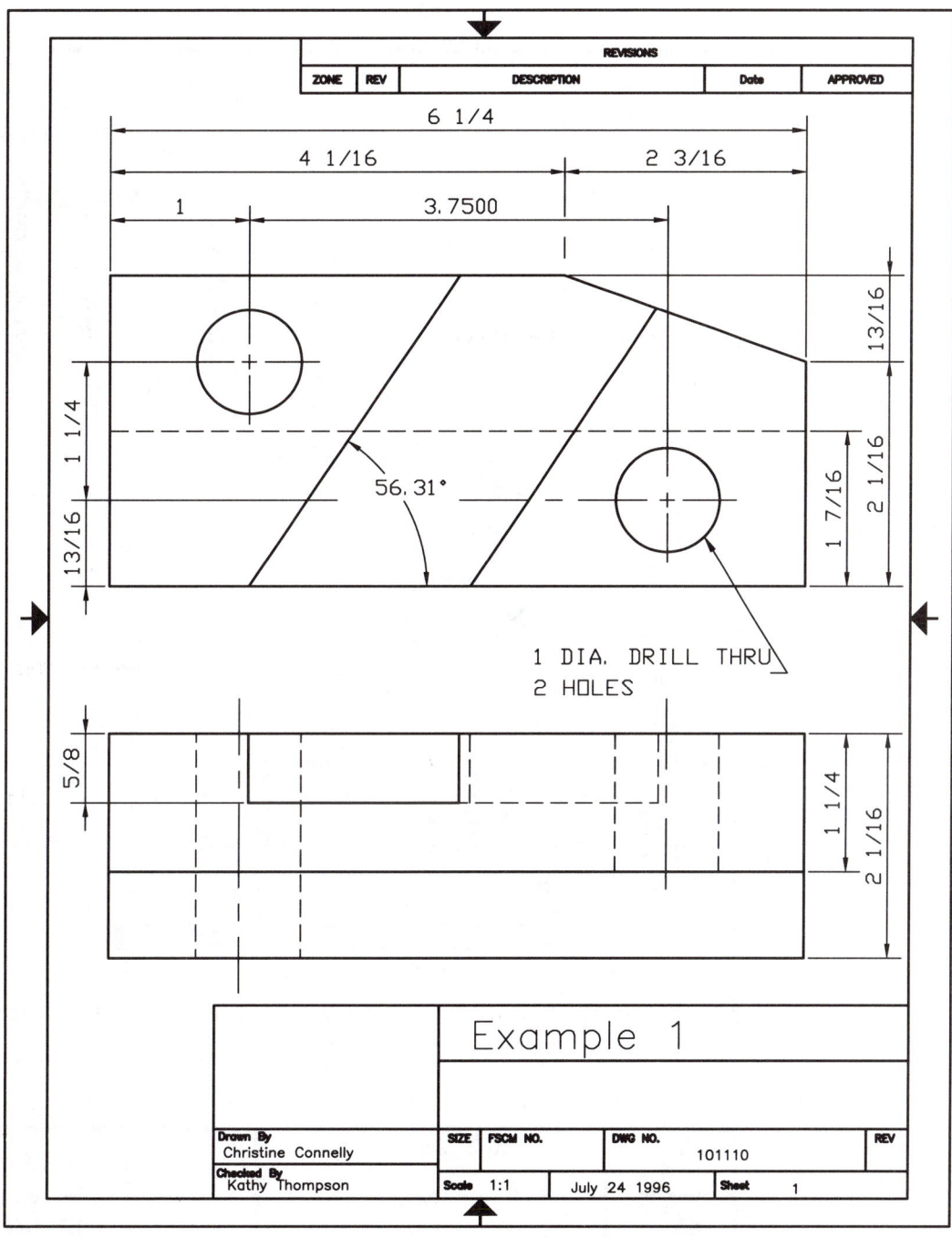

Exercise A11

Quickly draw up the staircase. Then use PSPACE to create two viewports, one for the full view and one for the detail. Add the title block in PSPACE and plot.

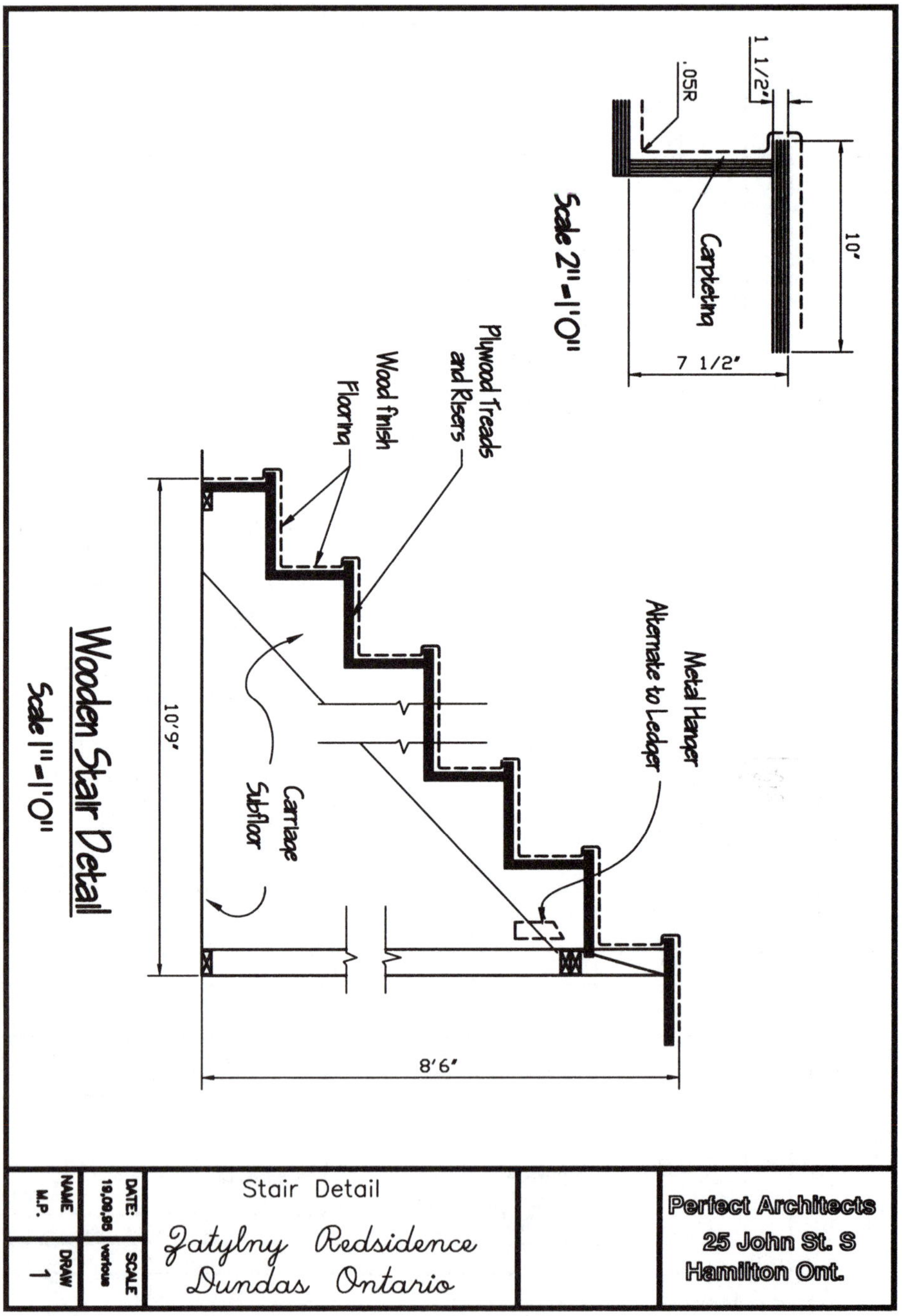

Exercise C11

Draw in the footing. Then turn tilemode off and add two viewports, one for the front and top, the other for a detail view. Use model space to create the dimensions in the appropriate dimscale, then add the views in paper space.

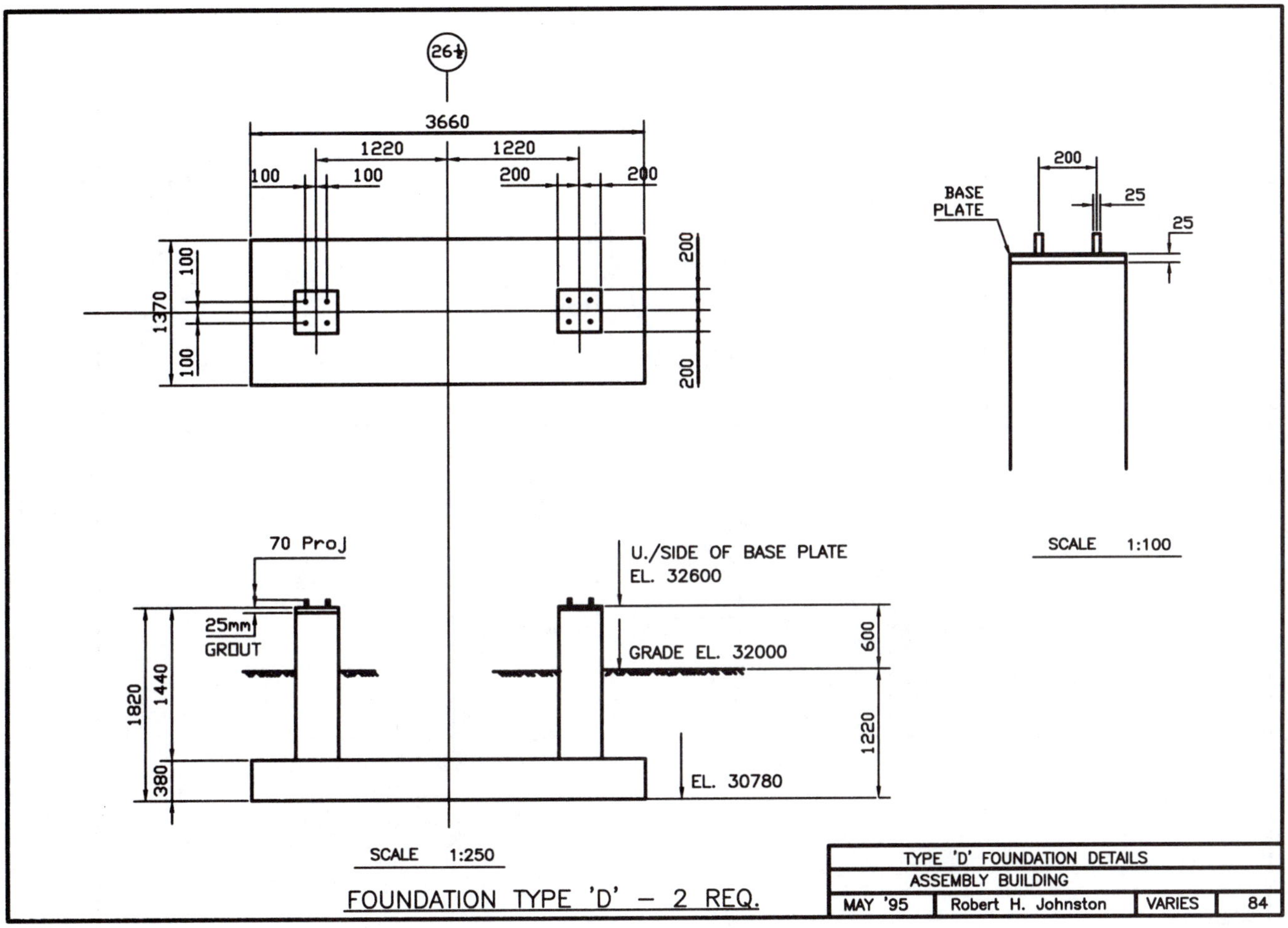

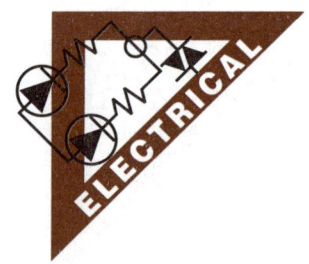

Exercise E11

Draw in the circuit, then use MVSETUP to place it on a metric title block.

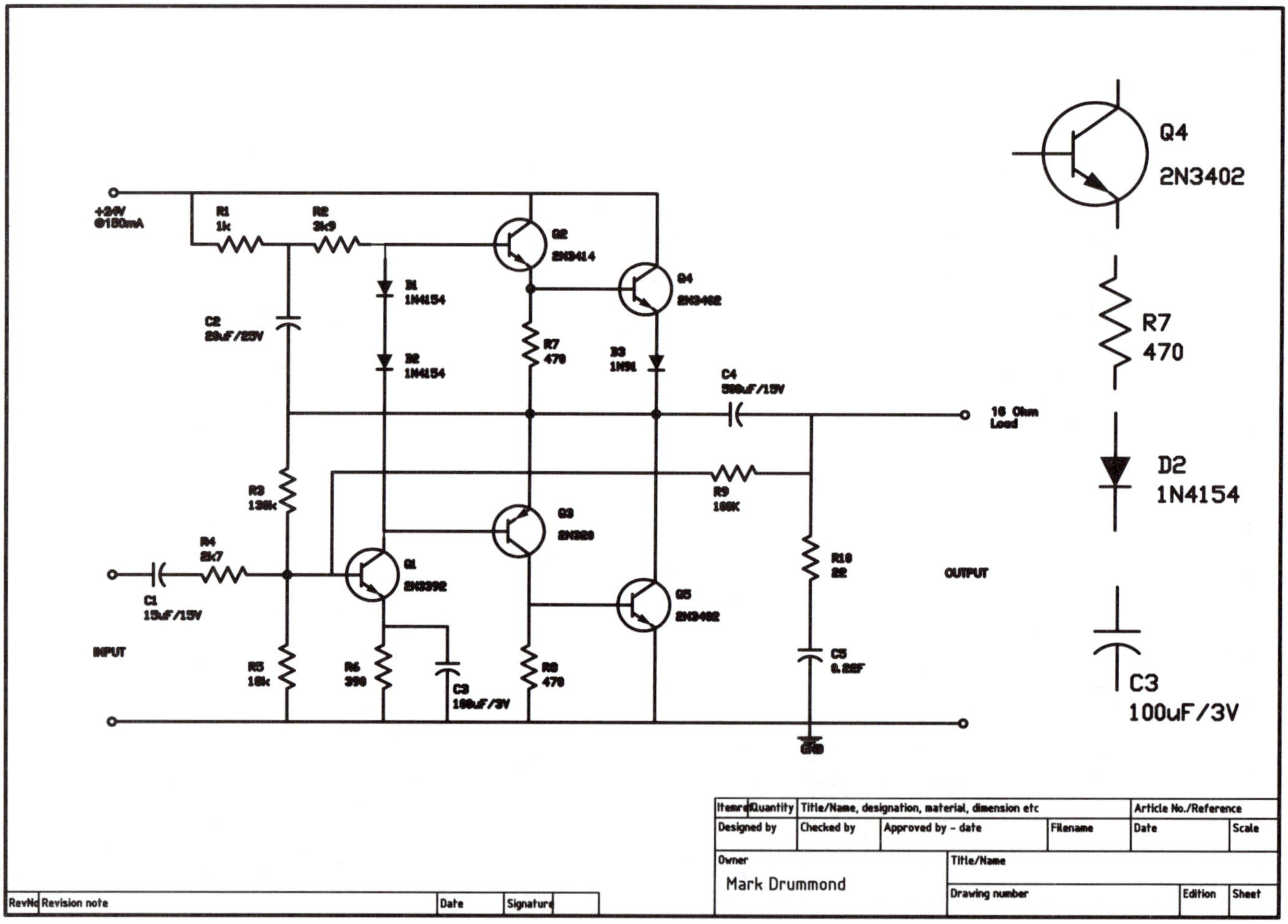

Exercise M11

Draw the body of the part and the pin at a scale of 1:1. Then create two viewports in paper space and dimension the parts in model space. Add a title block and print.

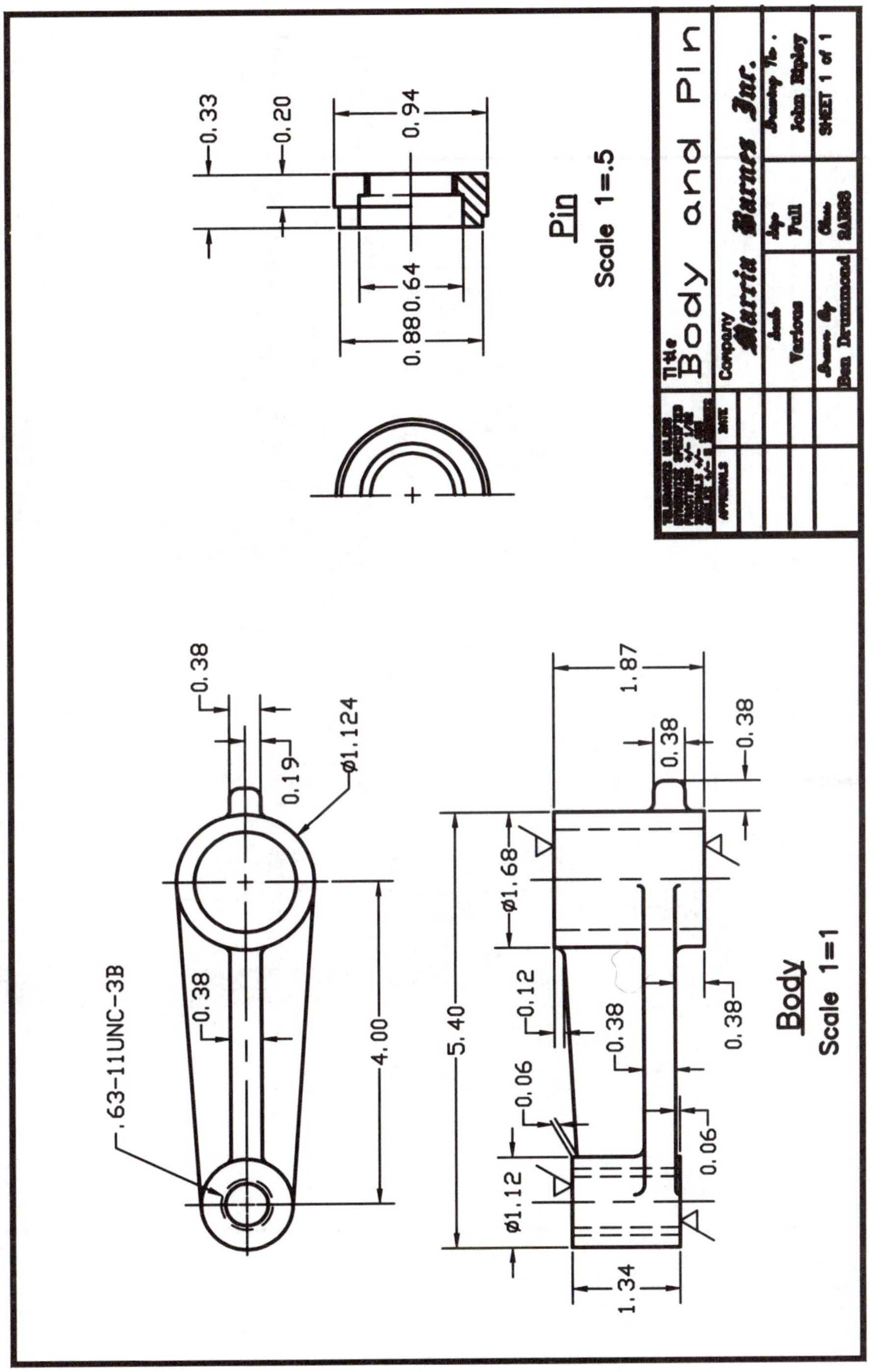

Challenger 11

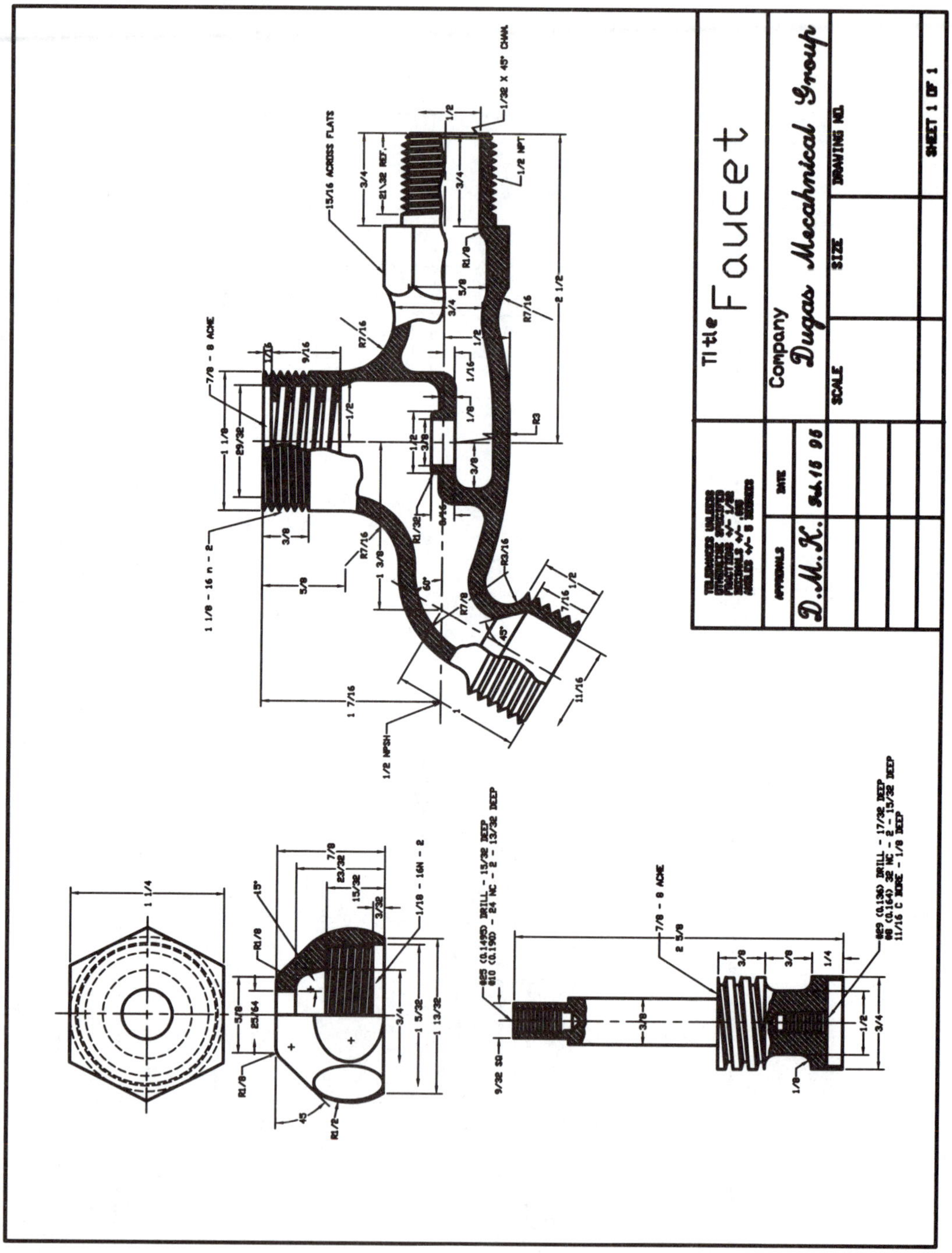

12 2D Review and Final Drawings

1. Review of creation of 2D drawings
2. Final practice in creating large drawings

OBJECTIVES

Each of the preceding eleven chapters has introduced a different aspect of AutoCAD software. You should now be capable of creating 2D drawings and plotting or printing them.

The real problems with any CAD system begin when you start compiling large drawings and multiview drawings. The drawings at the end of this chapter are included to help you practice creating large drawings.

Most people learn about file management the hard way. Most CAD users can tell you the tragic story of the first time they lost a very important drawing. In order to avoid joining the list of people who have lost their final drawings just before completion, stick by the following rules.

1. Always keep two backups of your files, not one, on separate disks.
2. Save the file every hour; this way, you will never lose more than one hour's work. You can also set the variable SAVETIME, for example to 15 minutes or 30 minutes, and AutoCAD will save the file automatically.
3. Save your drawings to the hard drive, then copy them to the two floppies. This will prevent file size errors.
4. Keep your disks away from magnets — cell phones and digitizers can contain magnets.
5. Never trust a computer.

If your files on the hard drive are too large to copy onto a floppy, it is usually because you have unwanted HATCH data or unwanted database records. To make the files smaller:

1. PURGE the file, then QSAVE it.
2. OPEN the file again.
3. QSAVE the file again.

Now list your files. You may find that the .DWG files are very much smaller than the .BAK files.

AutoCAD became the most popular CAD package in the world because it offered the most flexibility. The *only* way to become really proficient with the software is to practice. If you have time, it would benefit you a great deal to create *all* of the drawings in this chapter.

2D Review and Final Drawings **309**

Problem 1

Draw the desk and panel unit shown, starting with the 0,0 at point A. Set Units to Architectural with a one-inch readout.

Refer to the drawing of the desk to answer the following questions.

1. What would be a reasonable "First point of mirror line" to create the desk unit in dash-dot lines? Be sure that you avoid creating duplicate panels or duplicate lines. Once you have identified the point, perform the MIRROR command.

    a. 0,1.5"
    b. -1.5,0"
    c. 0,3"
    d. -3,0"

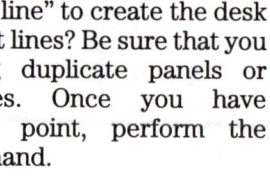

2. What would be a reasonable "First point of mirror line" to create the desk units in hidden lines? Perform the MIRROR command.

    a. 0,-1.5"
    b. -1.5,0"
    c. 0,-3"
    d. -3,0

3. What would be the absolute coordinate value of the midpoint of line B?

    a. 3'3",1'9"
    b. 2'3",2'3"
    c. 2'6",2'6"
    d. 2'3",2'6"

4. What would be the absolute coordinate value of the midpoint of line C?

    a. 3'3",1'9"
    b. 2'3",2'3"
    c. 3'3",1'9"
    d. 2'3",2'6"

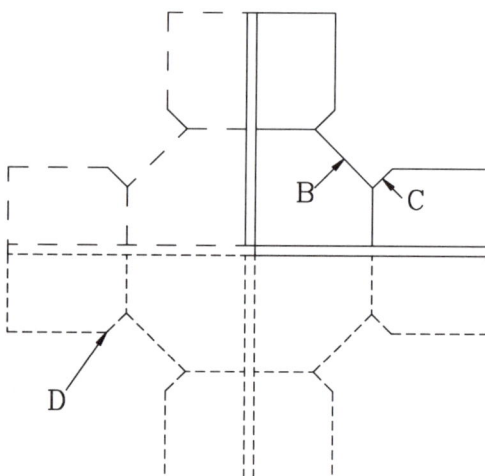

5. What would be the absolute value of point D?

    a. -3'9",-2'3"
    b. -4'3",-2'0"
    c. -3'6",-2'3"
    d. -3'9",-2'0"

6. If drawn so that each desk unit is 3' in length and each connecting panel is 6' in length, how many lines are there?

    a. 52
    b. 48
    c. 44
    d. 40

Problem 2

Draw the object shown, locating the center of the 3.250 radius circle E at 5.000,5.000.

Use the UNITS command to set the units to decimal, with the number of digits to the right of the decimal point at 3.

Refer to the drawing on your screen to answer the following questions.

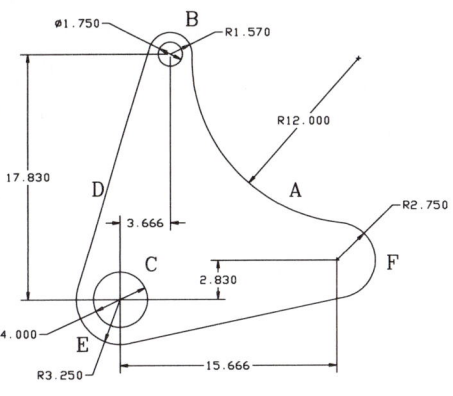

1. The total length of line segment D is:
   a. 18.127
   b. 17.239
   c. 17.238
   d. 16.986
   e. 16.374

2. The absolute coordinate value of the center of the 4.00 radius arc is:
   a. 24.245,23.456
   b. 23.464,22.259
   c. 22.231,22.497
   d. 21.321,22.497
   e. 20.437,23.267

3. The length of the 2.750 radius arc segment F is:
   a. 5.790
   b. 5.970
   c. 7.590
   d. 7.950
   e. 9.750

4. The distance from the center of the 3.250 radius arc E to the center of the 0.750 radius arc A is:
   a. 24.557
   b. 24.957
   c. 25.975
   d. 25.759
   e. 26.349

5. The circumference of the 4.00 diameter circle C is:
   a. 11.476
   b. 11.957
   c. 12.560
   d. 12.750
   e. 13.254

6. Using the center of the 12 unit radius arc A as a base point, scale the object in size by a scale factor of .333. The absolute value of the center of the 1.570 radius circle B is:
   a. 13.676,20.467
   b. 15.897,21.349
   c. 16.947,22.393
   d. 17.714,22.608
   e. 18.354,22.538

Quiz

Following are 104 questions that should provide a good review of the work performed so far.

Choose the response that is *most* correct.

1. Who makes the AutoCAD software?

    a. AutoCAD

    b. AutoDESK

    c. AutoSHADE

2. What does ORTHO do?

    a. SNAPs to a preset integer

    b. Allows only straight lines

    c. Allows only vertical and horizontal lines

3. What function key do you use to see your position with relation to the origin?

DOS	Windows
a. F7	a. Status bar, double-click ORTHO
b. F8	b. Options menu, double-click ORTHO
c. F6	c. Snap Angle ORTHO box

4. What function sets your point picks to a preset integer?

    a. GRID

    b. SNAP

    c. ORTHO

5. How would you draw a line 3 units in length at 45 degrees?

    a. @3<45

    b. @3>45

    c. @45<3

6. What would be the final point in this line?

    From point: 3,4

    To point: @4,0

    To point: @0,3

    To point: @-2,0

    a. 9,7

    b. 7,9

    c. 5,7

7. What does the Object SNAP QUADrant do on a circle?

    a. SNAP to the top, bottom, and far left or right sides

    b. SNAP to a tangent

    c. SNAP to the center point

    d. SNAP to the end point

8. What does the C stand for in the CIRCLE command?

    a. Center

    b. Close

    c. Clip

9. What does the C stand for in the ZOOM command?

    a. Clip

    b. Crossing

    c. Center

10. What does the R stand for in the Erase command?

    a. Replace

    b. Remove

    c. Restore

11. How do you cancel a command in progress?

DOS	Windows
a. QUIT	a. QUIT
b. ESC	b. ESC
c. ^C	c. ^C
d. F1	d. F1

12. Can you use the Object SNAP TANgent to create a line between two lines?

    a. Yes

    b. No

    c. Sometimes

13. How do you edit a polyline?

    a. CHANGE

    b. PEDIT

    c. EDIT

    d. CHPROP

14. How do you get a fully solid circle?

    a. SOLID

    b. FILLET Rad 0

    c. DONUT

15. Can you change the grid size once the drawing has started?

    a. Yes

    b. No

    c. Sometimes

16. Can you use the cursor to pick the points from the screen for the bottom left and top right of your limits?

    a. Yes

    b. No

17. Can you turn the limits off?

    a. Yes

    b. No

    c. Sometimes

18. Does a ZOOM All affect the size of your limits?

    a. Yes

    b. No

    c. Sometimes

19. What does **U** ⏎ do?

    a. Erases the last line

    b. Edits the last erase

    c. Undoes the last command

20. How can you set AutoCAD to reference the ENDpoint of existing objects for more than one command?

    a. Running Object SNAP under Options

    b. OSNAP on the command line

    c. LISP routines only

    d. All of the above

    e. a. and b. above

21. What does the L stand for in selecting objects?

    a. LINE

    b. Last

    c. Lost

    d. Link

22. What command places multiple copies of objects in patterns that are either polar or rectangular?

    a. ROTATE

    b. ARRAY

    c. TWIST

23. What does the F stand for in BREAK?

    a. First point

    b. Freeze

    c. Find

    d. Flip Screen

24. How do you make many copies of an item at random placing?

    a. COPY Multiple

    b. ARRAY

    c. ROTATE

    d. MIRROR

25. EXTEND helps you to elongate items to reach what?

    a. A boundary

    b. A cutting line

    c. A corner

26. What do you need to create an OFFSET?

    a. An object, a distance, and a side to offset

    b. An object, a cutting line, and an edge

    c. An object, a center point, and an edge

27. What command allows you to break off and delete any overhangings beyond a cutting edge that you specify?

   a. TRIM

   b. BREAK

   c. EXTEND

28. Can you use a Crossing window to identify the cutting lines in the TRIM command?

   a. No

   b. Yes

   c. Only in paper space

29. Under what circumstances does the cursor turn into a box?

   a. When the zoom factor is too low

   b. When an item selection is needed

   c. When a prompt is missing

30. Can you use incremental values (@3,4) to describe the displacement in the MOVE command?

   a. Yes

   b. No

31. What two commands may automatically perform a REGEN on your file?

   a. ZOOM and PAN

   b. DISPLAY and EDIT

   c. MOVE and BREAK

32. What do the angle brackets (< >) in a command string indicate?

   a. A default value

   b. A preset integer

   c. The suggested size

   d. The current size

33. What does CP mean in the "select objects" prompt?

   a. Crossing Polygon

   b. Create Polygon

   c. Center of Polygon

34. Can you set two concurrent Running Object Snaps?

    a. No

    b. Yes

35. What does the command PURGE do?

    a. Exits AutoCAD and saves the file

    b. Cleans the file of all unused blocks, layers, and linetypes

    c. Erases WBLOCKS

36. What does .AC$ stand for?

    a. A temporary file

    b. A backup file

    c. An erased file

    d. All of the above

37. Can you change the width of a pline once it has been entered?

    a. Yes

    b. No

    c. Sometimes

38. Must your mirroring line in the MIRROR command always be part of the object?

    a. Yes

    b. No

    c. Sometimes

39. What happens if you enter **UNDO 5**?

    a. The last five erases will be undone

    b. The last five commands will be undone

    c. The last five lines will be undone

40. When you use SAVE or QSAVE, where does the file get saved to?

    a. The root directory

    b. The directory that the file is on

    c. The A: drive

41. In the filename A:AC$EF.$A, what does A: stand for?

    a. The drive

    b. The directory

    c. The extension

42. In AutoCAD, which command do you use to get a listing of your files on the A: drive?

    DOS   *Windows*

    a. DIR A: a. List File Utility

    b. LIST A: b. FILES

    c. UTILITY A: c. DIR A:

    d. FILES   d. File Manager

43. In DOS, how would you copy a backup file on the A: drive into a drawing file in the B: drive?

    a. COPY A:NAME.BAK B:NAME.DWG

    b. COPY A:NAME.DWG B:NAME.DWG

    c. COPY B:NAME.DWG A:NAME.DWG

44. In DOS, how would you erase a drawing file from your disk?

    a. ERASE A:*.DWG

    b. ERASE A:NAME.DWG

    c. ERASE A:*.*

    d. None of the above

45. Why is it important not to remove your floppy disk from the drive while you are writing to it in AutoCAD?

    a. You may lose the address of the temporary file

    b. You may insert the wrong diskette later

    c. The RAM will not be able to retrieve the file

46. Why is DISKCOPY an incorrect command when you have a 360 K floppy in the A: drive and a 1.2 meg floppy in the B: drive?

    a. Because the files will be stretched to fit the larger format

    b. Because the DISKCOPY command reformats the target disk

    c. Because you cannot use two different sizes of floppy drives

47. Can you rename a layer once it has been entered?

    a. Yes

    b. No

    c. Sometimes

48. When using the STRETCH command, what is the system automatically set to if you pick the command from the screen menu?

    a. Crossing

    b. Window

    c. Last

    d. None of the above

49. If you stretch a circle, does it turn into an ellipse?

    a. Yes

    b. No

    c. Only with ORTHO on

50. If you change the rotation of GRID and SNAP, does the ORTHO rotate as well?

    a. Yes

    b. No

51. How would determine the layer of an object?

    a. Undo

    b. LIST

    c. CHPROP

52. Why must you press ↵ once you have selected all of the objects that you need?

    a. To signal the end of the object selection

    b. To exit OSNAP

    c. To reset the cursor

53. If you have selected a Window full of objects for editing, but one item was picked up that you do not want to edit, how would you release it from the list?

    a. Release

    b. Remove

    c. Undo

54. If you have an item in one layer, but you want it to be on another layer, how can you alter it?

    a. CHPROP

    b. LIST

    c. DDLMODEs

55. Can you rename the layer 0?

    a. Yes

    b. No

    c. Sometimes

56. How do you change the color of an object once it has been entered?

    a. Change the color in DDLMODE

    b. Use CHPROP to change the color

    c. Turn the layer off, then reenter

    d. All of the above

    e. Just a. and b. of the above

57. Can you delete a layer once it is in the file?

    a. Not if there are objects in it

    b. Yes, if it is empty, and with PURGE

    c. Not while it is current

    d. All of the above

58. Of what use is OK within the layer format?

    a. To exit DDLMODE

    b. To thaw layers

    c. To change colors

59. What wildcard can you use to load all the linetypes?

    a. ^F1

    b. *

    c. ?

60. If you have a set linetype, will this override your current layer linetype?

    a. Yes

    b. No

    c. Sometimes

61. What is the default setting for radius in FILLET?

    a. 0

    b. .5

    c. 1.0

62. How many colors are available with Release 13 of AutoCAD?

    a. 8

    b. 16

    c. 256

    d. More than 256

63. How do you change the size of the dashes in the hidden-line display?

    a. LINETYPE

    b. LTSCALE

    c. DIM VARs

64. What is the advantage of using QSAVE over END?

    a. There is no advantage

    b. The QSAVE command cleans up the drawing database by deleting all marked records

    c. The END file will be larger

    d. Both c. and b.

65. If you create several styles under the same basic dimension style, what have you created?

    a. A file data replacement

    b. A dimension style family

    c. A dimension style triad

66. If you want to change any of these settings, what do you need to access?

    a. The Geometry dialog box

    b. The Dimension Style dialog box

    c. The Dimension Style icon menu

67. How do you use different text fonts in the dimension text area?

    a. Set to the new font before creating the dimensions

    b. Set a new font in the dimension style menu

    c. Use APPLY to set a style from the font menu

68. What command do you use to make an image appear at 70% of its current screen size for purposes of dimensioning?

    a. SCALE .7

    b. ZOOM .7X

    c. ZOOM 70

    d. ZOOM .7XP

69. Can you use more than one basic dimension style in a drawing?

    a. Yes

    b. No

    c. Sometimes

70. What command do you use if you have changed the unit readout and want to have your existing dimensions revised according to the current setting?

    a. APPLY

    b. OVERRIDE

    c. UPD

    d. All of the above

71. Why does text already entered become highlighted when the command TEXT or DTEXT is reentered?

    a. In case you want to replace it

    b. To indicate where to place the next paragraph

    c. To indicate where the text will default to line up

72. What is the purpose of QTEXT?

    a. QTEXT allows you to enter the text more quickly

    b. QTEXT allows for quicker REGENs

    c. QTEXT allows you to read the text as it is being entered

73. Can you pick the height with your cursor when entering text, or do you need to use a numeric entry?

    a. A numeric entry is needed

    b. Both cursor and numeric entries are allowed

    c. Only a numeric entry is allowed

74. When will the TEXT command not offer you a height option?

    a. When you have changed the aspect ratio in the Style

    b. When you have used your cursor to pick the first height

    c. When you are using QTEXT

75. What command do you use to change the type of letters that you would like to use?

    a. Text Style

    b. MTEXT

    c. DTEXT

    d. Object Creation

76. What does an obliquing angle do?

    a. Sets the angle of a string of text

    b. Sets the angle of a character

    c. Makes lines oblique

77. What command would you use to alter the spelling within your text string?

    a. CHPROP

    b. DDEDIT

    c. CHANGE

    d. Two of the above

78. Name two ways of repositioning text once entered.

    a. MOVE and CHANGE

    b. MOVE and COPY

    c. COPY and CHPROP

79. Which of these lettering styles takes up the most room on the disk?

    a. Gothic English

    b. Roman Simplex

    c. Monotext

    d. Standard

80. How do you change the position of a paragraph of text?

    a. Properties

    b. CHANGE

    c. MOVE

81. How do you create an associative hatch?

    a. BHATCH

    b. ASOHATCH

    c. AHATCH

82. Which standard are the hatch patterns derived from?

    a. ASCII

    b. ANSI

    c. CSA

83. Can you get a hatch pattern to ignore text?

    a. Yes

    b. No

84. How can you access the standard AutoCAD hatch patterns?

    a. Use HATCHSTYLE

    b. Use HATCH ?

    c. Use the Boundary Hatch dialog box

85. If you are using a User-defined hatch pattern, what will it be made up of?

    a. Lines

    b. Lines and points

    c. Lines and circles

86. What are the ISO standards?

    a. Isometric Ortho hatch patterns

    b. International Standards Organization

    c. Internal Set Organizer

87. Where does your drawing go to if you have been silly enough to crash your disk by not putting your HATCH in properly?

    a. The AutoCAD directory

    b. The root directory

    c. The directory that the drawing was in

88. Once hatches are in, how can you get them off the screen so your REDRAWs will not take so long?

    a. BLOCK them

    b. Turn the HATCH LAYER off

    c. Freeze the HATCH LAYER

89. If you stretch the boundary of the hatch, can you set the hatch to go with it?

    a. No

    b. Yes

90. Can you move a hatch?

    a. No

    b. Yes

91. What is an internal block?

    a. A block that is part of an existing file

    b. A block that exists on a floppy disk as a .DWG

    c. A block that is contained within another block

92. What symbol can you use to have the wblock placed on a file as separate entities?

    a. ^

    b. *

    c. #

93. How do you insert an external drawing file?

    a. WBLOCK

    b. BLOCK

    c. INSERT

94. How do you modify an entity on the block once it has been inserted?

    a. CHANGE the entity

    b. EDIT the entity

    c. EXPLODE the block

95. With what command do you erase blocks from your file that are not part of the file, but only part of the memory?

    a. DBLIST

    b. ERASE in DOS

    c. PURGE

    d. WBLOCK *

    e. c. and d.

96. Under what condition can you not SCALE a wblock instance?

    a. If the aspect ratio is changed

    b. If the entities have been changed

    c. If the object was inserted with MINSERT

    d. Both a. and c.

97. What happens to the LTSCALE of your external file block instance when it is inserted onto another drawing?

    a. It takes the LTSCALE of the original drawing

    b. It takes the LTSCALE of the current drawing

98. How can you get object lines on a plot thicker than dimension lines without using pline?

    a. Change the pens so that one is thicker

    b. Change the DIMLIN and UPD

    c. Change the LTSCALE

99. How do you change a drawing currently in architectural units into mechanical?

    a. Change the UNITS

    b. Change the DIMALT

    c. Change the drawing size

100. If you WBLOCK a model in paper space, must you be in paper space in the current model in order to insert it?

    a. Yes

    b. No

101. If you have the viewport frames or borders in the wrong layer in paper space, can you change them?

    a. Yes

    b. No

102. Can you have paper space active and have TILEMODE on at the same time?

    a. Yes

    b. No

103. Can you keep two drawings of the same model in the same paper space environment?

    a. Yes

    b. No

104. What command do you use to get viewport-dependent layer display?

    a. ZOOM XP

    b. VPLAYER

    c. VDLAYER

Final Drawings

Final Drawings (cont.)

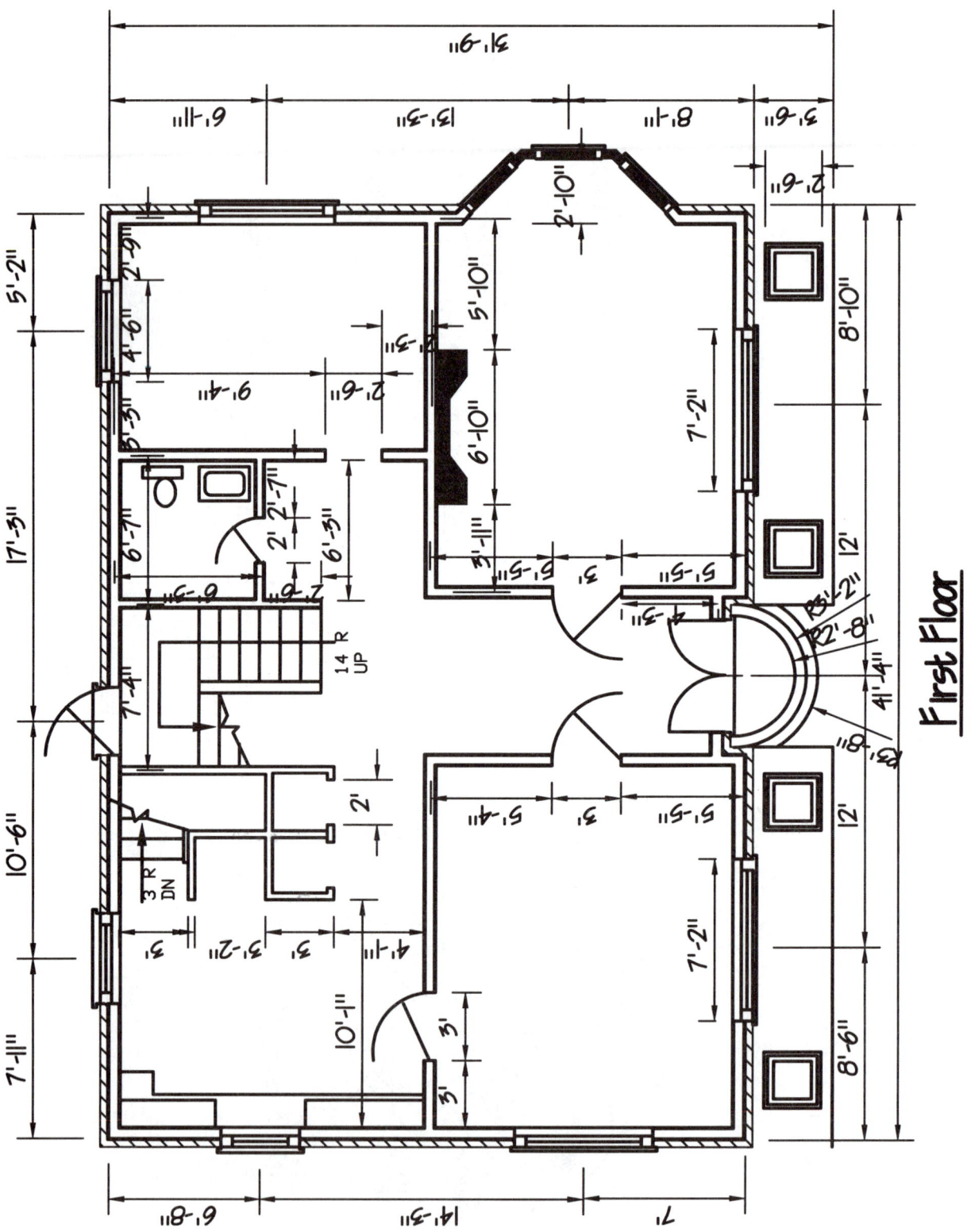

First Floor

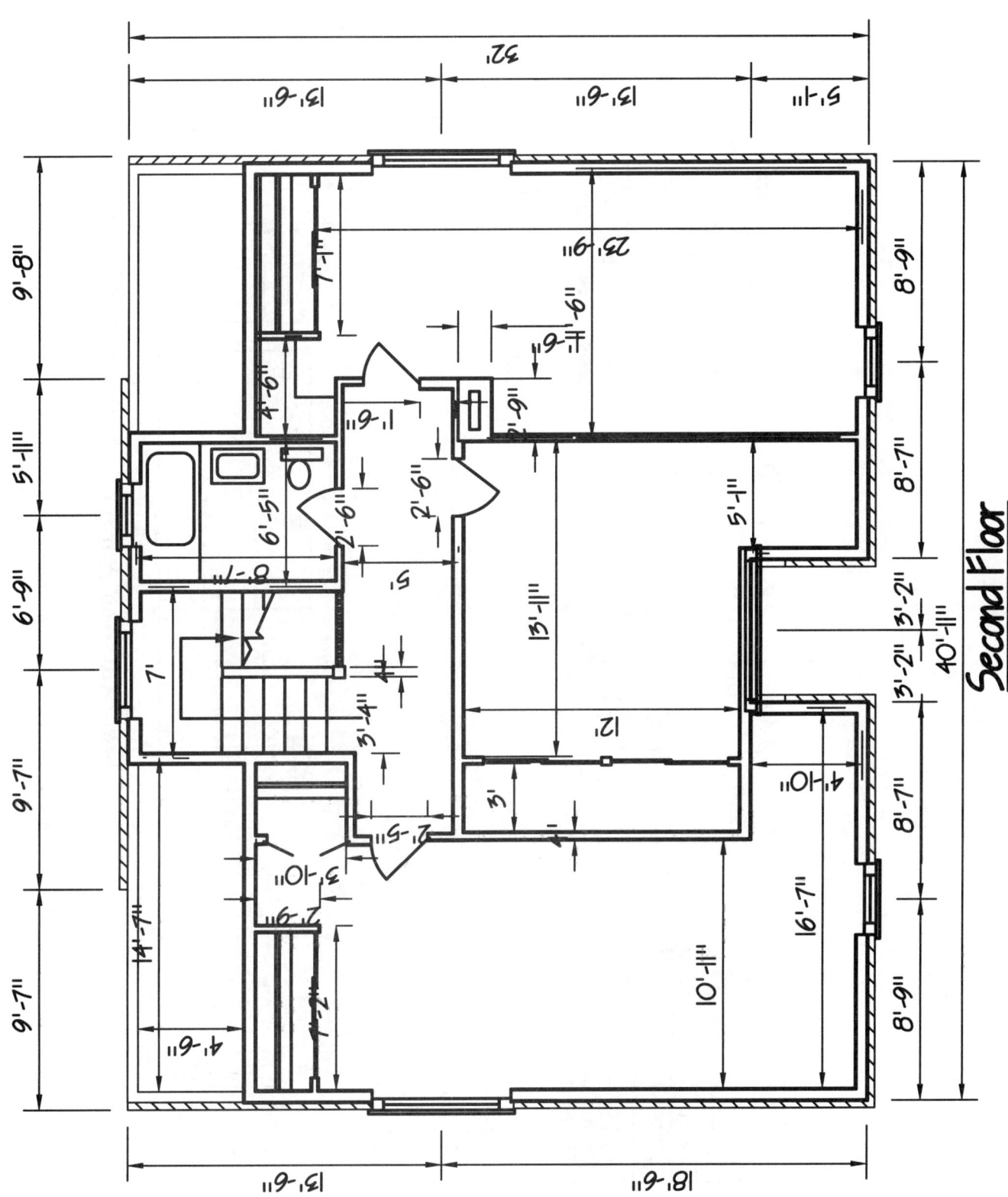

Final Drawings (cont.)

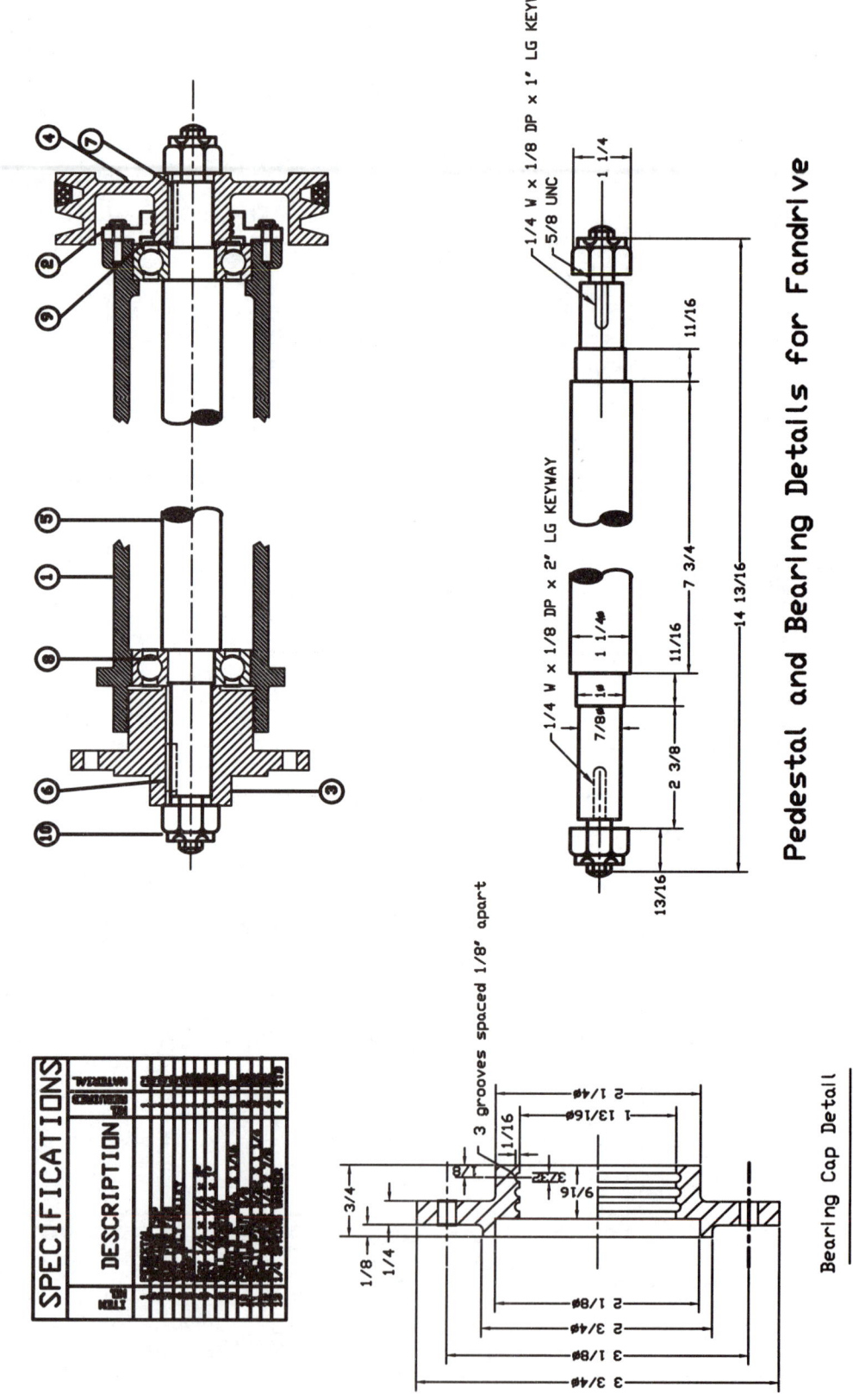

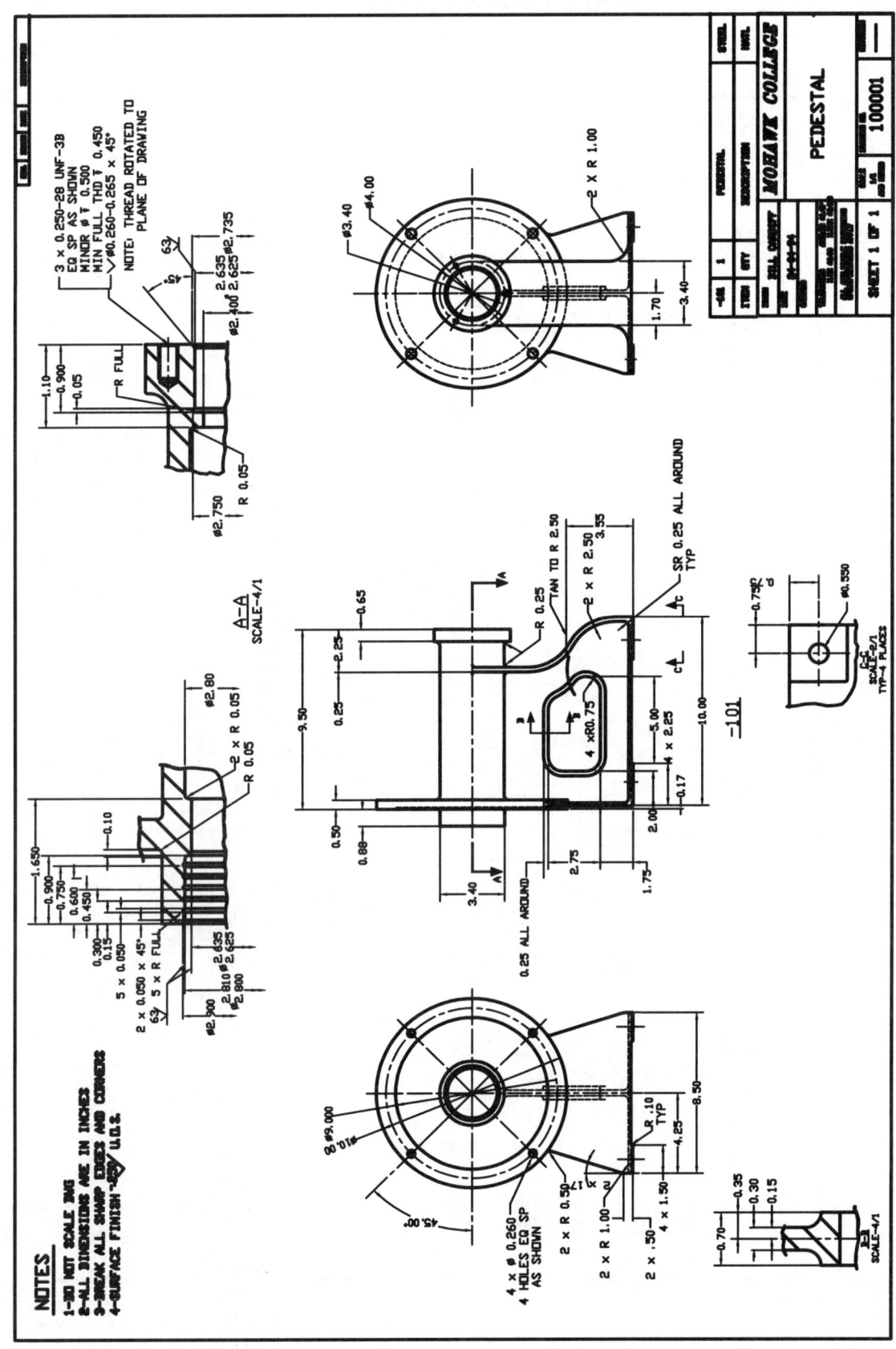

Final Drawings (cont.)

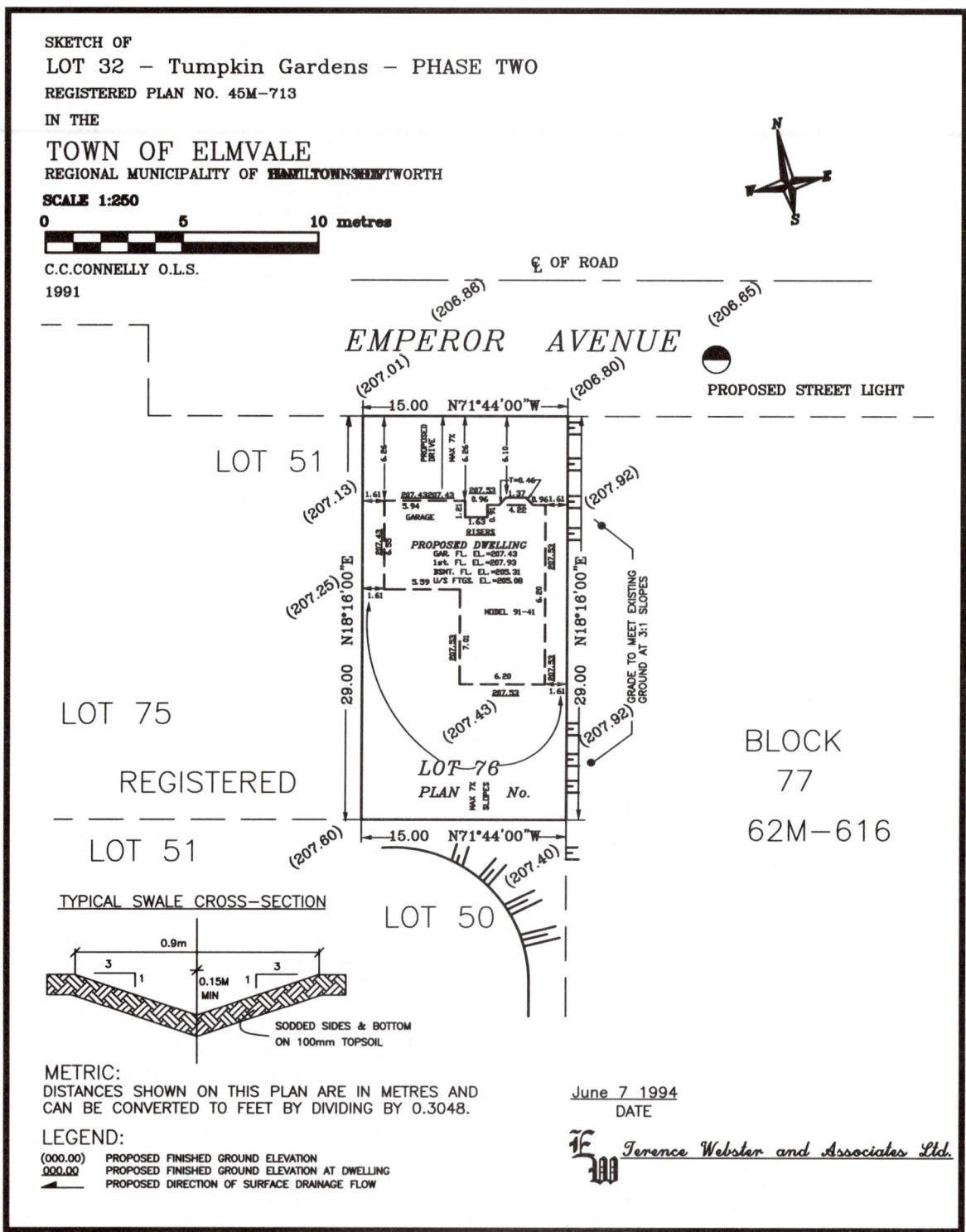

332 CHAPTER TWELVE

13 POINTS, DIVIDE, MEASURE, INQUIRY, and System Variables

Upon completion of this chapter, you should be able to:

1. Change the PDMODE and PDSIZE
2. Use DIVIDE and MEASURE to place points and/or multiple objects where desired within a model
3. Check the parameters and properties of objects within the model
4. Check the overall size of the model
5. Calculate the area of a model with AREA or Boundary
6. Use the time management facility
7. Use the system variables
8. Use the SPLINE command
9. Use MLSTYLE and MLEDIT commands

Point Display or PDMODE Options

Points are used in spline generation, in many 3D applications, and as node or reference points which you can snap to or offset from. When you divide or measure an object, points are used to show the divisions. You can set the style of the point and its size either relative to the screen or in absolute units.

> **Windows** and **DOS** From the Options menu, choose Display, and then Point Style.

The command line equivalent for the dialog box is **DDPTYPE**.

To set the point size or style without using the dialog box, use PDMODE for the style and PDSIZE for the size.

Point Size

In PDSIZE, a positive number will represent the actual size in drawing units of the point. A negative number is taken as a relative percentage of the screen and a difference in the zoom factor will have no difference in the size of the point display.

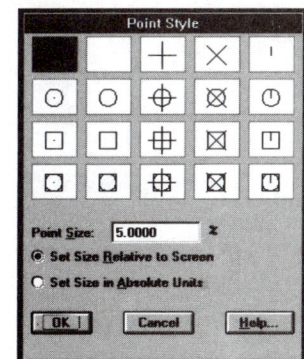

Figure 13-1

First choose whether you want the point to be relative to the screen or in absolute values. AutoCAD then stores the point size in the PDSIZE system variable. All points will be added relative to the new size, and all existing points will be updated according to this size upon the next regeneration.

Points, either created by MEASURE and DIVIDE or entered using the POINT command, can be accessed with the OSNAP option NODE. This is particularly important when entering blocks at specific points, or, in 3D, when finding centers for fillets, etc.

The points will then be added in at the points that are determined by either the DIVIDE or the MEASURE command.

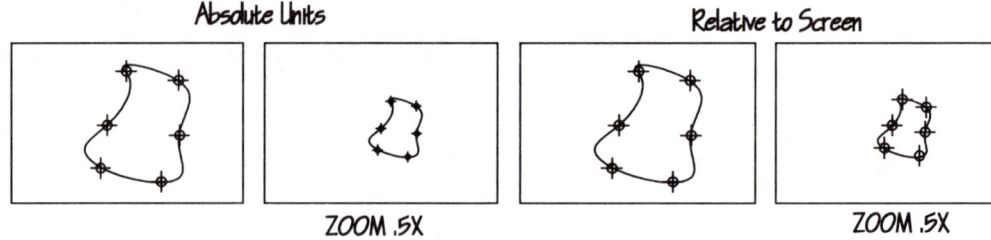

Figure 13-2

Using Divide and Measure

If you have an object or space that needs to be cut into equal pieces or portions, you can use DIVIDE or MEASURE.

DIVIDE will visually divide any linear element — an arc, a circle, a line, or a pline — into a specified number of equal parts.

MEASURE will visually measure a linear element into segments of a specified length.

The DIVIDE Command

DIVIDE places equally spaced point objects or blocks along the length or perimeter of an object.

> **Windows** From the Draw toolbar, choose the Point flyout, then Divide.
>
> **DOS** From the Draw menu, choose Point, then Divide.

The command line equivalent is **DIVIDE**.

DIVIDE with Points

Using the Point Style dialog box, set the current point to the one illustrated.

```
Command:DIVIDE
Select object to divide: (pick the object)
<Number of segments>/Block:6
```

The selected object is not altered in any way, but there are points in the current style at regular intervals. The points become objects on the current model. They can be accessed with the OSNAP NODE.

Figure 13-3

DIVIDE Using a BLOCK

DIVIDE can also be used with blocks. In the following example, a mullion block is used to divide a window into equally spaced sections. Both a curved window and a block called MULLION will be needed.

In Figure 13-4 we have a curved window and a block called MULLION, which is the shape of a mullion. The insertion point on the block is the middle of the bottom line.

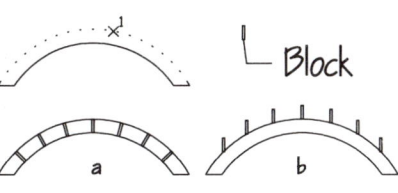

Figure 13-4

The command will place this block at regular intervals along the window. The base point of the block is very important.

```
Command:DIVIDE
Select object to divide: (pick 1)
<Number of segments>/Block:B
Block name to insert:MULLION
Align block with object?<Y>: (Y for (a), N for (b))
Number of segments:8
```

The MEASURE Command

The MEASURE command is very similar to the DIVIDE command in that it divides a specified object into a series of equal portions. The difference is that the equal portions are given a specific length, and thus there may be a portion of the object "left over" when the command is finished.

Again, the MEASURE command works on lines, arcs, circles, and pline;, and again, the markers can be either points or blocks.

> **Windows** From the Draw toolbar, choose the Point flyout, then Measure.
>
> **DOS** From the Draw menu, choose Point, then Measure.

The command line equivalent is **MEASURE**.

MEASURE with Points

In the following example, use point display 4 or the vertical line in the dialog box, and change the length to be 18 m with PDSIZE 18. Then use MEASURE to divide a road illustrated by a pline, (a), into equally spaced lots, (b). The pline representing the road will be needed.

Type in the system variables **PDMODE** and **PDSIZE**.

```
Command:PDMODE
Select new point mode<1>:4
Command:PDSIZE
Enter new point size<1.000>:18
Command:MEASURE
Select object to measure: (pick (a))
   <Segment length>/Block: 15
```

We now have a road that is divided into equal portions of 15 m lengths with a depth of 18 m each. Notice that the road is created with pline and thus has line and arc segments.

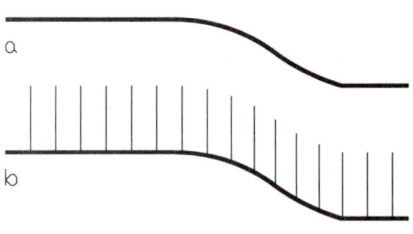

Figure 13-5

MEASURE Using BLOCKs

In the next example, we will place a block of a toilet along an existing wall. Both a LINE representing a 19' wall and a BLOCK that represents a toilet will be needed.

Create a toilet that has an interior space of 3 × 5 feet.

Now BLOCK the toilet, making sure that the insertion base point leaves enough space for a 2 inch wall on the back and on the sides.

```
Command:BLOCK
Block name:TOILET
Insertion base point: (pick 1)
Select objects:WINDOW
```

Now use MEASURE to place the toilet along a 19 foot wall.

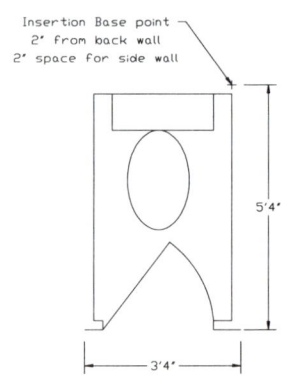

```
Command:MEASURE
Select object to measure: (pick
  1) (take the left side of the 19' wall)
<Segment length>/Block:B
Block name to insert:TOILET
Align block with object<Y>:⏎
Segment length:3'4"
```

Figure 13-6

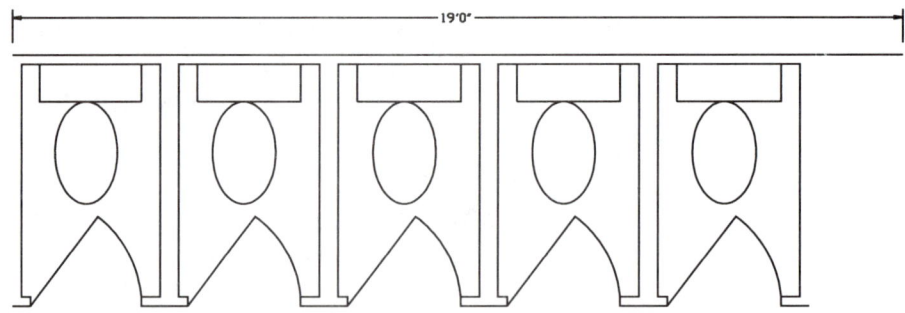

Figure 13-7

When creating the block to be used in a MEASURE command, be sure that you have no overlapping items.

If you use points in either the DIVIDE or the MEASURE command, you will have to change the PDMODE to be able to see the displayed points. Once placed, these points become objects in the file and will be affected by editing commands such as ERASE, MOVE, COPY, etc. If you do not change the PDMODE, these may be difficult to see.

The SPLINE Command

In Chapter 3 we looked at polylines and how to edit them into splines. There is also now a command to create splines or smooth curved lines without accessing the PLINE command.

AutoCAD used the Nonuniform Rational B-Spline (NURBS) formula to describe the splines entered. A NURBS curve produces a smooth curve between control points; this spline can be either quadratic or cubic.

If you are creating a large drawing with multiple splines for mapping or airfoil design, a drawing containing splines uses less disk space and memory than a drawing with polylines. To access the SPLINE command:

> **Windows** From the Draw toolbar, choose the Polyline flyout, then Spline.
>
> **DOS** From the Draw menu, choose Spline.

The command line equivalent is **SPLINE**.

```
Command: SPLINE
Object/<Enter first point>: (pick 1)
Enter point: (pick 2)
Close/Fit tangency/<Enter point>:
   (pick 3 through 9)
Enter start tangent: (pick 10)
Enter end tangent: (pick 11)
```

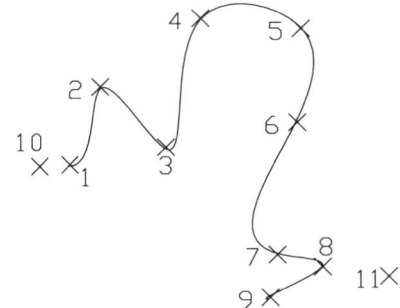

Figure 13-8

Entering Points

Enter points in to add additional spline curve segments until you press ⏎. Like the Line and PLINE commands, a **U** for Undo will remove the last entered point.

As the points are entered you can see the spline being created.

Start and End Tangency

The "Enter start tangent:" prompt specifies the tangency of the spline at the first point; the "Enter end tangent:" prompt does the same for the end point. You can specify tangency at both ends of the spline, and you can use either point, TANgent, or PERpendicular object snaps to make the spline tangent or perpendicular to existing objects.

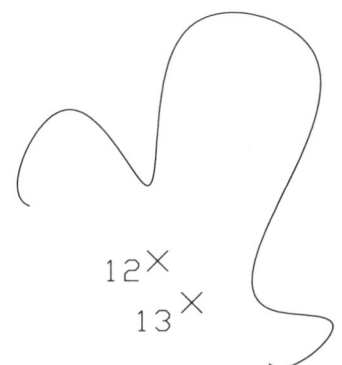

Figure 13-9

```
Enter start tangent: (pick 12)
Enter end tangent: (pick 13)
```

Close

Like the POLYLINE command, the Close option defines the last point as coincident with the first and makes it tangent there.

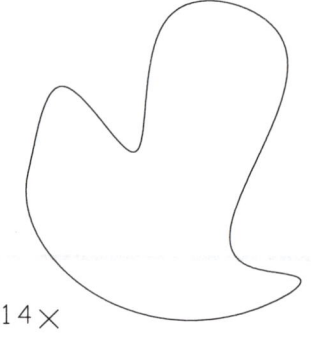

14×
Figure 13-10

Fit Tolerance

This changes the tolerance for fitting the spline through the points. The number is higher or lower depending on how you want the spline to fit through the points.

Object

This option converts either 2D or 3D polylines into splines.

The SPLINEDIT Command

The SPLINEDIT command edits the spline object.

> **Windows** From the Modify toolbar, choose the Polyline flyout, then Splinedit.
>
> **DOS** From the Modify menu, choose Splinedit.

The command line equivalent is **SPLINEDIT**.

The options for this edit command are similar to those of PEDIT.

Inquiry Commands

The Inquiry commands are used to see and list the parameters of a model.

AREA   computes the area of a closed polygon

DIST   computes the distance between two points

LIST   lists the position and properties of a specific object or group of objects

ID   identifies a point

STATUS   displays a listing of all the statistics of a file along with other information

The AREA Command

AutoCAD offers built-in area computational abilities which also display the perimeter of the object calculated. This can be extremely useful for calculations of plines and irregular shapes.

AREA Using an Entity

On the right we have a pline that has been fit with a spline curve, and the area and perimeter are to be calculated. In this case an entity is chosen for the area option.

```
Command:AREA
<First point>/Object/Add/Subtract:O
Select objects: (pick 1)
Area = 16.1434 Perimeter = 19.6824
```

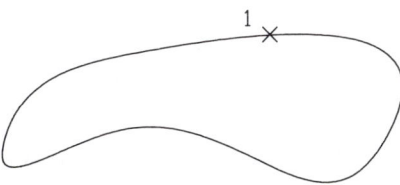

Figure 13-11

AREA Using Lines

When calculating a series of straight lines, the Point option is used, as in the following command sequence. OSNAPs are needed for accuracy.

```
Command:AREA
<First point>/Object/Add/Subtract:END of (pick 1)
Next point:END of (pick 2)
Next point:END of (pick 3)
Next point:END of (pick 4)
Next point:END of (pick 5)
Next point:⏎
Area = 67.5000 Perimeter = 31.1131
```

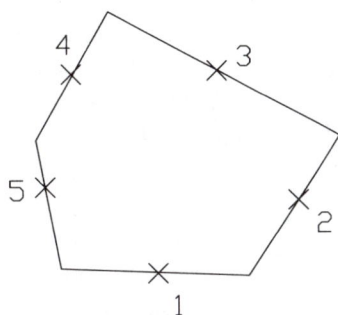

Figure 13-12

Subtract and Add with AREA

To calculate the net floor area of a bathroom, a drawing of a bathroom will be needed. Draw in a bathroom 8'6" × 5'6" as shown.

Set the osnap to ENDpoint and calculate the total floor area.

```
Command:OSNAP
Object snap modes:ENDpoint
```

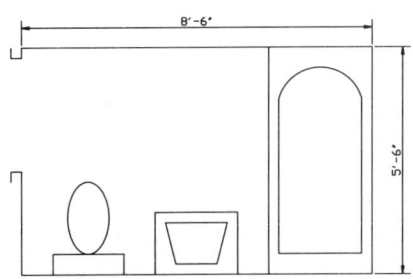

Figure 13-13

POINTS, DIVIDE, MEASURE, INQUIRY, and System Variables

```
Command: AREA
<First point>/Object/Add/Subtract: A
First point: (pick 1)
Next point: (pick 2)
Next point: (pick 3)
Next point: (pick 4)
Next point: ⏎
Area = 6732.00 square inches
(46.7500 square feet)
Perimeter = 28' 0"
Total area 6732.00 square inches
(46.75 square feet)
```

Figure 13-14

Without leaving the command, subtract the fixtures.

```
<First point>/Object/Add/Subtract: S
(SUBTRACT mode)<First point/Object/Add: (pick 5)
(SUBTRACT mode) Next point: (pick 6)
(SUBTRACT mode) Next point: (pick 7)
(SUBTRACT mode) Next point: (pick 8)
(SUBTRACT mode) Next point: ⏎
Area = 1980 square inches
(13.7500 square feet)
Perimeter = 16' 0"
Total area 4752 square inches
(33.0000 square feet)
```

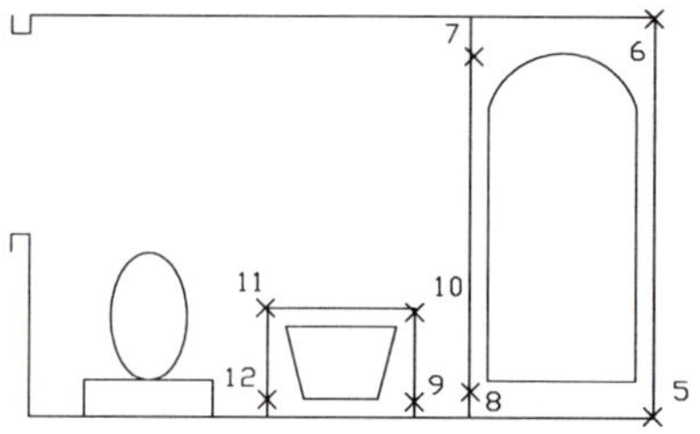

Figure 13-15

Notes

Your numbers might differ marginally depending on how you drew the diagram.

```
(SUBTRACT mode)<First point>/Object/Add: (pick 9)
(SUBTRACT mode) Next point: (pick 10)
(SUBTRACT mode) Next point: (pick 11)
(SUBTRACT mode) Next point: (pick 12)
(SUBTRACT mode) Next point: ⏎
Area = 432 square inches (3.00 square feet) Perimeter = 7' 0"
Total area 4320 square inches (30.0000 square feet)

(SUBTRACT mode)<First point>/Entity/Add: (pick 13)
(SUBTRACT mode) Next point: (pick 14)
(SUBTRACT mode) Next point: (pick 15)
(SUBTRACT mode) Next point: (pick 16)
(SUBTRACT mode) Next point: ⏎
Area = 126 square inches (0.8750
   square feet) Perimeter = 4' 6"
Total area 3852 square inches (26.75
   square feet)
```

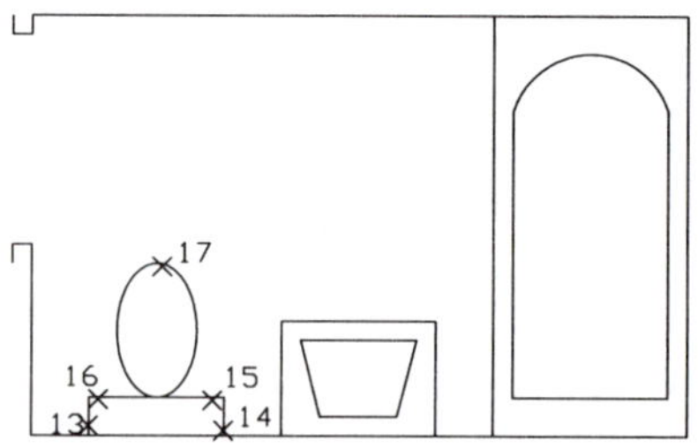

Figure 13-16

```
(SUBTRACT mode)<First point>/Object/Add: O
Select objects: (pick 17)
Area = 260 square inches (1.81 square feet) Length = 4' 2"
Total area 3592 square inches (24.94 square feet)

(SUBTRACT mode) Select objects: ⏎
```

As you can see from the example above, points and Objects can be used together. With the SUBTRACT mode, subsequent area calculations are subtracted from the accumulated area. Use Add to enter the first area, then Subtract for the subsequent areas.

With the Add option, subsequent areas will be added to the accumulated area.

In 3D calculations, all of the points used for calculation must be on the same plane. The UCS can be changed to allow for this. The extrusion direction of entities must also be parallel with the current UCS.

AREA and PERIMETER are also system variables and can be viewed but not changed by displaying them. Use the SETVAR command to access the system variable. If you calculated an area, but neglected to jot it down, type **AREA** as the variable name and the last calculated area will be displayed.

To find the length or circumference of a circle, arc, or pline, use the LIST command.

Be sure to change the units to the desired values before entering the AREA command; otherwise the readings may be inaccurate.

The BOUNDARY Command

The BOUNDARY command creates a region or a polyline of a closed boundary. Specifying a boundary set can produce the boundary more quickly, because AutoCAD examines fewer objects when a boundary set is identified. For an example of how BOUNDARY works, see Prelab 13.

> **Windows** From the Draw toolbar, choose the Polygon flyout, then Boundary.
>
> **DOS** From the Construct menu, choose Boundary.

The command line equivalent is **BOUNDARY**.

The DISTance Command

The DISTance command will give you the actual distance between two points. Remember that, when dealing with points in AutoCAD or any other CAD package, you can specify the points in three ways:

1. If they are associated with objects, by finding them with OSNAPs such as ENDpoint, MIDpoint, etc.
2. By picking on them on the screen, with or without the use of SNAP
3. By specifying their coordinates

When calculating DISTance, you can identify the points by any combination of the above. First enter the DIST command, then respond with the two points that identify the distance you want to know. It is better that you specify the exact points with either an OSNAP or a coordinate entry. As with the other commands, you will be offered the information relative to the current units specified.

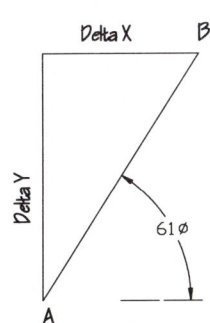

Figure 13-17

You will note that the readout contains not only the distance, but also the angle from the *X-Y* plane, and the length and width of the object.

```
Command:DIST
First point: (pick point A)
Second point: (pick point B)
Distance= 10.2956, Angle in X-Y Plane = 61, Angle from X-Y
    Plane = 0 Delta X = 5.0000 Delta Y = 9.0000 Delta Z = 0.0000
```

The LIST Command

We looked at LIST briefly in the first section of this book to establish the layer, color, and linetype properties of objects that had already been entered. In addition to this, LIST is an extremely useful command for the following purposes:

1. Cleaning up objects before hatching and editing
2. Finding out the angle of an existing LINE
3. Finding out the radius of a specified circle
4. Determining text fonts and sizes
5. Finding out if existing items were put in incorrectly or without the use of SNAP
6. Checking for incorrect dimensions

In 3D, LIST can be used to find out the current Z depth and extrusion length of an object as well as the properties listed above. LIST can also be used to show the number of objects within a specified window to check to see that the objects follow good CAD practice. In the following example, we will see how LIST can also be used to tell us if the dimension for an object is real or "fudged."

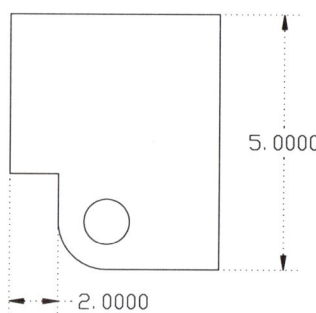

Figure 13-18

In Figure 13-18, there are two associative dimensions listed. To the eye, they look the same. They are identified by being highlighted in the DOT linetype.

Note the difference in the readouts after the LIST command has been used.

```
Command:LIST
Select objects: (pick 1, 2)
                    DIMENSION Layer: dim
                    SPACE: Model Space
                    Handle = 1D14
type: horizontal
1st extension defining point:   X= 2.0000 Y= 4.5000 Z= 0.0000
2nd extension defining point:   X= 4.0000 Y= 3.5000 Z= 0.0000
dimension line defining point:  X= 4.0000 Y= 2.5000 Z= 0.0000
default text position X= 3.875 Y= 2.5000 Z= 0.0000
default text
dimension style *UNNAMED
                    DIMENSION Layer: dim
                    SPACE: Model Space
                    Handle = 1D14
```

```
                 type: vertical
                 1st extension defining point:   X= 6.0000 Y= 8.5000 Z= 0.0000
                 2nd extension defining point:   X= 6.0000 Y= 3.0000 Z= 0.0000
                 dimension line defining point:  X= 6.5000 Y= 2.5000 Z= 0.0000
                 default text position X= 6.500 Y= 5.7500 Z= 0.0000
                 dimension text modifier: 5.0000
                 dimension style *UNNAMED
```

You can see that the first dimension has been entered properly, and the text for the dimension is the default text. In the second dimension, however, the dimension text has been modified to read 5.0000. If you subtract the second extension defining point from the first in the Y value, you will notice that the actual distance should read 5.5000, but the dimension text has been altered before it was entered.

In addition to checking your own work to be sure that it is all entered correctly and that there are no overlapping items, with the LIST command you can check to see that the dimensions associated with the drawing you have on file are correct and not altered in any way. If the dimensions have been altered, check to see that *all* of the necessary dimensions have been altered.

Another great advantage of LIST is that it enables you to determine if there are overlapping lines when making hatches and dimensions. Use LIST, then Crossing, to see the number of objects overlapping.

The ID Command

The ID function is similar to the LIST function, in that it gives an exact position. The difference is that it gives an exact position of a point rather than an object. This command is used in 3D modelling more frequently than in 2D modelling, to determine the Z depth of items. In 2D this can be useful for finding out the exact position you are looking at on a ZOOMed screen, or for simply "getting your bearings." Keep in mind that, if you are picking a point in space, a snap can give a more usable readout.

The location of the point will be relative to the origin or 0,0,0 of the model. If a point in space is chosen, the Z depth will be the current elevation; if an object is chosen with an OSNAP, the actual Z depth of the object will be used.

ID can be used as a reference point for the next point entered using @:

```
Command:LINE
From point:@2,0 (starts the line 2 units in X from the ID
   point)
```

There are basically two ways of using ID. The first is to find the parameters of a point on the screen.

```
Command:ID
Select point: (pick a point)
X = 34.6375    Y = 24.8758    Z = 0.0000
```

This will give the coordinates of a point in space in the defined units. The second is to locate a point by typing in the coordinates.

```
Command: ID
Select point: 23,4,0
```

If BLIPMODE is on, you will get a blip on the screen at the exact location of the point entered. The blip will disappear with the next REDRAW.

The STATUS Command

Another useful command for determining what is happening is STATUS. As you become proficient with AutoCAD, you will find the STATUS command more and more useful, because it displays a listing of all the statistics of a file.

The STATUS command will also offer information on memory and the partition on the hard drive where your temporary file or .AC$ file is being stored. If you run out of space in this partition (in a classroom this is often the A: drive), the program will terminate after first saving your file.

STATUS is most often used by beginners to determine if the color setting is overriding the layer color setting.

Enter **STATUS** at the command prompt and read the status of your file.

Creating Multiline Styles

As seen in Chapter 3, multilines are multiple parallel lines. Multilines consist of between 1 and 16 parallel lines called elements. You can create and save multiline styles, or use the default style shown in Chapter 3. The color and linetype of each element of the multiline can be set, and the ends of multilines can be set with caps.

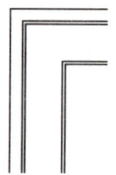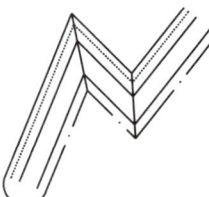

The MLSTYLE Command

Like dimension styles and text styles, multiline styles can be named and saved to control the elements and properties of each element.

> **Windows** From the Object Properties toolbar, choose
>
> **DOS** From the Data menu, choose Multiline Style.

The command line equivalent is **MLSTYLE**.

The LINETYPE command loads the linetypes from the ACAD.LIN file. MSTYLE also loads the multilines from an external file called ACAD.MLN.

The dialog box displays the multiline style names, makes them current, loads them from a file, and saves, adds, and renames them.

Current   sets the current multiline style.
Name   creates a new multiline style. First define the elements, then enter a new name on this line and pick Save.
Description adds a description of up to 255 characters to the name.
Load   Loads a style from the library.
Add   adds the multiline style in the Name text box to the current list

Here is an example of how to load a multiline style. We will use a typical northern exterior wall as an example.

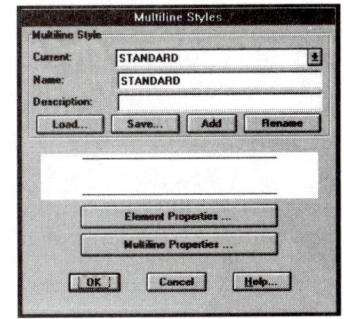

Example

Step 1 Open the MLSTYLE dialog box.

Step 2 Pick the Element Properties button, and Add, using the Offset option to load the 5 line elements.

When using the offset option, the offset starts from 0.0 and continues both up and down from there. Offset the first line at 2.5 from 0 by highlighting it, then changing the number in the offset box, and highlighting again. Then add the offset of .5 by typing .5 in the offset box, then picking the Add button. To add a line, use Add. To change an offset, highlight, change and re-highlight.

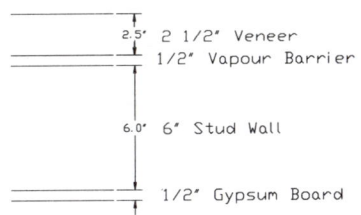

Step 3 Once the lines are entered, use the color button to change the colors of the lines to correspond to the line widths of your plotter. These changes are the same as the color changes in layers. Assume that a red line is the thinnest, a yellow line is medium, and a green line is the thickest object line width. Pick OK when finished.

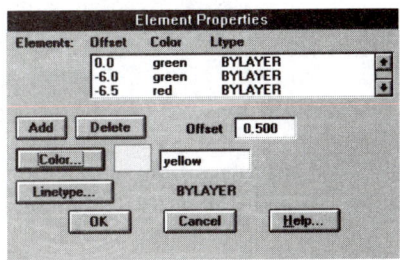

Step 4 In the MLSTYLE dialog box, pick the box beside the word, Name. Type in Extwall for your multiline style. Use Add to add the name to the current list.

Step 5 Use the new MLINE style to create an exterior wall.

Your MLINE will default to being justified at the outside of the veneer. Many residential designers prefer to have their home designs justified at the outside of the studwall. This will allow for greater accuracy when dimensioning the part for the framers. To do this, add a line in the dot linetype at a distance of positive 6.5. Then use Zero as the Justify option of the MLINE command. The dot line will not print, and the line will be justified at the center of the MLINE, or the edge of the wooden frame.

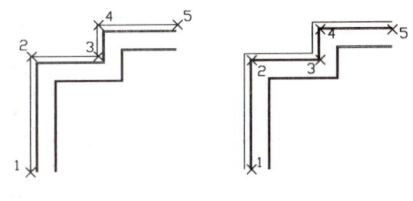

Step 6 To add straight line caps onto the wall, re-enter the MLSTYLE dialog box, and use the multiline properties button to add caps to the outer edges of the multiline. Use Save to save the MLINE style.

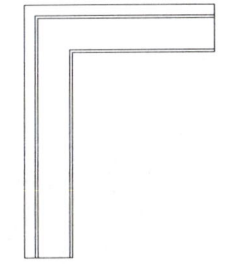

Editing Multilines

Once a multiline is entered, it is considered to be single unit. Like polylines, multilines can be edited by adding and deleting vertices. In addition, you can control the display of corner joints and the intersection of the multilines. You can also edit multiline styles to change the properties of individual line segments, or the end caps and background fill of future multilines.

The MLEDIT Command

The Multiline Edit Tools dialog box controls intersections between multilines.

Windows From the Modify toolbar, choose the Special Edit flyout, then

DOS From the Modify menu, choose Edit Multiline.

The command line equivalent is **MLEDIT**.

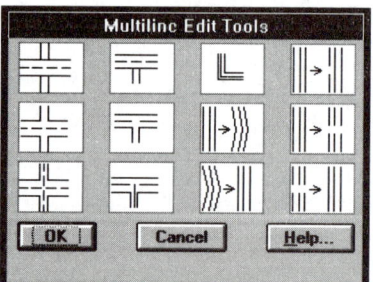

Once the multilines are created, the dialog box is used to change the intersection or to add or delete a vertex or break.

The first column works on multilines that cross. The second on multilines that form a tee, the third on corner joints and vertices, and the fourth on multilines to be cut or welded.

The first two columns work on pairs of multilines, the second two work on single multilines.

In the first two columns, you can edit more than one intersection. You will be prompted as follows:

```
Select first mline: (pick one mline)
Select second mline: (pick an intersecting mline)
Select first mline: (pick another mline or U to erase the
   first)
```

Once you have selected the first MLINE, you will be prompted for the second MLINE. If you enter U, AutoCAD undoes the closed cross intersection and displays the "Select first MLINE" prompt.

Closed Cross creates a closed cross intersection between two multilines

Open Cross creates an open cross intersection between two multilines, breaking all elements of the first multiline and only the outside element of the second multiline

Merged Cross creates a merged cross intersection between two multilines

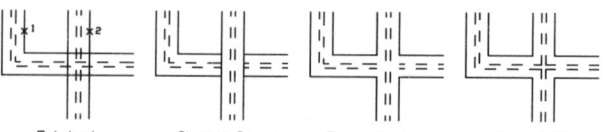

Closed Tee   creates a closed intersection between two multilines, trimming or extending the first multiline to its intersection with the second multiline

Open Tee   creates an open tee intersection between two multilines, trimming or extending the first multiline to its intersection with the second multiline

Merged Tee   creates a merged tee intersection between two multilines, trimming or extending the first multiline to its intersection with the second multiline

Corner Joint   creates a corner joint between multilines, trimming or extending the first multiline to its intersection with the second multiline

Add Vertex   adds a vertex to a multiline

Delete Vertex   deletes a vertex to a multiline

Cut Single   cuts a single element of a multiline

Cut All   cuts a multiline in two

Weld All   rejoins multiline segments that have been cut

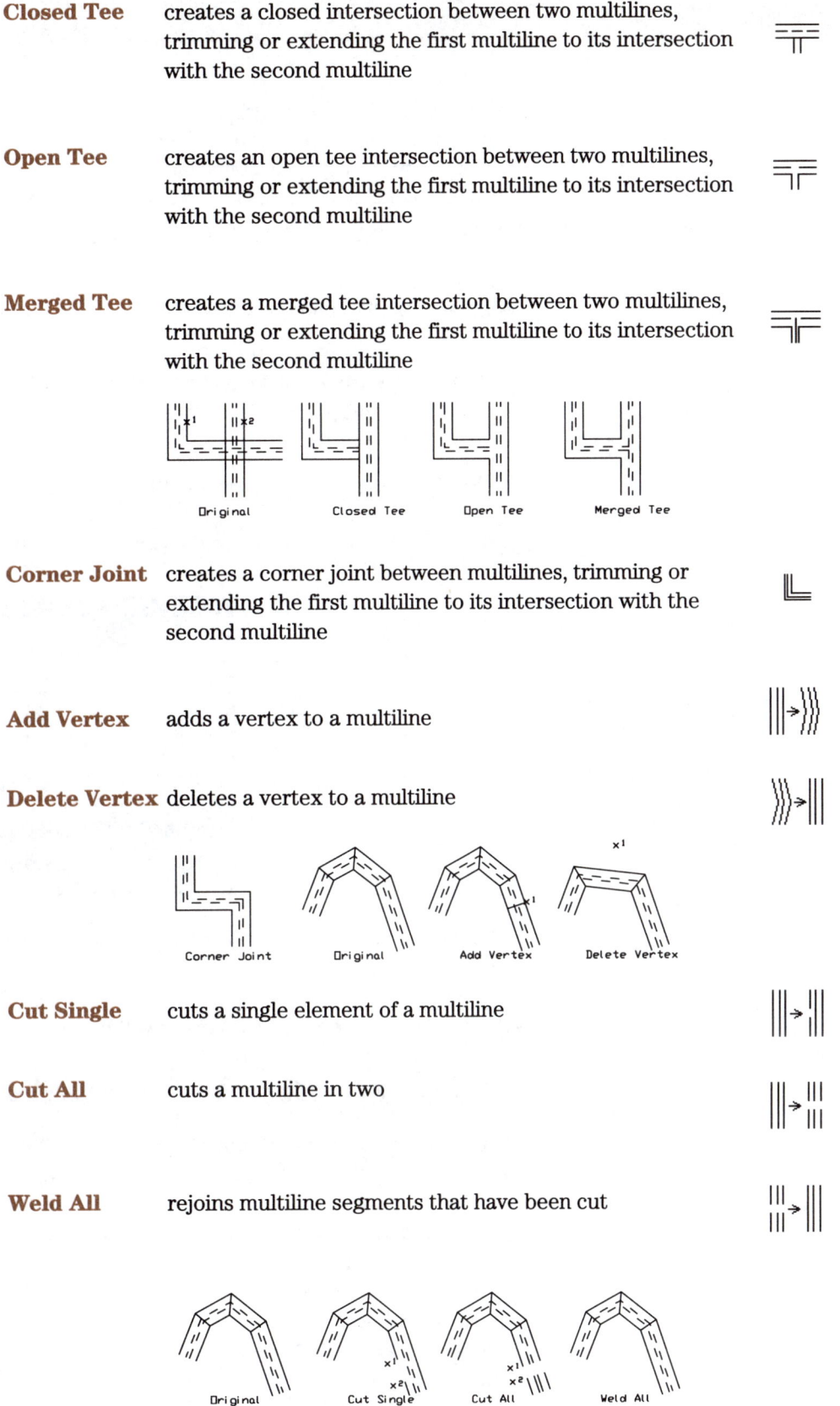

POINTS, DIVIDE, MEASURE, INQUIRY, and System Variables

Prelab 13 Point Style, SPLINE, and Inquiry

This top view of a corner gusset is drawn using both a complex spline and a series of lines. In order to get the spline so that it can be edited later if necessary, we will put in the points as POINTs and generate the spline through the nodes. In order to see the points you will need to change both the PDMODE and the PDSIZE.

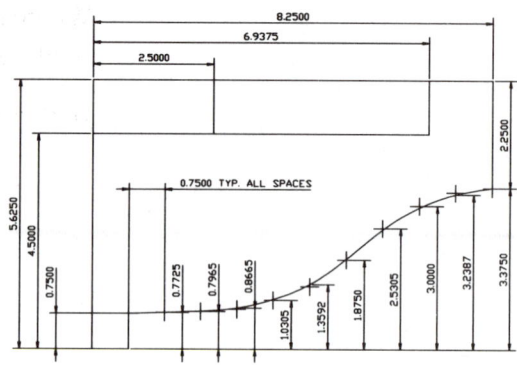

Step 1 Set up three layers, for POINTS, OBJECT, and DIMENSIONS. Make POINTS current. Then add points.

> **Windows** From the Options menu, choose Display, then Point Style.
>
> **DOS** From the Options menu, choose Display, then Point Style.

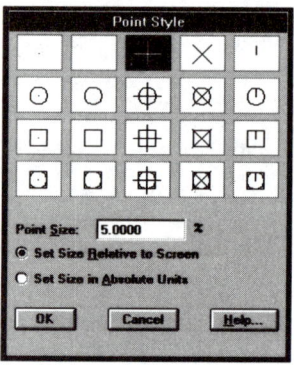

Make the top center or third point active, and change Size Relative to Screen to 5%.

Step 2 Add the points. The lower left corner will be 0,0.

> **Windows** From the Point flyout on the Draw toolbar, choose Point.
>
> **DOS** From the Draw menu, choose Point.

```
Command:POINT
Point:.75,.75
Command:POINT
Point:1.5,.7725
Command:POINT
Point:2.25,.7965
Command:POINT
Point:3.0,.8665
Command:POINT
Point:3.75,1.0305 etc.
```

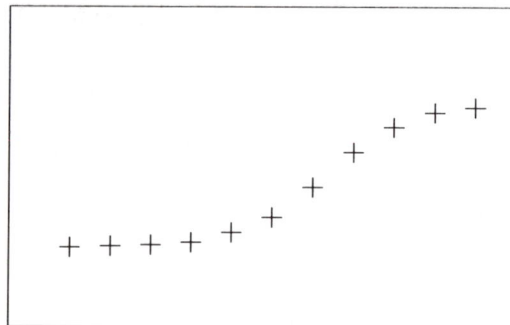

348 CHAPTER THIRTEEN

Step 3   Change your OSNAP to NODE and use SPLINE to create a spline through all of the points.

> **Windows** From the Object Snap flyout on the standard tool bar, choose Running Object Snap.
>
> **DOS** From the Options menu, choose Running Object Snap.

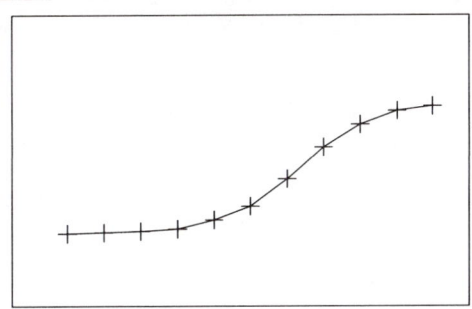

```
Command:OSNAP
Object snap modes:NODE
```

Now use SPLINE to create a spline through the identified points.

> **Windows** From the Draw toolbar, choose Spline.
>
> **DOS** From the Draw menu, choose Spline.

```
Command:SPLINE
Object/<Enter first point>:
Enter point: (pick the first point)
Close, Fit Tolerance/<Enter point>: (pick the points in
   sequence)
Close, Fit Tolerance/<Enter point>: (pick the last point)
Close, Fit Tolerance/<Enter point>:⏎
Enter start tangent: (pick the first point)
Enter end tangent: (pick the last point)
```

Step 4   Using the drawing given at the beginning, add the outside lines.

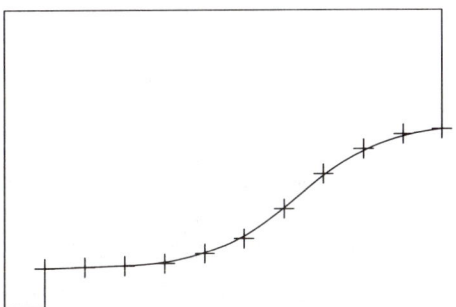

POINTS, DIVIDE, MEASURE, INQUIRY, and System Variables

Step 5 Now use the Boundary option to help determine the area of the top of the part.

> **Windows** From the Draw toolbar, choose the Polygon flyout, then choose Boundary.
>
> **DOS** From the Construct menu, choose Boundary.

From the Boundary Creation dialog box, choose Make New Boundary Set.

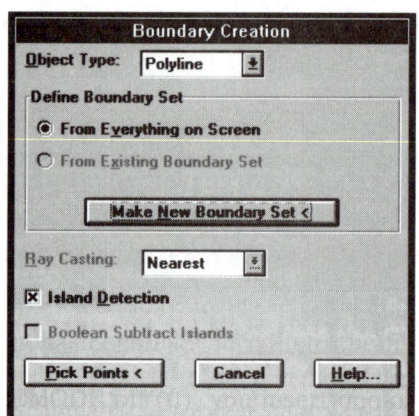

Now pick the lines and the spline.

Press ↵ when you are finished and you will return to the Boundary Creation dialog box.

From the dialog box, pick "Pick points." You will be prompted to pick the internal point; choose the inside of the part.

A region will be created and you will have the area for the part.

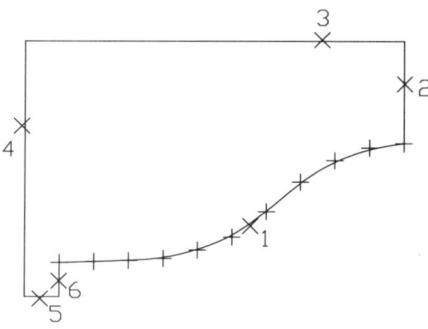

If your system doesn't allow this, enter the AREA command, use the Object option, and **L** for LAST object.

Step 6 Now list the properties of the spline and one of the point nodes.

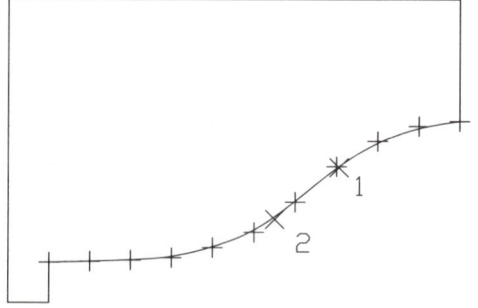

> **Windows** From the Object Properties toolbar, choose the Inquiry flyout, then this button:
>
> **DOS** From the Assist menu, choose Inquiry, LIST.

You will get a listing of the properties.

Step 7 Now add some points along the top edge of the part that correspond to the *X* measurement of all of the points. This is to double-check that all of your points are entered correctly. Use DDPTYPE to change the point style to the vertical line as shown.

> **Windows** From the Draw toolbar, choose the Point flyout, then this button:
>
> **DOS** From the Draw menu, choose Point, then Measure.

```
Command:MEASURE
Select object to measure:
   (pick 1)
<Segment length>/Block:.75
```

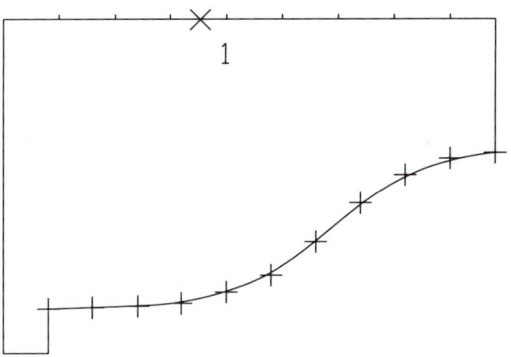

Use Undo to remove the points once they have been used to double-check the lower points.

Step 8 Now add the remaining lines and dimensions to this view.

Step 9 Pan the screen over and, using the same method, draw in the front view of the gusset. The points will be much easier to enter if you make 0,0 the upper left corner. Move the first view up and out of the way.

When you are done, add a title block and notations to complete the drawing.

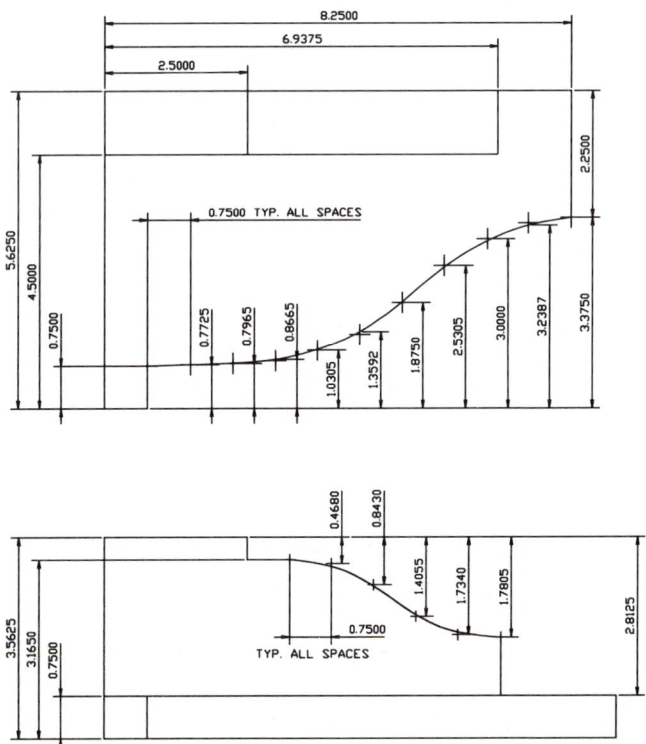

Command and Function Summary

AREA calculates the area and perimeter of objects or defined areas.

BOUNDARY creates a region or polyline of a closed boundary.

DIST measures the distance and angle between two points.

DDPTYPE specifies the display mode and size of point objects.

ID displays the coordinates of a location.

LIST displays database information for selected objects.

MLEDIT edits multiline intersections and vertices.

MLSTYLE sets multiline styles.

POINT creates a point object.

SPLINE creates a quadratic or cubic spline (NURBS) curve.

Exercise A13

Step 1 Use DDPTYPE to load the point shown, then create nine points at a regular distance to form a decorative wall.

Step 2 Use SPLINE with the NODE OSNAP to join the points together into a spline element.

Step 3 Offset the spline at 8" and use LINE to close the ends so that the left end is perpendicular and the right end is horizontal.

Step 4 Use AREA with the Boundary option to figure out the area of the decorative wall.

Step 5 Create a rectangle 1" × 8". Block it making sure that the insertion base point is the middle of the bottom line.

Step 6 Use MEASURE to place the blocks at 8" intervals to illustrate decorative brick.

Step 7 Draw in the remainder of the room using MLSTYLE and MLEDIT where needed.

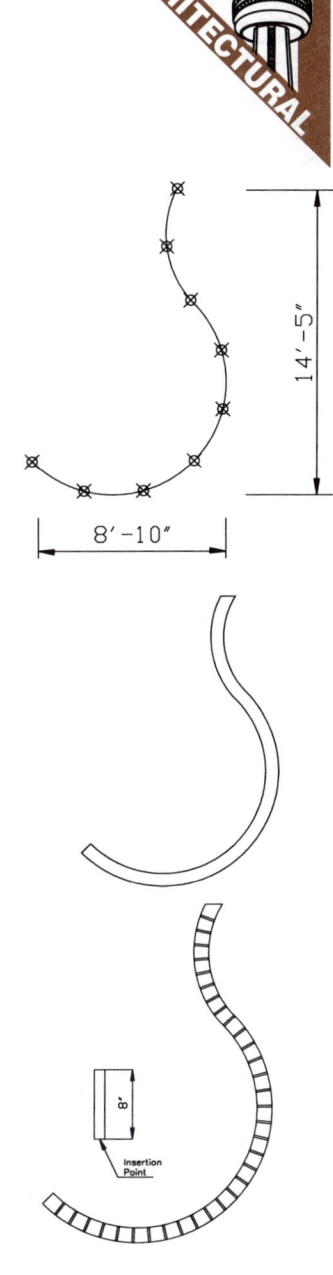

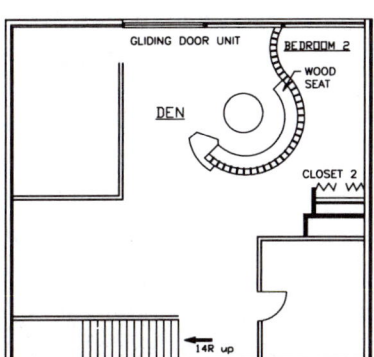

POINTS, DIVIDE, MEASURE, INQUIRY, and System Variables

Exercise C13

Step 1   Use SPLINE to create a riverbank fit through the following points.

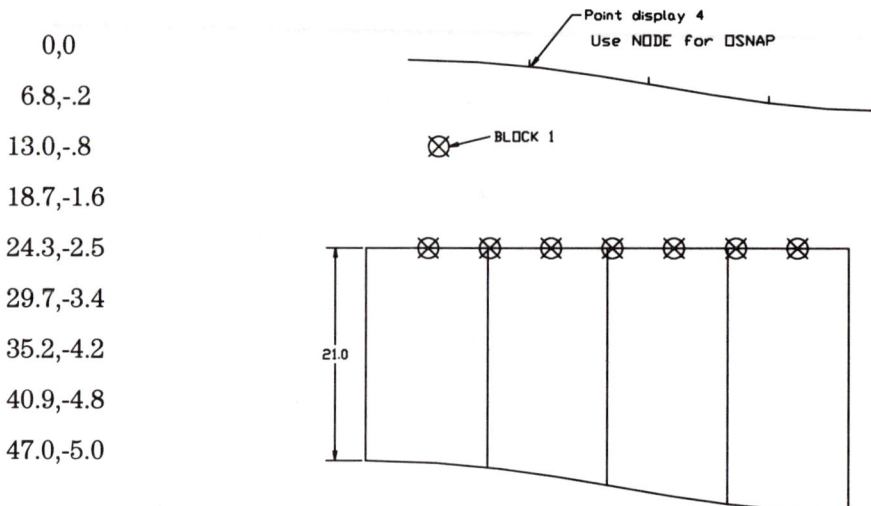

0,0
6.8,-.2
13.0,-.8
18.7,-1.6
24.3,-2.5
29.7,-3.4
35.2,-4.2
40.9,-4.8
47.0,-5.0

Step 2   Create a light as shown in the illustration and BLOCK it under the name LIGHT.

Step 3   Using DIVIDE, divide the riverbank into four equal sections. Use the OSNAP NODE to create lot lines between these sections and the road 21 m to the north of the river.

Step 4   Use MEASURE to place the lights along the roadway at spaces of 7 m each.

Step 5   Use the AREA command to calculate the total area of each lot. Make sure that you have the correct area, not an area with a straight line from the intersection of the lot line and the plines.

Step 6   Use LIST to find the length of road along the northern edge of these four lots. Use DIST to find the total length in a straight line from the southwest end of the first lot to the southeast end of the last lot.

If the southwest corner of the lots is at 0,0, what is the exact point at the centre of the four lots?

What is the total area taken up by the lots?

How much space have you left in your partition or RAM?

Exercise E13

Step 1 Use DIVIDE and MEASURE to create this partial drawing of electrical connectors in an oscilloscope. Use DIVIDE and MEASURE to place connections in both a circular and a linear fashion. In the linear example, the block was not rotated; in the circular, it was.

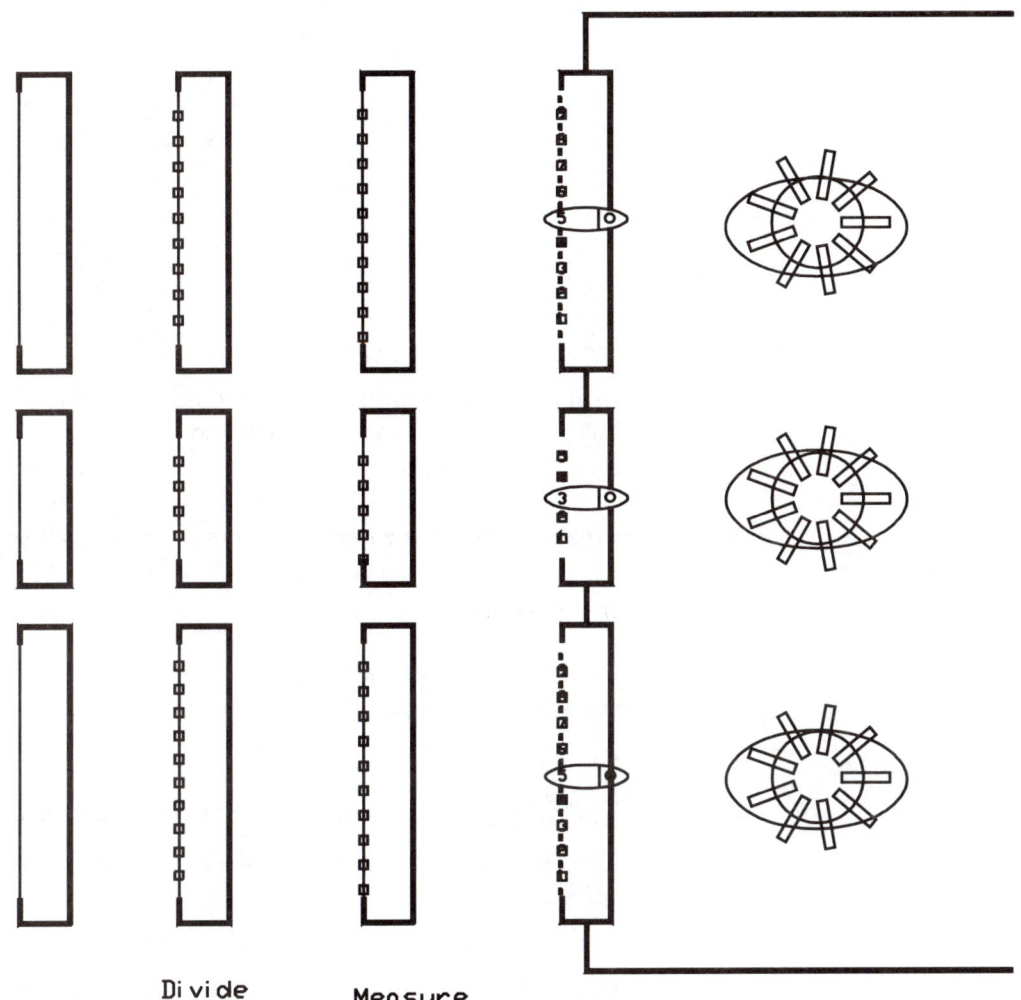

Step 2 Use LIST to find the length of the connections.

Use DIST to find the height of the schematic.

Exercise M13

Step 1 Draw the outlines for the sprocket shown. If the SNAP is set at .25 it will be quite simple.

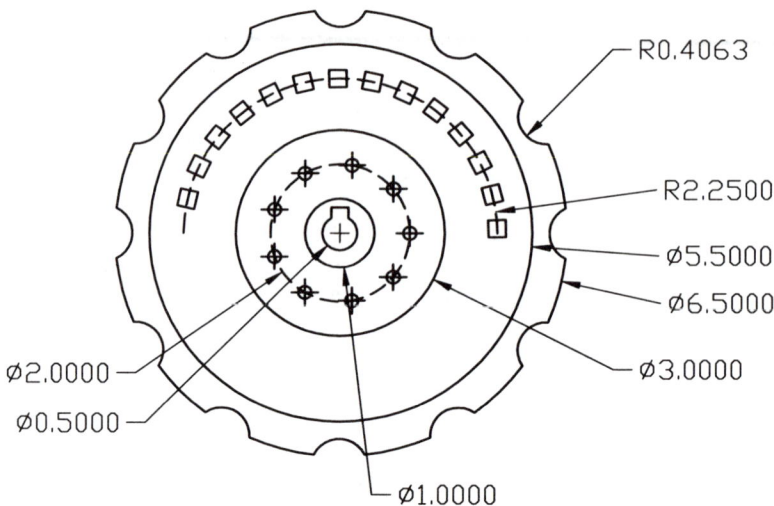

Step 2 Change the point display to 34 and use DIVIDE to place nine points to indicate the tap holes on the 1 unit radius centre line.

Step 3 Create a block of a square .25 units by .25 units. Use MEASURE to place this block at a distance of .5 units along the 2.25 radius arc.

Step 4 Use the AREA command to calculate the total area of the sprocket, making sure to subtract the keyway.

Step 5 Use LIST to find the length of arc along the perimeter of the sprocket.

Use DIST to find the minimum distance between one square and another on the sprocket.

Use PEDIT and Join to find the outer perimeter of the entire object.

If the center of the sprocket is at 0,0, what is the exact location of the left side of the keyway opening?

What is the total area taken up by the squares?

How much space have you left in your partition or RAM?

Challenger 13A

Step 1   Use the SPLINE command to place the outline of this part. The points will need to be placed perfectly to get an even, undulating curve. Don't forget to use Close. If you find this too frustrating, draw the part using arcs, then change it to a spline (not a pline, but a spline).

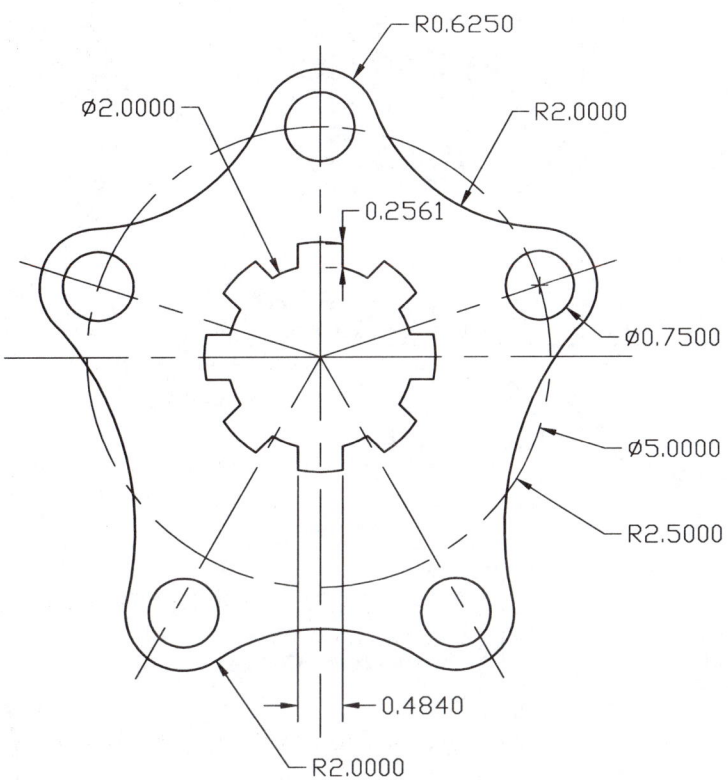

Step 2   Use Boundary and Area to discover the area of the surface of the part.

POINTS, DIVIDE, MEASURE, INQUIRY, and System Variables   **357**

Challenger 13B

This illustration can be added to pages 220, 250, and 457 to create a complete commercial stair layout.

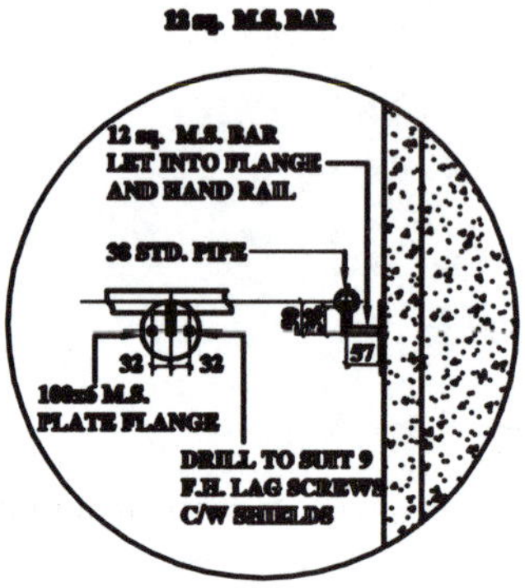

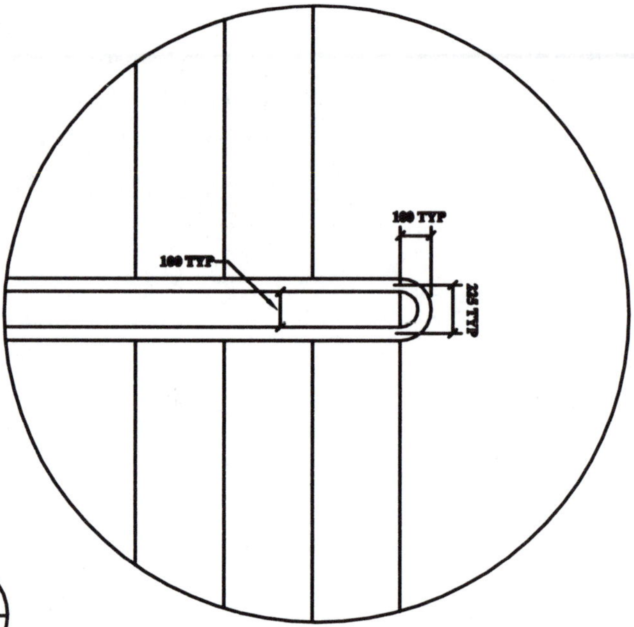

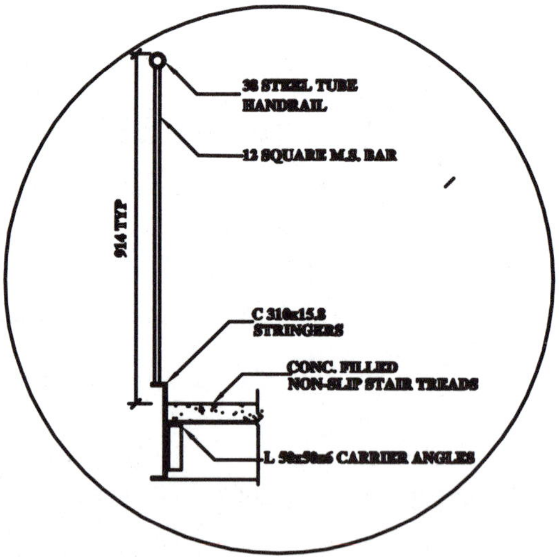

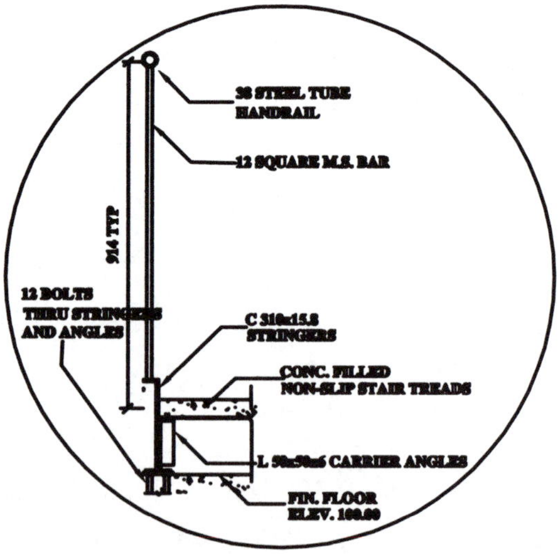

14 Creating Attributes

Upon completion of this chapter, you should be able to:

1. Define a series of attributes
2. Block the attributes
3. Insert the attributes onto a drawing
4. Use the ATTribute DIAlog boxes

OBJECTIVES

Introduction

In addition to creating geometry and drawings, AutoCAD allows for the generation of nongraphic information which can be accessed in the form of bills of materials, schedules, parts lists, and other data that is cross-referenced on the drawing. This data is not necessarily displayed on the drawing, but is filed with the drawing. This nongraphic intelligence is called an ***attribute***. Attributes provide a label or tag that lets you attach text or other data to a block. Attributes can also be used to generate templates for fill-in-the-blanks situations such as on drawing notations and title blocks.

You can define *constant* attributes which have the same value for every occurrence in the block, and *invisible* attributes that are not displayed nor plotted.

When you define an attribute, you are creating a program to prompt the user for data. You decide what the user must enter by creating the prompts for the entry of data.

Attributes for Title Blocks and Notations

In a standard title block there are areas identified for such information as date, part number, drawing number, etc. These would be printed on the title block. The user would then fill in the information that pertains to the part being drawn.

If this were a computer-generated title block, without attributes, the user would need to zoom into each area, adjust the TEXT size and style, then enter the current data using the TEXT command.

With attributes the user is prompted for the current information when the block is inserted. All of the sizes for the inserted text are determined at the time the attribute is defined and blocked. So are the placement of the text and the lettering font. All the user needs to do is add the missing information for the customer, name of part, date, name of designer, etc.

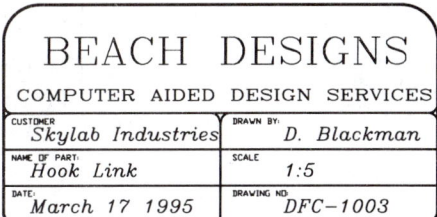

Figure 14-1

In the title block in Figure 14-1, all the information relative to the current drawing has been added using attributes. If this title block had attributes, you could expect the following prompts in the INSERT command:

Creating Attributes **359**

```
Command:INSERT
Block name (or ?):TITLE
Insertion point:0,0
X scale factor<1>/Corner/XYZ:⏎
Y scale factor (default = X):⏎
Rotation angle<0>:⏎ (the same as block inserts so far)
Customer name<Skylab Industries>:⏎
Name of Part<>:Hook Link
Enter the current date (DD/MM/YY):17/03/96
Drawn by<E.C.Jones>:⏎
Scale:<1:5>:⏎
Drawing no. DFC-####<>:DFC-1003
```

Notice that the INSERT command starts out the same as in Chapter 10, but there are prompts at the end that are specific to this title block. The prompts and the related information are what you create with the ATTDEF command or attribute dialog box.

Defining the Attributes

Prior to creating the attributed block, think through the information you are likely to need. How much is needed and where? A bit of forethought can save hours of editing.

In a title block the text font for the block titles (Name, Date, Drawn By) should be different from the text font for the current information. Before creating the attribute definitions, load the new text font.

To define the attribute use ATTDEF or the Attribute dialog box.

The ATTDEF Command

The ATTDEF command allows you to define an attribute. We will create the attribute definition for the date. Access the ATTDEF command through the command line.

```
Command:ATTDEF
Attribute modes -- Invisible-N, Constant-N, Verify-N, Preset-P
Enter (ICVP) to change, RETURN when done:⏎
```

Change the modes if you want the attribute to be constant, invisible, preset, or verified. In this case no change is necessary.

```
Attribute tag:DATE
```

The *attribute tag* is a one-word summary of the subject of the attribute. In this case, you want to enter the current date onto an existing title block. The word DATE is used to summarize what you want from the attribute.

```
Attribute prompt:Enter the current date (DD/MM/YY)
```

The attribute prompt is what is actually going to appear on screen. This prompt asks you for the required information, in this case the date, and it should offer any further information that may be required in order to enter that information. With regard to the date, you probably want a standard format such as Day/Month/Year so that an entry such as 03/04/94 will be April 3 and not March 4.

```
Default attribute value:↵
```

The default attribute value is what the user will usually want to use. There is no default required for the date because, in most cases, it changes daily.

```
Justify/Style/<start point>:J
```

Now that the you have identified what your attribute is going to say, you need to place it on the title block. Like the TEXT command, this prompts for the position of the attribute.

```
Align/Center/Fit/Middle/Right/TL/TC/TR/ML/MC/MR/BL/BC/BR:M
Middle point: (pick a point in the center of the area reserved
   for the date)
Height<0.1800>:.25 (this sets the size of lettering. Make sure
   this is both logical and readable!)
Rotation angle<0>:↵
```

Text in the Standard font was used to define the text for the title block. The tag for the date attribute is in the Italic Complex Font. This is where the current date will appear.

Prelab 14A describes in detail the construction of a fully attributed title block.

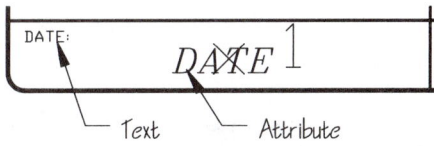

Figure 14-2

Using the ATTribute DIAlog Box

Windows From the Attribute toolbar, choose Define Attribute.

DOS From the Construct menu, pick Attribute.

The command line equivalent is **ATTDEF**.

In this dialog box you can set the modes, tag information, prompt, location, and text style as described above. When Pick Point is chosen, the dialog box will disappear until the point is chosen, then reappear once the selection is made.

Any number of attributes can be added to or associated with each block, as long as the tag is different.

Use the **TAB** key or the mouse to change fields.

The Value box means default value.

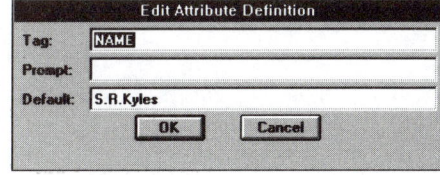

Figure 14-3

Editing Attribute Definitions

If you have made an error in defining the attribute you can change the attribute definition with the CHANGE command or with the Edit Attribute Definition dialog box.

```
Command:CHANGE
Select objects: (pick the definition you wish to change)
Select objects:↵
```

Creating Attributes **361**

```
Properties/<Change point>: (complete your changes)
```

You will be prompted for changes in the tag, prompt, and default values.

> **Windows** From the Special Edit flyout on the Modify toolbar, choose Edit Text.
>
> **DOS** From the Modify menu, choose Edit Text.

The command line equivalent is **DDEDIT**.

```
Command: DDEDIT
<Select a TEXT or ATTDEF object>/Undo: (pick the ATTDEF)
```

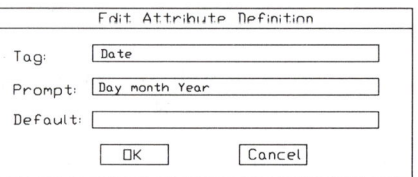

Figure 14-4

This will offer you three input boxes to change the TAG, PROMPT, or DEFAULT value within the dialog box. Press OK to complete the command.

BLOCKing the Attributes

Once the attributes are all defined, you must BLOCK the information in the order that you want your prompts to appear in by picking each attribute. If you use Window, you may be prompted in the reverse order, depending on where the Window starts.

Attributes can be used with both the BLOCK and the WBLOCK command.

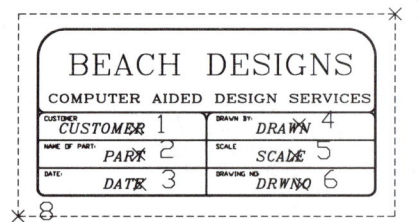

Figure 14-5

Saving the file also saves the attribute information without using the BLOCK or WBLOCK commands.

Inserting the Attributed Blocks

Having created the attributed block, you can now use INSERT to place it on your drawing. Once inserted, the block will contain the values for each attribute in that particular instance or application.

Using the ATTribute DIAlog Box

The Attribute dialog box will give you an automatic on-screen list of all the attribute data you have entered as soon as the INSERT command has been activated. You can then make any changes to the data on-screen. To activate the command, type ATTDIA at the command prompt, and make the variable active.

```
Command: ATTDIA
New value for ATTDIA<0>: 1
```

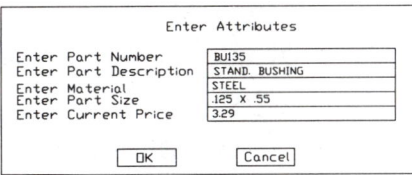

Figure 14-6

With ATTDIA at 0, the box does not display; with ATTDIA set to 1, it does.

The information in the box represents the default values. Just indicate which values you want to change by using the cursor arrow to highlight them. Then type in the new values and press OK to have the attributes display on the screen where placed.

This dialog box can be an advantage if there are a number of attributes that need to be coordinated.

Changing Attribute Definitions

While inserting the block, you may notice some errors in the block definitions. To change the attribute definitions at this point, use EXPLODE and then CHANGE. When you EXPLODE the attributed block, it will revert back to the tag format.

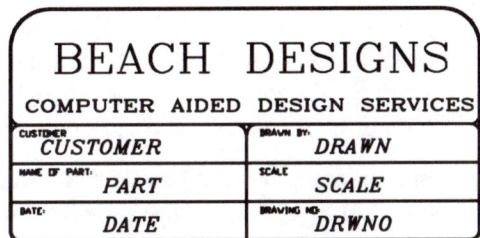
Block with tags

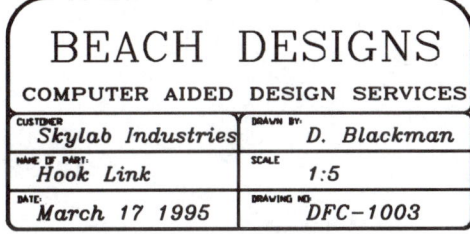
Attributed block

Figure 14-7

In Figure 14-7, we have the block with attribute tags on the left, and the inserted block with the instance values on the right.

If you use EXPLODE, you will return the attributed block to the attribute definitions, and only the tags will be shown. You can now use the CHANGE command or the Edit Attribute Definition dialog box to update the attribute definitions. Once changed, the attribute can be reblocked and reinserted.

```
Command: EXPLODE
Select block reference, polyline, dimension, or mesh: (pick 1)
Command: CHANGE
Select objects: (pick the block)
Select objects: ⏎ (etc.)
```

Displaying Attributes

When attributes are added to objects, they can be made universally invisible by setting the Invisible mode in the ATTDEF command, or you can make them invisible later using the ATTDISP command.

The ATTDISP Command

ATTDISP controls the display of attributes.

> **Windows** and **DOS** From the Options menu, choose Display, then Attribute Display.

The command line equivalent is **ATTDISP**.

```
Command:ATTDISP
Normal/ON/OFF<current value>:OFF
```

Where: **ON** = all attributes visible

OFF = all attributes invisible

Normal = normal visibility set individually

Attribute Modes

If you want to INSERT the attributed blocks without having to turn the display off, you can have the attribute mode set to invisible. You can set these as follows to make attributed block insertion easier:

INVISIBLE hides data associated with a BLOCK that does not need to be seen; for example, on a desk layout you may want to have the name of the current occupant for reference, but to provide specs for a decorator this is not necessary

CONSTANT makes an attribute uneditable; for example, the president's desk is always the president's desk even though the president may change

VERIFY allows you to take a final look at what you have entered prior to having it added to the drawing

PRESET is used when creating attributes that will always have the same value. This lets you insert a block that has an attribute or a set of attributes automatically attached to it. No user response is needed, because the attribute never changes. An example of this would be a part that would always maintain the same order number. The number would be inserted with the attribute, but the user is not prompted for any change

Creating Attributes for Data Extraction

Attributes can be used both for title blocks and drawing notations and for occasions where the information can be downloaded to a price list or bill of materials. The attributes are all defined in the same way, and the extractions take place once the attributed blocks are all inserted.

Attributes are always associated with blocks. If you want an attributed block, first create the geometry for the final block, if there is any, then add the attributes. You will need to have lines and text as well as attributes in a title block, but you will not need geometry in an attributed block that is meant to extract room colors.

Prelab 14B gives an example of attributes used for extraction. Extracting attribute data is covered in Chapter 15.

Prelab 14A Attributes for a Title Block

Step 1 Use LINE and PLINE to create the design for a standard title block. Use TEXT to enter the headings for each area. Use the parameters of this title block, or retrieve your title block from Chapter 8 to start.

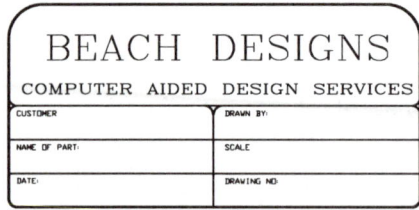

Step 2 Add the attribute definition for the part number.

```
Command:ATTDEF
Attribute modes -- Invisible-N, Constant-N, Verify-N, Preset-P
Enter (ICVP) to change, RETURN when done:↵
Attribute tag:Part
Attribute prompt:Enter name of part
Default attribute value:HOOK
Justify/Style/<start point>:J
Align/Center/Fit/Middle/Right/TL/TC/TR/ML/MC/MR/BL/BC/BR:R
Right point: (pick 1)
Height<5.0000>:.25
Rotation angle<0>:↵
```

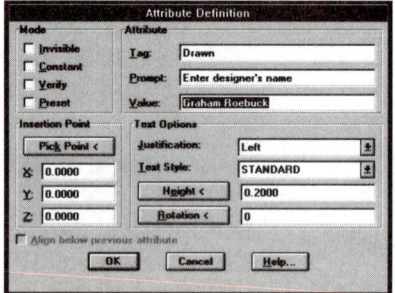

Step 3 Now enter the attribute for the designer using the Attribute Definition dialog box.

> **Windows** From the Attribute toolbar, choose Define Attribute.
>
> **DOS** From the Construct menu, pick Attribute.

The designer who created the block can use the default here, but if someone else wants to use this title block as well, they would simply enter their initials without using the default.

366 CHAPTER FOURTEEN

Step 4 Define the other attributes so that each area has an attribute tag.

```
BEACH DESIGNS
COMPUTER AIDED DESIGN SERVICES
CUSTOMER: CUSTOMER    DRAWN BY: DRAWN
NAME OF PART: PART    SCALE: SCALE
DATE: DATE            DRAWING NO: DRWNO
```

The block is now finished. We can block it and insert it.

Step 5
```
Command: BLOCK
Block name (or ?): TITLE
Insertion base point: (pick the corner)
Select objects: (pick attributes 1-6, then use Crossing to
   select the rest)
```

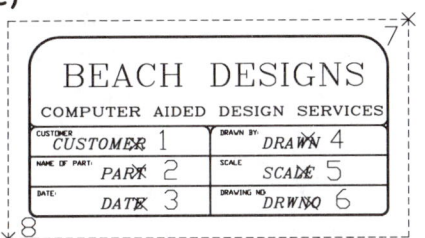

Remember that the attributed block will disappear once it is blocked.

Step 6 Insert the attributed block and check to see that you have specified the proper size of lettering and that prompts are clear. You can use ATTDIA if you like.

```
Command: INSERT
Block name (or ?): TITLE
Insertion point: 0,0
X scale factor<1>/Corner/XYZ: ↵
Y scale factor (default = X): ↵
Rotation angle<0>: ↵
Customer name<Skylab Industries>: ↵
Name of Part<Hook>: Hook Link
Enter the current date (DD/MM/YY): 17/03/93
Drawn by<E.C.Jones>: ↵
Scale<1:5>: ↵
Drawing no. DFC-####<>: DFC-1003
```

Step 7 A good way of checking your attributed block is to have a fellow student insert the block and fill in the details of the attributes without your help. If he/she has any problem understanding the prompts, go back and edit it. While you may understand your prompts now, the fact that someone else doesn't understand them may indicate that you won't either after a few days.

Other areas you may want to adjust are the attribute text sizes. Use EXPLODE and then CHANGE or DDEDIT to adjust them.

```
Command: EXPLODE
Select block reference, polyline, dimension, or mesh: (pick 1)
Command: CHANGE
Select objects: (pick 1)
Select objects: ⏎
```

Once changed, reblock the title block and reinsert it. To have the title block accessible for all files save the file as ATTITLE or use WBLOCK and take TITLE as the block.

Chapter 15 will illustrate how to edit the inserted attributed block.

Prelab 14B Defining, Blocking, and Inserting Attributes

In this tutorial, you will create a jig with a bushing and jigleg. In a manufacturing environment, it is usual to have items such as bushings purchased rather than made on-site to save both time and money. If you were to create a drawing of this part using these items, you could also have the ordering information stored as attributes with the block before it is inserted into the drawing.

Step 1   Draw up these very simple parts on different areas of your screen.

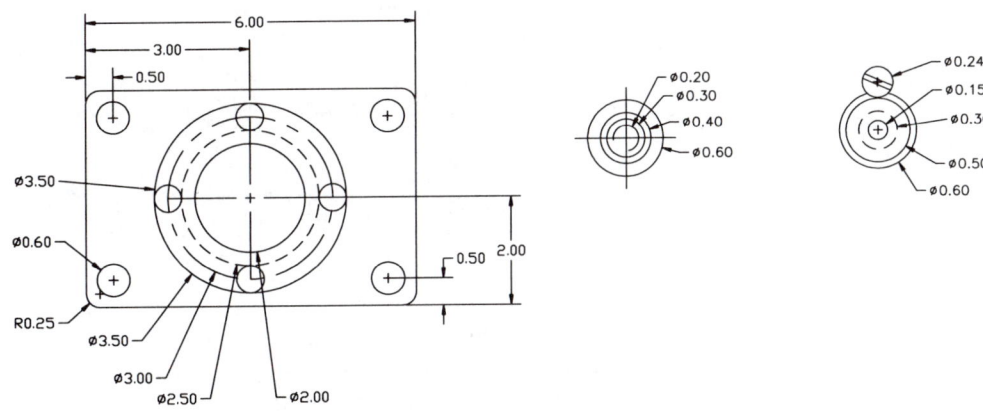

Step 2   If you wanted to create a bill of materials for these items, the information could include:

- Part Number
- Description
- Material
- Size
- Price

Then you would add the quantity and order the parts. The information listed above can be added as attributes to the block before it is blocked and then extracted later.

PARTNO
DESCRIPT
MATERIAL
SIZE
PRICE

The words in this illustration are attribute tags, not text. No text is needed on this exercise.

Add the attributes for the part using these default factors:

Tag	Prompt	Default Value
Partno	Enter Part Number	JIG125
Descript	Enter Part Description	Jigleg
Material	Enter Material	Steel
Size	Enter Part Size	.60 × 1.25
Price	Enter Current Price	5.26

Enter this information with the ATTDEF command or with the Attribute Definition dialog box.

> **Windows** From the Attribute toolbar, choose Define Attribute.
>
> **DOS** From the Construct menu, pick Attribute.

```
Command:ATTDEF
Attribute modes -- Invisible-N, Constant-N, Verify-N, Preset-P
Enter (ICVP) to change, RETURN when done:⏎
Attribute tag:PARTNO
Attribute prompt:Enter Part Number
Default attribute value:JIG125
Justify/Style/<start point>: (pick a spot beside the part)
Height<5.0000>:.10
Rotation angle<0>:⏎
```

When entering the next line for the attribute, the attribute definition will line up with the last entered line if you press ⏎ at the text justification line or simply choose OK on the dialog box.

```
Command:ATTDEF
Attribute modes -- Invisible-N, Constant-N, Verify-N, Preset-P
Enter (ICVP) to change, RETURN when done:⏎
Attribute tag:DESCRIPT
Attribute prompt:Enter Part Description
Default attribute Value:Jigleg
Justify/Style/<start point>:⏎
Command:ATTDEF
Attribute modes -- Invisible-N, Constant-N, Verify-N, Preset-P
Enter (ICVP) to change, RETURN when done:⏎
Attribute tag:MATERIAL
Attribute prompt:Enter Material
Default attribute value:Steel
Justify/Style/<start point>:⏎
```

Continue with the final two definitions until all five are complete.

Step 3 When all of the information has been added, create a block of the part with a new name to help you identify it as an attributed block.

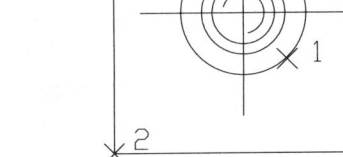

```
Command: BLOCK
Block name (or ?): ATTJIGLG
Insertion base point: CENter of (pick 1)
Select objects: (pick 2, 3)
```

Remember that the block will disappear.

Step 4 Do the same with the bushing. Use the values listed below.

Tag	Prompt	Default Value
Partno	Enter Part Number	BU125
Descript	Enter Part Description	Stand. Bushing
Material	Enter Material	Steel
Size	Enter Part Size	.125 × .55
Price	Enter Current Price	3.29

Block the new file under the name ATTBUSH.

Step 5 Use ZOOM All to get the original file back onto the screen. Now insert four jiglegs and four bushings onto the file.

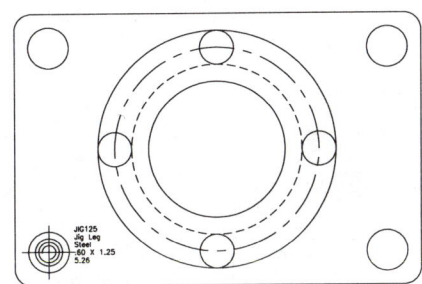

```
Command: INSERT
Block name (or ?): ATTJIGLG
Insertion point: CEN of (pick 1)
X scale factor<1>/Corner/XYZ: ↵
Y scale factor (default = X): ↵
Rotation angle<0>: ↵
Enter Part Number<JIG125>: ↵
Enter Part Description<Jigleg>: ↵
Enter Material<Steel>: ↵
Enter Part Size<.60 x 1.25>: ↵
Enter Current Price<5.26>: ↵
```

Creating Attributes

To get all four of these jiglegs placed, you could insert the object three more times or you could copy the existing block.

Step 6   Use the ATTribute DIAlog box to enter the final four parts.

```
Command: ATTDIA
New value for ATTDIA<0>: 1
Command: INSERT
```

Rotate the bushings by 90 when inserting.

When all four bushing blocks have been entered, your file will look like this:

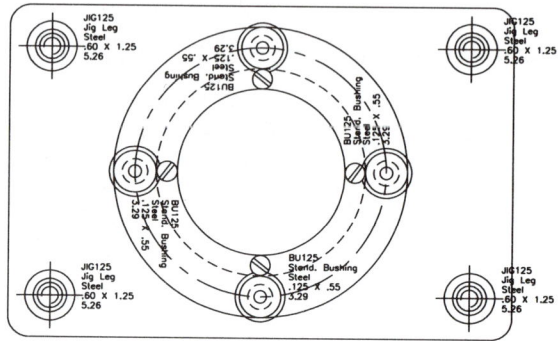

Notice that when you insert the bushings, the attributes will be rotated along with the geometry. This is of no importance, because the attributes will not be part of an overall drawing. In an electrical drawing, however, the attributes must be visible and legible; in this case, create attributed blocks for both directions. When attributes are added to parts, they are not usually printed or plotted.

Step 7   If you were just making a drawing, it would never look like this. Use ATTDISP to turn all of the attributes off.

```
Command: ATTDISP
Normal/On/Off<current value>: OFF
```

Command and Function Summary

ATTDEF creates an attribute definition which can be blocked and added to a drawing.

ATTDIA uses a dialog box to enter an attribute.

ATTDISP controls the display of attributes.

DDATTE edits inserted attributed blocks.

Exercise A14

Title Block — Retrieve your title block from Chapter 8 or create a new one. Use ATTDEF to define the attributes.

Project		Drawing	Scale	Project
Project		Drawing info	Scale	Pro
			Date	
		School	Drawn	Drawing
		School	Drawn	DR#
			Class	

Block the attributes and title block. Insert it to see how it works.

Project		Drawing	Scale	Project
Elgin Residence		Foundation and Plan	1" = 1'0"	1
			Date 17/3/92	
		School	Drawn G.G	Drawing
		Mohawk College	Class 1AT12	2

Room Finishes — Create one block containing the indicated attributes. No geometry is needed with this block. Draw a simple floor plan, and work out the color scheme by inserting the block in all of the rooms.

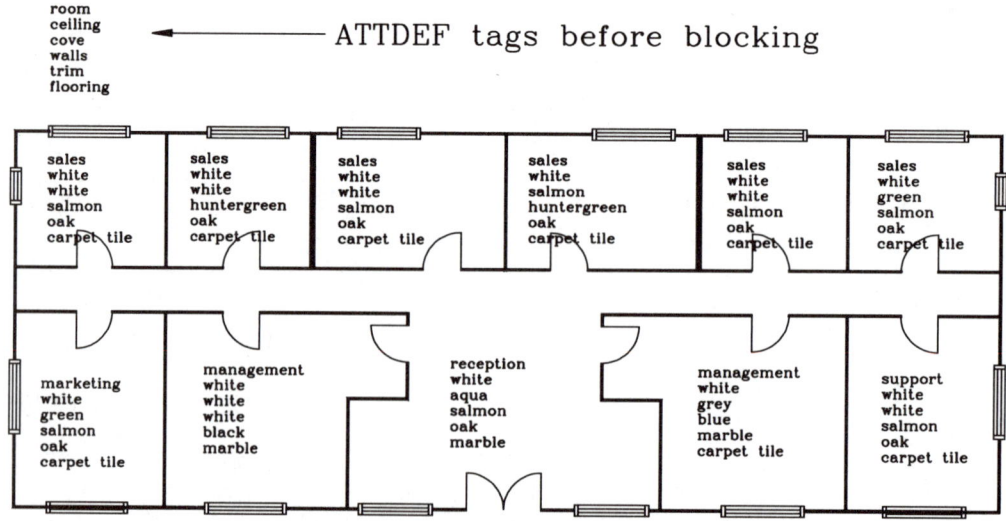

Exercise C14

Title Block Retrieve your title block from Chapter 8 or create a new one. Use ATTDEF to define the attributes.

Block the attributes and title block. Insert it to see how it works.

A Survey Draw a simple survey. Create attributes for the desired services and particulars as illustrated below. Block the attributes, then insert the block into each lot indicating the different lot numbers and services for each lot.

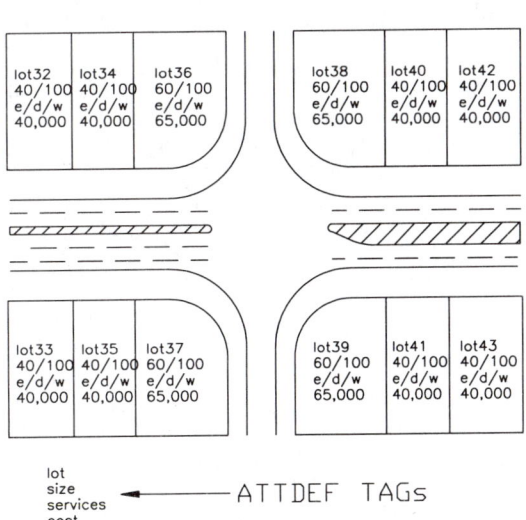

Creating Attributes 375

Exercise E14

Title Block — Retrieve your title block from Chapter 8 or create a new one. Use ATTDEF to define the attributes.

Project		Drawing	Scale	Project
Project		Drawing info	Scale	Pro
			Date	
		School	Date	Drawing
		School	Drawn	DR#
			Class	

Block the attributes and title block. Insert it to see how it works.

Project		Drawing	Scale	Project
Elgin Residence		Foundation and Plan	1"=1'0"	1
			Date 17/3/92	
		School	Drawn G.G	Drawing
		Mohawk College	Class 1AT12	2

Wiring Devices — Create four wiring devices with attributes containing color, price, and finish. Insert the blocks with attributes where necessary.

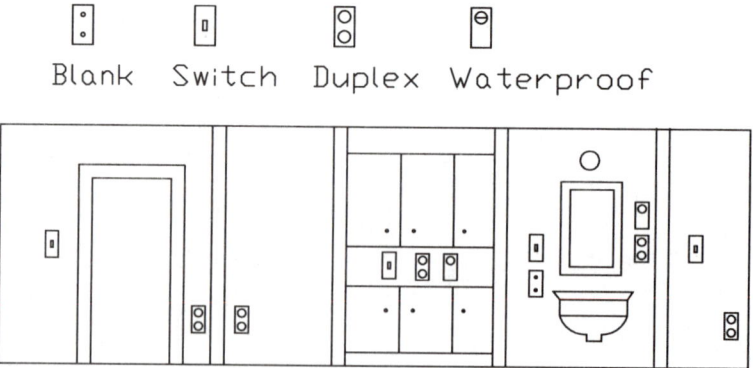

Exercise M14

Title Block Retrieve your title block from Chapter 8 or create a new one. Use ATTDEF to define the attributes.

```
┌─────────────────────────┬──────────────────────────┐
│ TOLERANCES UNLESS       │ Title                    │
│ OTHERWISE SPECIFIED     │                          │
│ FRACTIONS +/- 1/32      │      Title               │
│ DECIMALS  +/- .005      │                          │
│ ANGLES    +/- 5 DEGREES │                          │
├──────────────┬──────────┼──────────────────────────┤
│ APPROVALS    │ DATE     │ Company                  │
│              │          │     Company              │
│              │          ├────────┬──────┬──────────┤
│              │          │ SCALE  │ SIZE │ DRAWING NO. │
│              │          │ Scale  │ Size │  DRW#    │
│              │          │        │      │ Numbering│
└──────────────┴──────────┴────────┴──────┴──────────┘
```

Block the attributes and title block. Insert it to see how it works.

```
┌─────────────────────────┬──────────────────────────┐
│ TOLERANCES UNLESS       │ Title                    │
│ OTHERWISE SPECIFIED     │                          │
│ FRACTIONS +/- 1/32      │    Crane  Hook           │
│ DECIMALS  +/- .005      │                          │
│ ANGLES    +/- 5 DEGREES │                          │
├──────────────┬──────────┼──────────────────────────┤
│ APPROVALS    │ DATE     │ Company                  │
│              │          │     Miller  Inc          │
│              │          ├────────┬──────┬──────────┤
│              │          │ SCALE  │ SIZE │ DRAWING NO. │
│              │          │  1=1   │  A4  │  ch-48   │
│              │          │        │      │ Sheet 1 of 1│
└──────────────┴──────────┴────────┴──────┴──────────┘
```

Oiler Tray This is an assembly drawing for an oiler tray. Create one wire support rod. Add the part, part number, etc. as attributes. Then insert it six times on the drawing where needed.

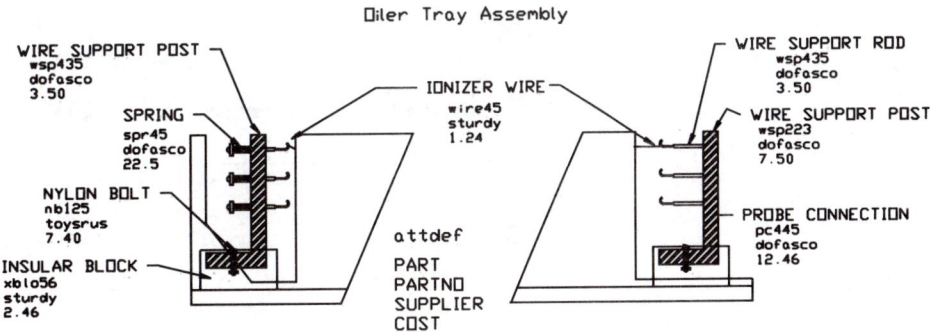

Do the same for each separate part. Create the geometry, add the attributes, and block them. Then insert them to complete the drawing.

Creating Attributes   **377**

Challenger 14

1. Draw the outline of a typical small office.
2. Draw two chairs, one slightly more complicated than the other, and two desks, one slightly larger than the other. Use the dimensions in Challenger 4.
3. Use ATTDEF to define the product number, description of the part, color, trim color, and price for the chairs, the desks, and the wall panels or dividers. Block each desk and each chair with their associated attributes separately.
4. Insert the windows and furnishings to make a full office layout.
5. Use ATTEXT to extract the product listing.

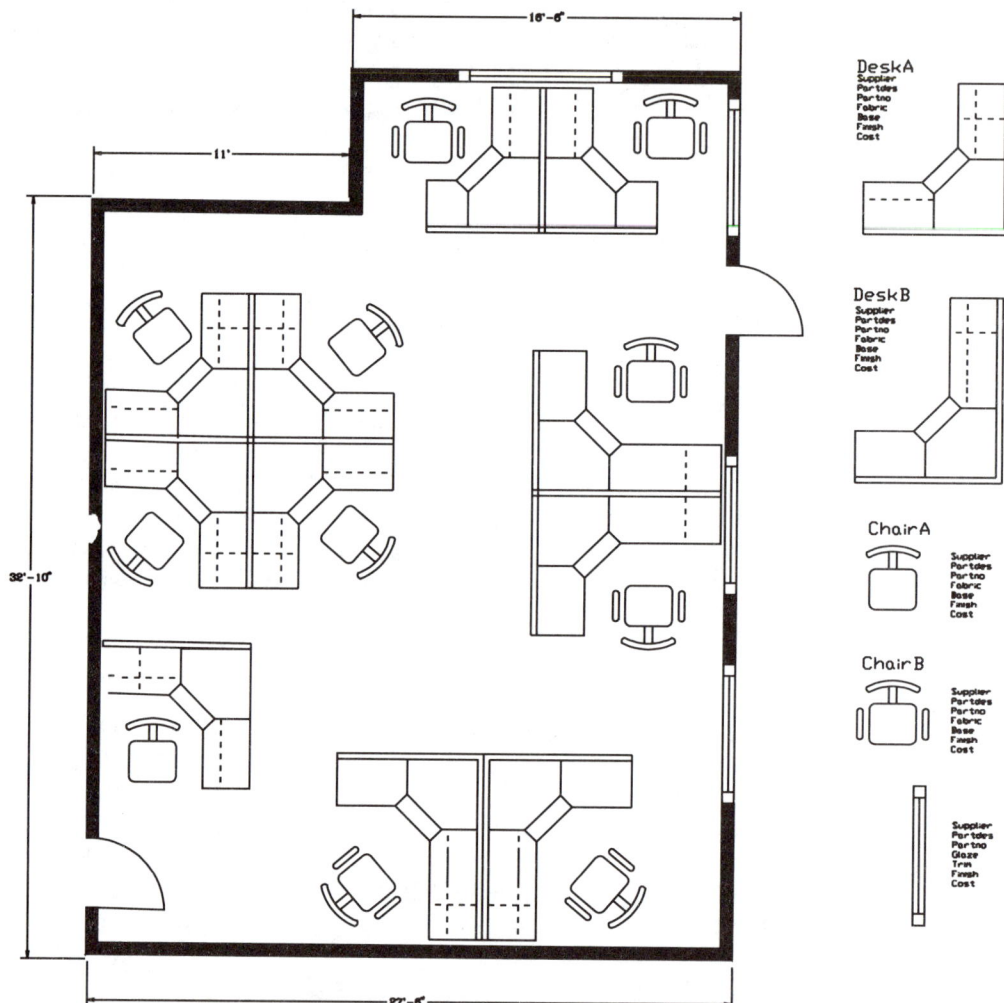

6. Take this attribute extract file into a spreadsheet or processing package that can produce a sum total for the price. Add applicable taxes, shipping of 10%, and $1200 installation charges to the total.

If you are using Lotus, you will need to PARSE the information before you can perform equations on it.

Then using the same office, change the color scheme of both the chairs and the offices. Change the trim to wood. This will reflect a different price for the entire office. Create another extract file and another quotation of the desks.

For a mechanical part of similar difficulty, see Challenger 15.

15 Editing and Extracting Attributes

Upon completion of this chapter, you should be able to:

1. Edit attributes
2. Extract attribute data to an ASCII format

Editing Attributes Attached to Blocks

As we saw in Chapter 14, you can use DDEDIT or Edit Text to edit an attribute definition before it is associated with a block.

Once the attributed block is in the drawing, you may want to edit the attributes that are already attached to a block and inserted in a drawing. To do this, use ATTEDIT.

CHANGE alters the characteristics of existing attribute definitions before they are blocked. Using the ATTEDIT command, you can change both the string value and the other characteristics of a blocked, inserted attribute.

The ATTEDIT Command

Windows From the Attribute toolbar, choose Edit Attribute.

DOS From the Modify menu, choose Attribute, then choose Edit.

The command line equivalent is **ATTEDIT**.

You will be prompted to choose the block to edit.

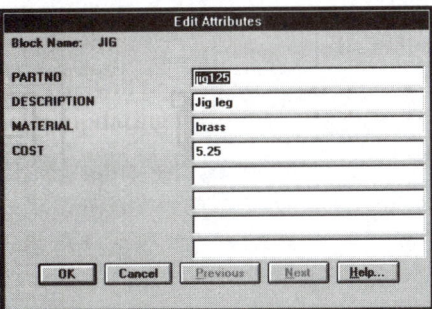

Figure 15-1

Use this dialog box the same way that you would use the Text Edit box. Change what you need to change, then pick OK.

Editing Individual Attributes

For example, to change the string value and color of just one attributed block, enter ATTEDIT at the command line, then choose to edit one by one.

```
Edit attributes one by one?<Y>:
```

You will be notified as to how many of the selected attributes have been chosen and an X will appear on the attribute you are editing. You will be prompted for filtering of tags, block names, and value specifications; just press to accept all the attributes chosen.

Then you will be prompted for the following possible changes:

```
Block name specification<*>:
Attribute tag specification<*>:
Attribute value specification<*>:
Value/Position/Height/Angle/Style/Layer/Color/Next<N>:
```

This will highlight the string values one by one and offer the user the possibility of changing each one.

Where: **VAL** = attribute value (the text string or what it says)
      POS = position of the text
      Hgt = text height
      ANG = angle of the text
      Style = text style
      Lay = layer
      Color = color
      Next = next

The response to this option is similar to that of CHANGE. Be sure you are answering the question posed and you will be fine.

The most common change required is the value. This will change what the attribute actually reads. Accept the default <N> for the next attribute, and scroll through until you have the top attribute, which will be your part number. Change the Value and Color of this attribute with the following.

```
Value/Position/Height/Angle/Style/Layer/Color/Next<N>:V
Change or Replace:R
New string:JIG135
Value/Position/Height/Angle/Style/Layer/Color/Next<N>:C
New color:BLUE
Value/Position/Height/Angle/Style/Layer/Color/Next<N>:N
```

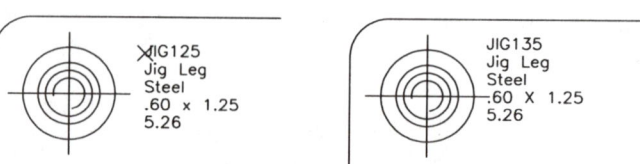

Figure 15-2

You can continue changing the attribute values, or use **Ctrl-C** to exit the command. The **N** option allows you to scroll onto the next attribute.

Editing Attributes Globally

When you are editing attributes with ATTEDIT one by one, you can change any of the values (orientation, text height, location, or string value); but when you perform a global edit, only the string value changes. A *global edit* is an edit that changes all of the attributes at once. A *string value* is the text actually created by the attributed block instance; a string of text is simply a line of characters.

To access global editing:

> **Windows** From the Attribute toolbar, choose Attedit, then edit globally.
>
> **DOS** From the Modify menu, choose Attedit, then edit globally.

The command line equivalent is **ATTEDIT**.

```
Command:ATTEDIT
Edit attributes one by one?<Y>:N
```

A positive response (**Y**) will identify and edit each attribute one by one. A negative response (**N**) will edit all of the attributes at once.

AutoCAD then prompts for the parts of the attributes to be edited.

```
Block name specification<*>:
Attribute tag specification<*>:PARTNO
Attribute value specification<*>:
```

The * is a "wildcard" and indicates that all of the items will be changed. You can accept this default and change string values in all the blocks, under any tags and values. If you want to change just attribute value in one type of block or one tag, specify the desired tag or value to avoid cycling through all the rest. You can use the block name, the tag name, and the string value to limit the number of attributes that will be edited. In the example above, only the PARTNO will be changed.

It is easier, though not necessary, to have the attributes displayed when you are editing them.

Finally, you are asked to select which attributes to change. Window or Crossing can be used to identify the attributes, but these must be preceded by **W** for Window or **C** for Crossing at the "Select attributes:" prompt.

Notes: When changing the string value, be sure to specify upper- or lowercase letters when needed.

```
Select attributes:W
First corner: (pick 1)
Other corner: (pick 2)
5 attributes selected.
```

You will now be asked for the string to change.

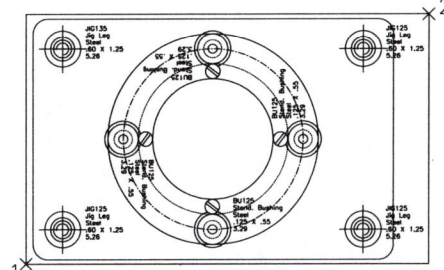

Figure 15-3

```
String to change:JIG125 ("jig125" will not work)
New string:JIG135
```

The ATTEDIT command changes the properties or value of the attribute once it has been inserted. The CHANGE command changes the tag, prompt, or default value of the attribute definition before it is blocked.

ATTEDIT for Global Edit — An Example

Attributes can also be copied and changed, which can save a lot of time.

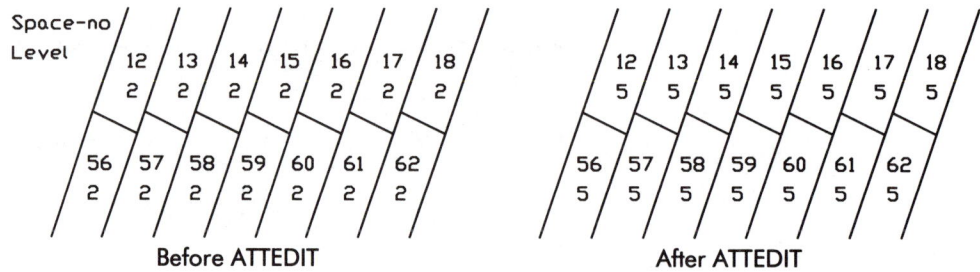

Figure 15-4

Step 1

Draw a series of parallel lines as in the illustration on the left. Create an attribute to define a space number and a level. Make the tag for the level LEVEL, and block the attributes under the name PARK. Insert the attributes as in the illustration on the left, making a series of different space numbers and keeping the level constant by accepting the default.

Step 2

Copy the parking lot over to the right. All of the attributed block instances will be copied as well.

Step 3

The second parking lot will be on level 5. The space number information will remain the same but the level will change.

Use ATTEDIT to quickly change all of the level specifications.

```
Command:ATTEDIT
Edit attributes one by one?<Y>:N
Global edit of attribute values.
Edit only attributes visible on screen?<Y>:↵
Block name specification<*>:PARK
Attribute tag specification<*>:LEVEL
Attribute value specification<*>:2
Select attributes:W
First corner: (pick 1)
Other corner: (pick 2)
14 attributes selected.
String to change:2
New string:5
```

Be careful to choose a string justification that will allow editing of the text, because the text string will be repositioned when the number of characters changes.

Data Extraction

Now that the data is entered correctly, you can have it printed out onto a spreadsheet or materials list by using the command ATTEXT. This operation does not change the drawing in any way, but takes the attribute data to applications such as dBASE, Lotus 1-2-3, or your favorite word processor.

You must create a template file to tell AutoCAD how to structure the file that contains the extracted information. Once this is completed, use ATTEXT to extract the information.

Extract File Formats

There are three main formats for data extraction.

- **CDF** Comma-delimited format
- **SDF** Space-delimited format
- **DXF** Drawing interchange file

CDF (comma-delimited format) takes each attributed block and extracts the attributes into a record where each attribute is separated by a comma. The extract file would look like this:

```
JIG135,Jigleg,Steel,.60 x 1.25,5.26
```

SDF (space-delimited format) will, again, provide one line for each occurrence of each block. In this format, there are no commas, the template file must specify the length of field for each extract, and any data that exceeds this limit is truncated. The extract file should look like this:

```
JIG135    Jigleg    Steel    .60 x 1.25    5.26
```

Note the difference in the format of this file as opposed to that of the CDF file. Note also that the area where the description appears must be large enough to contain all the letters.

Both the CDF and the SDF file must be made with a template file.

DXF (drawing interchange file) is the format used by many third-party programmers for drawing enhancements, analysis programs, and related applications. The DXF format is also used for exchange of engineering data, geological data, and nesting routines, as well as the extraction of purely spatial information. The extract file will contain all of the information in the block, such as insertion point, rotation angle, and *X,Y,Z* values. This creates a file with a .DXX extension. On MS-DOS and PC-DOS systems, you can also specify a filename of .CON to send the attribute extract directly to the screen, or .PRN to send it to a printer. Make sure your printer is connected before you use .PRN.

Template Files

A ***template file*** identifies the structure of the extracted information. It specifies which attributes are to be extracted, what information is to be included, and what the extract file will look like.

The template file must have the extension .TXT. Each line of the template file specifies one field to be written in the extract file. The extract file will contain the information in the order given in the template file.

For the jigleg example using an SDF format, your extract file might look like this:

```
JIG        Jigleg           Steel    .60 x 1.25    5.26
JIG135     Jigleg           Steel    .60 x 1.25    5.26
BU135      Stand. Bushing   Steel    .125 x .55    3.29
BU135      Stand. Bushing   Steel    .125 x .55    3.29
```

The template file might look like this:

```
PARTNO C008000
DESCRIPT C015000
MATERIAL C008000
SIZE C008000
PRICE C080000
```

The template file is set up with the tag plus the code:

```
PARTNO C008000
```

Where: **PARTNO** = TAG
       C = character
       008 = 8 characters maximum in the jig name (C008000)
       000 = number of decimal places (of course none are needed)

If your entry consists exclusively of numbers, as in the case of the price, you can use an N, which will signify a numeric field. If there is anything other than numbers, you must use a C. If the entry contains both numbers and letters, use C. If you want the block name and the *X,Y* factors, use BL:.

	Field Name	*C or N,* *Field Width of C or N,* *Decimal Places if N*
Block name	BL:JIG1	C008000
Insertion point	BL:X	N007001
	BL:Y	N007001
	PARTNO	C008000
	DESCRIPT	C015000
Attribute tags	MATERIAL	C008000
	SIZE	C008000
	PRICE	C008000

Creating the Template File

Create the file using one of the following methods:

> **Windows** From the Accessories program group, choose Notepad.
>
> **DOS** At the command line, enter **SHELL** or **EDIT**.

```
Command: SHELL
DOS Command: EDIT TEMPLATE.TXT
```

Type in the template file as outlined above using either the tags or the blocks. Then use ATTEXT to extract the file.

> **Windows** On the command line, enter **DDATTEXT**.
>
> **DOS** From the File menu, choose Export, then choose Attributes.

The command line equivalent is **ATTEXT** or **DDATTEXT**.

Creating the Extract file

Once the template file has been created, you can extract the data using the command ATTEXT. This will ask for the name of the template file and the name of the extract file.

```
Command: ATTEXT
CDF, SDF, or DXF Attribute extract ? (or entities)<C>: SDF
Template file<default>: TEMPLATE (this is the template file you
   created)
Extract file name<JIG>: ⏎ (this is the extract file)
8 records in extract file.
```

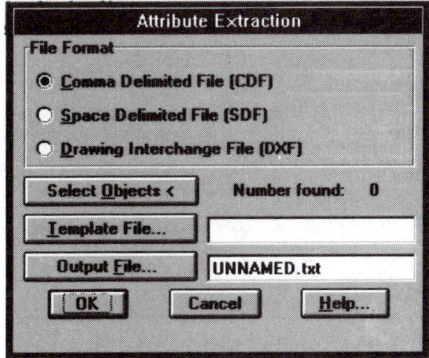

Figure 15-5

What Can Go Wrong with an ATTEXT

Field Overflow

On the readout after the ATTEXT command you may get a message saying:

```
Field overflow on line 3
```

This means that there are not enough spaces for characters in the template file. To remedy this situation, you must return to the template file and allow more characters in the field. For example, you may have allowed for only ten characters for the DESCRIPT tag:

```
DESCRIPT C010000
```

In this case the string value "Stand. Bushing" will be truncated, because it contains more than ten characters, and there will be a field overflow.

Extra Lines in File

Danger

The new file created by the ATTEXT command has the extension .TXT. Be sure that you do not have the same name for the template and the extract file; otherwise you will lose one of the files. There is no prompt to tell you that you are overwriting the first file.

When you are entering your template file, make sure that there are no extra lines in the file. Use a hard return on the last line of the text, but do not use a hard return after the final entry.

```
PARTNO   C008000[HRt]
DESCRIPT C015000[HRt]
MATERIAL C008000[HRt]
SIZE     C008000[HRt]
PRICE    C080000[HRt]
^Z       (only Ctrl-Z, no hard return)
```

Flow Chart for Creating ATTribute EXTracts

Start by creating the drawing and the attributes, and then create the template file. Alternatively you can start by creating the template file, as long as both the template and the attributed block instances are created when you use ATTEXT.

If you *start* by creating your template file, you can decide what information you need and then create the attributes to provide this information.

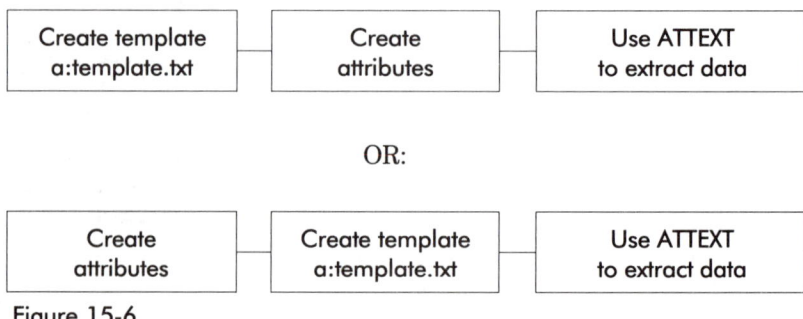

Figure 15-6

Prelab 15 Editing and Extracting Attributes

Step 1 Retrieve your file from Prelab 16B. If your attributes are turned off, turn them back on with ATTDISP.

Step 2 Use the following command to change all of the part numbers for the bushing.

> **Windows** From the Attribute toolbar, choose Attedit.
>
> **DOS** From the Modify menu, choose Attedit.

The command line equivalent is **ATTEDIT**.

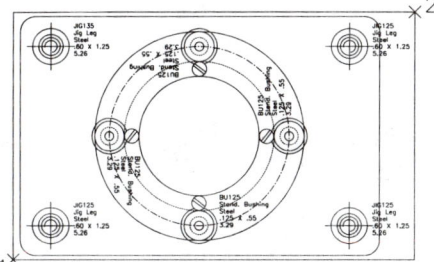

```
Command:ATTEDIT
Edit attributes one by one?<Y>:N
Global edit of attribute values.
Edit only attributes visible on
    screen?<Y>:↵
Block name specification
    <*>:ATTBUSH
Attribute tag specification<*>:
Attribute value specification<*>:
Select attributes:W
First corner: (pick 1)
Other corner: (pick 2)
20 attributes selected.
String to change:BU125 ("BU" must be uppercase)
New string:BU135
```

Use **INSERT ?** to find the name of the block if you think it may be different from ATTBUSH.

Notice that the prompts in ATTEDIT change with regard to the options chosen. In this example, it is a global edit, so *all* attributes will be changed at once. Since the attributes will not be highlighted, you are asked if the invisible attributes will be changed as well.

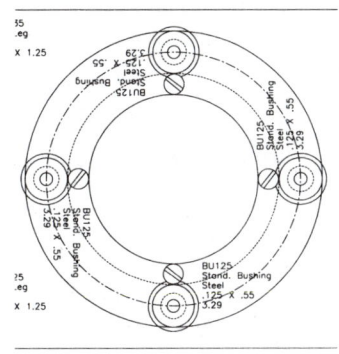

Step 3 Make a note of the tags of your attributes. If you have forgotten what they are, use LIST. Note the name of the tag exactly (no spelling errors are allowed).

Step 4 Now create a template file.

> **Windows** From the Accessories program group, choose Notepad. Once finished under the File menu, choose SAVE, then Minimize.
>
> **DOS** At the command line, enter **SHELL**, then **EDIT**.

```
Command:SHELL
DOS Command:EDIT TEMPLATE.TXT
Command:EDIT
File to edit:TEMPLATE.TXT
  PARTNO   C008000
  DESCRIPT C020000
  MATERIAL C010000
  SIZE     C010000
  PRICE    C008000
^Z (Ctrl-Z)
Command:
```

The template file will be saved on your disk as TEMPLATE.TXT. TEMPLATE is the name of the file and can be changed; .TXT is the extension and must not be changed.

You have now written a template file for an extract. It uses the template file TEMPLATE.TXT and the attribute data to merge the files.

Step 5 Back at the Graphics screen, enter the ATTEXT command to extract the data into a separate file.

> **Windows** On the command line, enter **DDATTEXT**.
>
> **DOS** From the File menu, choose Export, then choose Attributes.

The command line equivalent is **ATTEXT**.

```
Command:ATTEXT
CDF, SDF, or DXF Attribute extract ? (or entities)<C>:SDF
Template file<default>:TEMPLATE
Extract file name<jig>:PRELAB13
8 records in extract file.
```

AutoCAD will tell you how many extracts you have. The extract filename will default to the drawing name unless changed. Make sure it is PRELAB13.

Step 6 Now take a look at your file in the editor. The file should be PRELAB13.TXT.

```
Command:SHELL
Command:EDIT
File to edit:PRELAB13.TXT
```

```
JIG125      Jig Leg          Steel     .60 x 1.25    5.26
JIG125      Jig Leg          Steel     .60 x 1.25    5.26
JIG125      Jig Leg          Steel     .60 x 1.25    5.26
JIG125      Jig Leg          Steel     .60 x 1.25    5.26
BU135       Stand. Bushing   Steel     .125 x .55    3.29
BU135       Stand. Bushing   Steel     .125 x .55    3.29
BU135       Stand. Bushing   Steel     .125 x .55    3.29
BU135       Stand. Bushing   Steel     .125 x .55    3.29
```

Step 7 There are often a few problems with the file the first time through. Either your fields were not large enough, or you have forgotten one field. In this case, the description area could be a little larger.

> **Windows** From the Accessories program group, choose Notepad, then open the file TEMPLATE.TXT.
>
> **DOS** Use EDIT (or SHELL) to adjust your template file. Then try ATTEXT again.

```
Command:EDIT
File to edit:TEMPLATE.TXT

   PARTNO C008000
   DESCRIPT C020000  (change to C025000)
   MATERIAL C010000
   SIZE C010000
   PRICE C008000
```

Step 8 Now use ATTEXT to extract the file again, and look at it to make sure that the description area has been made larger.

Bringing Up the ASCII File into Other Software Packages

You have created a document that is written in ASCII format. If you want to bring this up in WordPerfect, Microsoft Word, or any other program, you can simply enter it as you would any text file. If you want to bring this file into Lotus, use Import and then Parse the file to create a columnar format.

Save the current file as JIG and exit the file.

Command and Function Summary

ATTEDIT edits the value of inserted attributes.

ATTEXT extracts information from attributes in your drawing file to a CDF, SDF, or DXF format.

DDATE opens a dialog box that allows you to edit inserted attribute blocks.

EDIT allows DOS users to create and edit a text file using the DOS editor.

Notepad allows Windows users to create text files.

Exercise A15

Part 1 *Room Finishes*

In your floor plan layout, change all the carpet tile to linoleum. Change the black trim to navy.

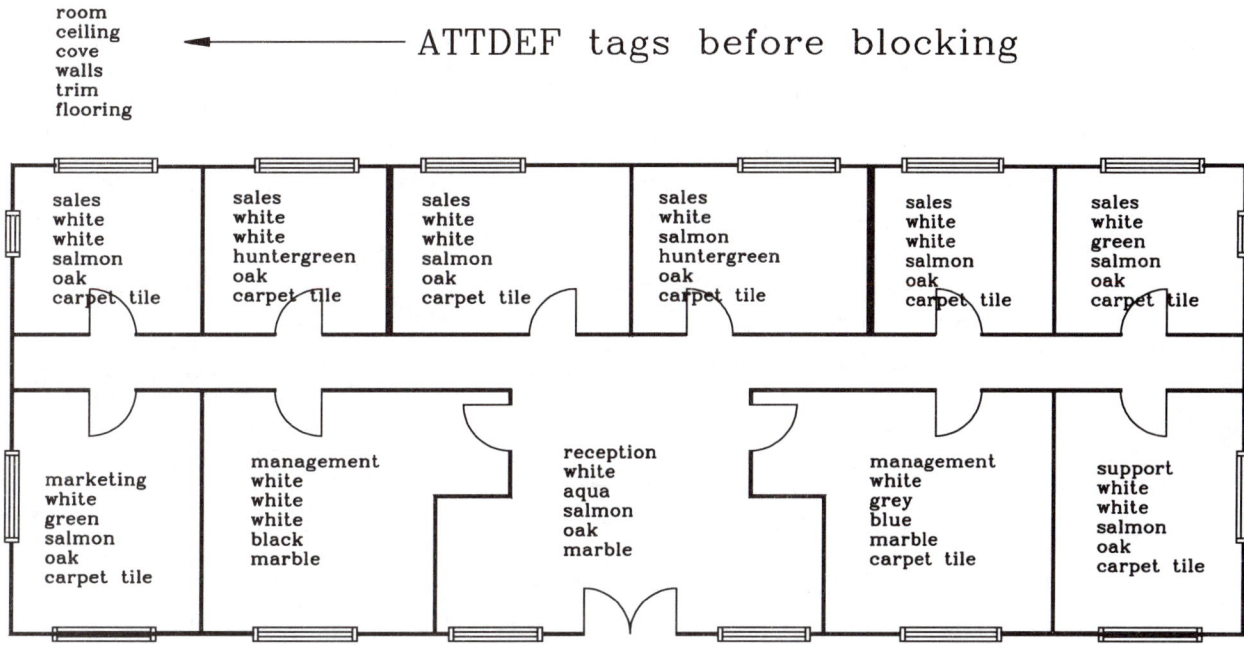

Part 2

Then extract the attribute data onto an ASCII file, retrieve it into a word processing package such as WordPerfect, and print your color schedule.

Exercise C15

Part 1 **A Survey**

On your survey, change the price of the 40,000 lots to 42,000. Then change the 65,000 to 67,000.

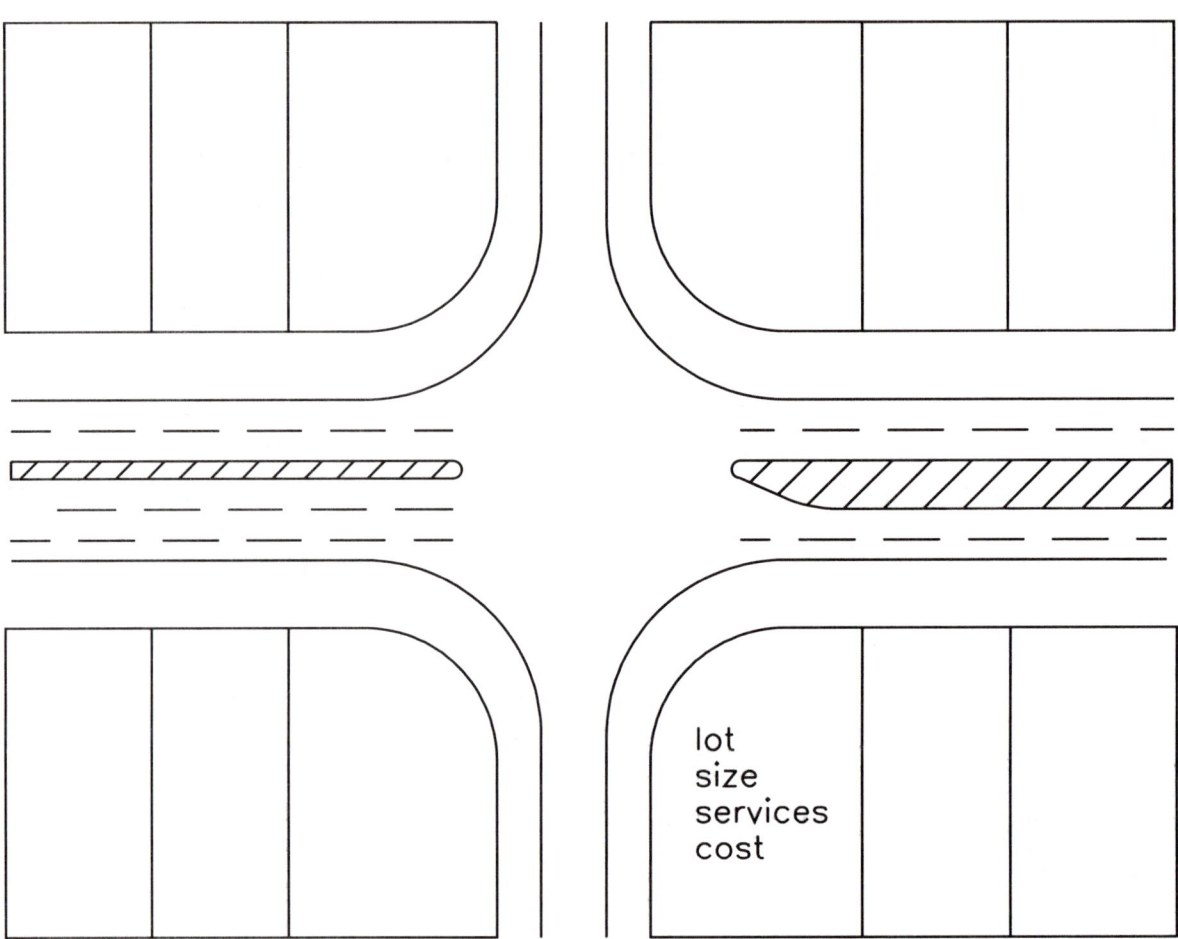

Part 2 Now extract the data into an ASCII file. Bring the file up into WordPerfect or a similar package, and create a listing of the lots and their values. Print the listings.

Exercise E15

Part 1 **Wiring Devices**

On your wall layout, change the color of all of the white fixtures to bone. Now change the prices by adding .15.

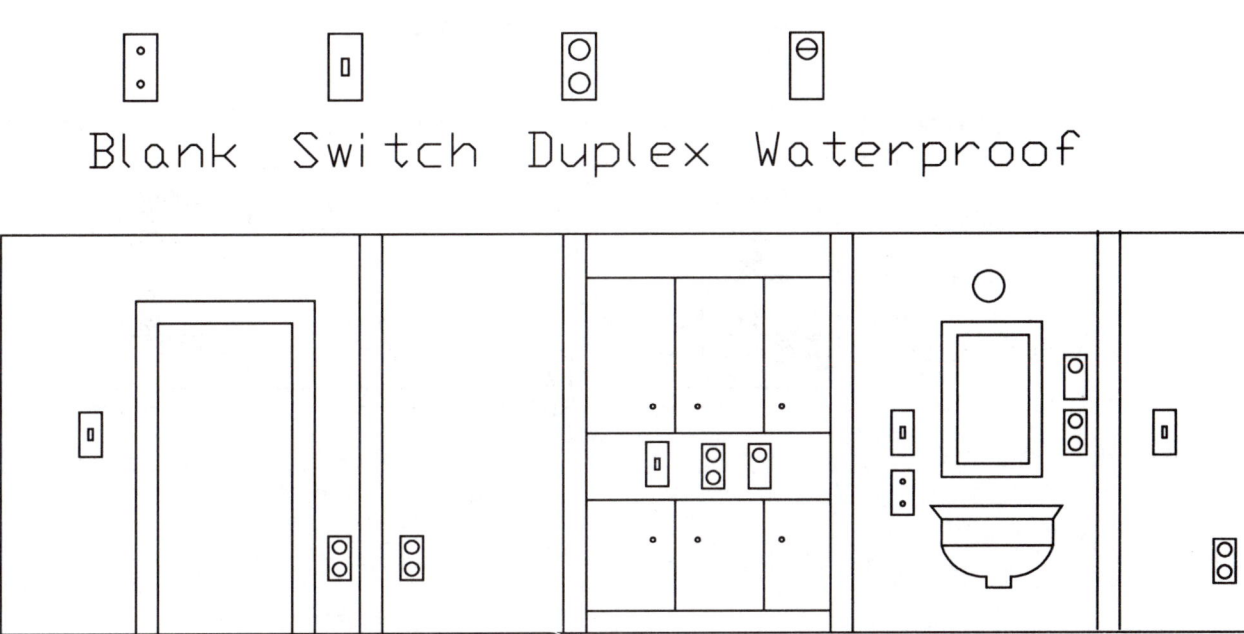

Part 2 Now extract this data into an ASCII format. Bring up the extract file into WordPerfect or a similar application. Create a full bill of materials.

Exercise M15

Part 1   *Oiler Tray*

Change the price of the wire support rod to 4.50. Change that of the nylon bolt to 8.25.

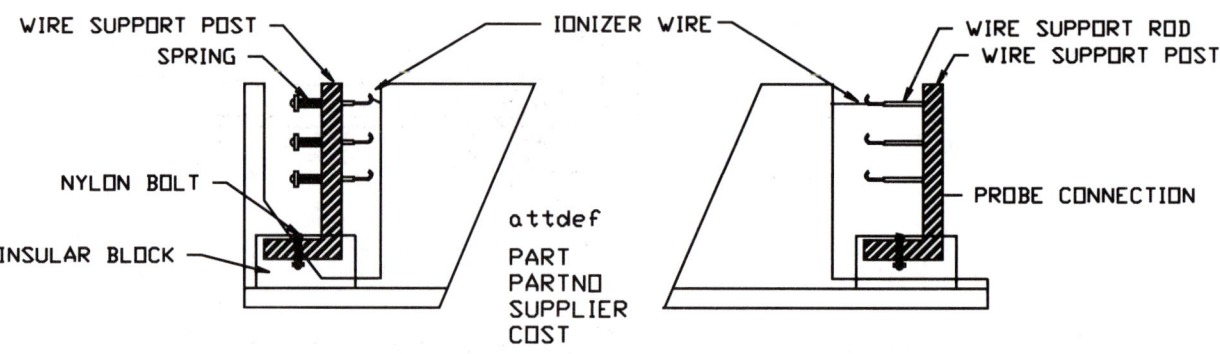

Part 2

Now extract the attribute data from all of the parts on this assembly into an ASCII file.

In Lotus or a similar package, create a spreadsheet for the costs of the parts. Print out a bill of materials.

Challenger 15A

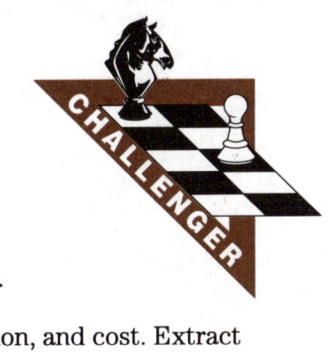

1. Draw this airbrush nozzle. For each individual part, create attributes for part number, description, material, and cost. Block the attributes with the parts. Insert them back into the assembly.

2. Create a template file for the part numbers, material, description, and cost. Extract the file and create a bill of materials.

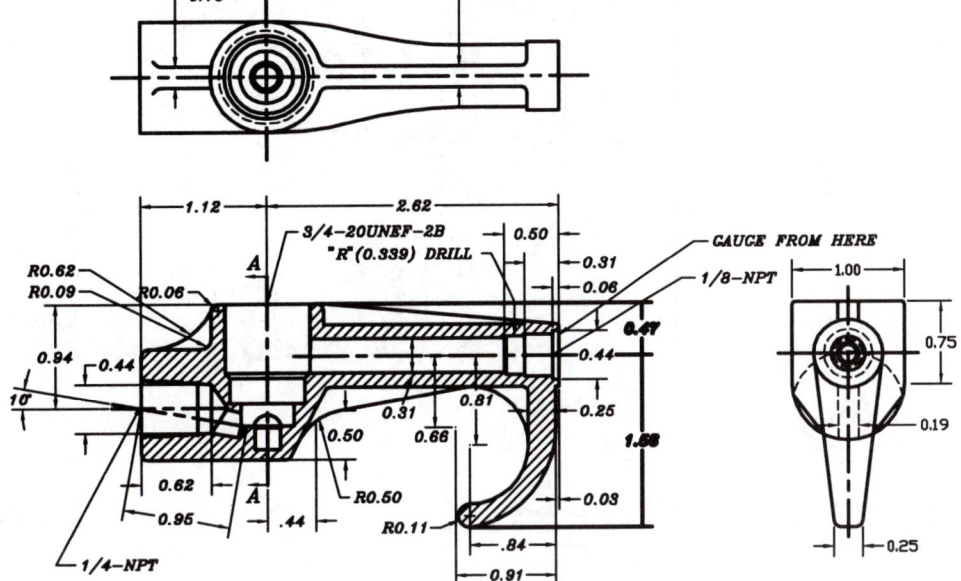

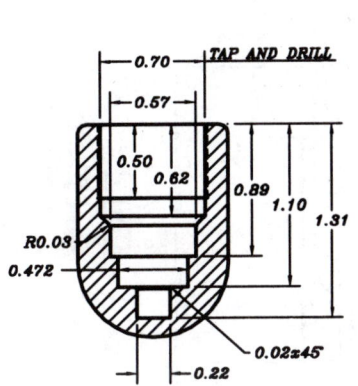

SECTION AA

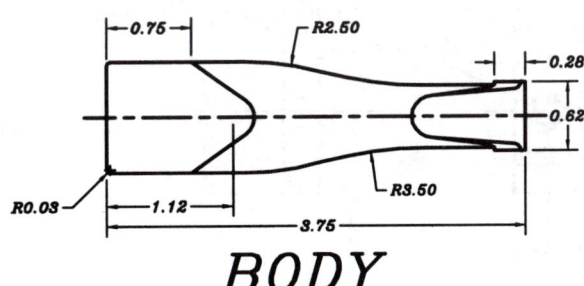

BODY

Challenger 15B

Draw the part as shown. Block each object as a separate attributed file, and insert back into the drawing. Create a parts list from the various parts.

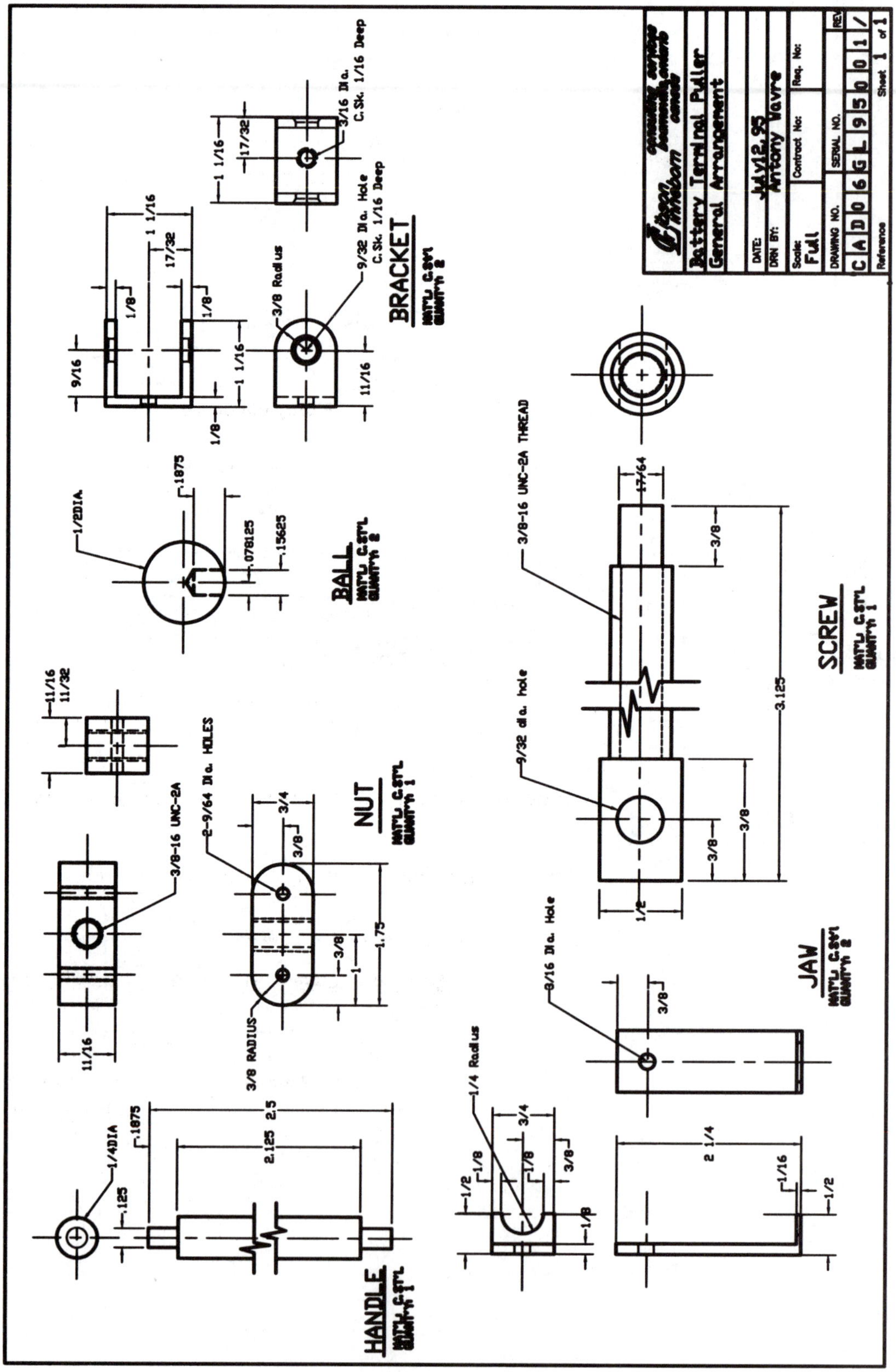

16 Isometric and Orthographic Drawings

Upon completion of this chapter, you should be able to:

1. Change SNAP to isometric to create isometric views
2. Use ISOPLANE to align arcs and circle to a certain plane
3. Change the obliquing angle of the text to create isometric text
4. Create orthographic views in 2D

OBJECTIVES

Isometric Views

In AutoCAD, it is possible to create rotatable 3D models by means of more advanced commands. However, if you are only concerned with a 2D representation of a 3D model, similar to what you can get by drawing with pencil on paper, you will find it satisfactory to change SNAP and use the ISOPLANE commands. The results are not 3D, but look 3D. This method has the advantage of enabling you to create an image much more quickly than you could with full 3D modelling techniques.

Changing the SNAP to Isometric

Remember from Chapter 1 that SNAP sets a spacing for point entries, and GRID places a dot grid on the screen.

SNAP allows for both rotated and isometric drawings to be entered. To access SNAP use either the command line or the Drawing Aids dialog box.

> **Windows** From the Options menu, choose Drawing Aids, then Drawing Aids.
>
> **DOS** From the Options menu, choose Drawing Aids.

The command line equivalent is **SNAP**.

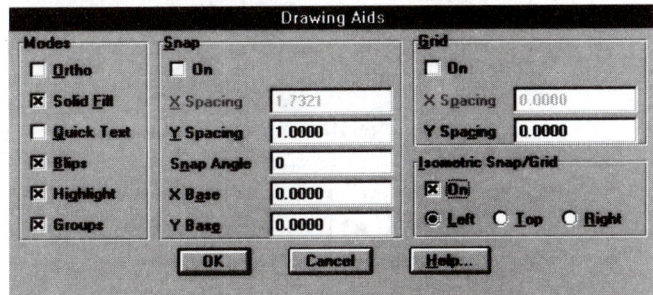

```
Command:SNAP
Snap spacing or ON/OFF/Aspect/Rotate/Style<1.0000>:S
Standard/Isometric<Standard>:I
Vertical spacing<0.4>:↵
```

Once you set SNAP you will notice that the grid is set at a 30 degree angle as well. The GRID size will follow the SNAP size unless changed. In addition, the crosshairs will be viewed at a 30 degree angle and ORTHO will also be isometric.

The easiest way to test the isometric snap is by drawing a simple cube on the screen with the aid of SNAP and GRID. The easiest way is to simply pick the points from the screen with the SNAP set to 1.

Follow the lines of the grid.

```
Command:LINE
From point: (pick a point)
To point: (pick following the grid)
```

Now that you've seen how easy this is, how do you think coordinate entry reacts?

```
Command:LINE
From point: (pick a point)
To point:@2,0
To point:@0,2
To point:⏎
```

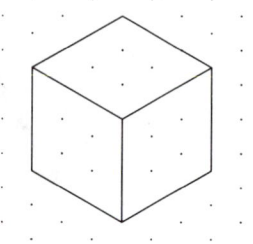
Figure 16-1

Notice that your line entries when using points from the screen SNAP to an isometric plane, but line entries using coordinates are not isometric.

Using ISOPLANE

Once you have entered the cube you will notice that the crosshairs line up with only one plane. Use ISOPLANE to get the crosshairs to line up with all three planes. Change the ISOPLANE through the Drawing Aids dialog box or by typing it in.

You can toggle to the next plane by using **Ctrl-E**. ISOPLANE is not available unless you have set SNAP to Isometric.

```
Command:ISOPLANE
ISOPLANE Left/Top/Right/<Toggle>:R
```

ISOPLANE will act as a toggle switch that toggles through the three planes and then off. ORTHO is useful when using ISOPLANE to make rectangular shapes.

Figure 16-2

Drawing Circles in Isometric

Lines can be entered quickly without ISOPLANE and with the use of the SNAP, but circles and arcs are entered relative to the actual X,Y plane, much like the line coordinate entry. In Figure 16-3(a), CIRCLE was used to create a circle, which on the isometric plane looks like an ellipse.

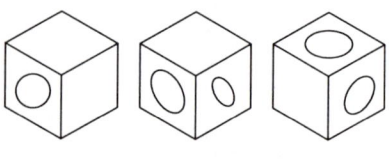
(a)   (b)   (c)
Figure 16-3

You will need to use ISOPLANE and ELLIPSE with the Isocircle option to create circles in the three standard isometric planes.

On the left cube, (a), a circle is drawn using the CIRCLE command.

The central circles shown on cube (b) are drawn relative to the left plane using the ELLIPSE command. The planes on cube (c) are toggled before the circles are added.

```
Command:ELLIPSE
<Axis endpoint 1>/Center/Isocircle:I
Center of the circle: (pick 1)
<Circle radius>/Diameter: (pick 2)
```

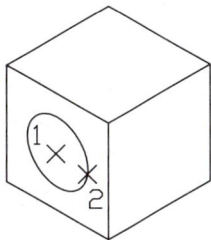

Figure 16-4

The current isoplane governs the plane in which the circle will be drawn; change the ISOPLANE to change the orientation of the circle.

Drawing LINEs at Angles

Because this is a 2D drawing, lines are drawn relative to the X-Y plane.

```
Command:LINE
From point: (pick a point)
To point: @1<0
To point:
Command:LINE
From point: (pick a point)
To point: @1<30
To point:
```

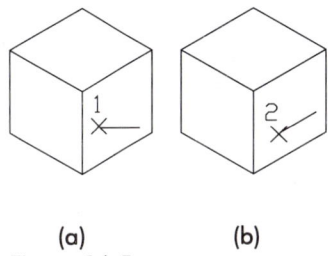

(a)   (b)

Figure 16-5

Thus, lines drawn at angles will not be rotated relative to the ISOPLANE. In this illustration, the lines are drawn as rotated normal to the X-Y plane.

In order to add a line at an angle relative to the isoplane, add 30 degrees to the angle that you need.

```
Command:LINE
From point: (pick a point)
To point:@1<75
```

This will give you a line of 45 degrees relative to the current plane.

Prelab 16A An Isometric Example

This example will illustrate how to change SNAP and ISOPLANE, and how to use ELLIPSE with ISOCIRCLE to generate this simple part.

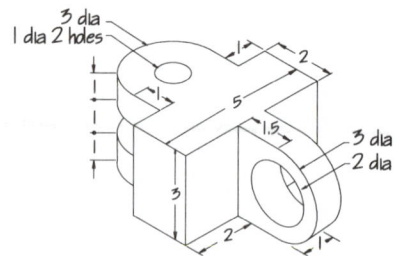

Step 1   Start by setting SNAP to isometric with a value of .5, and GRID to a value of 1. Then put in the lines as shown. Setting the coordinate readout to relative coordinates (F6) will help to attain the correct dimensions.

> **Windows** From the Standard toolbar, choose Drawing Aids, then Drawing Aids.
>
> **DOS** From the Options menu, choose Drawing Aids.

The command line equivalent is **SNAP**.

```
Command:SNAP
Snap spacing or ON/OFF/Aspect/Rotate/Style<1.0000>:S
Standard/Isometric<Standard>:I
Vertical spacing<1.0>:.5
Command:GRID
ON/OFF/<1.0>:⏎
Command:LINE
From point:
```

Add the lines as shown.

Step 2   Make sure that ISOPLANE is set to Left and add a circle using ELLIPSE.

```
Command:ISOPLANE
ISOPLANE Left/Top/Right/<Toggle>:L
Command:ELLIPSE
<Axis endpoint 1>/Center/Isocircle:I
Center of the circle: (pick 1)
<Circle radius>/Diameter: (pick 2)
```

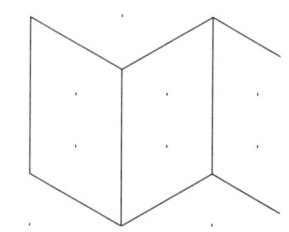

Now TRIM the left side of the circle to the two lines.

```
Command:TRIM
Select cutting edge(s) ...
Select objects: (pick 3)
Select objects: (pick 4)
Select objects:↵
<Select object to trim>/Undo: (pick 5)
<Select object to trim>/Undo:↵
```

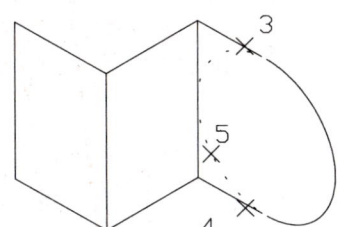

Step 3 Add the circle in the center of the arc, and then copy the arc, the circle, and the line onto the back plane. The dimensions are shown in the figure at the start of this Prelab.

```
Command:COPY
Select objects: (pick the arc, circles,
  and line)
Select objects:↵
<Base point or displacement>/Multiple:
  (pick 6)
Displacement: (pick 7)
```

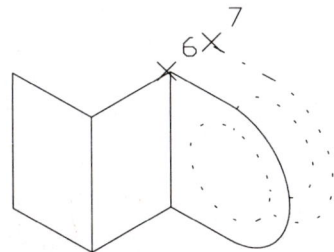

Step 4 The circle on the back plane needs to be trimmed behind the front plane. Remember that these objects are still only in 2D, so the TRIM command will work as it normally does.

```
Command:TRIM
Select cutting edge(s) ...
Select objects: (pick 8)
Select objects:↵
<Select object to trim>/Undo: (pick 9)
<Select object to trim>/Undo:↵
```

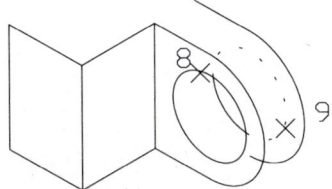

Step 5 With SNAP still on, create a line tangent to the first circle at a length of 1 unit at an angle of 30 degrees.

```
Command:LINE
From point:TAN to (pick 10)
To point:@1<30
To point:↵
```

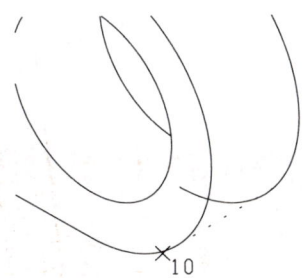

Step 6 Now trim the back arc to the line that you just entered.

```
Command:TRIM
Select cutting edge(s) ...
Select objects: (pick 11)
Select objects:↵
<Select object to trim>/Undo: (pick 12)
<Select object to trim>/Undo:↵
```

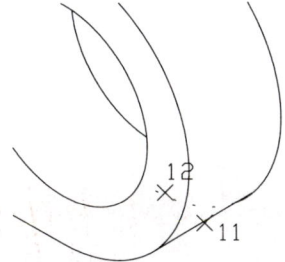

Step 7 Now using the crosshairs and SNAP, add the lines that make up the top and sides of the part.

The line visible through the hole in the part can be put in between two points, then trimmed.

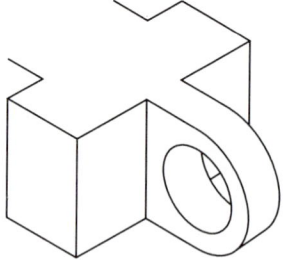

Step 8 Now change ISOPLANE to the top and add the circle along the top as shown.

```
Command: ISOPLANE
ISOPLANE Left/Top/Right/<Toggle>:T

Command: ELLIPSE
<Axis endpoint 1>/Center/Isocircle:I
Center of the circle: (pick the center
   point)
<Circle radius>/Diameter: (pick the end of
   the line)
```

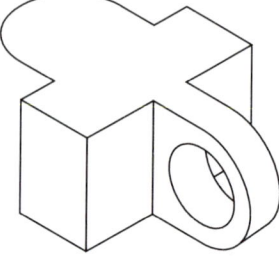

Trim the circle as shown.

Step 9 Now copy the arc and line down, place vertical lines, and trim the part until it is complete.

The vertical lines will be 1 unit in length and can be started tangent to the arcs.

Step 10 SNAP and ISOPLANE are set to isometric, but text will still be placed as circles would be placed, relative to the *X-Y* plane unless the obliquing angle is changed. We used isometric text in Chapter 8.

> **Windows** From the Data menu, choose Text Style.
>
> **DOS** From the Data menu, choose Text Style.

The command line equivalent is **STYLE**.

```
Command:STYLE
Text Style name<Monotxt>:NEG
Font File:CITYBLUEPRINT
Height<0.00>:
Width<1.00>:
Obliquing factor:-30
Backward<N>:
Upside Down<N>:
Vertical<N>:
```

Now add your text using the Fit option.

```
Command:TEXT
Justify/Style/<start point>:F
First point: (pick on the left)
Second point: (pick on the right)
Height<.2000>:.7
Text:Top
```

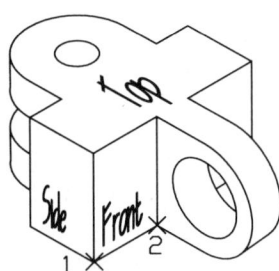

Create another text style called POS for positive, and change the obliquing angle to positive 90. Review Chapter 8 for more details.

When creating isometric text, use one of the fonts that is not already slanted such as Monotext, Standard, or any of the Roman numerals. The obliquing angle is calculated relative to the slant of the side.

Dimensioning isometric drawings can be quite tricky. Create an oblique text font, save it, and make it part of your dimension style. Use TEDIT if necessary.

Creating Orthographic Views

If you are drawing an object that needs to be seen in a number of views, you could — if you used advanced AutoCAD techniques — draw it in 3D, then generate multiple 2D views to fully define it in drawing format. Should you wish to create orthographic views starting in 2D, however, AutoCAD has many tools that can help you.

Traditionally, orthographic drawings include a set of views from mutually orthogonal lines of sight. The three most commonly drawn views are Top, Front, and Right Side as seen at right.

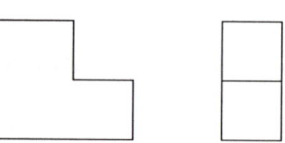

The top view is drawn above the front view because the two views share the dimension of width. The side view is placed beside the front view because the two views share the dimension of height.

In AutoCAD, point filters and layers can make this process simpler.

Using Layers with Construction Lines

You can use construction lines to line up the views. Point filters are quicker, and in many ways preferable, because they make better use of the computer; but until you become comfortable with point filters, use construction lines or ray line options.

Place construction lines on a separate layer so that you can freeze them later. Construction lines transfer the depth by projecting this value from the top view to the side view through a 45 degree angle.

Orthographic Projection — An Example

Here we will illustrate the procedure for determining the side view of this object with just the front and top views.

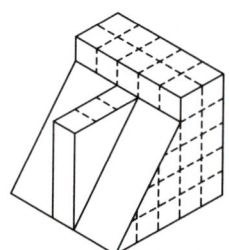

Set the drawing limits to 24 × 18. It is easiest to draw if the SNAP is set to 1 unit and GRID is on. ZOOM All will expand the file to the drawing limits.

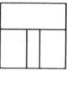

The dimensions can be taken from the isometric view above.

To create the depth for the part, you can use a 45 degree *mitre line* (dotted line) and construction lines (hidden lines) to create construction lines to make sure they line up.

Adding Construction Lines from the Top View

The construction lines extend from the end of the lines on the top view, straight across to the diagonal line. This, in most cases, will not lie on a snap point, so use ORTHO to make the line straight; then make another line from the intersection of the diagonal line and the construction line straight down to line up with the bottom of the front view.

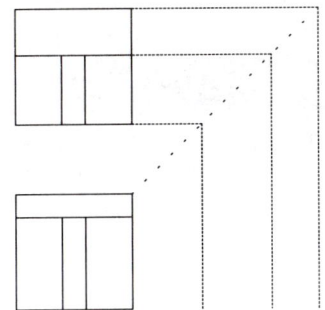

Adding Construction Lines from the Front View

On the front view, extend the construction lines straight across from the vertical points through to the far side of the existing construction lines.

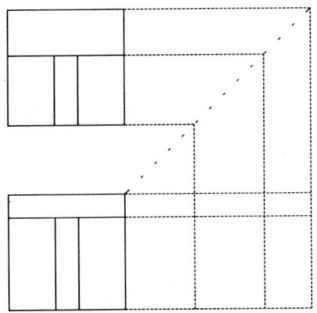

Finally, change your layer back to construction lines and draw in the outline of the part where the construction lines overlap.

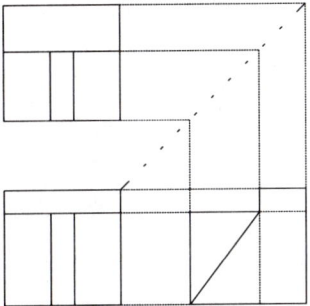

Turn the construction lines off and you will have the right side view.

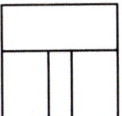

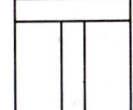
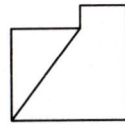

Isometric and Orthographic Drawings   **405**

Notes: Using OFFSET/FILLET and TRIM can be simpler.

You can now change SNAP to Isometric and draw in the isometric view. It is easiest to draw it in at a scale of 1:1. This will be far too large on the final drawing, so once it has been entered, scale the isometric to .66 of the original value.

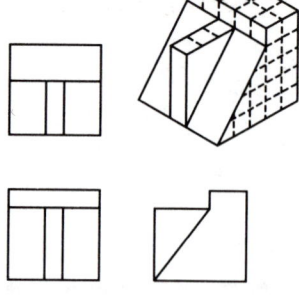

You now have a completed drawing.

Using another version of this rather simple formula, we will now create a top view from a front and side view.

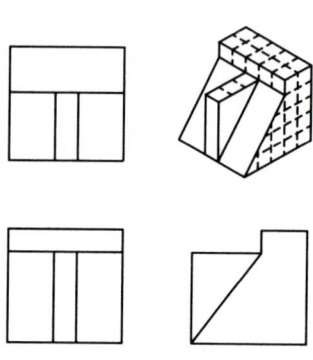

Prelab 16B Orthographic Projection

With the use of GRID and SNAP, orthographic projection of regular shapes is very simple. You may encounter trouble, however, when projecting arcs and ellipses as in the following example.

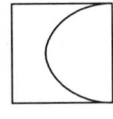

As you can see from the illustration, the arc in the right view is transformed into an ellipse in the Top view. You know the height of the arc from the front view, and the depth of the piece can be standardized with the use of construction lines. But how do you calculate the ellipse?

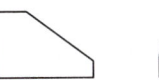

Step 1   First, draw in the part with simply the lines and the circles.

Step 2   Create a new layer for construction lines, and ensure that you change color and/or linetype. Make the new layer current. This is going to get a bit messy, so you might as well make the visualizing of the part as easy as possible.

Project the top horizontal and the right vertical lines on the front view to form a rectangle and draw in a line of 10 units at 45 degrees from the corner point of the right and top views.

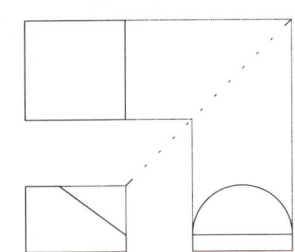

Draw in the rectangle of the outside of the top view using the construction lines, the SNAP, and the crosshairs.

Step 3   Now project a line from the right ENDpoint of the top horizontal line of the front view up through the top view (pick 1 and 2). Then project a line from the MIDpoint of the arc in the side view up through the 45 degree angle line (pick 3 and 4). Draw a horizontal line from the INTersection of the last line drawn (pick 5 and 6) and the 45 degree angle line to the first vertical projection line.

```
Command: (F8 to put ORTHO on)
Command: LINE
From point: END of (pick 1)
To point:  (pick 2)
To point: ↵
Command: LINE
From point: QUAD of (pick 3)
To point:  (pick 4)
To point: ↵
Command: LINE
From point: INT of (pick 5)
To point: PER (Pick 6)
To point: ↵
```

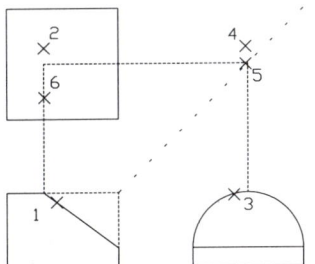

This will create your first point, the left central axis of the ellipse.

Step 4   Now create another point on the ellipse using the same method. Draw a line from a random point (use NEAR) along the side on the sloping surface directly horizontal to the side view. Where this intersects, draw a line vertically to the 45 degree angle mark. Where this intersects, draw a line horizontally back to the Top view. Then draw a line

Isometric and Orthographic Drawings

from the first point vertically. The intersection of these two points will be a correct point on the ellipse.

Make sure your ORTHO is still on.

```
Command:LINE
From point:NEAR (pick 1)
To point: (pick 2)
To point:↵
Command:LINE
From point:INT of (pick 3)
To point: (pick 4)
To point:↵
Command:LINE
From point:INT of (pick 5)
To point: (pick 6)
To point:↵
Command:LINE

From point:INT of (pick 7)
To point: (pick 8)
To point:↵
```

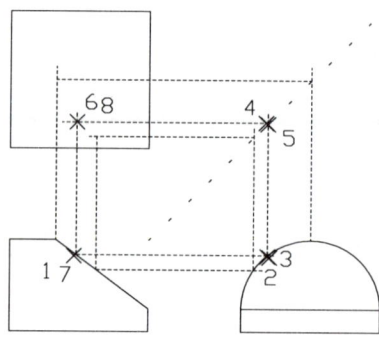

This will give you the second point of your ellipse. Identify another point in the same way.

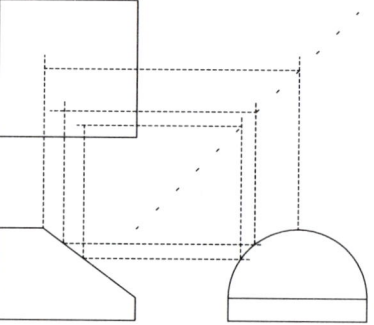

Step 5 Now use the POINT command to create points at the intersection of the lines on the top view. Change your layer to the geometry layer before you start and turn ORTHO off. Setting SNAP to intersection will help.

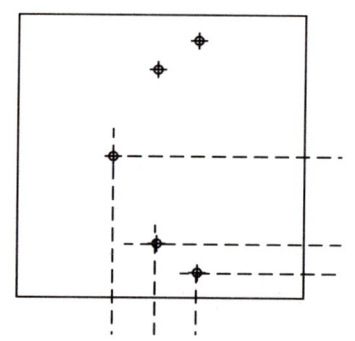

```
Command:OSNAP
Object snap modes:INTersection
Command:POINT (pick the
   intersections)
```

Now MIRROR the points through the midpoint of the vertical line on the right.

Step 6 Finally, use the SPLINE command to create the spline. Use OFFSET with the distance of 2 to create a line, the endpoints of which will be used for the start and end tangent points.

With SPLINE, set Running Object Snap to both NODE and Endpoint for easy access to endpoints.

Your part should be finished.

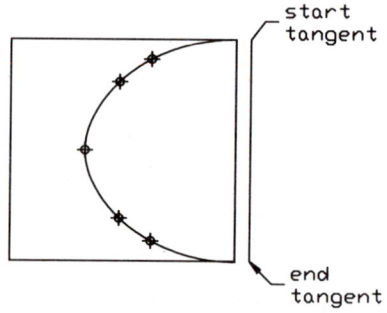

Command and Function Summary

ELLIPSE allows you to draw ellipses according to defined axes, or with the Isocircle option allows you to draw circles on receding planes.

ISOPLANE allows you to access any of the three planes in the SNAP's Isometric mode and set the axes to that plane.

SNAP with Style Isometric allows you to draw in isometrics. GRID, crosshairs, and **ORTHO** will change to isometric mode as well.

STYLE with a change in the obliquing angle allows text entry on a receding plane.

Practice Exercise 16

For the three parts shown, create four views: the front, the top, the side, and the isometric. Be sure to create different layers for construction lines, and project the lines properly. When you are done, dimension.

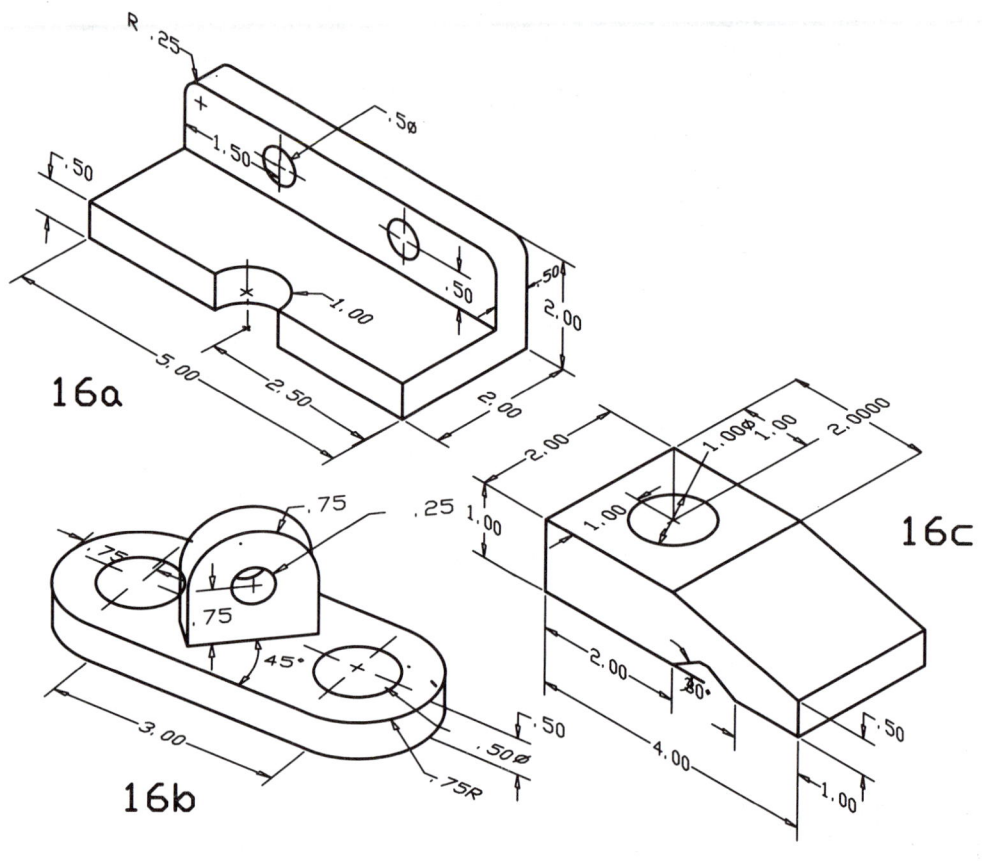

Create side and isometric view

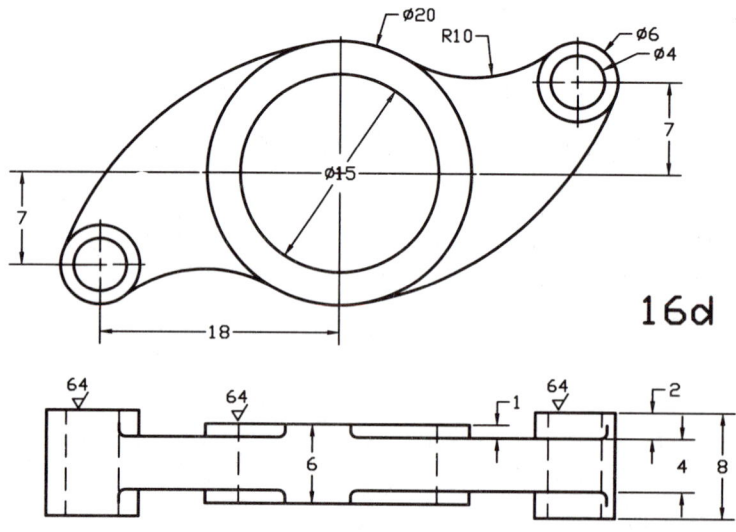

Exercise A16

Use SNAP Isometric and the ELLIPSE command with Isocircle to create this standard stair detail. Make sure that the distance between the ballisters matches current code requirements.

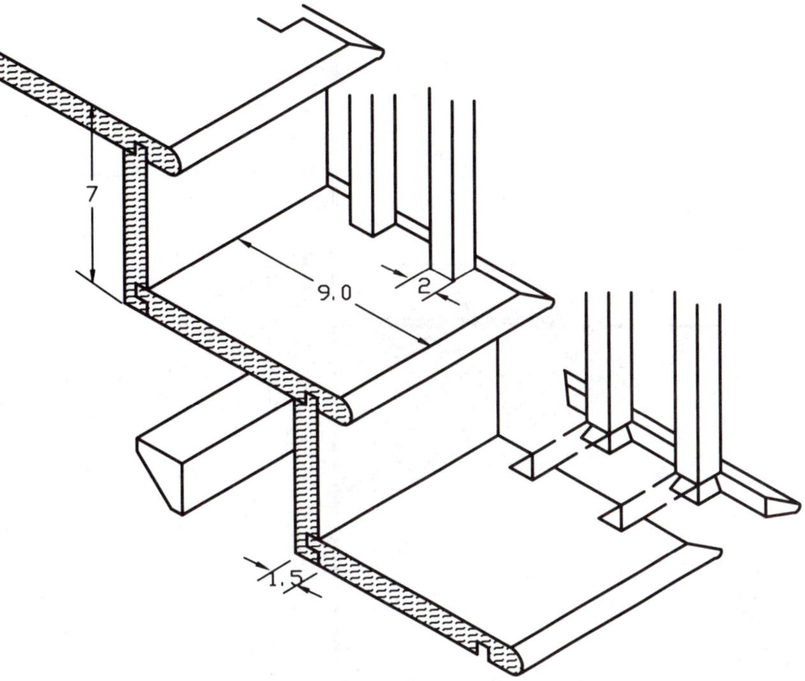

16a

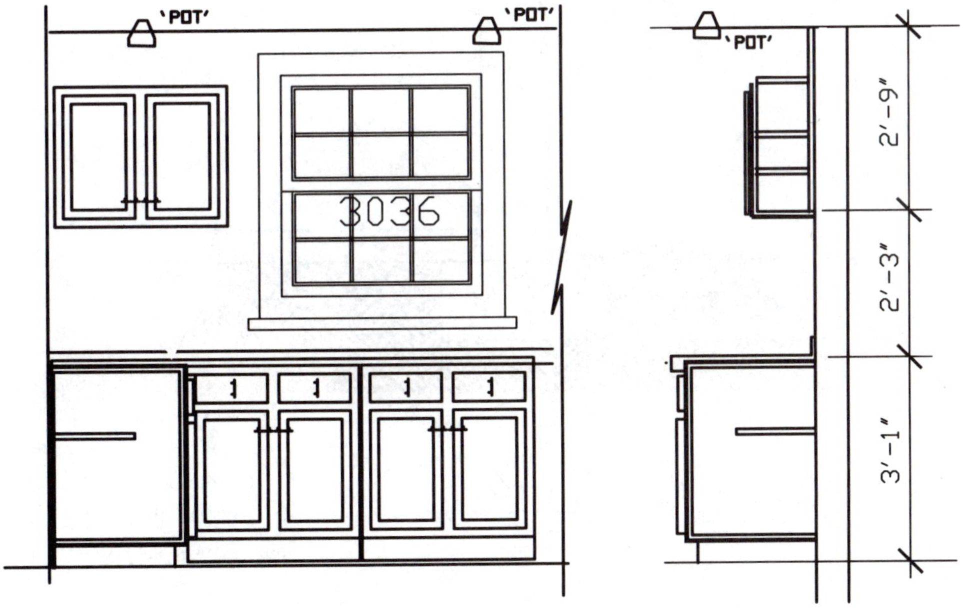

16b   KITCHEN CABINETS

SCALE: 3/8"=1'-0"

Exercise C16

Using this typical wall detail, create an isometric view of the footings of a house 24′ by 28′.

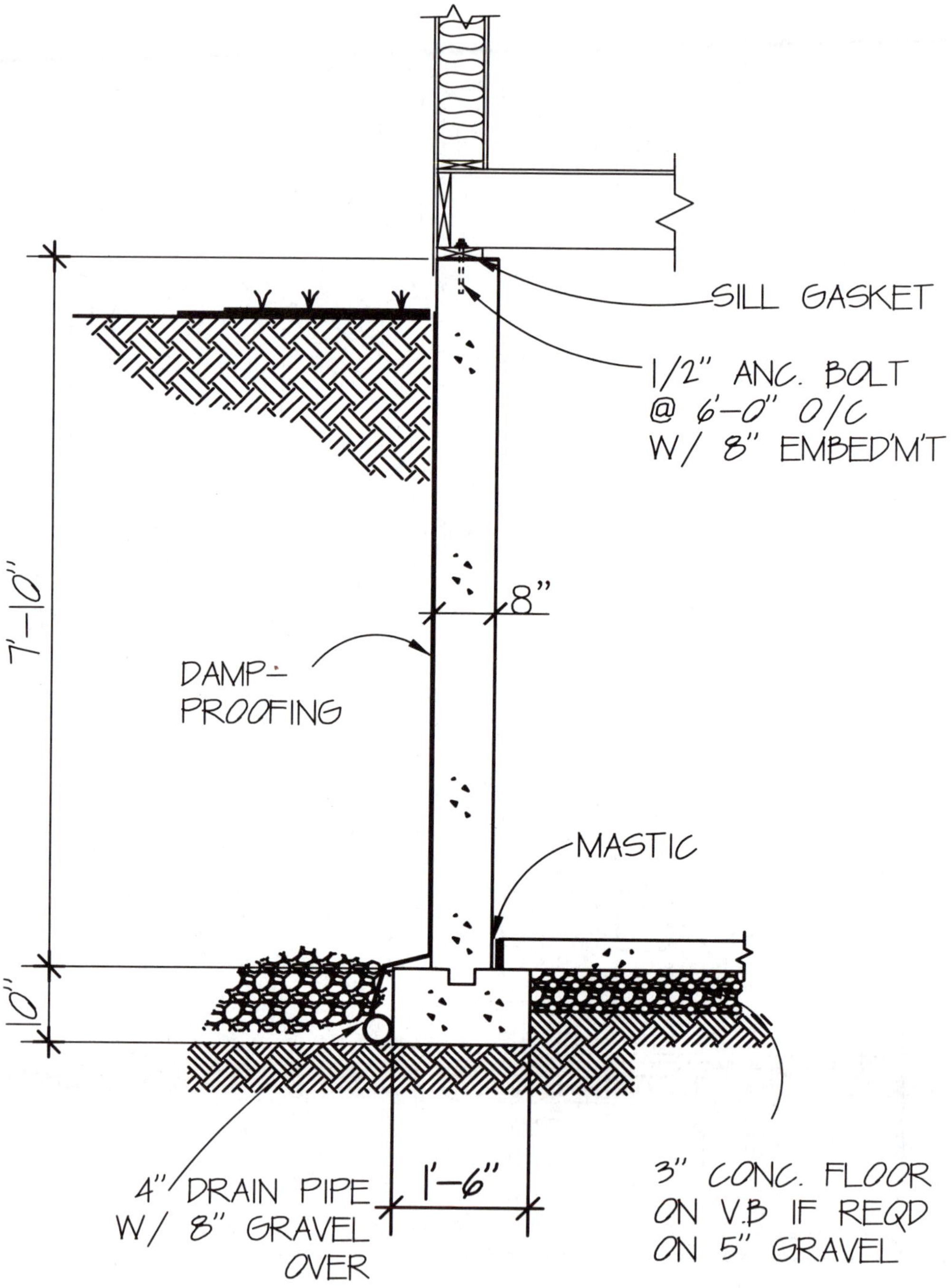

TYPICAL WALL SECTION
SCALE: 1/2"=1'-0"

Exercise M16

Take any or all of these and create the isometric views. Note that the difficult part of this object will, of course, be the fillets. Be sure to use different layers and dimension the part when it is finished.

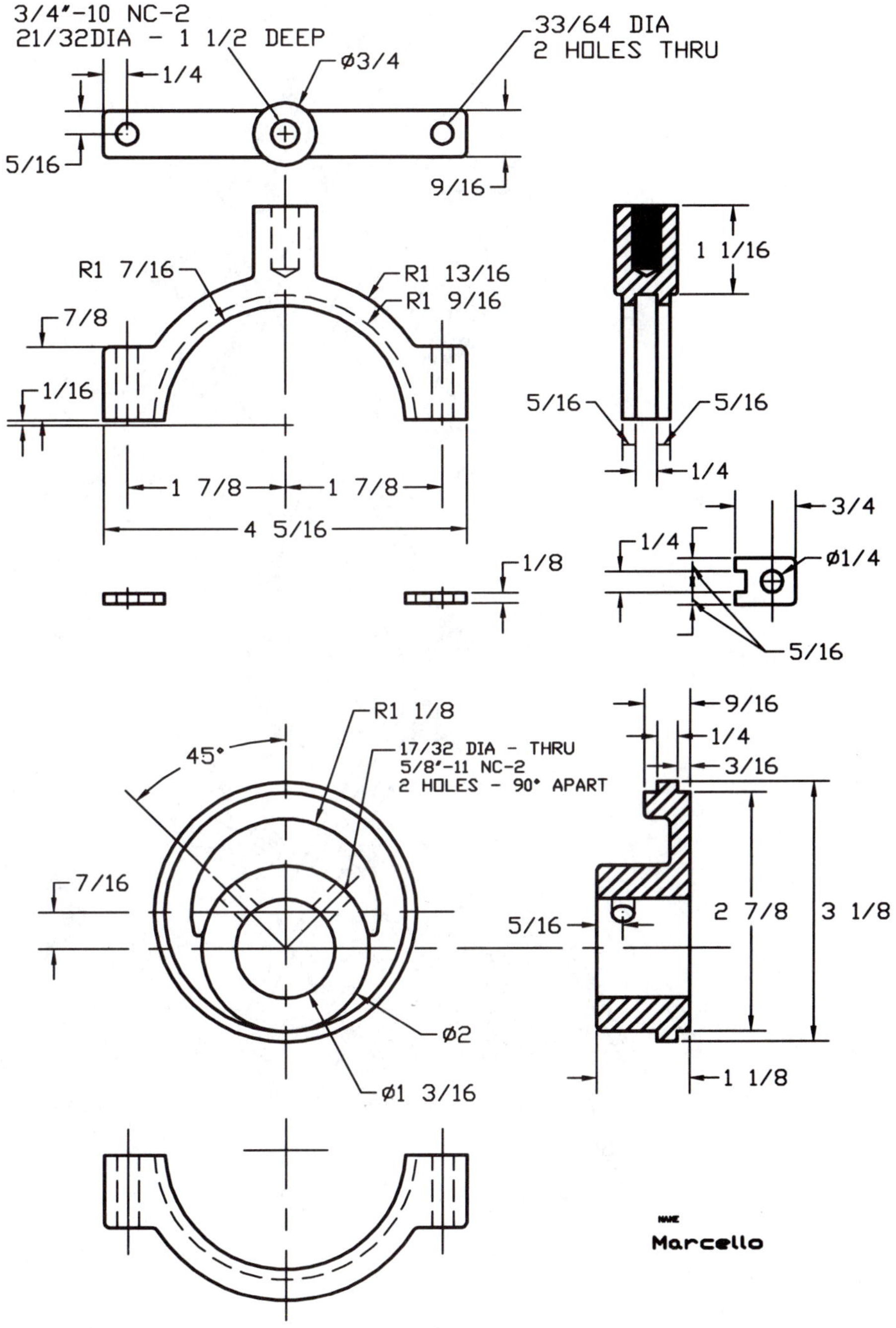

Isometric and Orthographic Drawings 413

Challenger 16

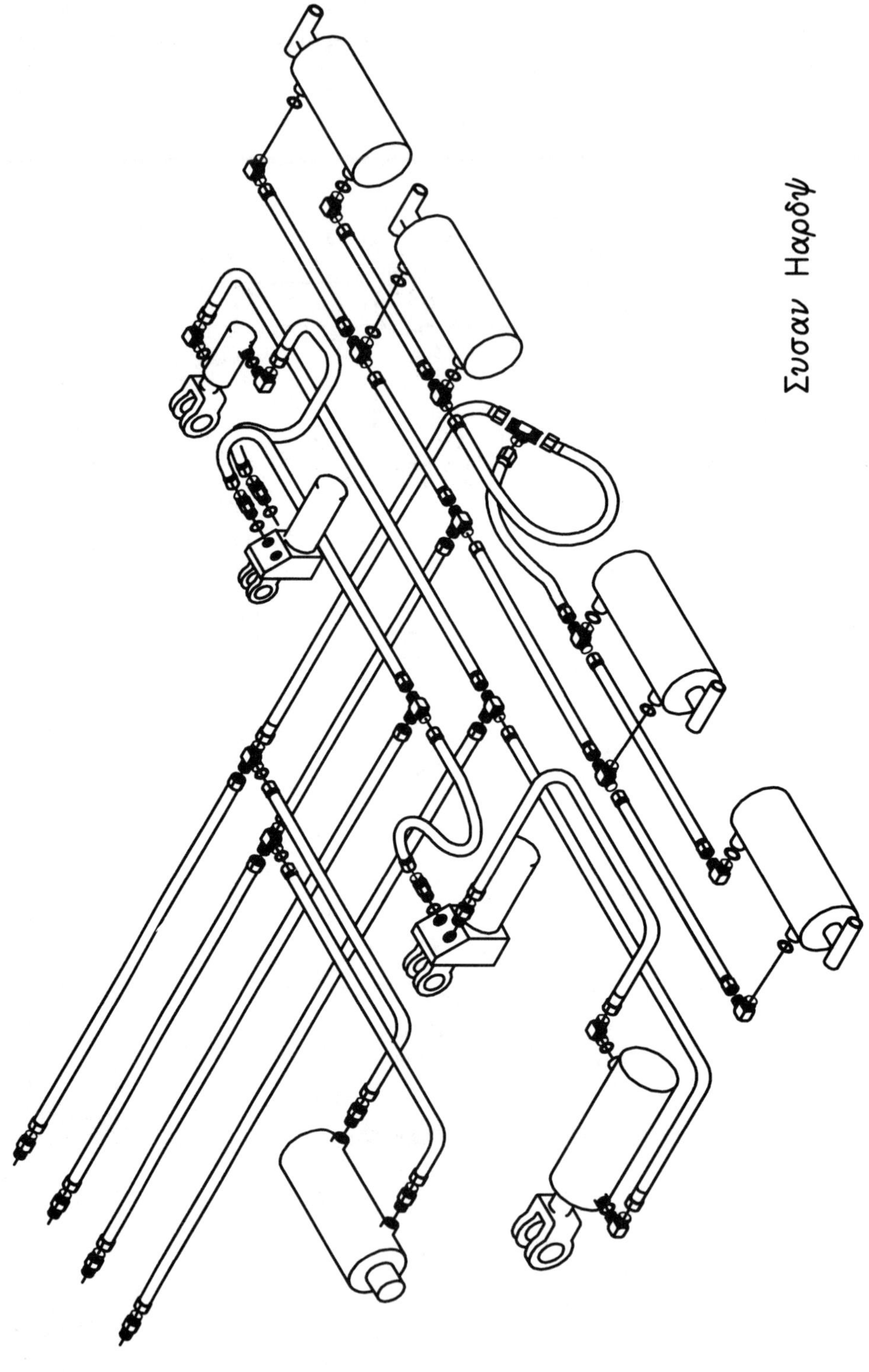

Σνσαν Ηαρδψ

17 File Formats and Management

Upon completion of this chapter, you should be able to:

1. Create slides from graphics data using MSLIDE
2. Access on-line slides using VSLIDE
3. Create a SCRIPT file for viewing slides and other purposes
4. Create files for export to other programs
5. Use the GROUP command
6. Use the FILTER command

What Are Slides?

In order to understand slides and script files, it is important to understand what is happening when you are working on a CAD program.

The first step in creating a CAD model or drawing is to use the available commands — LINE, ARC, CIRCLE — to create a *vector file* — one in which all of the geometry is calculated by vectors. The system relates all entered data to a fixed origin, 0,0,0.

Once the vector file is created, the image must be displayed on the screen. To do this the screen must be able to display the vectors.

Each screen is made up of a series of small, addressable spots called *pixels*. Each pixel can display a certain number of colors, depending on the quality of the screen.

The image is relayed to the screen starting from the top left corner and moving in rows down the screen itself, pixel by pixel. This same process displays the image on a television screen, except video images are not usually vector files. This part of the process is referred to as the pixel address format or bit map, because the image is being addressed to the screen as a series of pixels, not as a series of lines.

- **Step 1** The user enters the commands to create the geometry.
- **Step 2** The input data is made into a vector file.
- **Step 3** The image is relayed to the screen pixel by pixel in a bit map file.

Note also that the on-screen menus are addressed in the same manner, so a large part of the image space is taken up by nonimage data. The menus can be taken off the screen, leaving more room for the image.

Up until now you have been using vector files to create and manipulate images. In certain situations, you do not need to work on the CAD file, you only want to see it. Therefore, it is desirable to have just the pixel address file and not the other parts of the file. AutoCAD has the ability to make slides that can be viewed on the screen by extracting only the pixel address portion of the file for viewing.

Slides are useful for several applications:

- To have for easy access a "view" of part of a project; e.g. when working on an elevation of a building, the floor plans can be instantly accessed for consistency
- To provide a slide show of a particular product to clients
- To provide a slide show of your work to a prospective employer
- To have available on file an easily viewed set of images of your own BLOCKs, linefonts, title blocks, etc.
- To have an instructional tool to tell a story or explain a certain concept
- For exchanging images with other graphics and desktop publishing programs
- For creating icon menus

The MSLIDE Command

The procedure for making slides is very simple, but the concepts behind the command are sometimes difficult at first for those who have not worked extensively with computers.

- **Step 1** The first thing you must do is create the graphics for the slide. This is done in the regular Drawing Editor using standard drawing and editing commands.
- **Step 2** Once the drawing is complete, use ZOOM and PAN to centre the objects on the screen. The slide that you will be making will take up the entire screen; it is not defined by a Window. Be sure you have the view of the object that you want.
- **Step 3** Now make a slide of the geometry using the command MSLIDE. You need only enter a filename for the desired slide file, and AutoCAD will make a slide of the image on the screen.

MSLIDE creates a slide file of the current viewport. In model space, MSLIDE makes a slide file of the current viewport only. In paper space, MSLIDE makes a slide of a paper space display including all viewports and their contents.

> **Windows** and **DOS** From the Tools menu, choose Slide, then Save.

The command line equivalent is **MSLIDE**.

```
Command:MSLIDE
Slide file<>:A:ELEV
```

Enter a filename in the File Name box, or select a slide file (.SLD) from the list.

This will create a slide file (ELEV.SLD) in the A: directory. If you do not specify a specific drive and/or directory, the slide will be created in the same directory as the current drawing.

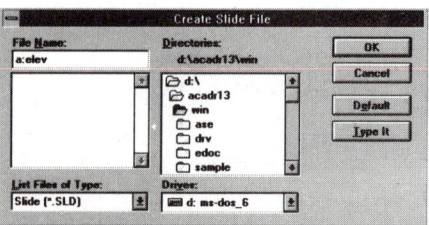

Figure 17-1

```
Command:MSLIDE
Slide file<FIRERATE>:D:\SLIDES\FIRERT1
```

Where: **<FIRERATE>** = the name of the drawing file
       D: = the directory
       \SLIDES = the slides subdirectory
       FIRERT1 = the filename of the slide

Check the directory (**DIR A:**). You should have a file with the extension .SLD; this is your slide file. The extension is added automatically with each MSLIDE command.

The slide is now ready to access, and can be accessed at any time in a drawing file by invoking the command VSLIDE. The slide will temporarily replace the information currently on your screen. You can draw on the slide, but when you change the display with either PAN, ZOOM, or REDRAW, your original drawing will reappear. Any objects added onto the slide will appear added to the current drawing.

> **Danger**
> Be careful when editing the information added on top of a slide, because you will also be editing the current drawing underneath the slide.

Editing Slides

You cannot edit slides. You must edit the original drawing and then recreate the slide.

Slide Resolution

When creating slides, always use a full screen set to the highest resolution. Otherwise, slides made in a smaller viewport or at lower resolution may show black lines.

The VSLIDE Command

You can view slides individually with the VSLIDE command or in sequence with the SCRIPT command. (See next section for the SCRIPT command.)

To view a slide, use VSLIDE with the slide filename.

The command line equivalent is **VSLIDE**.

> **Windows** and **DOS** From the Tools menu, choose Slide, then View.

Select a slide to view and choose OK.

The slide file image appears in the graphics area.

To remove the slide from the view, use REDRAW. Type in **R** at the command prompt, or choose REDRAW from the Standard toolbar in Windows or the View menu in DOS.

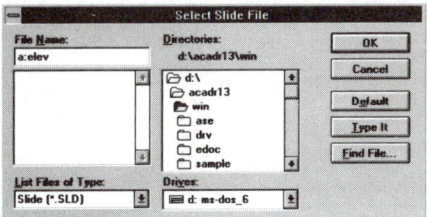

Figure 17-2

```
Command:VSLIDE
Slide file<default>:A:ELEV
```

> **Notes**
> You have not lost your current drawing; a simple REDRAW will bring it back.

A:ELEV will be retrieved over top of your current drawing. The slide will be displayed in the current viewport if more than one viewport is open.

Slides from Slide Libraries

Slide libraries allow you to organize your slides.

In the Select Slide File dialog box, enter a slide file (.SLD file) to display.

To display a slide in a slide library (.SLB file), choose Type it to see the following prompt:

```
Slide file: (enter the library file name and the slide file
   name as: library(slidename))
```

For example, enter GEAR(SECTION) to open the "section" slide which is stored in the "gear" library.

Slide Libraries

Managing slides in a specific slide library is an efficient way to avoid cluttering your disk. The library will have an extension of .SLB, and will contain all of your slides for either slide shows or other functions. To modify a slide in a slide library, you must recreate the slide and also recompile the slide library.

You can create slide libraries from slide files by using the SLIDELIB utility program supplied in the AutoCAD support directory.

At the command prompt, type **SLIDELIB**, then the name that you want to create.

Creating the SLIDE LIBrary — An Example

To create the slide library, follow these steps:

Step 1

First create the geometry for the electrical symbols shown. Make each symbol into a separate wblock, then make a slide of each symbol. Use MSLIDE to create the slides.

Once the blocks and sides are complete, exit to DOS. Find the SLIDELIB.EXE file. This should be under the ACAD\SAMPLE or ACAD\SUPPORT directory.

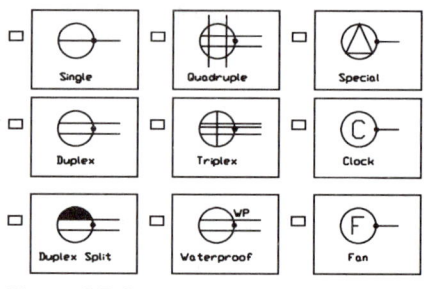

Figure 17-3

Once you have located it, copy it to the directory in which you would like to create your icon file. On the hard drive, create a directory called ICONS. Now copy SLIDELIB.EXE file to this directory.

```
c:\acad>MD\ICONS
c:\acad>CD\ACAD\SUPPORT
c:\acad\support>COPY SLIDELIB.EXE C:\ACAD\ICONS
```

If you are working with the A: drive, use the following:

```
c:\acad>CD\ACAD\SUPPORT
c:\acad\sample>COPY SLIDELIB.EXE A:
```

A subdirectory such as ICONS can be used if desired.

```
a:>MD\ICONS
a:>C:
c:\acad>CD\ACAD\SUPPORT
c:\acad\sample>COPY SLIDELIB.EXE A:ICONS
```

Step 2

Once SLIDELIB.EXE is loaded, you can create a slide library. Copy your slides into this directory. Your directory should contain the following:

```
c:\acad>DIR A:
SLIDE      EXE     24112     07-06-95     12:49a
Single     SLD       547     03-01-95     08:05a
Duplex     SLD       457     03-01-95     08:06a
Triplex    SLD      1138     03-01-95     08:12a
```

etc.

Step 3

The file SLIDELIB.EXE is used to execute a slide library within the directory that it is copied to.

At the DOS prompt, create the slide library using the command SLIDELIB as follows:

```
c:\acad>A:
a:>SLIDELIB ELEC
```

The library file will be called ELEC. After pressing ⏎ you should see the following:

```
SLIDELIB 2.2 (3/8/93)
(c) Copyright 1987-89 Autodesk, Inc.
All rights reserved.
```

Now you must enter the name of the each slide; the extension is not required.

SINGLE ⏎
DUPLEX ⏎
DUPSPL ⏎
TRIPLEX ⏎
SPECIAL ⏎
WEATHER ⏎
QUAD ⏎
CLOCK ⏎
FAN ⏎

To end the sequence:

^Z (use the F6 function) ⏎

Your slide library is now complete and ready to access. It should be on the drive that you copied the .EXE file to.

Script Files

A *script* is a simple text file that contains a set of AutoCAD commands that are executed in succession with a single SCRIPT command. Script files are used to automate or preprogram a process such as viewing slides or running batch plotting.

Creating Script files

Script files are created outside AutoCAD using a text editor, saved in text format, and stored in an external file, like a slide, with the extension .SCR. When invoked, each line of the file is read and executed as if it had been typed in at the command prompt.

As with the template file for ATTEXT, any ASCII-based file can be used to create a script file. The easiest files to use are created with Doskey or Editor in DOS and Notepad in Windows.

Scripts to Run in the Drawing Editor

Scripts can be run from within the AutoCAD graphics editor to enter a setup for a file such as the linetype and linetypescale, units, grid, etc., and many other simple routines.

To create a script file, you must be familiar with both the commands and your responses to provide an appropriate sequence of responses in the script file. Every keystroke, every blank space, is significant. AutoCAD accepts either a space or ⏎ as a command or data file delimiter.

For example, a script file to center a drawing called A:ONE.SCR and generate a plot might look like this:

```
ZOOM     (zoom command)
A        (all = ZOOM All)
ZOOM     (zoom command)
.95x     (makes the view a bit smaller than total screen area)
PLOT     (plot command)
D        (plot display)
N        (want to change anything?)
         (blank line represents a return to start plotting)
```

You must turn off the dialog box before using this. Filedia <1>:0

Once the text file is created, simply run the script.

There are many ways to customize your AutoCAD environment for the projects you will be completing. In many offices, only one standard drawing size is created. If this is true in your case, a prototype drawing is all you need. If there are several different types of drawings, you can create script files that will get you into the desired size and set up for your model or drawing.

For example, if you are working on a part that is 18 × 14, you may find a script file called A2.SCR useful. It could look like this:

```
UNITS     (units command)
2         (decimal linear units)
2         (two decimal places)
1         (decimal angular units)
          (space to accept angle default)
          (measure angles counterclockwise)
LIMITS    (limits command)
-1,-1     (lower limit)
20,16     (upper limit)
ZOOM      (zoom command)
A         (all)
GRID      (grid command)
2         (two unit spacing for grid)
SNAP      (snap command)
.5        (.5 unit snap spacing)
```

This can be called in as soon as you enter the Drawing Editor. It will give you all the sizes you may need to start the project.

A script file can easily take the place of a prototype drawing file if you find this method more suitable to your AutoCAD environment.

The script file can be very useful for creating standardized plots with similar layer/color and layer/linetype adjustments.

The SCRIPT Command

This command invokes the text file created as a script and executes the sequence of commands.

> **Windows** and **DOS** From the Tools menu, choose Run Script.

The command line equivalent is **SCRIPT**.

Enter the filename without the extension in the Select Script File dialog box or at the command prompt.

Scripts for Slide Shows

In creating a script file for the slide show, you will be preprogramming a series of VSLIDE commands that will allow you to view a series of slides without touching the keyboard. This is a popular technique for trade shows, etc., because it gives the impression that many drawing files are being accessed, one by one, while the station is unattended. The commands needed to perform the slide show are as follows. Slides 1 and 2 are precreated slides.

```
VSLIDE SLIDE1    (begins the slide show with slide 1)
VSLIDE *SLIDE2   (preloads SLIDE2)
DELAY 1000       (delays 1 sec for viewing (2000 = 2 sec, etc.))
VSLIDE           (displays the preloaded slide)
RSCRIPT          (will repeat the entire program)
```

Stopping the Script File

Ctrl-C (**^C**) in DOS, Esc in Windows will stop the script file. A backspace will temporarily halt it. Use **RESUME** to continue if the script file has been halted.

Invoking a Script While Loading AutoCAD

Scripts can be used from within the drawing editor and also when you start AutoCAD using a special form of the ACAD command. To invoke a script when you start AutoCAD, use the following command form:

```
drive>ACAD DRAWING-NAME SCRIPT-FILE
```

ACAD is your sign-on code. This may be different with regard to your system setup, e.g. ACAD13, ACAD13T, etc.

The script file must be the second file named on the ACAD program call line; a file type of .SCR is assumed. If AutoCAD can't find the script file, it reports that it can't open the file.

This format is useful for creating setups with grid, ltscale, snap, layer settings, and system variables.

Exporting AutoCAD Files

Word processing and desktop publishing programs can import AutoCAD files. You can also import raster images created by scanners, paint programs, and other applications into an AutoCAD drawing.

Exporting in Windows

One major advantage with Windows is the multitasking facility. To take a drawing image from AutoCAD to a word processing format, use Cut and Paste.

1. Open the drawing file in AutoCAD.
2. Under the Edit menu, choose Cut and identify the image to be exported.
3. Under the File menu, choose Minimize.
4. Open your word processing or desktop publishing document, position cursor at desired spot.
5. Under the Edit menu, use Paste to place the image.

In Windows, .SLD, .PLT, and other formats are not necessary.

Slides

Some programs will import .SLD files. Within the figure box, graphics box, or user box, try importing the .SLD file. If this is not compatible, try using .PLT files or .HPG files. (The latter extension indicates HPGL or Hewlett Packard Graphics Language files.)

When generating the graphic in AutoCAD, be careful that the sizing of text and dimensions will allow viewing on an 8 1/2" × 11" sheet. Generally, the space given for graphics is about 6" × 6". If your text is barely readable on the screen at 8 1/2" × 11", there is no chance you will see it on a WordPerfect sheet.

Plot Files or HPGL Files

A .PLT or .HPG file can be imported into many software packages. Simply generate a plot file (A:NAME.PLT) for the graphics by using the PLOT command. When in the PLOT command, make sure that you:

- Plot to a file
- Set the driver to an HP driver

Graphics must be generated in a Hewlett Packard plot-file format. Configure the plot for a Hewlett Packard plotter through the Device and Default Selection ... of the Plot Configuration menu.

The HP.DRV (plotter driver file) is standard on any AutoCAD release. If you need to reconfigure to add the plotter driver, use the CONFIG command and add any standard HP plotter driver (the HP 7420 for example) to the plotter options.

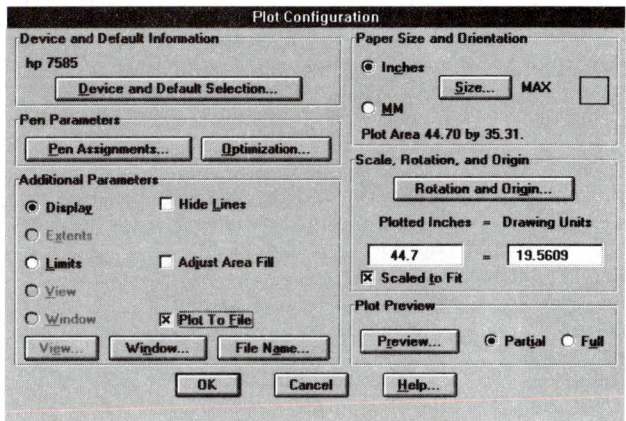

Figure 17-4

Use the PLOT command in AutoCAD to select the drawing or portion of a drawing to be exported. Make sure the plot is scaled to Fit; otherwise your graphics may be much too large.

Importing Files to AutoCAD

Raster images can be imported into AutoCAD. These images are not vector images and cannot be edited like a drawing file, but they can be useful for tracing or for viewing purposes. Because imported raster files can create large file formats, you should erase the raster image when you no longer need it.

AutoCAD can import the formats GIF (graphics interchange format), PCX, and TIFF (tagged image file format). To start the import, use the following:

> **Windows** At the command prompt, enter **GIFIN**, **PCIN**, or **TIFFIN**.
>
> **DOS** From the File menu, choose Import. Then choose Raster, then GIF, PCX, or TIFF.

Enter the path and name of file that you want to import.

AutoCAD displays the raster image. Drag the image into position, then specify the scale.

The command line equivalents are **GIFIN**, **PCXIN**, and **TIFFIN**.

Prelab 17A Creating Slides and Scripts

Step 1 This splatter file of a young girl blowing a bubble is very simple, but it should illustrate a sequence of events. It also takes up a very small amount of space, so it can be useful in testing out slides and scripts without using up too much time and disk space. If this is too violent for your liking, try another cartoon sequence, making sure that you have a distinct difference between each slide. (Use color 14 for grape bubble gum.)

You can either create this series of drawings and create slides of them, generate a similar image or set of images, or use a series of completed drawings to create a slide show of your drawings to date for a perspective employer or client.

Step 2

Create the first view using arcs. Copy the view over and make minor adjustments.

Copy both of the existing files down, and create a sequence of events.

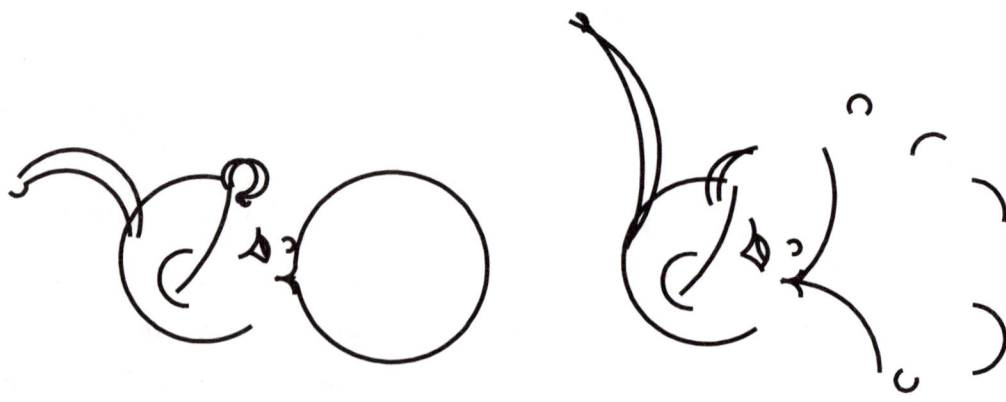

424 CHAPTER SEVENTEEN

Step 3 Now use ZOOM Window to bring each view separately onto the screen, then use MSLIDE to create a slide of each image in sequence. Be careful to number the slides in the proper sequence.

> **Windows** and **DOS** From the Tools menu, choose Slide, then Save.

The command line equivalent is **MSLIDE**.

```
Command:MSLIDE
Slide file<firerate>:A:BUBBLE1
```

You should have six slide files when this is completed.

Step 4 Now create a script file in the DOS Editor or Windows Notepad.

```
VSLIDE A:BUBBLE1
VSLIDE *A:BUBBLE2
DELAY 1000
VSLIDE
VSLIDE *A:BUBBLE3
DELAY 1000
VSLIDE
VSLIDE *A:BUBBLE4
DELAY 1000
VSLIDE
VSLIDE *A:BUBBLE5
DELAY 1000
VSLIDE
VSLIDE *A:BUBBLE6
DELAY 1000
VSLIDE
VSLIDE *A:BUBBLE1
RSCRIPT
Save the File as A:Blowup.SCR
Exit the Editor
```

Step 5 The command to load the script is SCRIPT; no extension is needed.

> **Windows** and **DOS** From the Tools menu, choose Run Script.

The command line equivalent is **SCRIPT**.

```
Command:Script
Script filename:A:Blowup
```

Managing Larger Files

The organization of a larger model can be a very difficult task even with the use of layers and blocks. AutoCAD has two other ways of organizing files: Groups and Filters.

Layers

There are two controlling factors that are utilized with the Layer function; the first indicates which layer is being drawn on, the second indicates which layers are being displayed and edited. Each object belongs to only one layer. Objects can also be editable or not editable by locking or unlocking the layers.

Filters

Filters allow the user to create a list of properties required of an object for it to be selected. Effectively, it acts like the Lock function of Layering in that it limits selectability of objects by property, such as color or linetype, or by object type, such as arc or polyline.

Filter identifies objects by linetype or color only when these properties have been set up independent of their layer. Filtering by object type is set up by Object Type and relative parameters such as Diameter.

Filter lists, like Dimension Styles and Multiline Styles, can be saved in a file and used repeatedly.

The FILTER Command

Filter is used at the command prompt to set up lists for later use at the Select Objects prompt. It can also be set up, like Layer Locking, within the editing command from the dialog box.

> **Windows** From the standard toolbar, choose the Select Objects flyout, then
>
> **DOS** From the Assist menu, choose Selection Filters.

The command line equivalent is **FILTER**.

The dialog box offers many of the same options as other naming lists.

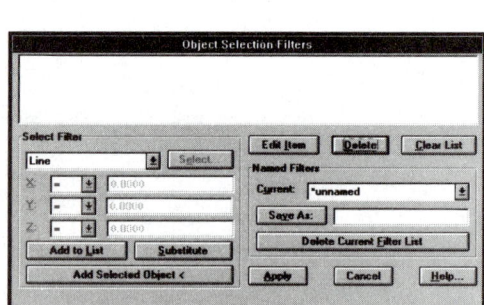

Delete	erases the selected filter from the list.
Clear List	erases the current filter list.
Named Filters	displays, saves, and deletes filter lists.
Edit Item	moves the selected filter into the Select Filter area for editing. To change a filter, select it and choose Edit item. Edit the filter and values and choose Substitute. The edited filter replaces the selected filter.
Select Filter	adds filters to the current list based on object properties. Additional parameter values can be added.

Relational Operators	Operator	Meaning
	<	Less than
	<=	Less than or equal to
	>	Greater than
	>=	Greater than or equal to
	=	Equal to
	!=	Not equal to
	*	Equal to any value

For example, you can have all circles selectable, or just those circles that have a diameter greater than 3, and are on the part of the file that is less than 25 in *X* and 25 in *Y*. Use the following relational operators to set your object filter.

```
Object = Circle
Circle Center   X<=25.00   Y<=25.00   Z<=0.00
Circle Radius  >=3.00
```

Grouping Operations selects filter sets by operands
Select displays a dialog box listing all items of the specified type in the drawing
Add to List adds the current Select Filter option to the filter list
Substitute replaces the selected filter with the one in Select Filter
Add Selected Object allows you to select one object in the drawing and add to the filter list

Groups

A group can be defined simply as a stored selection set. Objects within the model or drawing can belong to several groups, but only one layer. Groups are selection sets that are stored with the model.

Storing objects in groups allows the user to edit a large number of related objects while only selecting them once. The display is not affected by groups. An object can be a member of more than one group.

By picking one object in the group, the whole group is selected. Objects in locked layers are not edited with the other members of the selected group.

The GROUP Command

This creates a named selection set of objects.

Windows From the Standard toolbar, choose

DOS From the Assist menu, choose Group Objects.

The command line equivalent is **GROUP**.

```
Command:GROUP
?/Order/Add/Remove/Explode/REName/Selectable/<Create>:
```

Group Name	displays the names of existing groups
Selectable	indicates whether a selected object will be edited as an individual object or as a group. If the group is Selectable, the object selection pick will select the whole group
Group Indentification	when a group is selected in the group name list, the group's name and optional description appear in the Group Identification area
Group Name	displays the name of the selected group
Description	displays the description of the selected group
Find Name	lists the groups to which an object belongs
Highlight	shows the members of the selected group in the graphics area
Include Unnamed	specifies whether unnamed groups are listed

Create Group

New	creates a new group from the objects you have selected using the name in the Group Name text box
Selectable	specifies whether a group is selectable
Unnamed	indicates that you will accept a sequentially numbered identification for the group

Change Group

Remove	removes objects from the selection set
Add	adds objects to the selected group
Rename	allows the user to rename the group
Re-order	changes the numberical order of objects within the selected group. This is used for tool path generation. This option will invoke the order group dialog box

PICKSTYLE

If the PICKSTYLE system variable is set to 1 or 3, the objects in the group will be selected as a group. If the PICKSTYLE system variable is set to 0, the objects will be selected as separate objects.

Prelab 17B Drawing with Layers and Groups

In this example we will draw a bicycle wheel in three different layers, and save it as a group. Then we will draw in the rest of the bicycle, copy the wheel over as a group, and change just the circles to the hidden linetype.

Step 1 Draw in the wheel as shown. Make the tube of the wheel in layer Tube, and the spokes of the wheel in layer Metal.

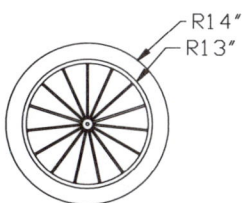

Step 2 Open the Create Group dialog box.

Under Group Identification, enter the word, wheel.

Under Select Group, specify that the group should be selectable.

Choose New.

In the drawing you will be prompted to select the objects. Pick the objects and press OK.

In the Object Group Dialog box, choose OK.

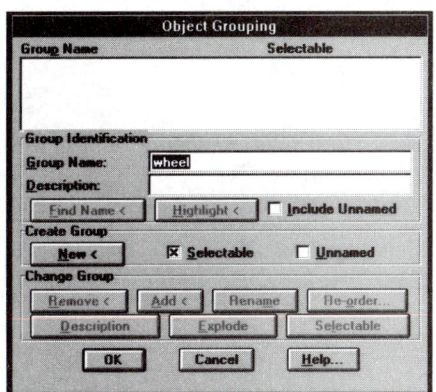

Step 3 At the command prompt, make sure that your PICKSTYLE is set to 1.

```
Command:PICKSTYLE
New value for PICKSTYLE<0>:1
```

Step 4 Draw in the remainder of the lines in layer Metal. Leave enough room for at least one wheel's width between the two wheel positions.

Step 5 Use the COPY command to copy the wheel from one end of the bicycle to the other.

```
Command:COPY
Select objects: (pick 1)
Select objects:⏎
<Base point>/or Multiple: etc.
```

Notice that the pick point picks the entire wheel, both layers.

Step 6 The circles and spokes are on group wheel. Now filter out only the circles on the drawing, and use Change Properties to change the linetype to hidden. First, you must change the PICKSTYLE system variable to 0 in order to have the individual objects selected.

```
Command:PICKSTYLE
New value for PICKSTYLE<0>:1
```

Step 7 Now invoke the CHPROP command, and use the Object Selection Filters dialog box to identify only the circles.

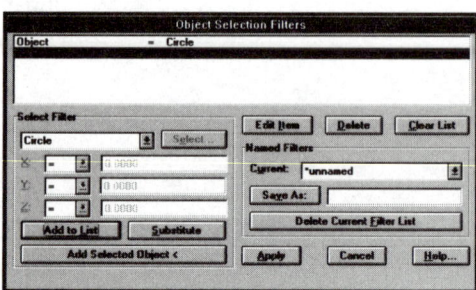

```
Command:CHPROP
Select objects:
```

Under Select Filter, select Circle.

Select Add to List.

Choose Apply. AutoCAD applies the filter list to whatever you now select.

```
Select objects: (pick a window around the entire bicycle)
146 found
136 were filtered out
Select objects:⏎
Select objects:
Exiting filtered selection. 10 found.
Change what property (Color/LAyer/LType/
   ltScale/Thickness):LT
New Linetype:HIDDEN
Change what property (Color/LAyer/LType/ltScale/Thickness):⏎
```

Is the new wheel considered a group with PICKSTYLE set to 1?

Command and Function Summary

FILTER creates lists to select objects based on properties.

GROUP creates a named selection set of objects.

MSLIDE creates a slide file of the current viewport.

SCRIPT executes a sequence of commands from a script.

SLIDELIB creates a slide library for use in icon menus and script files.

Text files are used to compile a series of AutoCAD commands that can be accessed through **SCRIPT**.

VSLIDE displays a raster image slide file in the current viewport.

Exercise A17

Attributed block
Building Section
BLOCK NAME: ATTBS

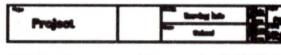

Attributed block
Horizontal Title Block
BLOCK NAME: ATTHT

Attributed block
Wall Section
BLOCK NAME: ATTWS

Block
North Arrow
BLOCK NAME: ARROW

Attributed block
Door
BLOCK NAME: ATTDOOR

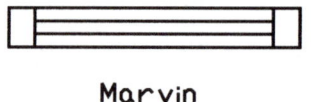

Attributed block
Window
BLOCK NAME: ATTWINDO

Part 1   Take six attributed wblock files. (Create them quickly if you do not have six on file.) Insert them onto a file and add notations as shown in the example. Zoom each one onto the screen individually, and create a slide using MSLIDE for each individual file. You can make the attributes invisible in the block.

Part 2   Generate a script file using either Notebook or the DOS Editor. Save the .SCR file.

Part 3   Use SCRIPT to view the slide show of your blocks.

Exercise C17

Attributed block
Road Section
BLOCK NAME: ATTRS

Attributed block
Horizontal Title Block
BLOCK NAME: ATTHT

Attributed block
Curb Cut Approach
BLOCK NAME: ATTCCA

Block
North Arrow
BLOCK NAME: ARROW

Attributed block
Turn Arrow
BLOCK NAME: ATTTARR

Attributed block
Stop Light
BLCOK NAME: ATTSTOP

Part 1   Take six attributed wblock files. (Create them quickly if you do not have six on file.) Insert them onto a file and add notations as shown in the example. Zoom each one onto the screen individually, and create a slide using MSLIDE for each individual file. You can make the attributes invisible on the block itself.

Part 2   Generate a script file using either Notebook or the DOS Editor. Save the .SCR file.

Part 3   Use SCRIPT to view the slide show of your blocks.

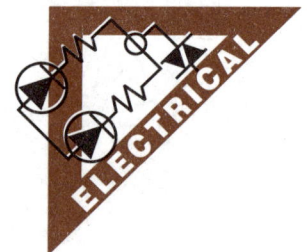

Exercise E17

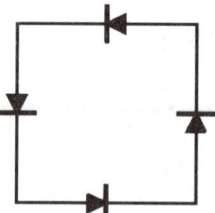

Attributed block
Rectifier
Full Wave Bridge
BLOCK NAME: FWB

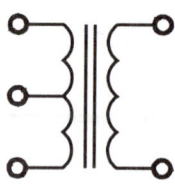

Attributed block
Transformer
Magnetic Core/Single
BLOCK NAME: MCST

Attributed block
Connecting
Dots
BLOCK NAME: DOTS

Attributed block
Connecting
Dots
BLOCK NAME: DOTS

Attributed block
Variable
Resistor
BLOCK NAME: VR

Attributed block
Switch
Single Pole
BLOCK NAME: SPSWITCH

Part 1 Take six attributed wblock files. (Create them quickly if you do not have six on file). Insert them onto a file and add notations as shown in the example. Zoom each one onto the screen individually, and create a slide using MSLIDE for each individual file. You can make the attributes invisible on the block itself.

Part 2 Generate a script file using either Notebook or the DOS Editor. Save the .SCR file.

Part 3 Use SCRIPT to view the slide show of your blocks.

Exercise M17

Attributed block
Section Arrow
BLOCK NAME: ATTSA

Attributed block
Horizontal Title Block
BLOCK NAME: ATTHT

Attributed block
Spring set
BLOCK NAME: ATTSS

Attributed block
Bolt
BLOCK NAME: ATTBOLT

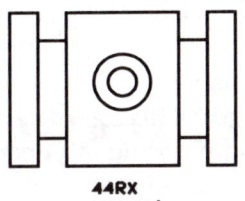

Slotted
.23xx

Attributed block
Centre Pin
BLOCK NAME: ATTPIN

Attributed block
Slotted
BLOCK NAME: ATTSLOT

Part 1   Take six attributed wblock files. (Create them quickly if you do not have six on file). Insert them onto a file and add notations as shown in the example. Zoom each one onto the screen individually, and create a slide using MSLIDE for each individual file. You can make the attributes invisible on the block itself.

Part 2   Generate a script file using either Notebook or the DOS Editor. Save the .SCR file.

Part 3   Use SCRIPT to view the slide show of your blocks.

Challenger 17

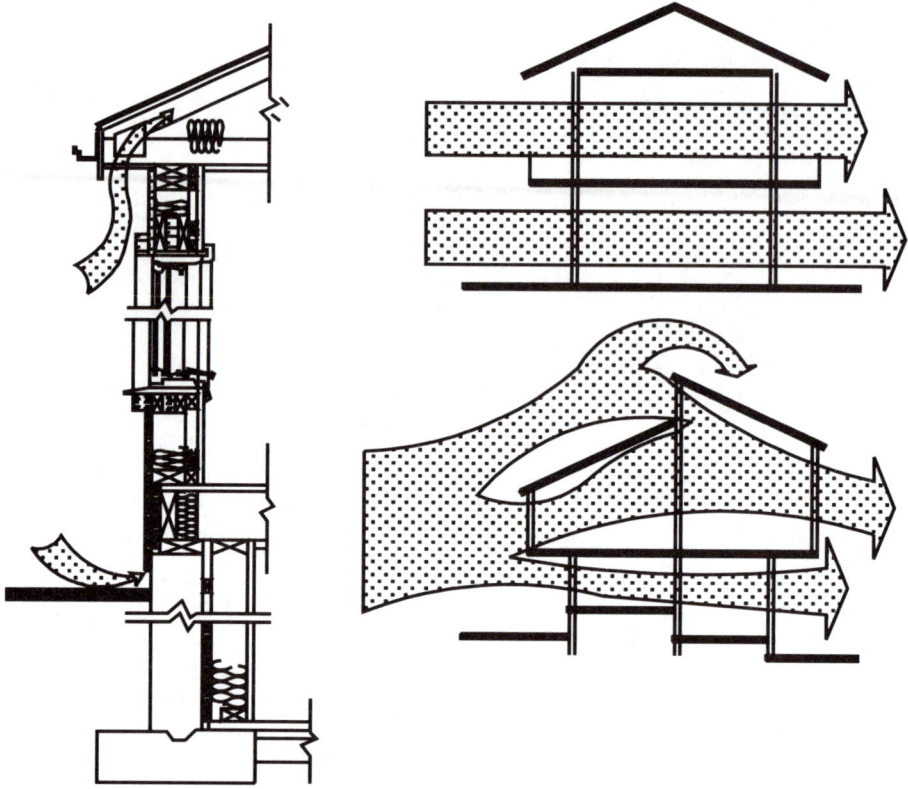

In this drawing of ventilation patterns, we can see how the first view illustrates an insulated house, and the ventilation going into it. The two houses demonstrate ventilation patterns through windows, doors, etc. Using a condition that exists in your discipline, generate a group of views that develop a certain thought process or project and create a slide show for presentation.

- Mechanical
    1. Movements of a robotic arm
    2. Drilling a piece of metal
    3. Turning gears
- Architectural
    1. Different framing techniques
    2. Movement of a window
    3. Renovation of a kitchen
- Civil
    1. Traffic flow problems
    2. Cut and fill before and after
    3. Installation of services in an area

18 Advanced Blocking and Xrefs

Upon completion of this chapter, you should be able to:

OBJECTIVES

1. Create nested blocks
2. Redefine BLOCKs
3. Substitute BLOCKs
4. Use XREFs to place BLOCKs
5. Use XBIND

Advanced Blocking

In addition to being used as an assembly tool and as a format for generating attributes, blocks can be compiled for further efficiency of design. One block can contain any number of other blocks so that it can be edited as a group of blocks instead of just a single block object.

Nesting Blocks

Nested blocks are blocks within blocks, or several interconnected blocks. An example would be a workstation in an office layout. In a large company there would certainly be some standards of what desk, storage, and chair arrangements could be used for each position. If each component of the workstation were stored as a separate block, the whole workstation could also be stored as a block and inserted separately as well.

In Figure 18-1, each individual item could be blocked, with or without attribute information, and the whole unit could be blocked for easy insertion into the office layout. Inserting one workstation would obviously be much easier than inserting all of the individual blocks.

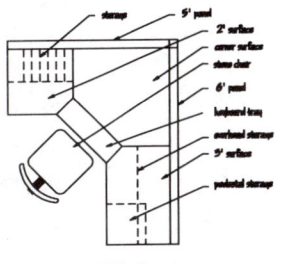

Figure 18-1

The blocks are created individually; then they are blocked as a unit to be part of a single large block. The large block is then inserted as one unit.

Layering, Color, and Linetype with Blocks

When you insert a block into a file, all of the layers associated with the block are inserted as well.

Each object of the block is stored on the layer that it was assigned. If you have planned correctly, you should have control of the objects that are inserted, and the color and linetype of each object should be the same as when it was created. Usually, the color and linetype will be set BYLAYER, but a COLOR or LINETYPE setting will override these.

All objects in layer 0 will take on the current color and linetype, unless they are entered using *BLOCKNAME or exploded. Objects will default to layer 0 when exploded.

When inserting blocks, it is often difficult to determine how best to set up and view the objects. In the example above, two or three manufacturers could be supplying the furnishings for the workstation. You may have the layers set up so that each manufacturer has its own layer. When inserting the workstations, however, you may wish to assign them a color to represent a department. On one floor there could be 12 secretaries representing 4 different departments, and you may wish to have the colors coordinate so that the departments, not the furniture manufacturers, are easily distinguished. If this is the case, the color of the nested block can be set BYBLOCK, and the current layer can be changed while inserting the 12 workstations so that the blocks, when inserted, will take on the color of the department. The layer for each department must be set up, and the BLOCK must be inserted while that layer is current. The block's color and linetype can be changed after it has been inserted by using CHPROP, and the color or linetype can be set BYBLOCK.

If the color and linetype objects are set BYBLOCK, they will take on the settings of the insertion layer at the time of insertion. Remember, if you insert a block on a layer that is frozen, the block will not be visible.

Block Redefinition

Once a block is inserted, you may encounter design changes that will make you want to substitute a different block at the same insertion base point. In the example are shown a coniferous tree symbol on the left and a deciduous on the right. If both trees are blocks, it is possible to substitute one type of tree with another on a layout.

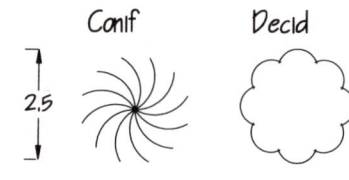

Figure 18-2

Both blocks must exist; then the equal sign (=) can be used to assign one block to another block's position.

```
Command: INSERT
BLOCK name (or ?)<LAST>: CONIF=DECID
BLOCK conif redefined
Regenerating drawing
Insertion point: ^C (to cancel)
```

The previous command has assigned a different block to each of the reference points on the blocks called CONIF. The graphic information of the block DECID will replace the graphic information of the block CONIF.

This function can be extremely useful when working on a large drawing or assembly where one of the components which has been entered as an external block has been updated or replaced.

If you have created an office layout with 125 chairs and management decides to change the manufacturer, simply redefine the blocks.

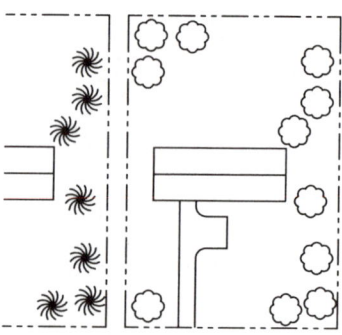

Figure 18-3

If you are working on an assembly with 150 components of the same type, simply redefine them if the component is upgraded.

Finally, if you are working on a large drawing that has a nested block or a large block with many objects, create a simple block, such as a rectangle, and substitute it for the large block to reduce visual clutter and improve redrawing speed. In this case you are redefining the block in order to substitute a simpler one temporarily. Use INSERT with the simple name equalling the complex block name.

```
Command: INSERT
BLOCK name (or ?)<CAR>: COMPLEX=SIMPLE
BLOCK complex redefined
Regenerating drawing
Insertion point: ^C
```

Remember to replace the simple block with the complex block before plotting.

Xrefs or External Reference Files

When you insert a block to a file, you are, in fact, adding all of the objects of the block into the current file. When you add an XREF to a file, you are bringing the latest version of a current drawing into your drawing as an external reference. The objects are not actually added to the file. They are referenced for the duration of your editing time and then released. The xrefs are reloaded every time the drawing is loaded. Xrefs don't significantly increase the size of the drawing, because they are not part of the drawing database.

The great advantage of having xref files as opposed to blocks is that the xref files are separate drawing files and thus are subject to change. Every time a file containing an xref is retrieved, it will load the most current version of the referenced file. Any changes in the components of a referenced drawing are automatically shown on the master drawing the next time it is retrieved. This feature is particularly useful when working on a networked system, because many people can access the same drawings for updating.

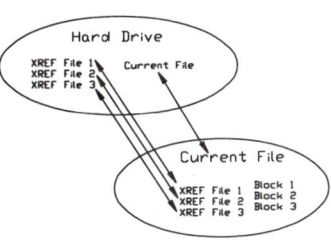

Figure 18-4

In addition, because the xref data is contained in another file, the file size is much smaller than it would be if data were inserted as the same amount of blocks.

Xrefs can be manipulated in exactly the same way as the blocks; the layers can be listed, the object can be edited as a single object, the objects can be used as OSNAP positions, and the colors can be changed. As with blocks, individual items of geometry within the xref cannot be changed. The only difference between an xref and a block is that the xref is not added to the file.

The XREF Command

The XREF command offers you seven options.

> **Windows** From the External Reference toolbar, choose a selection.
>
> **DOS** From the File menu, choose External Reference.

The command line equivalent is **XREF**.

```
Command:XREF
?/Bind/Detach/Path/Reload/Overlay/<Attach>:
```

Where: **?** = a listing of current xrefs
Bind = an ability to add the xref to the current file, thus making a BLOCK
Detach = a removal of xrefs from the current file
Path = the path that AutoCAD uses when loading one or more xrefs
Reload = an update of an xref without exiting the drawing editor
Overlay = an overlay of an XREF file onto the current file
Attach = an attachment of a new file or a copy of an xref that already exists in the file; much like INSERT

XREF Options

? will list all the xrefs in a file plus the path to the file. You can display a full listing of files, or just those from a specific path, or you can use * to obtain a wildcard listing with specific drawing names.

```
Command:XREF
?/Bind/Detach/Path/Reload/<Attach>:?

Xref name          Path

Axle               C:\ACAD386\1992
Tire               C:\ACAD386\1991

Total Xref(s): 2
```

BIND will change an xref into a block making it a permanent part of a file. This option is most often used when sending files to a client or another user to ensure that all the files are contained in the drawing. It is also useful when archiving files to make sure that all the files are included in the one master file.

When using the Bind option, you can bind the xrefs individually, or you can bind them all by using the * wildcard.

```
Command:XREF
?/Bind/Detach/Path/Reload/<Attach>:B
Xref(s) to bind:STENO1
    Scanning ...
```

DETACH When you erase an xref, the reference to the exterior file still exists, just as when you erase a block, the block reference still exists. To remove a reference completely from a file use Detach. When you retrieve a drawing with an xref that has been erased but not Detached it will bring the reference into the file but not display it. This takes up valuable space in memory.

Detach will affect an xref whether it is displayed or not, and it will also remove any nested references it finds. Again, the * wildcard can be used to detach all xrefs.

```
Command:XREF
?/Bind/Detach/Path/Reload/<Attach>:D
Xref(s) to detach:CORNER
      Scanning ...
```

If the xref that you are detaching has multiple insertions, AutoCAD will not perform the function.

PATH   If someone has updated or changed the directory of a referenced file, you will need to change the path to enable AutoCAD to locate and retrieve it. The Path option lets you change the path of loaded xrefs.

On a large model it is not unusual to have many or all of the xref source files in a separate directory. Path allows you to assign one or all of your source files to a separate directory. Use the * wildcard to list all of the files.

```
Command:XREF
?/Bind/Detach/Path/Reload/<Attach>:P
Edit path for which Xref(s):CORNER
      Scanning ...

XREF name: CORNER
Old path: C:\OFFICE\2NDFLR\CORNER
New path: C:\OFFICE\3RDFLR\CORNER

Reload Xref CORNER: C:\OFFICE\3RDFLR\CORNER
CORNER loaded.
```

The Path option prompts you to enter the names of the xrefs that you want changed. You are then given the old path name and prompted for the new path. You must enter the entire path name correctly. The problem most frequently encountered with this option is the spelling and nomenclature of the path. Pay careful attention to the use of backslashes (\) and colons (:).

RELOAD   There are two ways to reload your external files. The first way is simply by retrieving the file; the second is by using the Reload option.

Reload is used most frequently on a networked system where several people are working concurrently on the same project. If user 1 is updating a master file, and user 2 has made significant changes in one of the source files, user 1 may want to update the xref of user 2's file to see what effect the changes have on the master file.

If only one person is working on a project, Reload may be used when making significant changes on a portion of the drawing. If additions are made to a portion of a drawing that is an XREF, it may be WBLOCKed and then reloaded with the current changes. This function helps the file management process. Again, the * wildcard can be used to Reload all of the xrefs.

```
Command:XREF
?/Bind/Detach/Path/Reload/<Attach>:R
Xref(s) to detach:CORNER
      Scanning ...
```

```
Reload Xref CORNER: C:\OFFICE\3RDFLR\CORNER
CORNER loaded.
```

The files are first detached to make sure that all references to the block are removed.

OVERLAY attaches a file onto the current file, but doesn't allow the overlaid file to be attached or overlaid on a subsequent file. An overlaid file is not included in a drawing when the drawing itself is attached or overlaid as an xref to another drawing. Overlaid xrefs are designed for data sharing. (See next section, "Attaching Versus Overlaying.")

ATTACH is the default setting. It works much like the INSERT command in that it is used to place an existing xref on the current drawing. As with INSERT, the name of the last xref is used as the default.

```
Command:XREF
?/Bind/Detach/Path/Reload/<Attach>:A
Xref(s) to attach<CORNER>:A:24PANEL

Attach Xref 24panel:24PANEL
24PANEL loaded.
Insertion point:6,5
X scale factor<1>:↵
Y scale factor<1>:↵
Rotation angle<0>:↵
```

If you are using AutoCAD's Advanced User Interface, you can also use the dialog box by entering the tilde (~) character. This will display a list of all of the files already accessed.

Attaching Versus Overlaying

An overlaid xref is not included in a drawing when the drawing is itself attached or overlaid as an xref to another drawing.

In Figure 18-5 SCREW.DWG is attached to BUSHING.DWG, which is either attached or overlaid on WORKPART.DWG. All three files are visible, because SCREW.DWG was attached and, whether attached or overlaid, BUSHING.DWG was not more than one level of xref away from WORKPART.DWG.

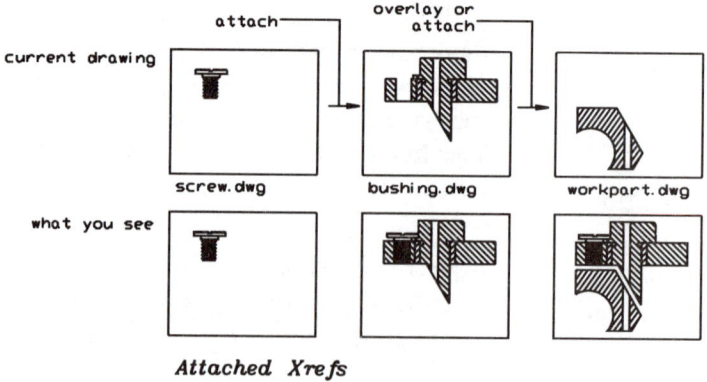

Figure 18-5

In Figure 18-6, SCREW.DWG is overlaid on BUSHING.DWG, which is either attached or overlaid on WORKPART.DWG. When you open WORKPART.DWG, SCREW.DWG is not visible, because it was overlaid, not attached, to BUSHING.DWG and is more than one level of xref away from WORKPART.DWG.

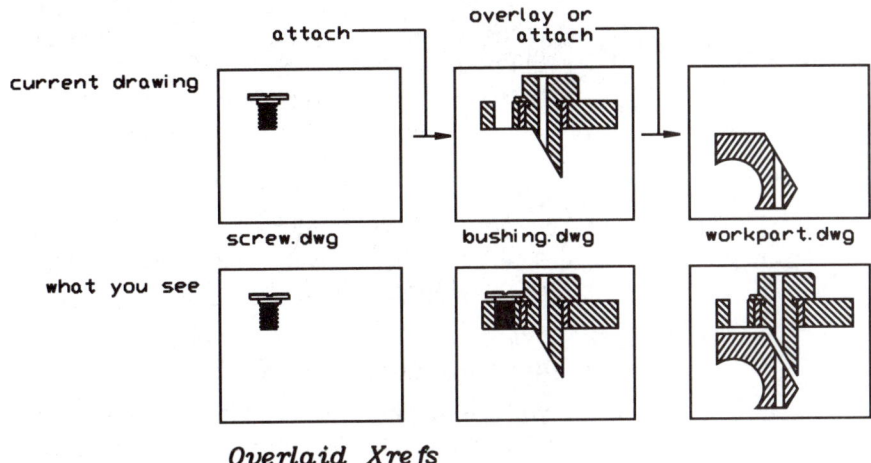

Figure 18-6

Using Xrefs

If you wanted to update a block, you could explode it, update it, and then, if the changes were significant and possibly useful on other files, you could WBLOCK the revised data and reinsert it. (Before inserting, you would need to purge the existing block; use the PURGE command.) If the block was entirely wrong, you could simply erase it.

You can erase an xref, but you can't explode it. Instead, you can access the external source file, update it, and reload it.

Notes

AutoCAD ignores block attributes and paper space-dependent information in XREFs.

The reinserted wblock would reflect all of the changes that were made on the one update. With xrefs, the revised source file will affect every drawing in which the file is referenced.

Like blocks as well, only the objects created in model space will be loaded into the drawing. No paper-dependent or viewport-dependent objects will be read.

If one or more of the xrefs attached to a drawing is moved or erased, AutoCAD will display an error message when the file is retrieved:

```
"C:\OFFICE\2NDFLR\CORNER" Can't open file
** Error resolving Xref CORNER
```

A block with the name of the xref will be entered at the position the xref was to occupy. This error can be corrected by updating the path name or reloading the referenced file into the original directory.

Xrefs and Layers

All of the properties of the objects of the referenced files will be added to the current file in addition to the objects themselves. Layers, linetypes, colors, and text and dimension styles will be added to the master file.

When these properties are added, they are renamed in the current drawing using a temporary name to avoid confusion and duplication.

When new layers are added, the new name will have the xref file name followed by the layer name.

```
Command: LAYER
?/Make/Set/New/ON/OFF/Color/Ltype/Freeze/Thaw:?
Layer name(s) to list<*>:↵

Layer Name      State       Color           Linetype
0               On          7 (white)       CONTINUOUS
DIM             On          1 (red)         CONTINUOUS
3RDFLR¦DIM      On          2 (yellow)      3RD FLR¦HIDDEN
?/Make/Set/New/ON/OFF/Color/Ltype/Freeze/Thaw:↵
```

The color and linetype of the referenced layers have the original file's properties.

Layer 0 is the exception to this layer renaming. Any objects on Layer 0 will be placed in Layer 0 of the current model, and all objects that are referenced will assume the settings associated with the current Layer 0.

Managing Blocks and Xrefs

If you are transporting a drawing to another machine or site, you must take along the xref files. If the file is not on the disk, it cannot be loaded. Similarly, when someone is backing up and erasing files from the hard drive, it is a good idea to provide him/her with a list of the files that you reference so that they are not backed up to an external hard drive and removed.

You can use the = symbol to reattach an xref that is already placed as a block on a current file.

```
Command: XREF
?/Bind/Detach/Path/Reload/<Attach>: A
Xref(s) to attach<CORNER>: A:36PANEL

**Error: A:36PANEL is already a standard block in the current
    drawing.                        *Invalid*

Command: XREF
?/Bind/Detach/Path/Reload/<Attach>: A
Xref(s) to attach<CORNER>: A:36PANEL2=A:36PANEL
```

This option is useful if you are trying to load different files which have the same name and are in different subdirectories.

When creating files, it is not a bad idea to record file information on paper — such information as:

1. Layers plus creation date
2. Blocks plus first insertion date
3. Xrefs plus first insertion date

If you do not note this information, you can guarantee that you will need it.

As xrefs are inserted, AutoCAD creates an external log (.XLG) file of the inserted file. This ASCII text file, located in the current drawing directory, records all relevant xref information concerning attached files and any subsequent xrefs added. It is updated every time an external reference file is attached. This file records a lot of the documentation explained above. However, because it can be deleted from the directory (and often is if there is a lack of memory space), it is still a good idea to keep a hardcopy list as well.

The XBIND Command

With the XREF command, an external source file is added to a current or master file for the duration of the editing period. When you edit the master file, the source file and all of its dependent "symbols," including linetypes, text styles, layers, and dimension styles, are returned to the original file. While part of the current file, the layers, linetypes, etc. are listed and referenced to the current file, but they are not accessible and remain part of the external source file. Therefore you can use a custom linetype from an attached xref.

The XBIND command allows certain of these "symbols" to be added to a current file when it is saved and can, therefore, be accessed for use. This would allow you to use the customized xref linetype on your current file. At the end of the editing session, the xref would return to its original source file, and the linetype would remain with the current file.

> **Windows** From the External Reference toolbar, choose Bind (reads All).
>
> **DOS** From the File menu, choose Bind.

The command line equivalent is **XBIND**.

The command offers five options.

```
Command:XBIND
Block/Dimstyle/LAyer/LType/Style:
```

Where: **Block** = adds a block to the current file
Dimstyle = adds a customized dimension style to the current file
LAyer = adds a layer to the existing file
LType = adds a customized linetype to the current file
Style = adds a customized text style to the current file

```
Command:XBIND
Block/Dimstyle/LAyer/LType/Style:LA
Dependent LAYER name(s):STENO
     Scanning ...
1 Layer bound
```

Once the layer is bound to the current drawing, the ¦ is removed from the name and replaced with 0. If the above layer, for example, was on the externally referenced file called 3RDFLOOR, it would appear as 3RDFLOOR¦STENO before it was bound and 3RDFLOOR0STENO after it was bound. This occurs so that a layer being bound to a current drawing will not take on the name of an existing layer and add all of its parameters to an existing layer. If the layer was called 3RDFLOOR¦DIM, for example, there is a good chance that there would already be a layer called DIM. The new layer is 3RDFLOOR0DIM to distinguish the two. The new name can, of course, be renamed using the layer command.

The linetype and color associated with the layer in the xref source drawing would remain with it when it is bound to the new drawing file.

The maximum amount of characters that can be added or bound to a new drawing is 31; if any number greater than that is attempted, the XBIND command will not work.

Prelab 18 Advanced Blocking and Xrefs

Step 1 Open the file called JIG from Chapter 14. Insert the block called JIGLEG on the right side of the file. If you haven't got this file, quickly draw up just the geometry from page 369.

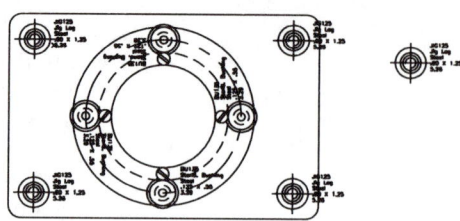

Step 2 Explode the block so that it is separate objects and attribute definitions.

Step 3 Use the ELLIPSE command to add two ellipses to the block. Then BLOCK the data using the same name.

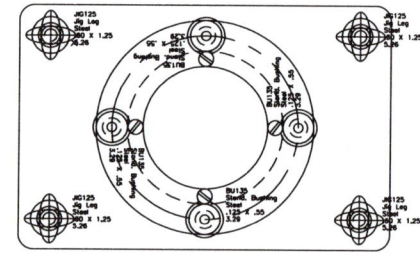

```
Command: BLOCK
Block name (or ?): JIGLEG
Block JIGLEG already exists,
   Redefine it?<N>: Y
Insertion base point: (pick the
   intersection of the lines)
Select objects: (pick the block
   and attributes)
```

The jiglegs on the part will be automatically updated.

Step 4 Open a new file *without saving the changes in JIG*.

You should have an attributed block file called ATTITLE from Chapter 9. If you don't have this file, quickly create a title block with attributes and WBLOCK it.

INSERT the file ATTITLE and fill in the attributes.

Step 5 Now use the XREF command to reference the JIG file from Chapter 14.

> **Windows** From the External Reference toolbar, choose Attach.
>
> **DOS** From the File menu, choose External Reference.

The command line equivalent is **XREF**.

```
Command: XREF
?/Bind/Detach/Path/Reload/Olay/<Attach>: A
Xref(s) to attach<CORNER>: JIG
Attach Xref JIG
```

```
JIG loaded.
Insertion point: (pick a point)
X scale factor<1>:⏎
Y scale factor<1>:⏎
Rotation angle<0>:⏎
```

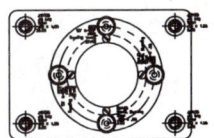

Xrefs can be manipulated in exactly the same way as the blocks: the layers can be listed, the object can be edited as a single object, the objects can be used as OSNAP positions, and the colors can be changed. As with blocks, individual items of geometry within the xref cannot be changed. The only difference between an xref and a block is that the xref is not added to the file.

Step 6   SAVE the file as XREF, then open your JIG file.

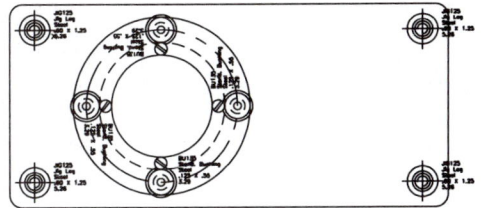

In the JIG file, use STRETCH to change the size of the JIG. Then use ATTEDIT to change the size of one set of attributes.

```
Command:ATTEDIT
Edit attributes one by one?<Y>:⏎
Block name specification<*>:⏎
Attribute tag specification<*>:⏎
Attribute value specification<*>:⏎
Select attributes:W
First corner: (pick 1)
Other corner: (pick 2)
5 attributes selected.
Value/Position/Height/Angle/Style/Layer/Color/Next<N>:H
New height<0.07>:1
Value/Position/Height/Angle/Style/Layer/Color/Next<N>:⏎
Value/Position/Height/Angle/Style/Layer/Color/Next<N>:H
New height<0.07>:1
```

Continue until all heights are changed.

Step 7   SAVE the file and OPEN the file called XREF.

Notice that the XREF has been updated automatically to reflect the new drawing. Both the size of the file and the size of the attributes have been altered.

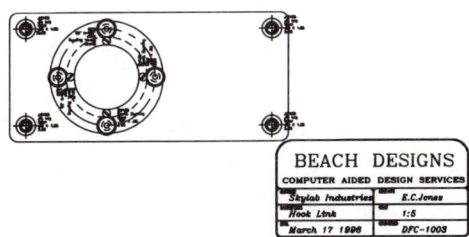

Advanced Blocking and Xrefs   **447**

Step 8 Now use the XBIND command to bind the jig to the XREF file.

> **Windows** From the External Reference toolbar, choose Xbind.
>
> **DOS** From the File menu, choose Bind.

The command line equivalent is **XBIND**.

```
Command:XBIND
Block/Dimstyle/LAyer/LType/Style:B
Dependent BLOCK name(s):JIG
     Scanning ...
1 Block bound.
```

Use SAVE-AS to create a new file called XREF1. Note the difference in the size of the file XREF and the file XREF1.

Command and Function Summary

XBIND binds a block, layer, linetype, or other object to a file.

XREF allows you to add files to your drawings as external references.

Notes

While doing the following exercises, it is suggested that you note on paper the names of the blocks and xrefs you are using as files.

Exercise A18

Step 1 Create the following files:

1. A plain steno chair, filename STENO1
2. A steno chair with arms, filename STENO2
3. A 5′ × 6′ workstation, filename 5X6WST
4. A 6′ × 7′ workstation, filename 6X6WST
5. An executive office, filename MGMT

(Use files from Challenger 4 if you have them.)

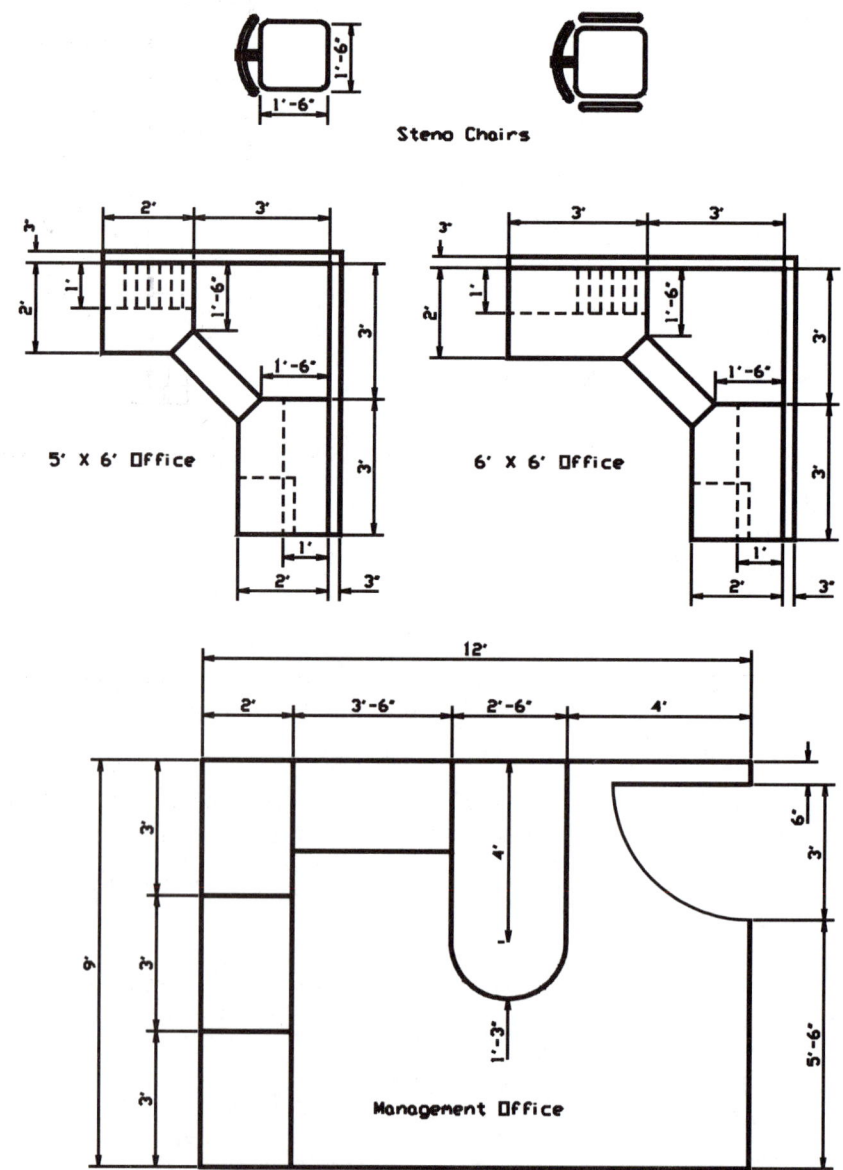

Step 2   Create an office that is 48' × 28' with a 6' double door.

Step 3   Create a bank of eight 5' × 6' workstations (filename 5X6WST) in the middle of the room by inserting them as BLOCKs. Add an armless STENO chair (STENO1) to each workstation.

Step 4   Using the XREF command, insert four management offices (MGMT). Insert a STENO chair with arms (STENO2) into each management office.

Step 5   SAVE the file, and note the time and the size of the file as follows:

Filename _____ Time _____ Size _____

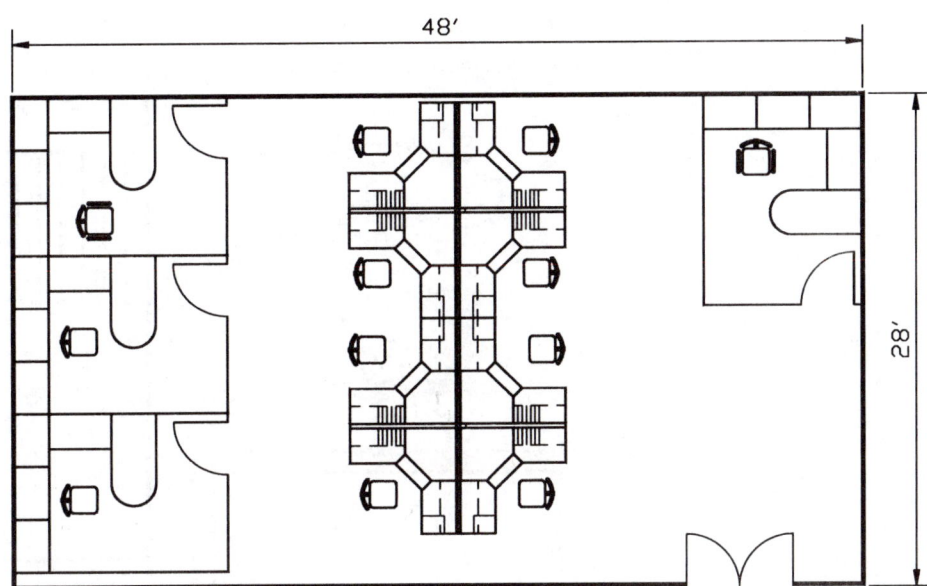

Step 6   Enter the management source file and change the size of the curved desk. Add some filing cabinets and a plant. END the file.

Step 7   Enter your office layout. Note the fact that the management offices have been automatically updated. Using the equal sign, change the 5 × 6 workstations (5X6WST) to 6 × 6 workstations (6X6WST). Make all the chairs armchairs (STENO2) using =.

Bind the management offices to the office layout. END the file and note its size:

File size _____

Reenter the file and erase the management offices. Detach them from your file and note the size of the file after you have SAVEd:

File size _____

Advanced Blocking and Xrefs

Exercise C18

Step 1   Create the following files:

1. A deciduous tree, filename DECID
2. A coniferous tree, filename CONIF
3. An 8 m × 8 m house, filename HOUSE
4. A 10 m × 12 m house plus deck, filename HOUSE2
5. A swimming pool, filename POOL

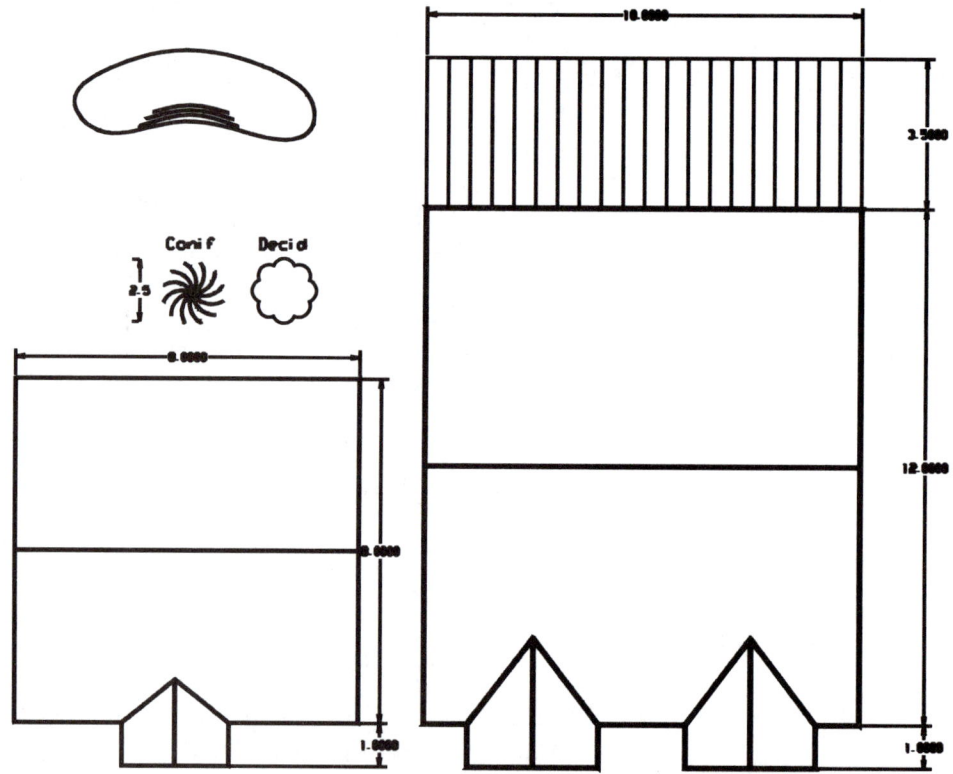

Step 2   Create a plot plan for a single lot that is 20 m × 30 m. In this lot, INSERT the first house (HOUSE1) and at least 12 coniferous trees (CONIF).

Step 3   Using the XREF command, Attach the pool in the backyard A:POOL.

Step 4 Using the = sign, change the first house with the second house HOUSE1=HOUSE2. Now change the coniferous trees with the deciduous trees CONIF=DECID.

Now SAVE and end the file and note the time the file was saved and the size of the file:

Time _____ Size _____

Step 5 While outside of the file, change the source file for the swimming pool (POOL). Add a diving board and some "natural setting" vegetation.

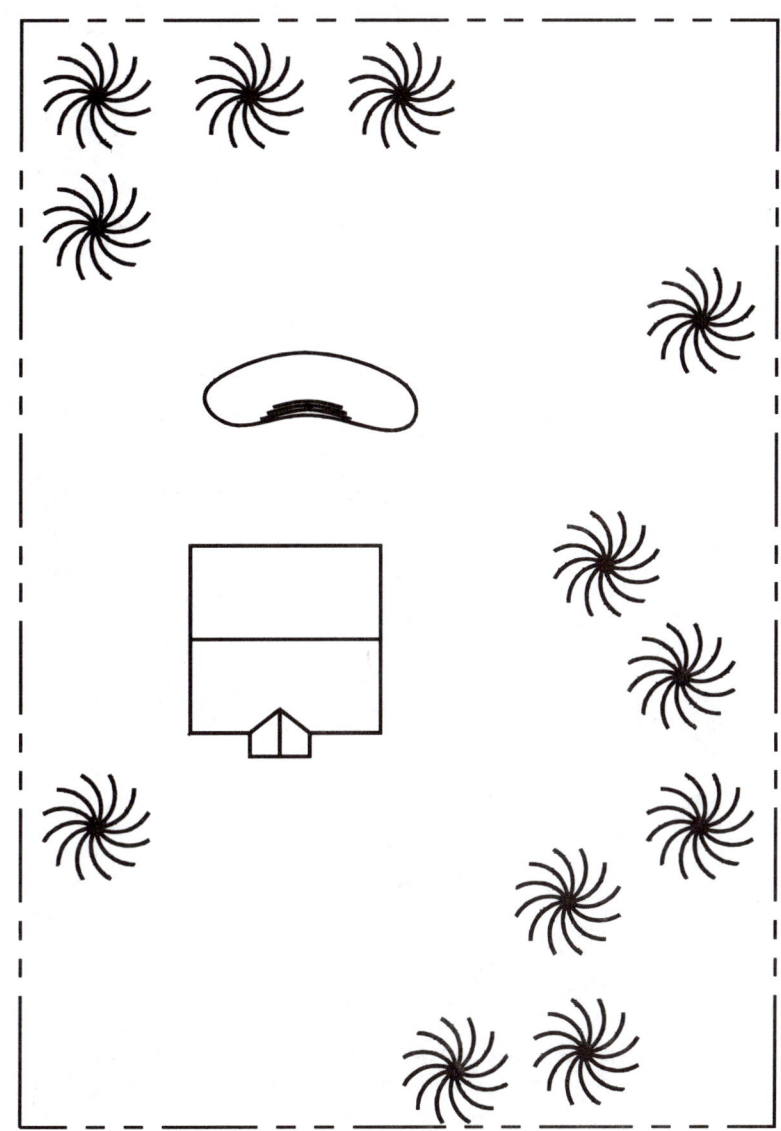

Step 6 Enter your site plan. Note the fact that the pool has been updated automatically.

Bind the pool to the site plan master file.

SAVE the file and note its size.

File size _____

Reenter the file and erase the pool. Detach it from the file and note the size of the file after you have SAVEed:

File size _____

Advanced Blocking and Xrefs

Exercise E18

Step 1 Create the following files:

1. A fuse, filename FUSE
2. A relay, filename RELAY
3. A temperature control overload, filename OL

Step 2 Using these external blocks, create this 3-phase supply diagram. Then create another file for the control circuit.

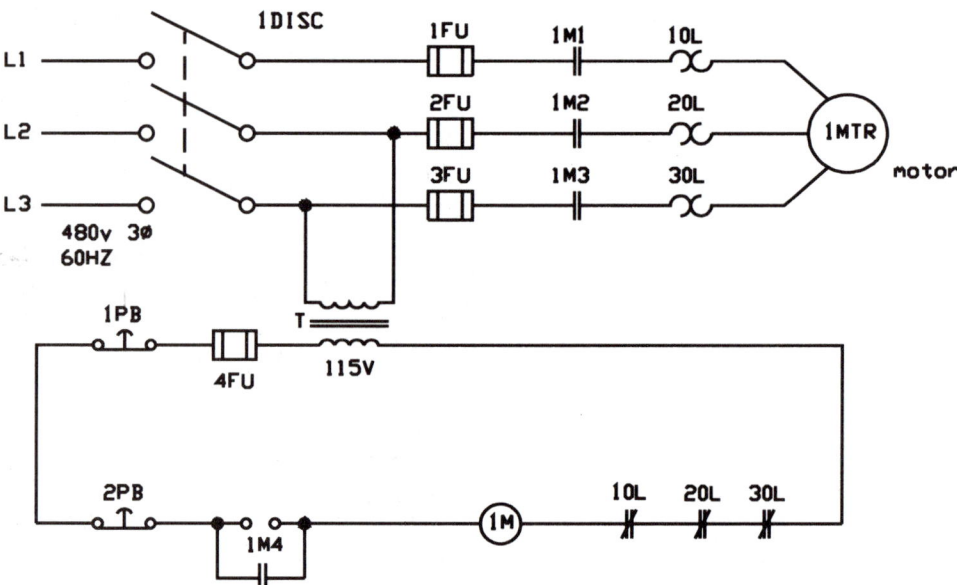

Step 3 Bring up the 3-phase supply diagram and explode one of the fuse blocks. Make some minor changes to the design, and reblock the fuse under the same name.

Step 4 Use XREF to place the control circuit into the 3-phase supply diagram.

Step 5 Save the drawing with the 3-phase supply, and bring up the control circuit file. Redesign the control circuit to a different sequence of operation. Save the file without changing the name.

Step 6 Bring up the 3-phase supply file and note how the file has been updated.

Exercise M18

Step 1   Create the following files:

1. An AutoCAD DOS station, filename ACADDOS
2. An AutoCAD Windows station, filename ACADWIN
3. A machine tool as shown, filename TOOL1
4. Icons for:

   Finite Element Analysis, filename FEA

   Materials, filename MAT

   Existing Parts, filename EXPARTS

   Standard Parts, filename STPARTS

   Machines, filename MACH

   Central Network or Database, filename DATA

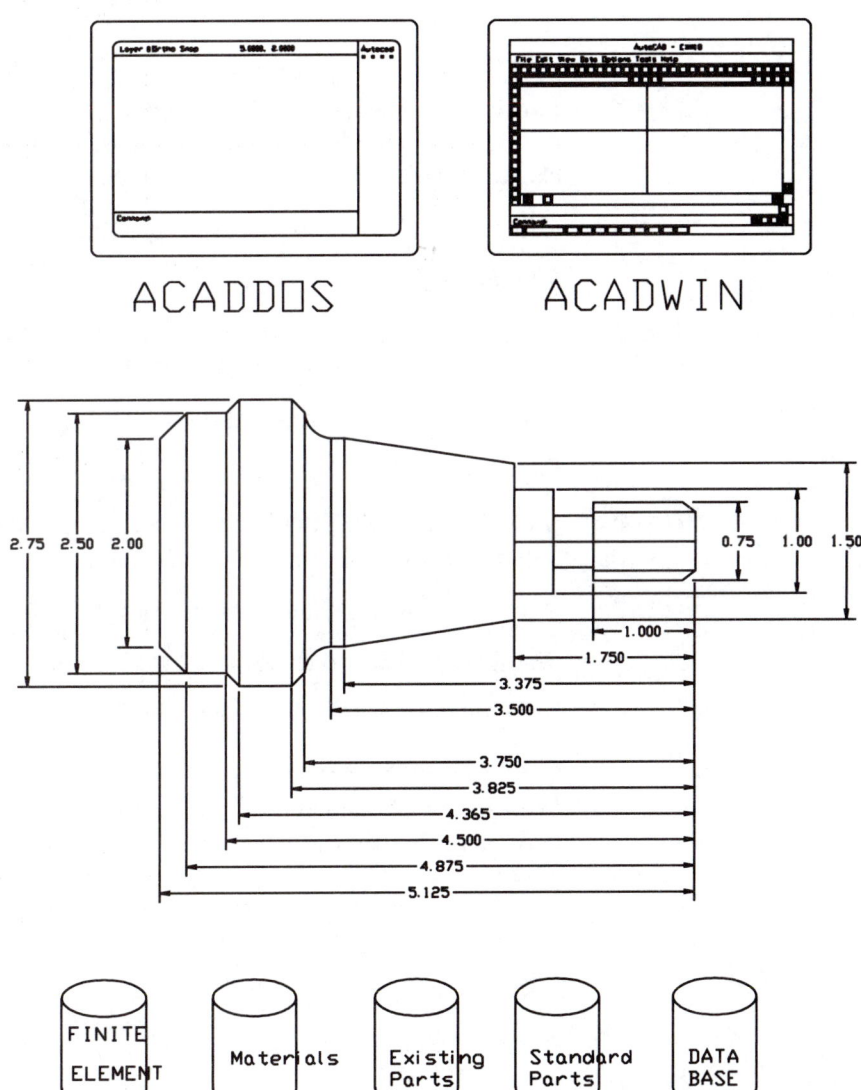

Advanced Blocking and Xrefs

Step 2 Create a data flow management chart as shown.

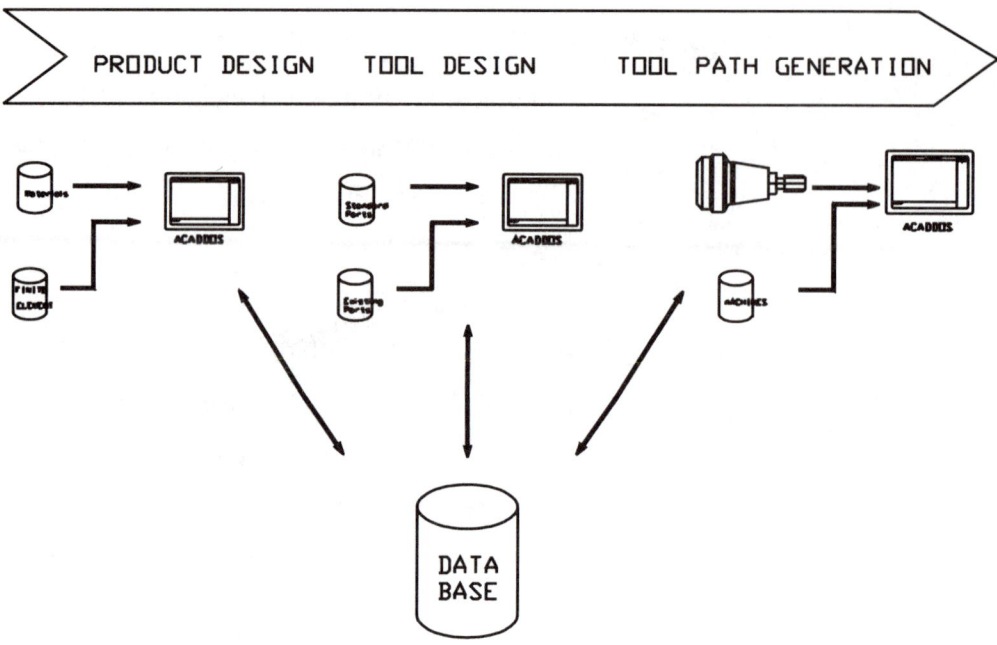

Step 3 For the first chart, use INSERT to place all of the ICONs. Use XREF to Attach to place the ACAD11 stations at each section of the data flow chart (A:ACAD11). Use XREF also to place the machine tool (A:TOOL).

Step 4 End the file and note the file size and time:

File size _____ Time _____

Step 5 Enter the file again and Bind the ACADDOS files to the Chart. Now, using the = sign, change the ACADDOS stations with ACADWIN stations.

Once again note the size and time:

File size _____ Time _____

Step 6 Enter the Tool file TOOL and use STRETCH to change some of the parameters of the tool. SAVE the file.

Step 7 Enter the flow chart and note the fact that the tool has been automatically updated.

Bind your tool file to the flow chart file.

END the file and note the size of the file:

File size _____

Reenter the file, erase the tools and ACAD stations and Detach them from your file. Make sure that they are Detached. Once more, note the file size:

File size _____

Challenger 18A

Using the preceding exercises, create attributes for each BLOCK that is added. Also create new LAYERs for each BLOCK.

Then create a data extract file to extract the files. When you exchange the first set of BLOCKs with the second, generate another data extract file.

Finally, create a file of five nested XREFed drawings (5 lots, 5 office floors, or 5 flow charts). Bind the drawings and save them. Note the size. In each case, use INSERT to create the same drawing and note any difference in the size of the files.

Create this drawing with attributes for wall, ceiling, handrail, floor and trim finishes for each separate floor. Block each floor separably, then compile into a finished floor.

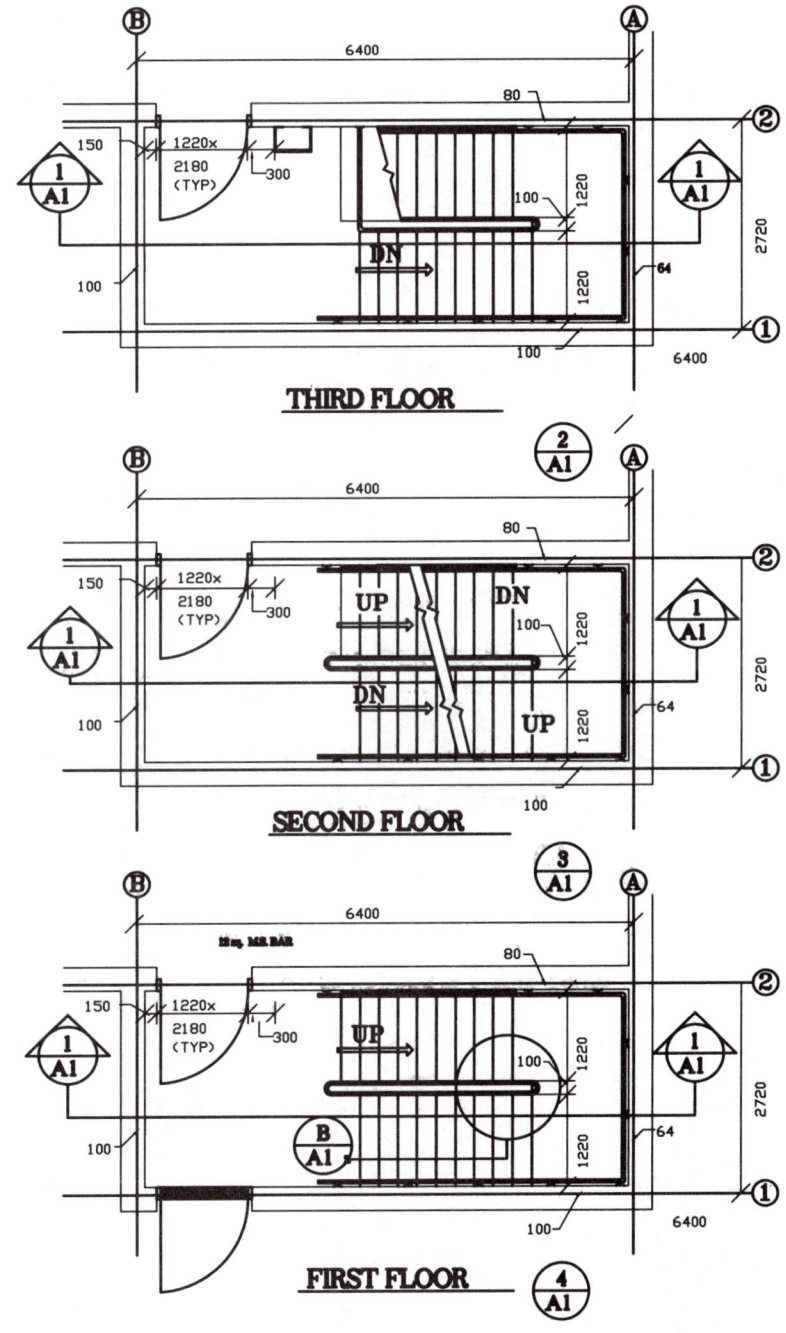

Challenger 18B

For this corner layout, create an attributed A2 title block. On a separate file draw the intersection to scale. Insert the block onto a new drawing, then attach the intersection as an XREF.

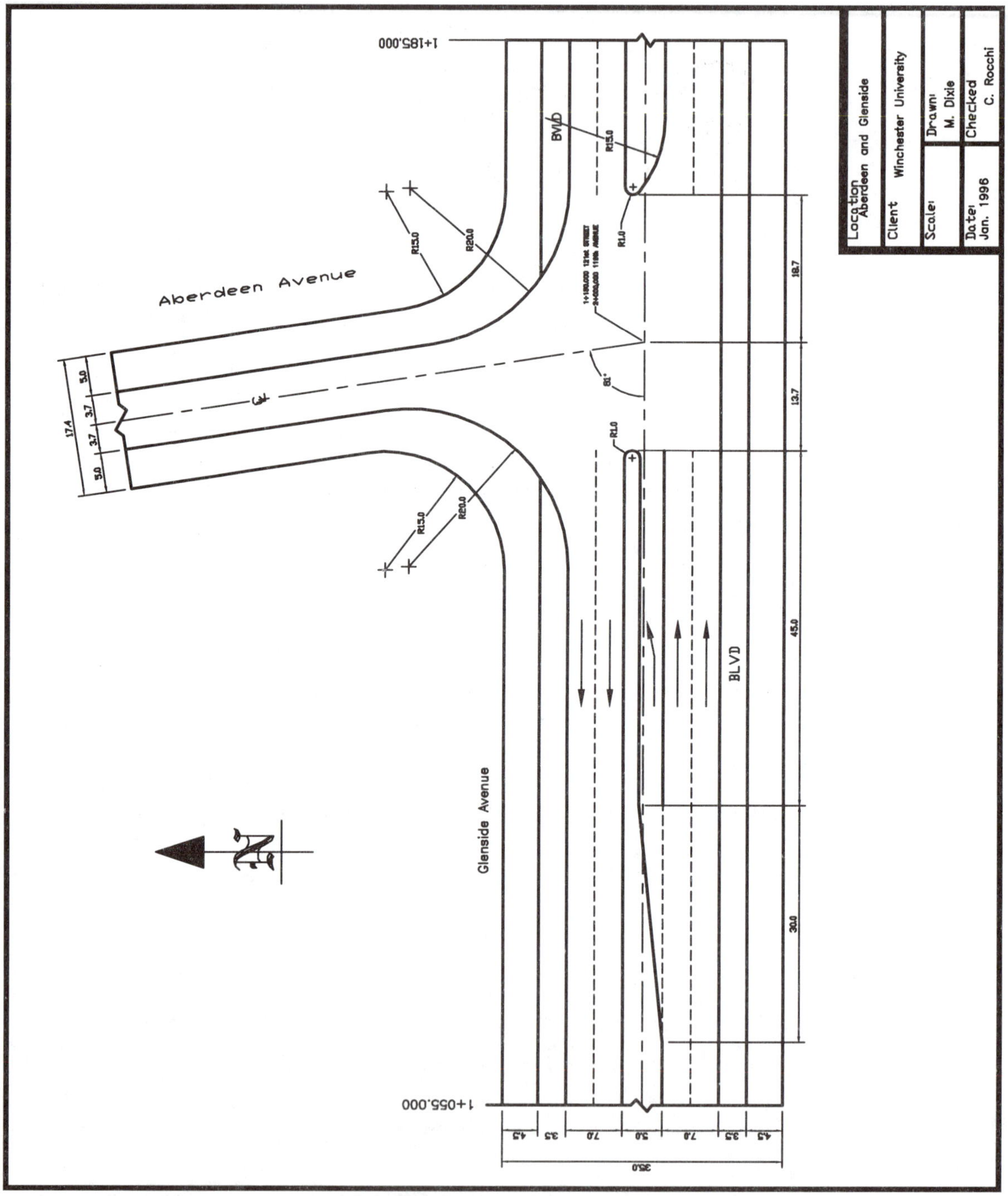

19 Final Tests and Projects

1. On-screen review
2. General review
3. Final drawing projects

OBJECTIVES

The next few pages offer two on-screen problems, 70 questions on the second six chapters of this book, and four final projects. Your instructor may offer you these or similar questions as a final exam. If you are not required to write a final exam, these questions will be nonetheless useful in showing you how much you understand.

Section 1: On-Screen Problems

This section is difficult to combine into a Final Project, because there is so little graphic information. Most of this material covers the creation of graphics data and exchanging this data with other types of computer documents.

Final projects are offered after Section 2, to ensure that your understanding of that section is complete and to help you combine all of the skills learned in this section.

Problem 1

Set SNAP to Isometric and draw in the part shown. Use the nominal value of each line — if it says 1.3 make the line 1.3 in length. Use the UNITS command to set the number of digits to the right of the decimal point to 4.

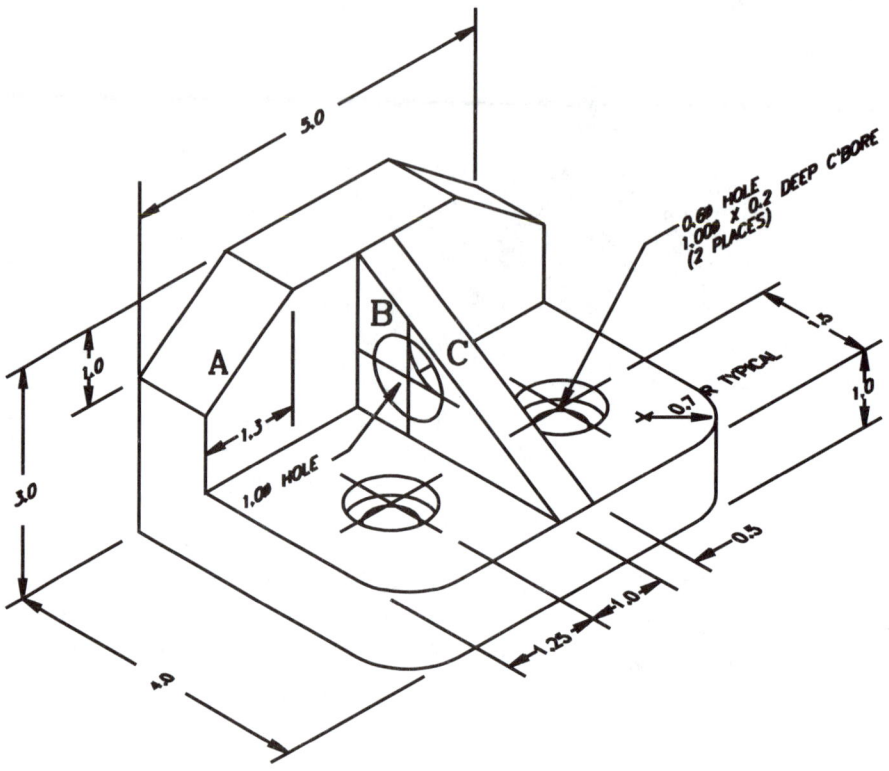

Begin the drawing by setting the lower left corner of the part at 0,0. Either move the UCS, PAN the screen over, or set your LIMITS so that the 0,0 is on screen.

Refer to your drawing to answer the following questions:

1. The total length of line segment A is:
   a. 1.8775
   b. 2.2015
   c. 1.9975
   d. 1.7500
   e. 2.2565

2. The absolute coordinate value of the center of the circle B is:
   a. 3.4641,2.0000
   b. 3.9672,2.2100
   c. 3.1462,1.5465
   d. 2.6574,2.0000
   e. 3.5837,1.9283

3. The total area of surface C is:
   a. 2.6574
   b. 1.9684
   c. 2.7947
   d. 2.1651
   e. 1.7483

Problem 2

Draw the electrical cover as shown. Start the drawing with the lower left corner at 0,0. Use PAN, UCS, or LIMITS to place the 0,0 on screen.

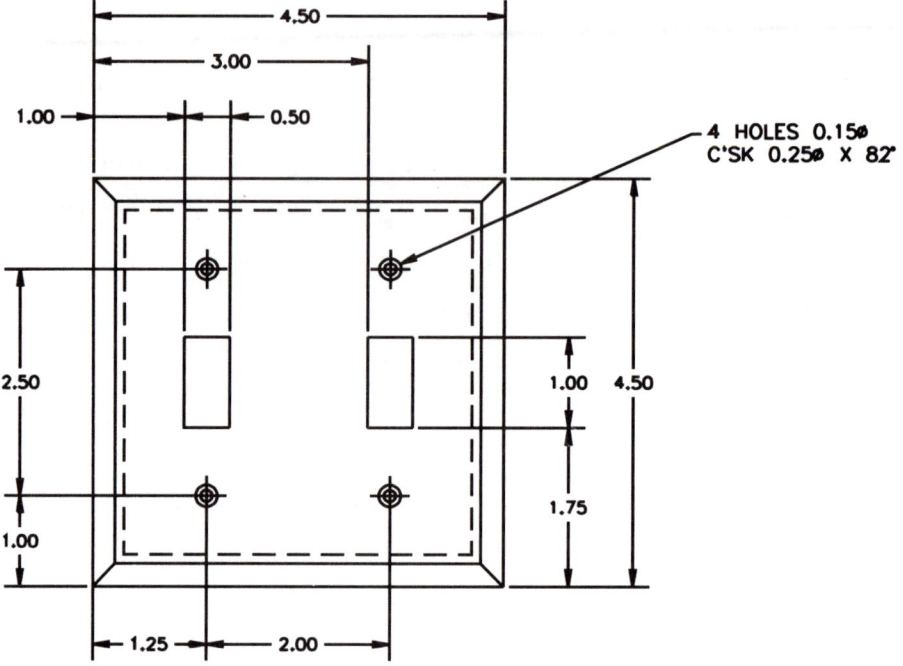

Use the drawing to answer the following questions:

1. The total surface area of the flat portion of the plate minus the bolt holes and rectangular holes is:

   a. 12.5430

   b. 14.8190

   c. 14.2340

   d. 13.3570

   e. 14.9055

2. The total area of the two rectangles and the bolt holes is:

   a. 1.1810

   b. 2.4570

   c. 1.7660

   d. 1.0045

   e. 3.9860

3. The circumference of one of the bolt holes is:

   a. .4520

   b. .8596

   c. .7500

   d. .6725

   e. .7540

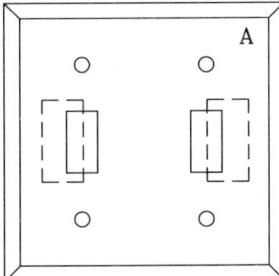

4. The absolute coordinate value of the end point of the line at A is:

   a. 4.25,4.25

   b. 4.00,4.00

   c. 3.75,4.00

   d. 3.75,3.75

   e. 4.25,4.00

5. Use the scale command to expand the two rectangles from the center point of the plate by a factor of 1.33. The new surface area of the plate minus the bolt holes and rectangles is:

   a. 15.0990

   b. 12.0422

   c. 14.0422

   d. 13.9586

   e. 15.0958

Section 2: Review

1. Before it can be inserted, an attribute must be:

   a. Blocked

   b. Edited

   c. Extracted

2. When using a Window to choose the series of ATTRIBUTEs:

   a. The first attribute defined is offered first.

   b. The attribute closest to the first Window position is chosen first.

   c. The attribute in the highest position in *Y* is chosen first.

3. What happens if you don't define the prompt in the ATTDEF command?

   a. The user will not be prompted.

   b. The tag is used as the prompt.

   c. The default is offered for the prompt.

4. What are you "writing" when you define an ATTRIBUTE?

   a. A program that will help the user define attributes

   b. A script file

   c. An attribute block

5. What happens when you accept the default by hitting ⏎ when asked for the default ATTRIBUTE value?

   a. The default value entered by the person writing the ATTDEF will be chosen.

   b. The system default will be chosen.

   c. The default will be .5.

6. Is an ATTDEF filed with the model even if it hasn't been blocked or inserted?

   a. Yes

   b. No

7. Within the ATTRIBUTE command, what does Invisible refer to?

   a. The lines around the block will be invisible.

   b. The tags will be invisible.

   c. The attributes when inserted will be invisible.

8. Can you change the LAYER of an attributed block?

   a. Yes

   b. No

9. What is the advantage of setting an ATTRIBUTE text height within a title block?

   a. The text will always be the same in all the title blocks.

   b. The text will always have the same justification.

   c. The height will not have to be changed in the STYLE command.

10. Once the ATTRIBUTE is defined, how can you change the prompts?

    a. CHPROP

    b. ATTEDIT

    c. CHANGE

11. Must the ATTRIBUTE be EXPLODEd before it can be moved?

    a. Yes

    b. No

    c. Sometimes

12. How can you modify the color of an ATTRIBUTE?

    a. This is not possible because it has nongraphics data attached.

    b. CHPROP

    c. ATTEDIT

13. If you would like to change the prompting sequence of the ATTRIBUTE instance, what must you do?

    a. CHPROP

    b. EXPLODE and reblock

    c. Change the default

14. What is a "tag"?

    a. The title you give to the attribute definition

    b. The word used to define a prompt

    c. The word for nongraphics data

15. How can an ATTRIBUTEd BLOCK be used in another file?

    a. ATTEXT

    b. By saving it as a WBLOCK

    c. By INSERTing it as an ATTDEF

16. What command do you use to define an ATTRIBUTE?

    a. ATTEXT

    b. ATTDEF

    c. ATTEDIT

17. When you want to attach an attributed BLOCK to a drawing, what command do you use?

    a. BLOCK

    b. INSERT

    c. PURGE

18. What does the term "global" mean when editing attributes?

    a. A global edit affects just the graphics.

    b. A global edit edits all the attributes selected.

    c. A global edit affects just the TAGs.

19. If you don't choose to select the attributes one by one, will all of the attributes on screen be selected?

    a. Yes, if you Window them all.

    b. Yes, they are all inserted from the same block reference.

    c. No, only those selected one by one will be edited.

20. If you have INSERTed many attributed blocks, how do you change the default value and then keep INSERTing it?

    a. It can't be done.

    b. ATTEDIT, then CHANGE

    c. EXPLODE, CHANGE, and reBLOCK

21. What command do you use to modify the value of an ATTRIBUTE instance?

    a. CHANGE and reBLOCK

    b. EXPLODE and CHANGE

    c. ATTEDIT

22. What command do you use to extract the values of all of the ATTRIBUTE instances to an ASCII-based format?

    a. ATTEDIT

    b. ATTEXT

    c. ATTASCII

23. What command is used to change the position of an ATTRIBUTE instance?

    a. ATTEDIT

    b. MOVE

    c. ATTEXT

24. Does QTEXT work on ATTRIBUTES?

    a. Yes

    b. No

25. Why does your attribute extract have to be ASCII-based?

    a. ASCII is an insertable code.

    b. ASCII is the accepted standard.

    c. ASCII has the alphabet that we use.

26. Can you MEASURE a SOLID?

    a. Yes

    b. No

27. What is the quickest method of making five equal portions out of a stretch of geometry that contains two arcs and a circle?

    a. BLOCK it and DIVIDE it

    b. PEDIT and DIVIDE it

    c. PEDIT and MEASURE it

28. Can you use the MEASURE command on an ELLIPSE?

    a. Yes

    b. No

29. What can you do to see the division markers created with the DIVIDE command?

    a. ZOOM Window

    b. Change PDMODE

    c. Change PDSIZE

30. What is the maximum size of the default point?

    a. 10 × 10

    b. No maximum

    c. One pixel

31. If you wanted to have a DONUT shape as opposed to a point displayed as a division marker in the DIVIDE command, how would you do it?

    a. Change the PDMODE

    b. Make the DONUT a BLOCK

    c. Change the PDSIZE

32. Why will the Crossing option not work in the MEASURE command?

    a. Because the LINEs must be picked in a contiguous order

    b. Because the ARCs must be selected first

    c. Because only one object can be selected at once

33. What is the key element when placing a BLOCK correctly within the MEASURE command?

    a. The scale factor of the block

    b. The insertion base point of the block

    c. The attributes of the block

34. What does the ID command do?

    a. Allows the user to add an ID number to an object

    b. Gives the position of a point relative to the origin

    c. Lists the LAYER, position, and block association of an object

35. If you wanted to find the angle of a LINE relative to 0, what command would you use?

    a. LIST

    b. ANGLE

    c. DBLIST

36. If you have UNITS set to two decimal points of accuracy (.00), will a LIST on the objects that you have created with this setting default to two decimal points of accuracy?

    a. Yes

    b. No

37. Can you use OSNAP when you are calculating a DISTance?

    a. Yes

    b. No

38. Does the AREA command calculate volume as well?

    a. Yes

    b. No

39. How would you change the time setting on your computer?

    a. TIME

    b. DATE

    c. TDCREATE

40. Where can you find the position of the last entered point?

    a. ID

    b. LIST

    c. STATUS

41. What does the term "elapsed time" mean?

    a. The time that has elapsed since the user signed onto a file

    b. The time that has elapsed since the user signed onto the system

    c. The time that has elapsed since the file was started

42. What is TDINDWIG?

    a. The time and date indicator in ROM

    b. The elapsed time setting value

    c. The total editing time value

43. What is an EPS file?

    a. Extract Printer Script

    b. Encapsulated PostScript

    c. Enhanced Printer Script

44. What is the difference between a raster file and a bit map?

    a. A raster file only works on a VGA screen.

    b. A raster file accesses the pixels more rapidly.

    c. There is no difference.

45. What is a pixel?

    a. The smallest addressable portion of your screen

    b. The center of your crosshairs

    c. The aperture setting

46. What is a vector file?

    a. A file that relates to the resolution of your screen

    b. A file that records the number of lines on your file

    c. A file that records the position of the objects in your file relative to a fixed origin

47. What three colors make up your screen display?

    a. Red, green, and blue

    b. Red, yellow, and blue

    c. White, black, and green

48. What kind of information is used to make a SLIDE file?

    a. Vector file

    b. Bit map

    c. Window file

49. What command would you use to get a list of slide files?

    a. DIR *.SLD

    b. DIR *.SCR

    c. DIR *.SWR

50. Why does a REGEN take longer than a REDRAW?

    a. A REDRAW will only redraw the vector file.

    b. A REDRAW will only redraw the bit map.

    c. A REDRAW involves the GRID.

51. What does the 1000 stand for in DELAY 1000?

    a. One minute

    b. One nanosecond

    c. One second

52. What does RSCRIPT do?

    a. Redraws the script

    b. Regenerates the script

    c. Repeats the script

53. What is a delimiter?

    a. The limit on the size of your file

    b. The code that indicates the end of an entry

    c. The code that indicates a change in direction

54. What code indicates a preloading of your slides in a script file?

    a. #

    b. *

    c. ^

55. What does the GL stand for in HPGL?

    a. Graphics language

    b. Gather line

    c. Grab line

56. What is a .PLT file?

    a. Perfect line type

    b. Pre-load type

    c. Plot

57. How can you make a .PLT file into an HPGL file ?

    a. Rename it (as long as the driver is correct)

    b. Copy it

    c. Extract it

58. What is an image file?

    a. A nongraphics file

    b. A vector file

    c. A bit map file

59. What command do you use to activate the isometric SNAP style?

    a. ISOPLANE

    b. ELLIPSE

    c. SNAP

    d. GRID

60. What pull-down menu is used to set the current drawing plane?

    a. DDLMODE

    b. DDRMODE

    c. DDIMODE

61. What command is used to load a circle on an isometric plane?

    a. ELLIPSE

    b. (Load Isocircle)

    c. ISOCIRCLE

62. What command is used to set the crosshairs to an isometric plane?

    a. F8

    b. SNAP

    c. GRID

    d. ORTHO

63. Can you use CEN to access the center of an isometric circle?

    a. Yes

    b. No

64. Can you draw a SOLID on an isometric plane?

    a. Yes

    b. No

65. What option needs to be changed to add text to an isometric drawing?

    a. Rotation angle

    b. Obliquing angle

    c. Style

66. What command would you use to replace 24 BLOCKs called one with 24 BLOCKs called two?

    a. INSERT ONE=TWO

    b. INSERT TWO=ONE

    c. EXPLODE ONE=TWO

67. Can a BLOCK contain XREFs?

    a. Yes

    b. No

68. Can an XREF contain BLOCKs?

    a. Yes

    b. No

69. If an XREF contains user-defined HATCH patterns, can these HATCH patterns be used on the file in which the XREF appears?

    a. Yes

    b. No

70. What option is used to add an XREF to the current file?

    a. ATTACH

    b. BIND

    c. BUILD

    d. ADD

Architectural Final

1. Create a floor plan and elevations for the commercial building (opposite page), using MLINE, MLSTYLE, and MLEDIT.

2. Create an ATTRIBUTEd BLOCK with the following:

   a. The area of each room

   b. The final floor finish

   c. The wall color

   d. The trim color

   e. The length of trim (floor trim) needed

3. Create another attributed block for the windows and doors containing such information as:

   a. Size of window/door

   b. Rough opening

   c. Glass finish (clear/smoked/triple glaze)

   d. Cost

4. Download this data onto a spreadsheet so that final amount calculations can be made.

5. Create an isometric rendering of the building using Isoplane and SNAP.

6. Document the time taken to create the drawing, the area of the interior, and the size of the file.

7. Accessing the isometric view as an XREF, compile the drawing including the plan, and elevations, and an attributed title block.

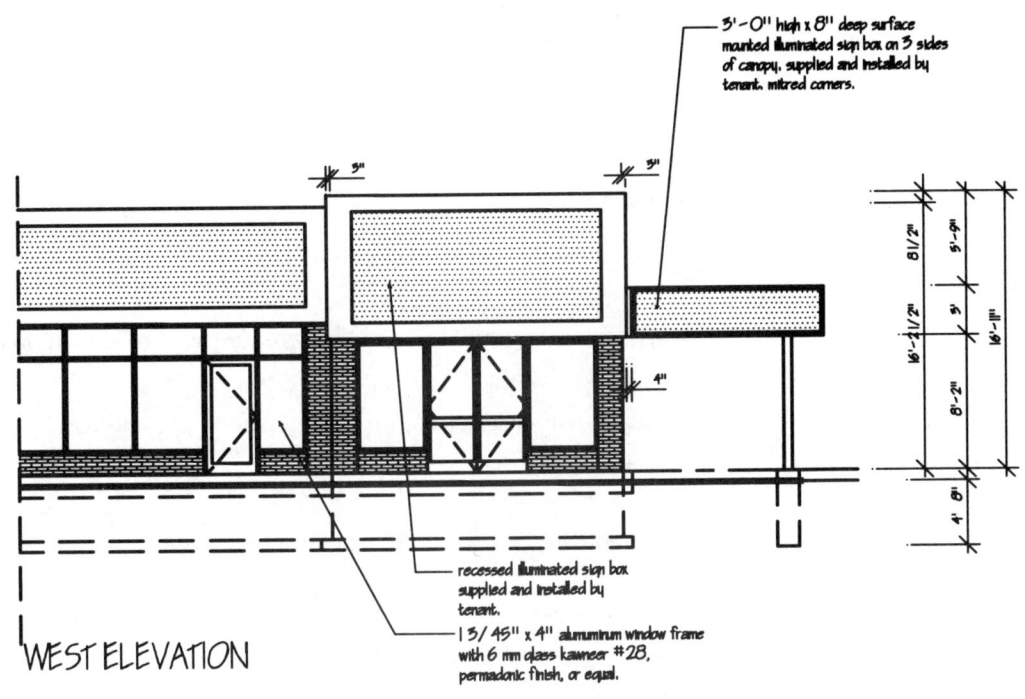

WEST ELEVATION

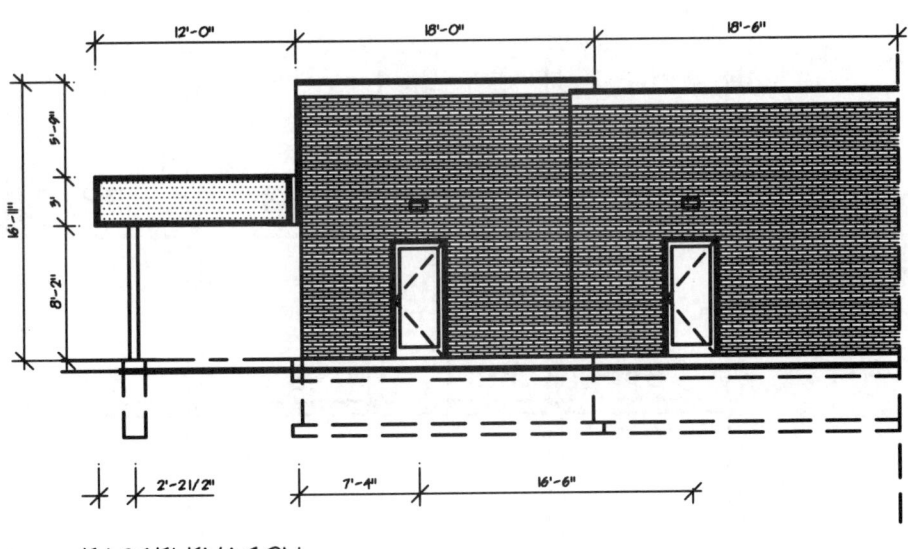

EAST ELEVATION

Final Tests and Projects 475

Civil Final

1. Create two beam details as shown in the drawing (opposite page) with top views as well as front views.

2. Create an ATTRIBUTEd BLOCK of the top view with the following:
    a. The beam title
    b. The material
    c. The finish
    d. The cost

3. Insert both attributed top views to create the roof shown on a third drawing.

4. Download the part data onto a spreadsheet so that cost for the beams can be calculated.

5. Document the time taken to create the drawing, the area of the interior, and the size of the file.

6. Accessing one file as an XREF, compile a drawing including both parts, the overall roof layout, and an attributed title block.

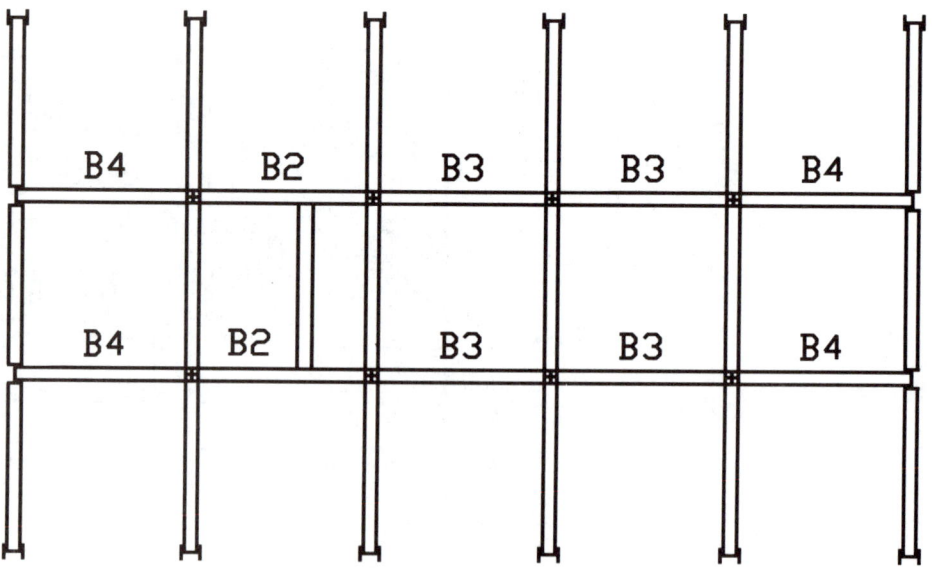

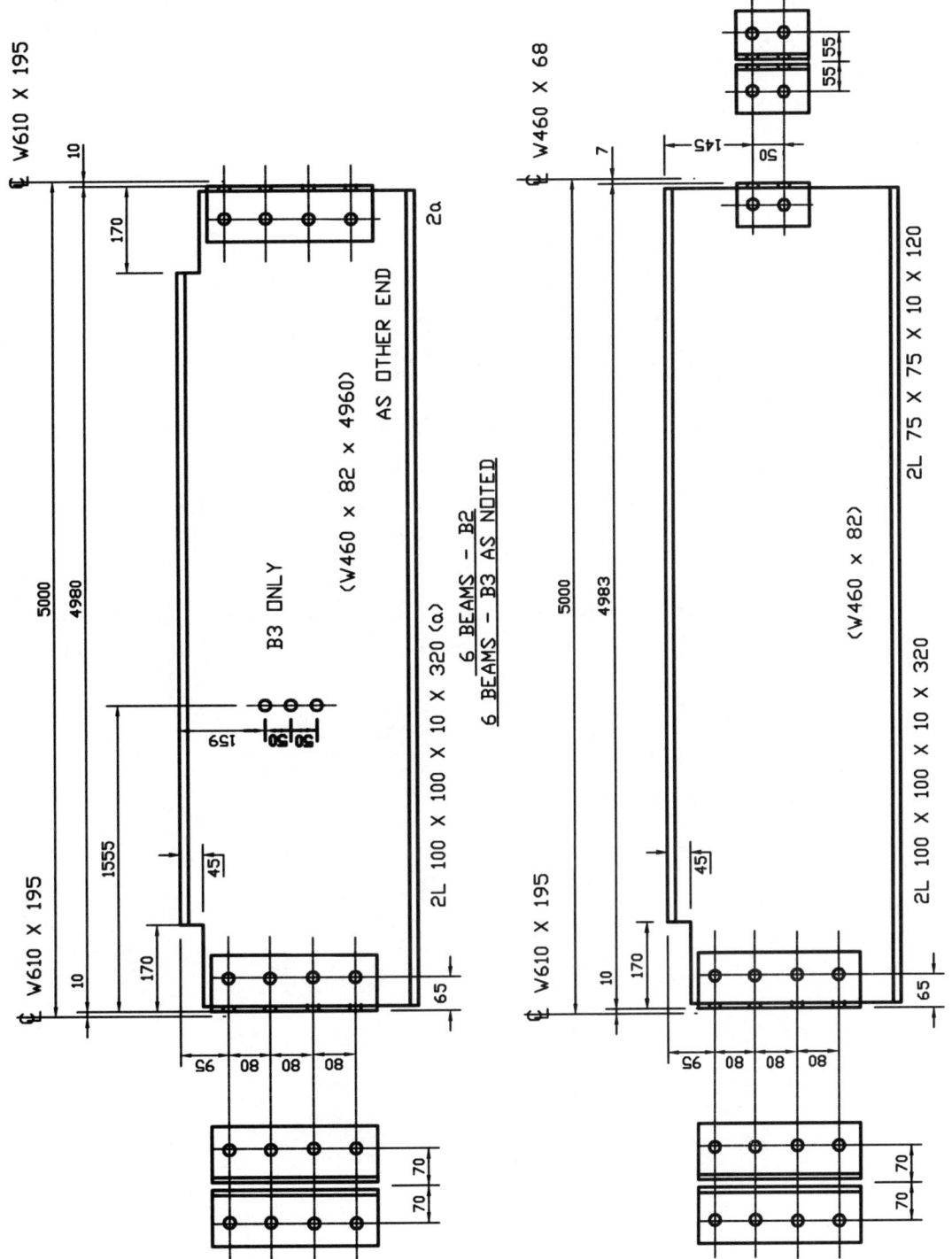

Final Tests and Projects

Mechanical Final

1. Draw two separate pulleys from the information provided. On these drawings show a front, a section, an isometric, and a sectioned isometric.

2. Create an ATTRIBUTEd BLOCK of the front view with the following:

    a. The product number

    b. The material

    c. The finish

    d. The cost

3. Insert both attributed front views on a third drawing.

4. Download the part data onto a spreadsheet so that the pulleys can be compared.

5. Document the time taken to create the drawing, the area of the interior, and the size of the file.

6. Accessing one file as an XREF, compile a drawing including both parts as well as an attributed title block.

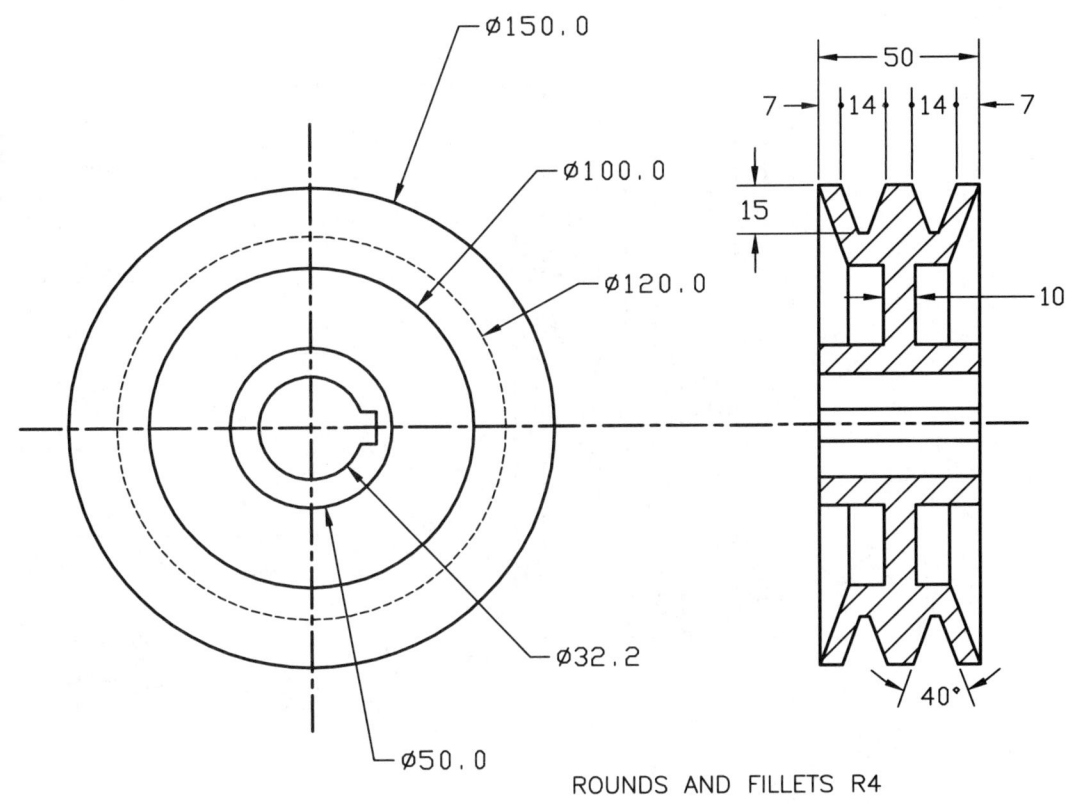

ROUNDS AND FILLETS R4
KEY SEAT 5D X 10W

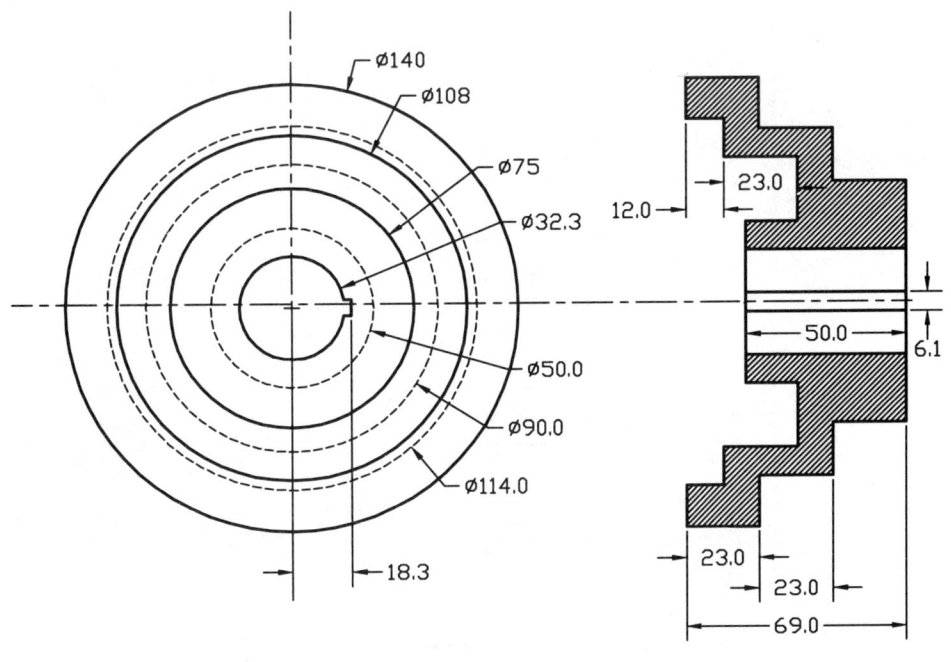

Challenger Final

Using the information on this and page 396, create an exploded isometric view of an assembly as seen in Challengers 10, 16, and 25. Add attributes to each part, and create a part list from that information. Note your time. Create a slide show of each individual part, and the assembly as a whole.

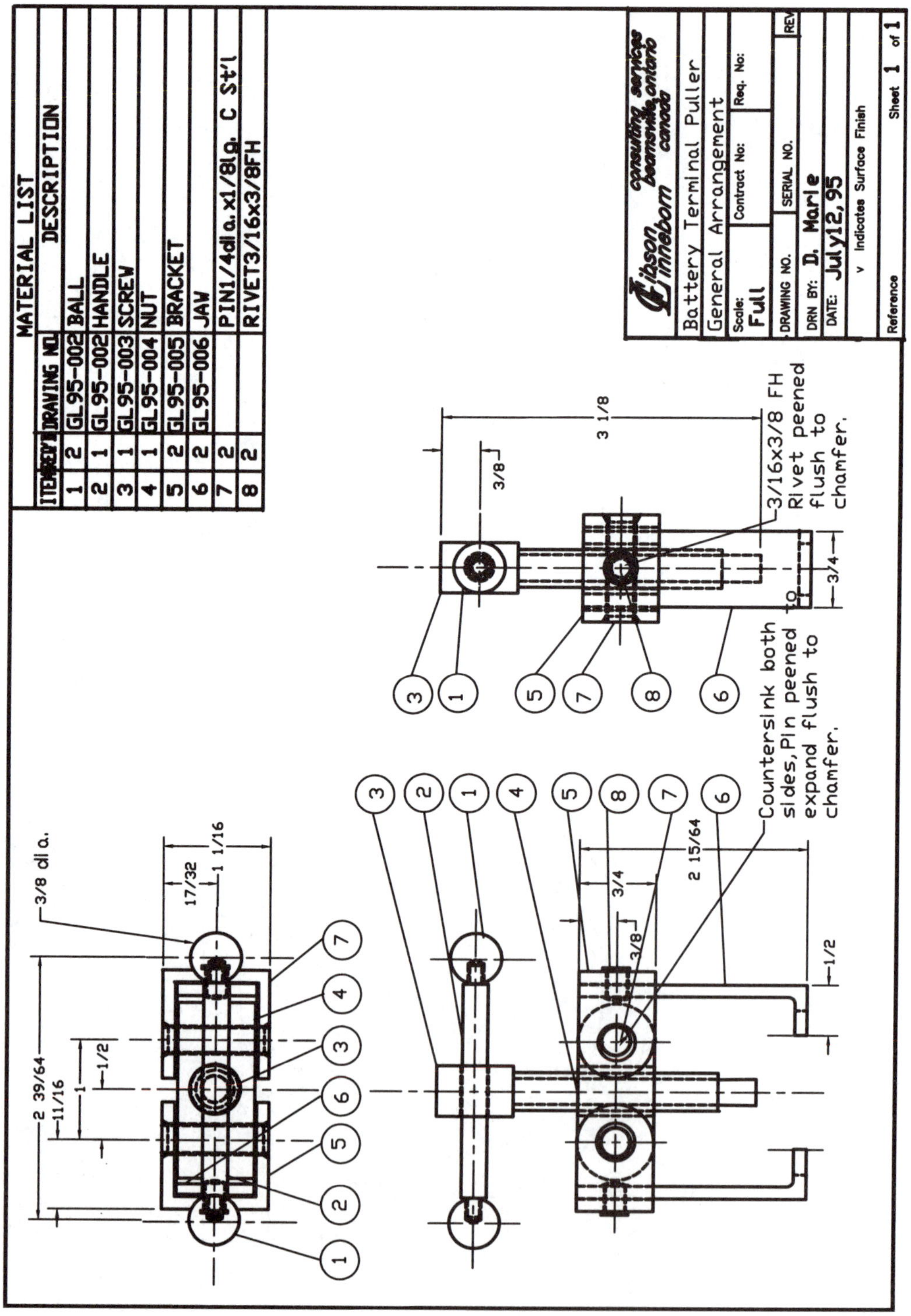

Appendix A
Glossary of Terms

a, A   **absolute coordinates**   Points located in space relative to the file's fixed origin, used to locate points on objects.

ADS (AutoCAD development system)   A programming interface that allows third-party developers to include applications written in high-level languages such as C.

alias   Shorthand for an AutoCAD command. (See Appendix C, Abbreviations and Aliases.)

alpha character   Any letter from A to Z.

alphanumeric   Any letter or number. Alphanumeric screens are ones that show only alphanumeric characters and no graphics.

annotations   Text, dimensions, tolerances, symbols, and notes.

ASCII (American standard code for information interchange) **code**   The code which describes each alpha, numeric, or special character in computer language. This code translates an M or a 2 or a * into a series of 0s and 1s that are understandable to a large percentage of computer languages.

aspect ratio   The image height-to-width ratio on a CAD display screen.

associative dimension   A dimension that adapts as the associated geometry is modified.

attribute definition   An AutoCAD object that serves as a template for assigning attribute values to drawing objects. (For attributes, see Chapter 14.)

AutoDESK   The company that produces AutoCAD.

AutoLISP   A programming language built into the AutoCAD program. It is an open program and users are encouraged to learn AutoLISP in order to create their own programs. To enter an AutoLISP program, use the following with ( ) and "".

```
Command: (load "d:\path\file name")
```

b, B   **basepoint**   1. In the context of editing grips, the grip that turns to a solid color when picked to specify the focus of the subsequent editing operation. 2. A point for relative distance and angle when copying, moving, and rotating objects.

Bezier curve A polynomial curve used in B-spline curve calculations, defined by a set of control points representing an equation of an order one less than the number of points being considered.

bit map A digital representation of a display image as a pattern of bits, where each bit maps one or more pixels. Multiple bit maps may be used in color graphics to assign values to each pixel, which are used as indices to the color look-up table, if one exists.

blip marks or blips Temporary screen markers displayed by AutoCAD when you indicate a point on-screen.

block The name, base point, and set of objects you create with the BLOCK command.

block reference Insertions of a block created by the INSERT command. Also called a *block instance*.

b-spline A mathematical representation of a smooth curve.

b-spline curve A curve defined by a set of control points. Also called a *NURBS curve*.

b-spline surface A mathematical description of a 3D surface which passes through a set of B-splines, e.g., Bezier, Coon's.

c, C

Cartesian coordinate system A coordinate system defined using three perpendicular axes (X, Y, Z) to specify locations in 3D space.

center line Line that radiates from the center mark of a dimensioned circle.

character An instance of a numeral, letter, or other linguistic, mathematical, or logical symbol.

character font The style of a character set.

chord A line segment joining two points in a circle or arc.

circular array Multiple copies of drawing objects around an arc or circle.

circular external reference An externally referenced drawing (XREF) that references itself directly or indirectly. The XREF that creates the circular condition is ignored.

command line A text area reserved for keyboard input, prompts, and special messages.

constructive solid geometry (CSG) The method of using intersection, union, and subtraction operations to construct composite solids.

coordinates Cartesian coordinates overlaid on the number space of the display screen. A pair of numbers (X, Y) or a triplet of numbers (X, Y, Z) that correspond to a point on a plane (X-Y) or in space (X, Y, Z).

crosshair Crossed horizontal/vertical lines representing a cursor, with the intersect being used to indicate desired device coordinates.

cross section A view of a part formed by the intersection of the part and a cutting plane.

cursor A symbol or a pair of intersecting lines on a video display screen that can be moved around to place textual or graphic information. Also called *graphics cursor*.

d, D   **database**   A comprehensive collection of information having predetermined structure and organization suitable for communication, interpretation, or processing.

default   Predefined value used for a program input or parameter, shown in AutoCAD within angle brackets <>.

definition points   Points for creating an associative dimension. AutoCAD refers to the points to modify the appearance and value of an associative dimension when the associated object is modified. Also called *defpoints*.

digitize   To enter graphical points into a computer from a data tablet with a puck or stylus.

digitizer   A device that tracks the relative position of the cursor for the purpose of recording the relative location or item.

diskette   A magnetic data storage device; also known as a disk.

dithering   To increase the variations of color or intensity on raster displays by trading picture resolution for patterns of pixel arrays.

DOS (disk operating system)   The Microsoft program that controls the CPU (central processing unit), output peripherals, etc.

drag   To move an object across the display screen using a puck, mouse, or stylus.

.DWG   File name extension of a drawing file.

e, E   **electrostatic printer/plotter**   A computer output peripheral that prints and plots by placing electrostatic charges on small areas of treated paper in desired patterns, upon which toner is spread and baked.

endpoints   Either of the points that mark the end of a line, arc, circle, or other primitive.

entity   Fundamental building blocks which the designer uses to represent a product—lines, arcs, ellipses, text, splines. Also known as an *item* or *object*.

f, F   **face**   A finite, planar, cylindrical, conical, spherical, or toroidal surface on a solid model.

fill   To fill an area of the display surface bounded by vectors with a solid color or pattern.

font   The style of a letter or character. A character set, comprising letters, numbers, punctuation marks, and symbols, of a distinctive size and design. Also called *typeface*.

freeze   To ignore the objects on specified layers when regenerating a drawing, thereby shortening regeneration time. Objects on frozen layers are not displayed, regenerated, or plotted.

function keys   A keyboard entry that saves typing time; a toggle switch.

g, G   **graphics screen**   The area of the AutoCAD screen used for creating and editing graphics.

gray scale   An ordered description of the tonal levels of an input image.

grid   Uniformly spaced points which create a visual drawing aid to determine distance.

grips   Small squares that appear on objects you select. After selecting the grip, you edit the object by dragging it with the mouse rather than editing commands.

h, H

half-space   The portion of 3D space that lies on one side of a surface. If the surface is planar, the half-space is known as a planar half-space.

handle   A unique alphanumeric representation of an object in the AutoCAD database.

hard copy   Any printed or plotted printout.

hatch   Regular pattern filling an enclosed area.

hidden lines   The line segments which would not be visible to a viewer of a 3D display item because they are "behind" other parts of the same or other display items.

hidden surfaces   Surfaces obscured by other surfaces from a specific viewpoint.

hole   1. A closed hatch boundary within another closed hatch boundary. 2. A feature of a solid.

i, I

icon   A graphic image of a function or facility to help either choose or recognize your position and options. The UCS icon in the bottom left corner of your screen is an example.

icon menu   A menu that contains multiple image tiles and can be customized by editing the ACAD.MNU file.

IGES (initial graphic exchange specification)   An ANSI-standard format for the digital representation and exchange of information between CAD/CAM systems.

image file   An AutoCAD command in pictorial format.

ink jet plotter   A plotter which uses electrostatic technology to first atomize a liquid ink and then control the number of droplets that are deposited on the plotting medium.

initial environment   The variables and settings for new drawings as defined by the default prototype drawing, such as ACAD.DWG.

instance   When used in conjunction with blocks, a copy of a BLOCK created and stored apart from the actual drawing or model. Every time the BLOCK is INSERTed, it is one instance or copy of the block, but the block remains intact in memory. This is like a rubber stamp, the impression of the stamp being an instance. Also called a *block reference*.

integer value   Many commands have options available by choosing a number such as 0, 1, 2, etc. These are integer values for the command.

interactive graphics   The use of a computer terminal to generate graphics command-by-command, as opposed to batch processing.

interpolation points   Points that a curve or surface pass through to define the curve or surface. Also called *fit points*.

island   An enclosed area within the hatch area.

isometric drawing   A drawing of an object with the X, Y, and Z axes spaced 120 degrees apart and the Z axis projected vertically.

k, K

key   Specifies a database location associated with a key value for searching the database.

l, L **laser plotter** A plotter which produces images through the use of a laser (light amplification through stimulated emission of radiation).

layer A logical grouping of data, like transparent acetate overlays on a drawing. You can view layers individually or in combination.

linetype Defines the display of a line or other type of curve. For example, a continuous line has a different linetype from a dashed line. Also called *line font*.

link An SQL connection between an AutoCAD object and an external database record.

lock file A binary file, created when you pen an AutoCAD data file, that contains permissions, which determine if other AutoCAD users can read or write to the opened file.

loop A closed statement that repeats.

m, M **macro** A single command made up of a group of commands.

mass properties Calculation of physical engineering information about a part, e.g. perimeter, centroid, volume, weight, and moments of inertia.

***M* direction, *N* direction** In the matrix that determines a polygon mesh, the *M* direction is established from the first to the second row, and the *N* direction is established from the first to the second column.

menu A list of options or functions displayed on the video screen.

mnemonic An abbreviated command entry scheme for simplifying command input. For example, CP can be a mnemonic for the COPY command and Z for the ZOOM command. (See also *alias*.)

model A geometrically accurate representation of an object. In AutoCAD, graphic data is often referred to as a model as opposed to a drawing because the data is not always used to create drawings, particularly in 3D applications.

model space The original position of the origin and axes with regard to the model. With the world coordinate system you are using model space.

mouse A data entry device that echoes the position of the cursor on the screen and helps the user to define a point or item that is desired. Much like a *puck*.

n, N **nesting** Embedding data in levels of other data so that certain routines or data can be executed or accessed continuously or in a loop.

node A point object. Used as an object snap in MEASURE and DIVIDE commands.

null response To accept the default by pressing the space bar or ⏎.

o, O **object** A primitive, such as a line, circle, or polyline, treated as a single element for creation, manipulation, and modification.

object snap (OSNAP) Modes for selecting commonly needed points on an object while you create or edit an AutoCAD drawing.

on-line documentation Information about the commands within the database.

operating system The microcomputer software program which controls the CPU and the input/output peripherals, and provides the working of programs.

origin The fixed *X*0, *Y*0, *Z*0 of the model. The point on the coordinate system whose values are all zero.

ortho mode   An AutoCAD setting that limits pointing device input to horizontal or vertical relative to the current snap angle.

p, P

pan   To move from one zoomed-in view to another without changing size. A horizontal translation.

paper space   An AutoCAD state for creating a finished layout for printing and plotting, as opposed to doing drafting or design work. Model space is the state for creating the drawing. Although both 2D and 3D objects can exist in paper space, commands that render a 3D viewpoint are disabled.

pixel   The smallest section of your screen; the dot resolution of an image. The discrete display element of a raster display represented as a single point with a specified color or intensity.

platform   A computer system, for example, the DOS platform, or Windows platform.

pline   Polynumeral line. A line composed of many different vertices.

polygon window   A multi-sided selection window for selecting objects in groups.

primitive   1. The simplest and most basic geometry you can create: LINEs, CIRCLEs, ARCS, etc., are all primitives. 2. A solid or region building block such as a box, wedge, cone, cylinder, sphere, and torus.

prompt   Any message or symbol from the computer system informing the user of possible actions or operations. A guide to the operator, indicating possible actions or options.

prototype drawing   A drawing file with preestablished settings for new drawings, such as ACAD.DWG, ACADISO.DWG. Any drawing can be used as a prototype drawing. (See also *initial environment*.)

puck   A mouse, the moveable cursor assembly used with a digitizing tablet to locate points accurately for input.

r, R

raster scan   Line-by-line sweep across the entire display surface to generate elements of a display image.

read-only   A readout that can be read but not edited.

redraw   To quickly refresh or clean up the current viewport without updating the file's database.

relative coordinates   Incremental coordinates. Coordinates specified relative to previous coordinates.

rendering   A shaded and hidden line image of solid objects.

resolution   The number of horizontal and vertical rows of pixels that can be displayed by a particular graphics controller or monitor. For example, a standard VGA graphics controller and color monitor has a resolution of 640 columns and 480 rows of pixels.

RGB color   A color described in terms of its red, green, and blue intensity levels.

right-hand rule   A method of determining which is the positive direction of rotation around the origin by using the right hand to point to the X (thumb), Y (index finger), and Z (middle finger).

ROM (read only memory)   Contains the commands that start the computer and address the various peripherals. This is a read-only chip and cannot be changed.

rubber band A line that stretches dynamically in conjunction with your cursor during many editing and drawing commands.

running object snap Setting an object snap mode so it continues for subsequent selections.

s, S

scale factor A number which multiplies the vector endpoint coordinates to produce scaling.

screen menu A list of commands you can display and execute beside the AutoCAD window while the graphics screen is active. The screen menu is automatically displayed in DOS but needs to be invoked with the ACAD.MNU in Windows.

script file A set of one or more AutoCAD commands executed sequentially with a single SCRIPT command. Script files are created outside AutoCAD using a text editor, saved in text format, and stored in an external file with the extension .SCR.

selection set One or more AutoCAD objects specified for processing as a unit.

selection window A rectangular area drawn in the AutoCAD graphics area to select objects in groups.

slide file A file that contains a raster image or snapshot of the display on the graphics screen. Slides have the extension .SLB.

slide library A collection of slide files organized for convenient retrieval and display. Slide library names have the extension .SLB and are created with the SLIDELIB.EXE utility.

snap A drawing aid function which allows you to place entities at a preset spacing.

standard toolbar In Windows, the portion of the AutoCAD user interface that appears under the title bar by default in the graphics window. It contains a layer control field, a coordinate display field, a box displaying the current color, and a series of macro buttons.

string A sequence of characters.

stylus A device analogous to a pencil which is used in conjunction with a data tablet to input coordinate information.

swap files Files needed by AutoCAD to create your temporary file, typically with hexadecimal numbers and the extension .SWR. Normally, these files are erased when you exit the program, but if the system locks for any reason, you can erase these files from your directory.

system variable A name that AutoCAD recognizes as a mode, size, or limit. Read-only system variables such as DWGNAME cannot be modified directly by the user.

t, T

tablet An input device which digitizes coordinate data indicated by stylus position.

temporary files Data files created during an AutoCAD session and deleted when you exit the file. If the session ends abnormally, either by power failure or by removing the diskette from the drive while addressing that drive, temporary files might be left on the disk.

tessellation lines Lines displayed on a curved surface to help you visualize the curved surface better.

text style A named, stored collection of settings that determines the appearance of text characters.

third-party developers Companies offering software enhancements to AutoCAD users that are based on AutoCAD software. AutoCAD has maintained an open policy toward such developers, offering a great deal of support to those who would like to customize their software for specific purposes.

toolbar Part of the AutoCAD interface containing icons that represent commands.

transparent command A command started while another is in progress. Transparent commands must be preceded with an apostrophe (').

u, U

UCS (user coordinate system) The three-axis coordinate system that can be rotated and placed at any location in order to help you with the creation of your model.

unit A user-defined distance, such as kilometers, inches, meters, and miles, used as a standard of measurement in a drawing. The default is inches.

v, V

vector A mathematical straight line which has both magnitude and direction.

vertex A topologically unique point in space used to define boundaries.

viewport A bounded area that displays some portion of a drawing's model space. The TILEMODE system variable determines the type of viewport created. (See *paper space*.)

virtual screen A pixel map of the current regenerated view. This is stored in memory and determines the speed of REDRAWs with regard to the percentage of the drawing that is regenerated in addition to the actual on-screen image.

w, W

window A selected rectangle for image display or processing.

wireframe A representation of a solid object that displays edges and tessellation lines.

world coordinate system (WCS) The original position of the origin and axes on the model.

WPolygon A multisided polygon window used to select objects within its borders.

wraparound The phenomenon whereby a vector which overflows the number space is continued on the opposite edge of the drawing.

x, X

Xref A file referenced to the current file but still on the main disk. (See Chapter 18.)

z, Z

Zoom The process of reducing or increasing the apparent magnification of graphics on the display screen.

Appendix B
File Extensions

.3ds	3D Studio files
.ads	Automatically loaded ADS application ACAD.ADS
.ahp	AutoCAD Help files ACAD.AHP
.bak	Drawing file backup
.bat	Batch file
.bin	Binary image file
.bxn	Emergency backup file
.c	ADS source code files
.cc	ADS source code files
.cfg	Configuration file ACAD.CFG
.com	Machine language command file
.dcl	Dialog box definition files ACAD.DCL
.dcc	Dialog box color control files (DOS only)
.dct	Dictionary files
.dwg	Drawing file ACAD.DWG (standard prototype drawing)
.dwk	File lock
.dxb	Binary drawing interchange file
.dxf	Drawing interchange file (ASCII or binary)
.dxx	Attribute extract file in DXF format
.eps	Encapsulated post script
.err	Error file ACAD.ERR (error log file)
.exe	DOS executable file ACAD.ERR
.exp	Driver files and ADS executable files (DOS only)
.hpg	Hewlett Packard Graphics

.hlp	Windows Help file ACAD.HLP
.lib	ADS library files
.lin	Linetype library file ACAD.LIN (standard line library)
.lsp	AutoLISP application files
.lpt	Line printer
.mnc	Compiled menu file (DOS only) ACAD.MNC
.mnu	Menu template file ACAD.MNU (standard menu template file)
.mnx	Compiled menu source file
.msg	Message file ACAD.MSG (message displayed by ABOUT command)
.old	Original version of converted drawing file
.pat	Hatch pattern library file ACAD.PAT (standard patterns)
.pgp	Program parameters file ACAD.PGP (standard parameters)
.pfb	PostScript font files
.pfm	PostScript font metric files
.plt	Plot file
.prp	ADI printer plotter output file
.pwd	Login file
.scr	Script file
.shp	AutoCAD shape file
.shx	Shape/font-definition source file
.slb	Slide library file ACAD.SLB (slide library file)
.sld	Slide file
.tif	Tagged image file
.ttf	TrueType font files
.txt	Attribute extract or template file
.unt	Units file ACAD.UNT (units conversion file)
.xlg	External references log file
.$ac	AutoCAD temporary file
.$a	AutoCAD temporary file

Appendix C
Abbreviations and Aliases

AutoCAD's .PGP file is an ASCII-based text file containing AutoCAD program parameters. Within this file you can abbreviate frequently used AutoCAD commands by defining aliases for them. You can make any command, device-driver command, or external command into an alias.

Any ASCII editor can be used to edit the ACAD.PGP file. Any AutoCAD, AutoLISP, ADS, Solids, operating system or graphics display-driver command can be abbreviated. Once you have altered the .PGP file, simply SAVE it and the commands will be accessible through the alias.

These two comma-delimited fields define a command alias in the ACAD.PGP file:

abbreviation,*command

abbreviation   The abbreviation of the command that you enter at the command prompt.

command   The command that it invokes.

For example, in the .PGP file, type in:

m,*move   This will make the MOVE command accessible from typing in only the letter m.

The following is a partial list of AutoCAD and Solids commands for creating abbreviations or aliases:

A,	*ARC	P,	*PAN
C,	*CIRCLE	PS,	*PSPACE
CP,	*COPY	PL,	*PLINE
DV,	*DVIEW	R,	*REDRAW
E,	*ERASE	Z,	*ZOOM
L,	*LINE		
LA,	*LAYER	SUB,	*SUBTRACT
M,	*MOVE	CONE,	*CONESOLID
MI,	*MIRROR	SOL,	*SOLIDIFY
MS,	*MSPACE	UN,	*UNION

Appendix D
Plotting and Printing

Once your drawing is finished and you want a hardcopy or plot, make sure that your title block is complete, then use the Print PLOT command to create either a plot or a .plt file that can be taken to a plotter. If your computer is linked directly to a plotter or printer, simply plot directly. If your computer is not linked directly to a plotter, create a .plt file on a floppy disk, and take it to load onto the plotter.

Plotters and Output Peripherals

Before we consider how to get your drawing or data actually plotted, it is a good idea to outline some of the common plotters to see what they are and how they are used.

There are many types of plotters and printers available, and prices are coming down while capabilities advance. The most common plotters and printers include pen plotters, electrostatic plotters, dot matrix, and laser printers.

At the high-end of plotters—those capable of 3D plotting—are mills, lathes and stereolithography, a process which manufactures a plastic prototype by using photo polymers and lasers. These will not be discussed here because much more software is needed to produce a final hardcopy product.

Pen Plotters

The most well-known vendors of plotters are Calcomp, HP (Hewlett-Packard), Houston Instruments, Roland, Versatec and Ioline. The criteria used to evaluate these types of plotters are:

RESOLUTION The measure of the smallest dimension the plotter can legibly draw. Typical values are in the 0.001" range.

REPEATABILITY The maximum distance the plotter would be out if it were to draw the same object twice in the same place. Typical values are 0.002"-0.005".

SPEED Usually the diagonal speed of the pen, measured in ips (inches per second). Typical speed is 14-18 ips.

BUFFER SIZE The plotter's built-in memory that frees the computer to do other things while graphics data are sent to the plotter. Graphics data are sent by processed vectors. Typical buffer sizes are 0-8 MB (megabytes). This may be the most important factor to consider because in many plotters, plots simply will not work if the drawing is too large for the buffer. Some plotters can split a file and plot it in sections, but this is very time consuming.

NUMBER OF PENS The size of the carousel and the number of pens it will hold. Typical size is 2-8 pens.

PAPER SIZE The largest size that can be plotted on large plotters is A-D size, on smaller plotters, B.

ACCURACY The measure of the difference between the length of a line being sent to the plotter and the length of line actually drawn. This value is usually quoted as a percentage, typically 0.1%.

PRICE Can range from $2,000 to $30,000, depending on size, performance, and number of pens. Other features to consider are single sheet feeders, continuous-feed paper, control panels, etc.

If your system cannot multitask, the computer's CPU (central processing unit) can be tied up while a drawing is being plotted. A "black box" can be used as a buffer to store the drawing until the plotter can process it, thus freeing the CPU.

When using pen plotters in the classroom, there are several variables that must be considered.

PENS There are several types of pens available, ranging in price from less than $1.00 to $75.00. Felt tip and ball-point are the most popular types. Regardless of the type of pen you are using, it needs to be capped when the drawing is completed, and in many cases, should be left in a horizontal position. Many students tend to overlook this, so you may need to supply your own pens to assure that they are always in working order.

Make sure the pens are appropriate for the type of paper you are using. Pens are also available for mylar and vellum. Note: these are not usually the same pens used in normal plots.

PEN SIZES In addition to having pens with different colors, you can also have different pen widths. Pen sizes can be programmed to create different line widths. Different line types can be assigned to different pen numbers.

PAPER A series of small wheels holds the paper on the plotter. The plotter itself often reads the size of the paper; thus, there is a certain portion of the paper held in place while the rest is being read. To create an A-size plot the paper will need to be slightly larger than A-size drafting paper. Before you start your plot, make sure that your paper is large enough to both hold your drawing and allow for the plotter's mechanics.

Paper quality is also important. If you are testing a plot before using good paper, try it out for size on a sheet of paper, but *never plot on blueprint paper*—one plot alone can effectively demolish a pen.

Electrostatic Plotters

This type of plotter can produce "photographic images," and is excellent for drawings as well as presentation graphics. Also known as a Xerographic™ printer, it forms an optical image on specially treated paper. Dark areas are charged and light areas are uncharged; ink adheres to the charged areas. The paper is heated, causing the ink to melt and adhere to heated areas.

Electrostatic plotters are not usually found in classrooms, but if you require a high-resolution image you can bring your floppy disk to a graphics store and have it plotted for a fee. Some large computer stores may also offer this service.

Dot Matrix Printers

Dot matrix printers form each character from a pattern of ink dots. The dots are printed by a print-head containing a line of seven or nine hammers that strike against a ribbon. Inside the printer, a ROM chip stores the pattern of ink dots associated with each character. The printer receives the ASCII character code for a letter, and the ROM chip "tells" the print-head which hammers to strike on the ribbon. A motor-driven belt moves the print-head back and forth as the paper is spooled forward.

Dot matrixes are used in both printers and plotters, allowing letter quality printing and reasonable graphics up to a C-size drawing. Dot matrix produces an adequate product at a reasonable cost.

When plotting on a dot matrix, the output is usually not to scale. It is important to make sure that the printer's storage or memory is sufficient for graphics. Dot matrix printers are useful for many reasons, but they are particularly desirable in classrooms because all that needs regular adjustment is the paper and the ribbon.

Dot matrix printers are usually used to create quick prints of drawings or to check plots. Though they are not to scale, they give an idea of what the model looks like.

Laser Printers

The laser printer operates on a principle similar to that of the electrostatic plotter. It uses a laser to create an image by charging the paper with electrostatic energy. With lasers though, the paper is not specially treated.

Fast and silent, this is a popular choice among low-cost printers. Laser output is usually impressive. Difficulties may be encountered, however, in making the plotter compatible with the existing system. When buying a laser for your home, make sure the store will take it back if it is not compatible. Again, when dealing with graphics, the size of the buffer is very important.

Get the largest buffer you can afford, and before purchasing the laser, test the size with a large plot.

Like a dot matrix printer, a laser does not produce drawings to scale (unless they are on an 8.5 × 11" sheet). But the final product is both pleasing to the eye and easily reproduced.

Pen widths can be programmed within the plot command.

The PLOT Command

While the drawing is on screen, access the PLOT or PRINT command:

WINDOWS From the Standard toolbar, choose

DOS From the File menu, choose Print.

The command line equivalent is **PLOT**.

Once you have accessed the PLOT command, you will get the main plot menu.

If you want to accept all of the defaults, press OK and your plot will be produced. In this screen, the plot will be produced on an HP plotter at a size of 8" × 11.5". The plot will be anything that is on the screen. The plot will be scaled to fit the paper.

Should you wish to change these defaults, pick the related button.

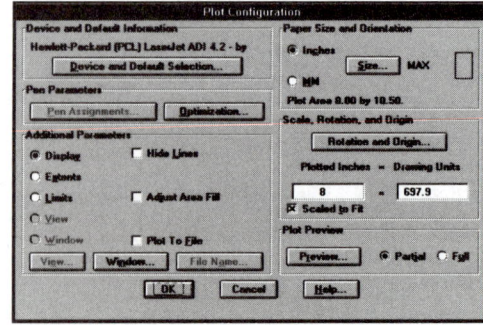

Device and Default Selection

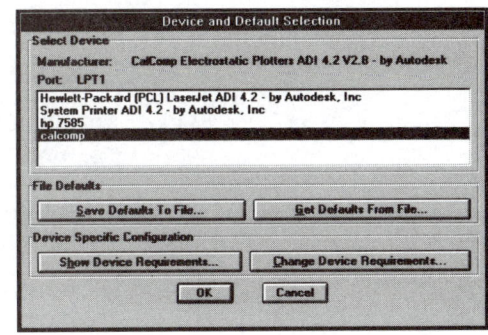

There is a large selection of plotters available. To access a new plotter, pick the Device and Default Selection button, then pick the device you require.

This screen will offer you a list of the loaded plotters. Pick one and continue.

Configuring a New Plotter

If the plotter you need is not on the list provided, use the CONFIG command to load a new one. At the command prompt, type in CONFIG. Then answer the questions for loading a new plotter driver. You will need to pick the driver you want, and provide a name for it. If there are questions you are not sure of, accept the default; it will probably work on the new plotter if it worked on the last one.

Note: Selecting another plotter may change the settings of other parameters in the plotting dialog box.

Pen Widths

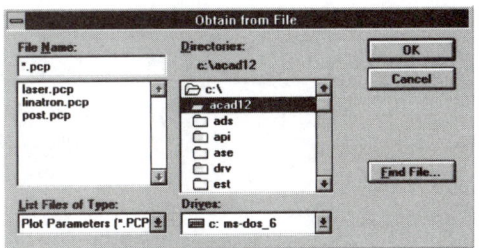

Each object in a drawing has a color associated with it. Depending on the pen plotter, you can plot each color with a different pen. If you are using a plotter that has different pen widths, these can be loaded with the pens themselves; or you may be able to change the pen widths in the pen parameters area.

If you have a list of pen widths on a .pcp file, use the Set Defaults From File button to pick the preset pen widths.

Similarly, if you have been "fiddling" for a while with your pen widths and finally find a set-up that works, save the pen settings with the Save Defaults to File button so you can have the same results for every plot you create.

Pen Parameters

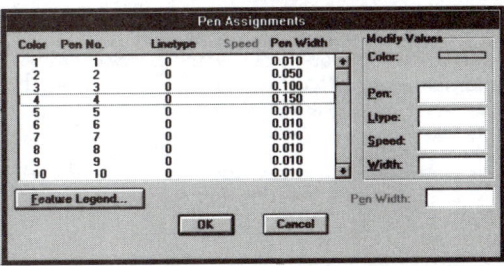

This controls the pen parameters for plotters with multiple pens, hardware linetypes, software-controlled pen speeds and pen widths.

Some plotters allow you to change pen widths, others do not. If you can change the pen width, use the Pen Assignments button to set the line width.

The size of the pens, once determined, can be set with the Save Defaults to File option under Device and Defaults Selection.

Additional Parameters

This area allows you to plot either a portion or all of the image on screen.

In this case, the default <D> means that the last person using the PLOT command plotted the Display. Make your choice:

Display takes everything on screen; in paper space from all viewports, in model space only from the current viewport.

Extents takes all portions of the drawing that contain entities. By using ZOOM Extent before plotting, you will be able to check that there is no "floating geometry" outside the limits which will be plotted.

Limits takes only that portion of the file contained in the limits.

Window allows you to select a rectangular portion of the file, a detail, or the whole view. It can only be used if you are in the Drawing Editor.

View takes a stored view of the data. Like Window, it can be all or a portion of the file. In order to plot a View, you must save a view prior to using the PLOT command. This option is particularly useful when you are plotting from the Main Menu because you can be sure of what you will get.

Hide Lines allows you to hide lines in a 3D drawing. The Hideplot option of MVIEW is usually much quicker.

Adjust Area Fill adjusts the area fill for polylines and solids. If you are getting a "pin stripe" effect, use this option to get a solid image.

Plot to File will create a .plt file for transfer into a word processing file using a HP driver. It is also used for transferring a plot file to a plotter as in an off-site Calcomp plotter. If your plotter is not attached to your computer, and you need a .plt file, use this option.

Paper Size and Orientation

First decide if you want to plot in inches or millimeters. Then choose the size of paper you need.

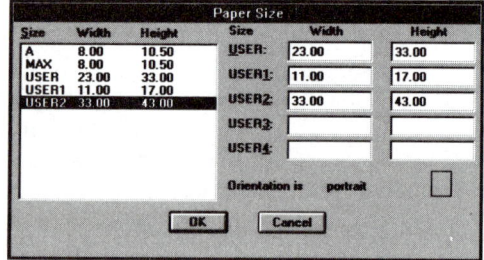

Depending on the size of drawing you require, pick the preset drawing size, or set a user-defined size.

Danger

This is where most students have problems. Do you know what size the paper is? Does the plotter require two or three inches to compensate for rollers?

The drawing sizes may have been set up during your software installation to agree with the available sizes of your plotter. The standard sizes should agree with national standards, and there will be an option for larger or custom size plots. Simply choose the letter designation in response to the plotting size prompt.

To create a user-defined size, select User, then enter the size of your paper.

Paper Size

It is a very good idea to have a block that includes the title block plus the actual size of the paper on which you will plot.

Insert this in paper space on the final drawing before plotting to make sure your drawing will fit nicely onto the paper.

Scale Rotation and Origin

Always do a plot preview before creating your plot file. If this shows only a portion of your file, use the rotation and scaling options to set the file correctly. If you pick Scaled to Fit, you will not be able to scale from the drawing.

Scale

The drawing will be set in terms of plotted inches : drawing inches or plotted mms : drawing units. Try to start plotting directly with the proper scale so that your plots will always be correct. Use Fit only when the actual scale of the drawing is not needed.

The plot origin is always assumed to be the bottom left corner of the paper. If you want to relocate the origin of the plot to another portion of the paper, reset this parameter by changing the X and Y values in the edit box. This is usually done to place a title block onto an existing drawing or to place a missing view onto an otherwise completed drawing. Some skill is needed here, so try placing the view or plot onto a test sheet before placing it on the final paper—especially if there are drawings on the paper that could be destroyed.

The default on a plotter is to have the drawing's X value perpendicular to the bar of the plotter, and the Y value parallel to the plotting bed (as the drawing usually looks on screen). For printers, however, the X value of the drawing is taken to be the 8.5″ edge of the paper, and the Y value is taken to be the 11″ edge of the paper. Therefore, you will probably *not* want to change this on a pen plotter, but *will* want to change it for a printout. The rotation you choose will become the default for the next plot. The final readout before printing will tell you the size of your plotting area, and will give you the option of checking the plot rotation.

Plot Preview

This will give you an idea of how big your plot is, and how it fits on paper. Use this option before you spend the time or money on a paper plot.

An Example

If you are creating a .plt file to be transferred to a Calcomp plotter not attached to your computer, use the following options. Prior to plotting, make sure that your text sizes are not too large and that your thin lines are all in one color (1); thicker lines should be in another color (2), and your border should be in another color (3). If you have points in white, and white is a very wide line setting, you will end up with blobs all over your drawing.

1. Change the plotter to a Calcomp plotter.
2. Change your pen assignments if they are not preset.
3. Pick Window to pick the outside of your drawing area.
4. Pick Plot to File, then File Name to set the .plt file name.
5. Choose the Size and Orientation of your drawing.

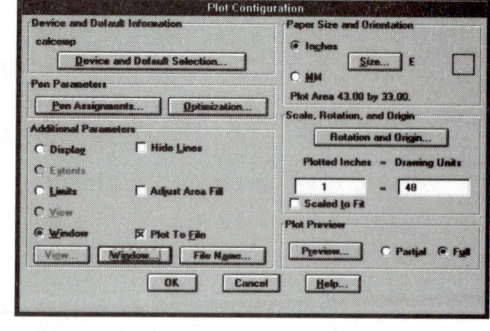

6. If you are plotting to a scale of 1/4'' : 1'0'', put your scale in as 1 plotted inch for every 48 drawing sizes. These scales can be found on page 278 of your text.
7. Do a Full preview of the plot to make sure that it is OK. If it is OK, continue. If not, change the options until it looks correct on the preview.
8. Then pick OK.

Trouble Shooting

Make sure you have a backup copy of the file before you plot (in case you get an error message on the disk). Remember to leave enough available memory on the disk; available memory must at least equal the size of the file you wish to plot. You cannot plot a file from a floppy disk that occupies more than 1/3 of the disk memory space.

Ideally, you should plot from a hard drive whenever feasible. But due to the size of many classes, this is not always possible.

Technical Difficulties

Plotter maintenance is mandatory. Pens will dry out if not recapped between plots. Cap them and put them away. In addition, pens run out of ink and, if felt-tipped, wear down. When making plots, remember that continuous lines improve plot speed. There is no point in having a hidden line if LTSCALE is too small to read. If your plot is taking a very long time, you may have forgotten to change the LTSCALE.

Practice at least one plot before your final project is due. Plots seldom work perfectly the first time.

Index

(period or dot) extensions, 489

A

Absolute coordinates, 6
acad.dwg file, 1
Add entry selection method, 114,
ALIGN command, **128**
Aligned dimensions, 168
ALL entry selection method, 83
Alternate units dimensioning, 176
Angle brackets, 38
Angles
    degrees/minutes/seconds, 20
    dimensioning, 171
    grads, 20
    input, 8
    overview, 8
    radians, 20
ANGULAR dimensions, 171
ANSI drafting standards, 275
Apparent Intersection snap, 45
APERTURE command, 47
ARC command, **11**, 24
Arcs
    add to polyline, 65, 66
    dragging, 12
    methods of specification, 24
    options, 12, 24
    radial dimensions, 170
AREA command, 338
ARRAY command, **125**
    array polar, 125
    array rectangular, 126, 127, 129
Arrow blocks, 172
Arrowheads, dimensioning, 172
ASCII, 420
Associative dimensioning, 180
ATTDEF command, 360
ATTDIA system variable, 363
ATTDISP command, 364
ATTEDIT command, 379
ATTEXT command, 383
Attribute definition, 360
Attribute extract, 364
Attributes
    blocking, 362
    defining, 360
    dialog boxes, 361
    displaying of, 364
    editing of, 379
    extraction of, 384
    options, 364, 380
    visibility, 364
    with Xrefs, 443
AutoCAD
    Release 12 converting files, iv, xx

B

BASELINE dimensions, 169
Begin AutoCAD session, 1, xi, xii
BHATCH command, **222**, 227, 236
    angle, 224
    advanced options, 227
    dialog box, 222
    editing, 228
    exploded option (*), 225
    overview, 222
    patterns, 223
    scale, 223, 224
    styles, 223
BLIPs, 17
BLOCK command, **252**
Blocks
    advantages, 251
    arrow, 172
    attributes and, 362
    color, 260, 437
    defining, 252
    editing, 258
    external, 255
    frozen layers with, 437
    insertion, 253
    instance, 254, 437
    layers, 260, 437
    linetypes, 260, 437
    listing, 254
    naming conventions, 261
    nested, 437
    output to disk, 255
    overview, 252
    place into drawings, 253
    redefining, 257
    removing from file, 261
    scaling, 257
    updating, 257, 259
    with XREFs, 439
Borders
    in PAPERSPACE, 290, 296
BOUNDARY command, 341, 350
Boundary
    Boundary Hatch dialog box, 222
    boundary set, 225
Boundary set, 225, 227
BREAK command, **46**
Browse Search dialog box, xxi
B-spline curves, 337
Buttons (mouse), xiii
BYBLOCK
    color, 260
    linetype, 260
BYLAYER
    color, 146
    linetype, 147

C

Calculating
    area, 338
    distance, 341
    point on a line, 15
Cartesian coordinates, 1
CDF attribute extract, 383
Center lines, 147
Center marks, 173, 184
CENter object snap, 42
CHAMFER command, **49**
CHANGE command, 149, **204**, 210
Change Properties, 149, 156
Changing Linetypes, 149, 205, 156
Changing colors outside of layers, 146
CHPROP command, **149**, 156
CIRCLE command, **11**, 25, 53
Circles
    creating, 6, 11
    isometric, 398
    filled, 69
    options, 11, 25
    solid filled, 69
    tangent to, 25
Circular (polar) arrays, 125
Closing lines, 9, 10
Closing polylines, 62
COLOR command, **146**
Color
    Blocks and, 261
    BYBLOCK, 261
    BYLAYER, 146
    changing, 149
    current, 146
Commands
    entering from keyboard, xv, xxiv
    entry, xv, xxiv
    repeated, 12
    transparent, 16
Command lines, xxii
Compatability with older versions of
    AutoCAD, vii

CONTINUE dimension, 169
Converting old drawings, xix
Coordinates
   absolute, 6
   display, 4
   entering, 6
   incremental, 7
   polar, 8
   relative, 7
   User Coordinate System, 2
   X,Y,Z point filters, 230
COPY command, **86**, 102
COPY Multiple, 86
Crosshatching, 221
Crossing polygon system variable, 48, 84
Crossing window entity selection method, 48, 84
Current Layer, 145
Current text style, 201, 202
Current UCS, 1
Cursor
   and paper space, 283
   target size, 47
Curves
   B-spline, 337
   fitting, 65
   spline, 337

D

Data entry
   angles, 8
   coordinates, 6, 7
   feet and inches, 18
   file names, xxi
   numeric values, 6, 7
   specifying points, 6, 7
   variables and arithmetical expressions, xi
   X,Y,Z point filters, 230
DDATTDEF command, 361
DDATTEXT command, 385
DDCHPROP command, 149
DDEDIT command, 205
DDGROUP command, 428
DDIM command, 171
DDINSERT command, 253
DDLMODES command, 143
DDLTYPE command, 147
DDLMODIFY command, 205
DDOSNAP command, 47
DDPTYPE command, 333
DDRMODES command, 41
DDUNITS command, 18
Default working environment, 1
Definition points, 166
DEFPOINTS layer, 152
Deleting
   line segments, 48
   multiline vertices, 347
   objects, 48
   toolbars, xvii
   unused blocks, 262
Dialog boxes, xxi
DIAMETER dimensions, 170
Digitizing, 9
Dimensioning, 165
   2D, 165
   aligned, 168
   alternate units, 176
   and paper space scaling, 286
   angular, 171
   arcs, 170, 183
   arrow blocks, 172
   arrows, 172
   baseline increment, 169
   basic, 171
   center lines, 173, 184
   commands, 165
   continued, 169
   definition points, 166
   diameter, 170
   dimension line, 166
   editing, 180
   entities, 166
   extension line, 166, 173
   format, 173, 183
   horizontal, 163
   leaders, 207
   Linear, 166, 181
   lines, 166
   mode, 166
   Radius, 170
   scale factor, 172
   selection grips, 95
   stretching, 180, 185
   styles, 171, 177, 179, 182, 186
   text, 173, 174, 175, 177
   text format, 172
   text location, 174
   tick, 172
   tolerances, 177
   units of measure, 176, 182
   updating, 178, 179
   vertical, 167
Dimensioning variables
   DIMALT, 176
   DIMALTD, 176
   DIMALTF, 176
   DIMAPOST, 175
   DIMASO, 180
   DIMASZ, 172
   DIMBLK, 172
   DIMBLK1, 172
   DIMBLK2, 172
   DIMCEN, 173
   DIMCLRD, 172
   DIMCLRE, 172
   DIMCLRT, 172
   DIMDLE, 172
   DIMDLI, 172
   DIMEXE, 172
   DIMEXO, 172
   DIMGAP, 172
   DIMLFAC, 172
   DIMLIM, 177
   DIMOVERRIDE, 178
   DIMPOST, 175
   DIMRND, 177
   DIMSCALE, 172
   DIMSE1, 173
   DIMSE2, 173
   DIMSTYLE, 178, 179, 182
   DIMTAD, 174, 175
   DIMEDIT, 180
   DIMTIH, 174
   DIMTIX, 174
   DIMTM, 177
   DIMTOFL, 174
   DIMTOH, 174
   DIMTOL, 177
   DIMTP, 177
   DIMTSZ, 172
   DIMTXT, 177
   DIMZIN, 176
Dimension styles, 178
Dimension text,
   angle, 174
   editing, 180
   styles, 177
   variables, setting, 177
Display
   Cleaning up of, 17
   grid, 40
   linetype, 147
   panning, 14
   zooming, 13
Displaying toolbars, xviii
DIST command, 341
DIVIDE command, 334
DONUT command, 69
Drafting standards, 275
Dragging
   Arcs, 11
   during mirror, 89
Drawing
   aids, 41
   compiling, 279
   conversion, vii, xx
   converting an old, vii, xx
   export of, 255
   extents, 5

inserting, 254
isometric, 397
opening existing, xix
prototype, 99
scale, 338
units, 17
DTEXT command, **197**, 209
duplicate items, see COPY, 86
Dynamic
  text, 197
  zoom, 13

E
Editing
  attributes in blocks, 379
  attribute definitions, 363
  commands, 83
  dimensions, 180
  floating viewports, 283, 293
  hatches, 228
  Multiline text, 346
  paragraph text, 205
  polylines, 65
  selecting objects for, 83
  snap angle, 40
  splines, 338
  text, 205
ELLIPSE command, 399
END command, xxi
ENDpoint object snap, 42
ERASE command, **48**
Exiting AutoCAD, xx, xxii
EXPLODE command, **258**
Exploding
  blocks, 258
  dimensions, 180
  polylines, 258
Exporting files, 255, 422
EXTEND command, **120**, 121, 131
Extend to Implied Intersection, 121
Extension line (dimensions), 173
External references, 439
Extracting attributes, 385

F
F1 key, 2
Feet and inches, 18
Fence, 85
File names, xxi
File
  ASCII, 420
  extensions, 489
  formats, 489
  managing, 309
  plotting to, 423
  recovery, xx
  renaming, xxi
  saving, xxi
  script, 420
  searching for, xxi
  slide, 416
  temporary, xxi, 490
FILLET command, **12**
Filleting
  arcs, 23
  circles, 101
  lines, 12, 25
  setting radius for, 25
FILTER command, **426**
Filters X,Y,Z, 230
Flip screen, 2
Floating
  command window, xvii
  toolbars, xvi
  viewports, 281, 290
Flyout properties, xvi
Font files, 203
  assigning, 202
  post script, 203
  True Type, 203
Fractional units display, 19
Freezing layers
  in single views, 144
  in paper space, 285
Frozen layers, 144
Function Keys, xiii

G
GIFIN command, 423
Graphics area, xii
Grid
  isometric, 40, 398
  standard, 1, 4, 21, 39
GRID command, **40**
Grips, 48, 94
GROUP command, 427

H
Hatch
  area, 225
  associative, 228
  boundaries, 225
  editing, 228
  from command line, 221
  patterns, 223
  previewing, 226
  selecting objects, 226
  styles, 223
HATCH command, **221**
HATCHEDIT command, **228**
HELP command, **32**
Hexagons, 67
Hidden lines
  in paper space, 275
Highlighting group members, 428
HORIZONTAL dimensions, 167
HPGL files, 423

I
ID command, 343
Implied intersection, 121
Importing images
  to AutoCAD, 423
  to word processing, 422
Inches, zero suppression, 176
Included angle, 126
Incremental point entry, 7
Inquiry commands, 338
INSERT command, **253**
Insertion
  base insertion point, 252
  multiple, 257
  scale, 253, 279
Intersection object snap, 42
Intersections
  editing multilines, 346
  trimming to implied, 121
Intervals, blocks inserted to, 334
Invisible attributes, 364
Isometric
  circles, 398
  grid and snap, 40
  text, 206
ISOPLANE command, 398

J
Joining PLINEs, 66
Justification of text, 70, 196

K
Keyboard entry, xi
Keys (function), x, xiv

L
Last entity selection method, 85
LAYER command, **143**
Layer control dialog box, 143
Layers
  "0" layer zero, 152
  Blocks and, 260
  changing, 145, 149
  colors, 144, 146, 149, 151
  controlling visibility, 151
  creating, 144, 153
  current, 144, 145
  DDLMODES, 143
  Defpoints, 152
  filtering, 152
  freeze and thaw, 144, 151
  freeze and thaw in paper space, 284, 290
  linetypes, 151

list, 152
lock and unlock, 144, 151
modifying, 145
naming conventions, 152
new, 151
on and off, 151
overview, 143
properties, 146
renaming, 145
thaw and freeze, 144, 151
turning off, 151
turning on, 151
visibility
and layers, 151
and viewports, 144
LEADER command, **207**
Left justified text, 70
Limits, 1, 3, 21
LIMITS command, **2**, 21, 39
.lin files, 97
Line width (PLINE), 63
LINE command, **6**
Lines
center, 147
chamfering, 49
extension, 173
filleting, 12
freehand, 229
multiple, 71
using and entering, 6
with undo, 10
LINETYPE command, 97, 98, **147**
Linetypes
and layers, 147
assigning, 147
by Block, 260
by object, 98
by Layer, 147, 149
changing, 98
current, 97
listing library files, 98
scaling, 99
scaling in paper space, 286
standard, 97
LIST command, **15**, 342
Listing
dimension styles, 171
frozen layers, 151
properties of objects, 15, 149
Loading
files, xix
linetypes, 97
Locking layers, 151
LTSCALE command, **148**

M
MEASURE command, **335**

Menu
bars, xiii, xvi
buttons, xiii, xvi
command entry, xvi, xviii
files, xviii
icon, xvi
pull-down, xviii
pop-up, xviii
screen, xviii
standard, xvii, 490
MENU command, xvii
MIDpoint object snap, 42
MINSERT command, **257**
Minus tolerancing, 177
MIRROR command, **89**, 100
Mirror grip mode, 97
Mirroring
and dimensioning, 189
existing objects, 89, 103
with grips, 97
MIRRTEXT system variable, 91
MLEDIT command, 346
MLINE command, **71**, 75
MLSTYLE command, 344
.mnu files, xvii, 490
Model space, 282
dimensioning in, 282
switching to, 281
Modes
grip, 95
grid, 40
ortho, 10
Snap, 39
Modify polyline, 65
Modify toolbar, xiii
Mouse, xiii
MOVE command, **87**, 100
Moving objects to 0,0, 88
Move grip mode, 97
Moving viewports, 283
MSLIDE command, **416**
MSPACE command, **282**
MTEXT command, **199**, 200, 213
Multiline editing, 346
Multiline Styles dialog box, 345
Multiline
adding vertices, 347
creating styles, 345
deleting vertices, 347
drawing, 71
editing, 346
elements of, 71
joints, 347
using existing styles, 346
Multiline text, 198
Multiple
copies, 86, 87

inserts, 257
linetypes, 347
viewports, 281
MVIEW command, 281
MVSETUP, 286, 298

N
Naming
groups, 428
layers, 144
NEARest object snap, 42
Nested blocks, 437
NEW command, xx
New File Name dialog box, xx
NODE object snap, 42
NONE object snap, 42
Null response text, 198

O
Object Grouping Dialog box, 428
Objects
changing properties of, 149
creation of, 6
editing of, 83
Object selection filters, 426
Object selection methods
Add, 144
All, 83
Crossing Polygon, 83, 85, 115
Crossing Window, 83, 38
Cycling, 47
Fence, 83, 85
Last, 83, 85
Multiple, 83
Previous, 83, 85
Remove, 113, 117
Single, 83, 84
Undo, 84
Window, 83, 84
WPolygon, 83, 85
Object snap
Apparent Intersection, 42, 45
CENter, 42, 44
disabling, 42
ENDpoint, 42, 44, 52
From, 42, 45
INSERTion, 42,
INTersection, 42, 46
MIDpoint, 42, 52
Multiple, 47
NEArest, 42
NODE, 42
NONE, 42, 46
PERpendincular, 42, 44
QUADrant, 42, 43, 50
QUICK, 42, 46
TANgent, 42, 43

OFFSET command, **122**
OOPS command, **48**
OPEN command, xxi
Origin of model or drawing, 1, 3
ORTHO command, **10**
Ortho mode, 10
Orthographic projection, 404
OSNAP command, **41**, 47

P
PAN command, **14**
Paper space
    dimensioning, 286
    floating viewports, 281
    freezing layers, 284
    objects, 263
    overview, 280
    scaling views in, 284
    switching to, 280
Paragraph text, 199, 200, 206
Parallel lines, 71
Parent dimension style, 177
PCX files, 423
PDMODE system variable, 333, 348
PDSIZE system variable, 333, 348
PEDIT command, **65**
Pen assignment for plots, 495
PERpendicual object snap, 42
Pick button, 9
PLINE command, **62**
Pline (see polyline)
Plot
    colors, 495
    configuration, 494
    display, 496
    extents, 496
    limits, 496
    origin, 497
    preview, 497
    rotation, 497
    scale, 497
    to file, **62**, 496
    to printer, 495
    view, 497
    window, 494
PLOT command, 423, 494
Plot dialog box, 62
Plus tolerancing, 177
POINT command, **348**
Point filters, 230, 234
Pointing devices, xx
Points
    display modes, 333
    entering, 41
Polar
    arrays, 125
    coordinates, 8

POLYGON command, **67**
Polygon window, 85
Polylines
    arc segment options, 65
    calculating area of, 338
    closing, 63
    corners, 64, 102
    drawing, 62, 72
    editing, 65, 66
    exploding, 258
    filleting, 65
    joining, 66
    splined, 66, 130
    width, 62, 64
PostScripts, 203
Preview
    hatch, 226
    plot, 497
Print dialog box, 494
Prototype drawing, 99
PSPACE command, **282**
Pull-down menus, xviii
PURGE command, 261, **262**

Q
QSAVE command, xx
QTEXT command, **204**
Quick object snap, 46
QUIT command, xxii

R
RADIAL dimensions, 170, 183
Radius
    arc, 24
    circle, 25
    setting for fillets, 25
ray casting, 227
Record (SKETCH option), 229
Rectangular arrays, 126
Redefining blocks, 259
REDO command, 94
REDRAW command, 16
REGEN command, 17
Relative coordinates, 7
Remove entity selection method, 113
Return key, x
ROTATE command, **91**, 92, 103
Rotate and copy, 92
Rotate grip mode, 96
Rscript, 421, 425
Running object snap, 47
Running object snap dialog box, 47

S
SAVE command, xxi
SAVEAS command, xxi
Save drawing as dialog box, xxi

Saving your work, xxi
SCALE command, **93**
Scaling
    dimensions, 172
    drawings, 93, 275, 276, 284
    hatches, 223, 275
    linetypes, 148
    plots, 497
    text, 70, 277, 278
    title blocks, 275
    viewports, 275
    with grips, 97
Screen
    DOS, xii
    Windows, xiii
Screen menus, xvii
.SCR files, 421
SCRIPT command, **421**
Script files, 420
Scroll bars, xvii
SDF attribute extract, 383
Sections, 232, 234, 236
Selection set (see object selection methods)
Setting up, 5, 6
SHELL command, 388
.SHP files 490
Sign on code, xi
Single entity selection methods, 42
SKETCH command, 229
SKPOLY variable, 230
SLIDELIB utility program, 419
Slider bars, xvii
Slides, 416
Slide shows, 421
Snap
    base point, 40
    isometric, 40
    rotation angle, 40
SNAP command, 1, 4, 21, **39**, 397
SOLID command, **68**, 74
SPLINE command, 337
SPLINETYPE command, 67
Spline curves, from polylines, 66
Starting AutoCAD, xi, xii, 1, 5, 6
Status line, xii
STRETCH command, **115**, 131
    dimensioning, 180, 185
Stretch grip mode, 95
STYLE command, **202**, 403
Styles
    dimensioning, 179
    hatching, 223
    isometric snap, 40
    linetype, 147
    text, 202
Suppressing zeros

in dimensions, 176
in units readout, 18
Symbol library, 263

T

Tangent with circles, 25, 51
TANgent object snap, 42, 43
Temporary files, xxi, 490
Text
    alignment options, 70, 196
    ASCII, 420
    dimension orientation, 175
    display, 204
    dynamic, 197
    editing, 204
    fonts, 201
    height, 70
    importing, 203
    justification, 70, 196
    mirroring, 91
    multiline, 198
    overscore, 198
    paragraph, 199
    special characters, 198
    styles, 201, 209
    underscored, 198
    vertical, 201
TEXT command, 69, **195**
Text editor, 201
Text files, 385
Text styles, 201
Thawing layers
    in paper space, 284
TIFF format, 423
TIFFIN command, 423
Tiled viewports, 281
TILEMODE command, **281**
Tilemode
    commands affected by, 281, 285
    MVIEW command, 281
    and paper space, 281
Title blocks, 275
Toggle keys, xiv
Toolbars
    accessing, xviii
    docking, xvi
    moving, xvi
TRACE command, **61**
Transparent commands, 16
TRIM command, 45, 51, 52, **118**, 119, 120, 130
Trim to implied intersection, 121
TrueType fonts, 203

U

U command, 10, 121
UCS command, 2

UNDO command, 121
UNITS command, **17**
Units of measure
    alternate, 176
    display format, 18
    drawings, 17
    setting style, 18
    surveyors, 20
Unlocking layers, 151
Updating dimensions, 180
User coordinate system, 1

V

Values
    absolute, 6
    relative, 7
Variables
    dimensioning (see Dimensioning variables)
Vector files, 415
Vertical dimensions, 167
Vertical text, 201
VPLAYER command, **285**
VSLIDE command, **417**

W

WBLOCK command, **255**, 443
WCS (see World Coordinate System)
Width of Polylines, 62
Window entity selection method, 48, 84
Windows
    switching from DOS to, xix
    using, xiii
World Coordinate System (WCS), 2
WPolygon entity selection method, 85

X

X Zoom scale factor, 15
XBIND command, 445
XP Zoom scale factor, 284
XREF command, 439
Xrefs
    adding to drawings, 442
    attaching to drawings, 442, 445
    binding to drawings, 440, 445
    layers, 439, 444
    overview, 439
    reloading, 441
    removing from drawings, 440
    updating, 441
X,Y,Z point filters, 230

Z

Zero suppression
    in dimensioning, 176
    in screen display, 18

Zoom
    full plot preview, 497
    in paper space, 284
    overview, 13
    scale factor X, 15
    scale factor XP, 284
    transparent, 16
ZOOM command, **13**
    All, 4, 14
    Center, 13
    Dynamic, 13
    Extents, 13
    In, 14
    Out, 14
    Previous, 13
    Scale, 13, 14
    Vmax, 13
    Window, 14
    XP, 284

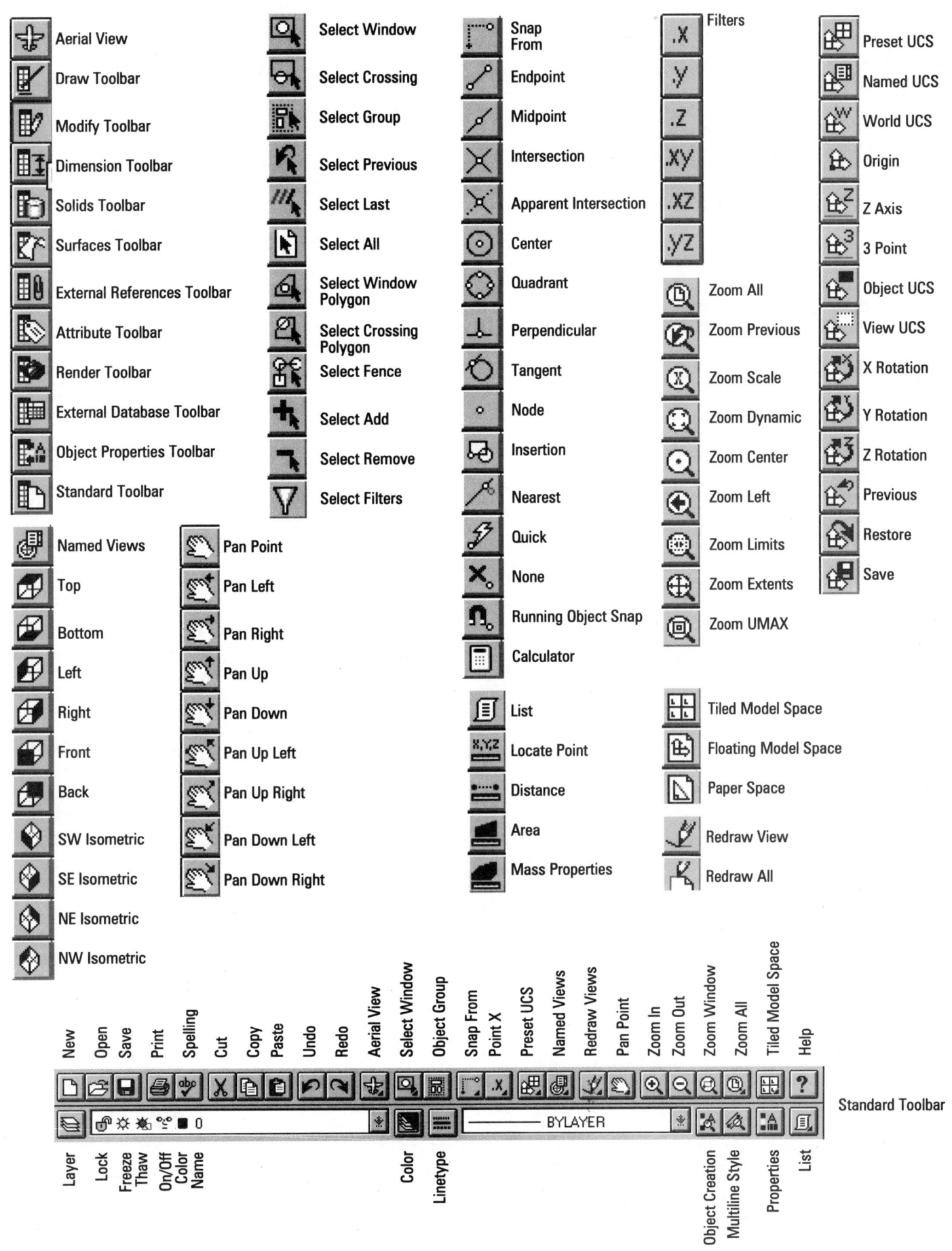

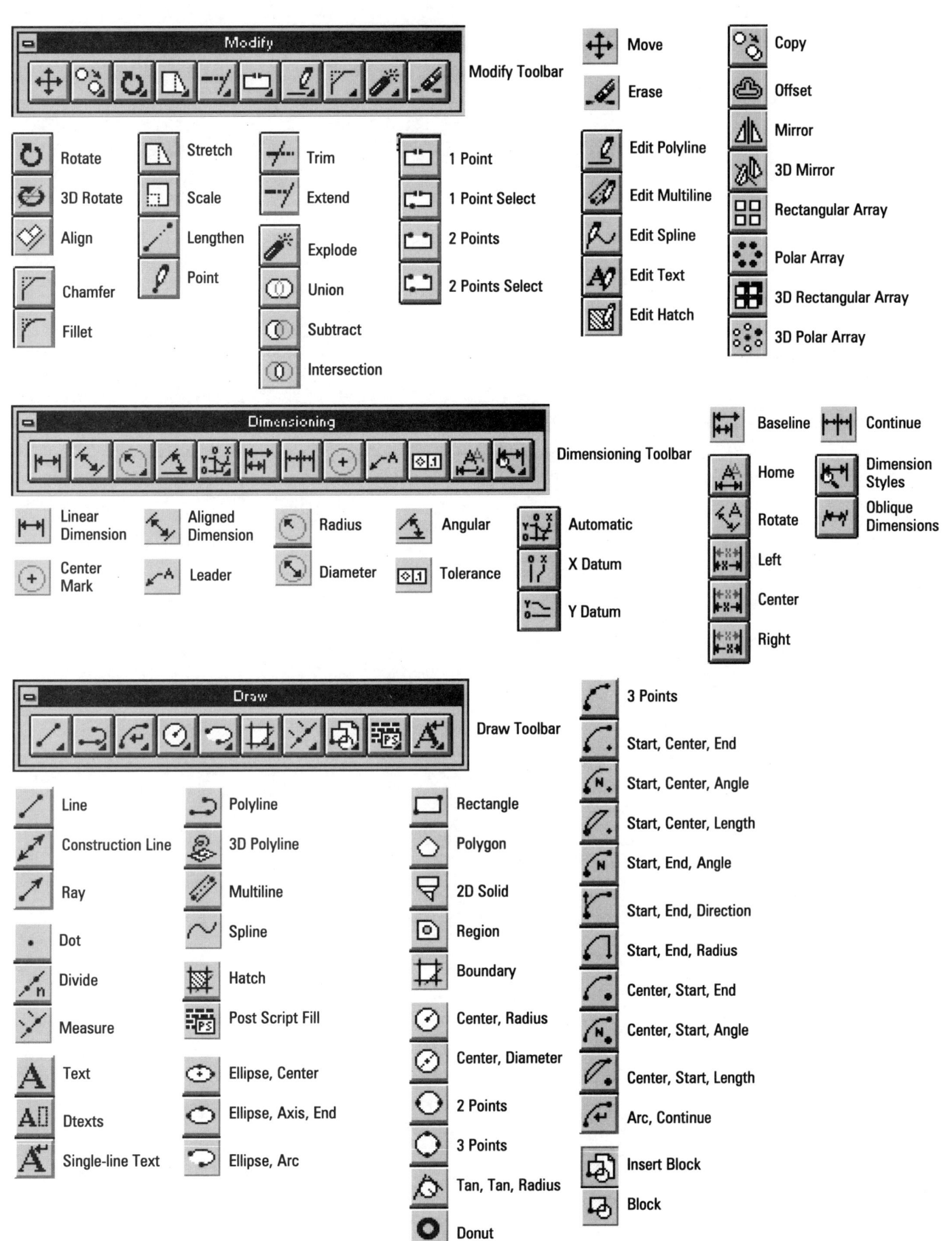

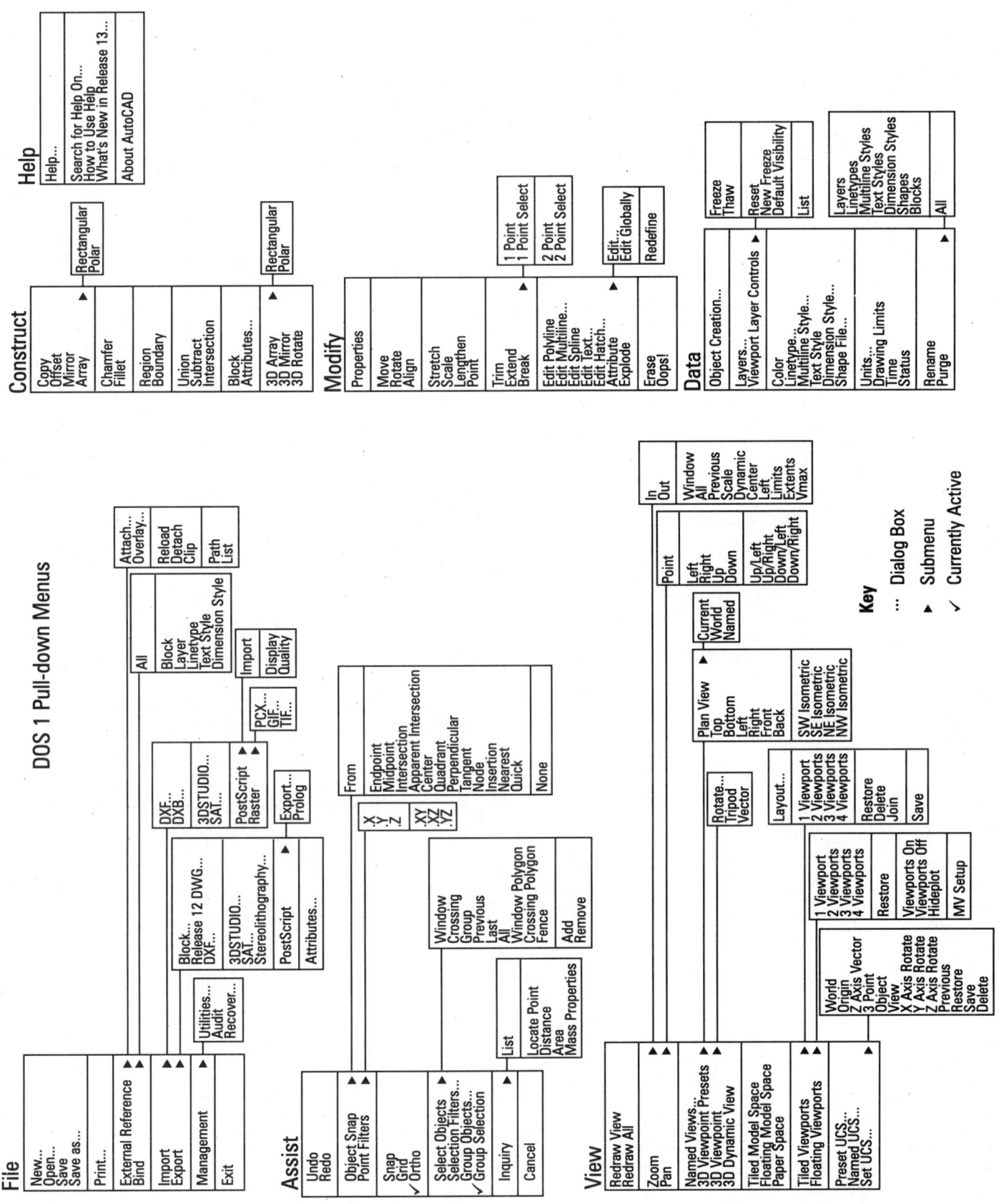

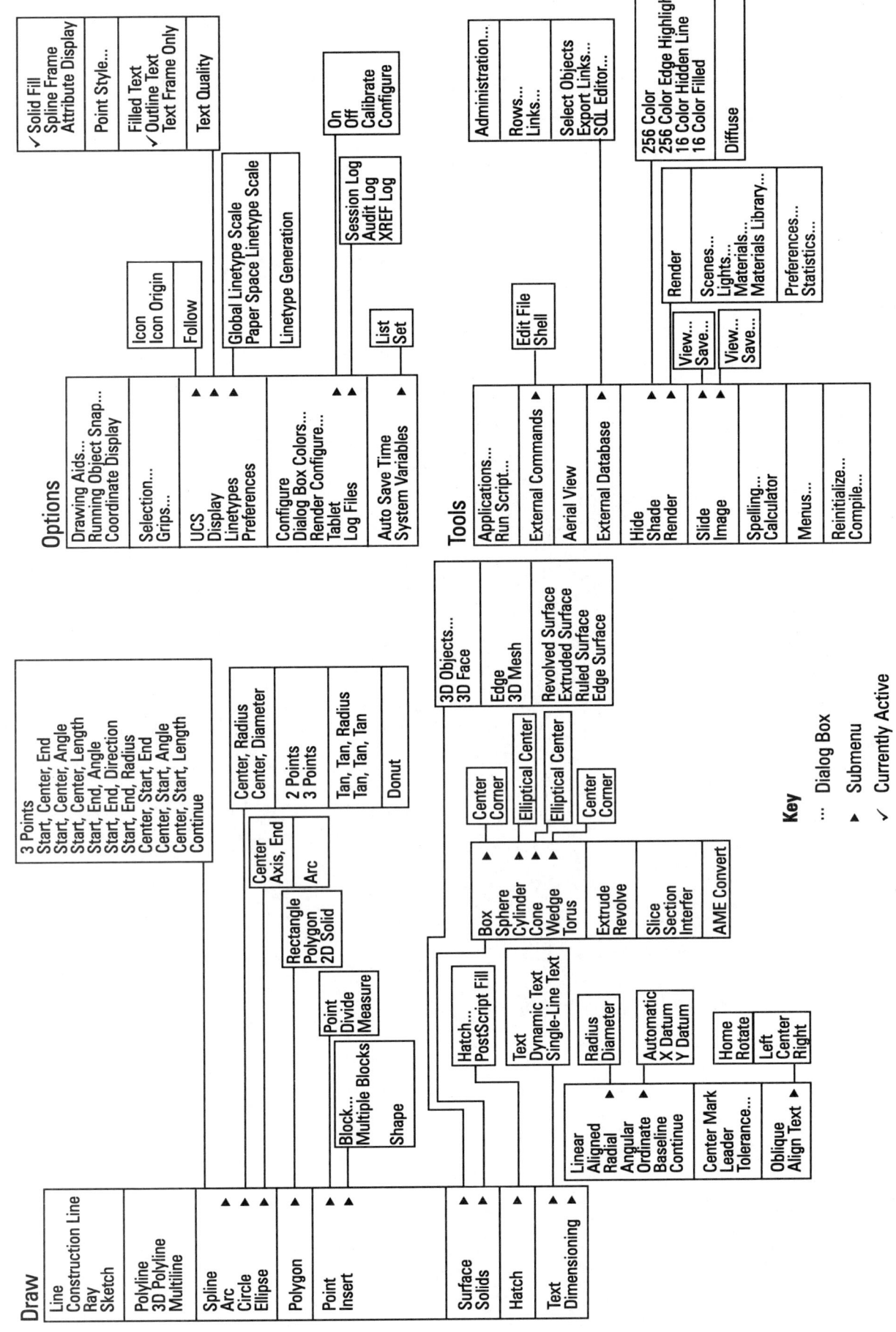